Perturbation Theories for the Thermodynamic Properties of Fluids and Solids

Perturbation Theories *for the* Thermodynamic Properties *of* Fluids *and* Solids

J. R. Solana

CRC Press
Taylor & Francis Group
Boca Raton London New York

CRC Press is an imprint of the
Taylor & Francis Group, an **informa** business

CRC Press
Taylor & Francis Group
6000 Broken Sound Parkway NW, Suite 300
Boca Raton, FL 33487-2742

First issued in paperback 2019

© 2013 by Taylor & Francis Group, LLC
CRC Press is an imprint of Taylor & Francis Group, an Informa business

No claim to original U.S. Government works

ISBN-13: 978-1-4398-0775-0 (hbk)
ISBN-13: 978-0-367-38025-0 (pbk)

Visit the Taylor & Francis Web site at
http://www.taylorandfrancis.com

and the CRC Press Web site at
http://www.crcpress.com

Contents

Preface

The aim of this book is to provide a comprehensive review of the present status of perturbation theories for the equilibrium thermodynamic and structural properties of classical systems. The scope of the book is much wider than it might appear at first sight, because the concept of "perturbation theory" is used here in a broad sense, embracing those theories based, in one way or another, in splitting the interaction potential into reference and perturbation contributions.

The starting point for any statistical mechanics theory of fluids and solids is the knowledge of the intermolecular forces, and, therefore, Chapter 1 includes some comments on the nature of intermolecular forces and describes a number of simple potential models for later reference in subsequent chapters. The book is intended to be accessible to a wide audience with a basic knowledge on thermodynamics and statistical mechanics, but, for completeness, Chapter 2 provides a summary of statistical mechanics concepts and formulae that are used throughout the book. A constant throughout the book is the analysis of the performance of the different theories. In most cases, this is carried out by comparing with simulation data. Therefore, a review on simulation techniques is given in Chapter 3 in order to provide to the reader a basic understanding of the subject. However, the ultimate goal of theories is their application to real substances, and so the capability of some theories for obtaining or fitting the experimental data of real substances is also discussed. As the thermodynamic and structural properties of the reference fluid are often determined from integral equation theories, Chapter 4 provides a not-so-short account on them, including integral equation perturbation theories. Most frequently, the reference system is the hard-sphere fluid or solid, and so Chapter 5 describes in detail several theoretical approaches for these systems, leading to analytical solutions for the thermodynamics and structure. Perturbation theories based on series expansions of the Helmholtz free energy for fluids and solids with simple spherically symmetric pair potentials are analyzed in Chapters 6 and 7 for monocomponent and multicomponent systems, respectively. Particular attention is paid in the latter chapter to the thermodynamic and structural properties of hard-sphere fluid mixtures, as they are usually the systems chosen as reference in perturbation theories for fluid mixtures with more realistic interactions. Monocomponent and multicomponent molecular fluids are the topics of Chapter 8, most of which is devoted to describe a number of theoretical approaches, both perturbative and nonperturbative, for the structure and thermodynamics of hard-body molecular fluids, as systems of election for reference in perturbation theories for molecular fluids. Density functional theories for inhomogeneous systems are the subject of Chapter 9; these theories are either perturbative in nature, or are closely related to perturbative approaches, or provide the thermodynamic and structural properties of inhomogeneous reference systems for use in perturbation theories for other inhomogeneous systems. The book concludes with Chapter 10, which deals with the study of several particularly challenging systems, as is the case of the fluids near the critical point because of the presence of large-scale fluctuations, or systems with complex interactions. Part of the contents of Chapters 3 through 10 is devoted to analyze the performance of the different theories considered, and to this end a considerable member of figures is included.

Therefore, although focused on perturbation theories, the book attempts to provide quite a complete account of the three main groups of theories for the equilibrium thermodynamic and structural properties, namely, integral equation theories, free energy perturbation theories, and density functional theories. The theories analyzed in this book have a sound theoretical basis so that they are susceptible of further refinements in a more or less systematic way. Therefore, although semiempirical corrections to certain theories are included, formulations with little or no theoretical basis are avoided. In this book, the emphasis is on practical applications, so that complex theoretical derivations are avoided as much as possible, guiding the interested reader to the appropriate literature for further details.

This book should be appropriate for both postgraduate students and experienced researchers. The former will acquire a global view, though quite detailed, on the subject. For experienced researchers, the book may be useful as a quick guide to help them judge in advance which theory might be more suitable for a particular application, or to gain insight on the directions along which a particular theory or group of theories are susceptible to new improvements and applications.

Author

J. R. Solana currently serves as full professor in the Applied Physics Department, Faculty of Sciences, University of Cantabria. He has taught thermodynamics, statistical mechanics, computer simulation, and liquid state physics among other subjects. He has coauthored more than 100 papers in the field of thermodynamics and statistical mechanics of fluids and solids, including a recent review on perturbation theories (Zhou, S. and Solana, J. R., Progress in the perturbation approach in fluid and fluid-related theories, *Chemical Reviews* 109, 2829–2858, 2009). He has also served as a referee for *Physical Review Letters*, *Journal of Chemical Physics*, *Physical Review E*, *Fluid Phase Equilibria*, *Molecular Physics*, and other journals.

Solana has been active in the field of theory and computer simulation for the thermodynamic and structural properties of fluids and solids. His research interests include computer simulation, perturbation theories, and integral equation theories, with application to simple fluids and solids, molecular fluids, mixtures, colloids, and aqueous protein solutions to obtain equations of state, phase equilibria, thermophysical properties, and pair correlation functions.

Notation

Latin Letters

a	In general, a parameter. In particular, the van der Waals parameter
b	In general, a parameter. In particular, the covolume
$B(r)$	Bridge function
B_2	Second virial coefficient
$c(r)$	Direct correlation function
$\hat{c}(k)$	Fourier transform of the direct correlation function
$c^{(n)}(\mathbf{r}_1, \mathbf{r}_2, \ldots, \mathbf{r}_n)$	n-particle direct correlation function
$\mathscr{C}(\mathbf{r})$	Effective one-body potential
$\mathscr{C}(\mathbf{r}_1, \mathbf{r}_2, \ldots, \mathbf{r}_n)$	Functional derivative of order $n-1$ of $\mathscr{C}(\mathbf{r})$ with respect to the density
e	Absolute value of the charge of the electron
f	Number of degrees of freedom of the system
F	Helmholtz free energy
$f(r)$	Mayer function
$f^E(\rho)$	Excess free energy per particle
$\mathbf{F}(r)$	Intermolecular force
$\mathscr{F}[\rho(\mathbf{r})]$	Free energy functional
$g(r)$	Radial distribution function
$g(\mathbf{r}_1, \mathbf{r}_2, \ldots, \mathbf{r}_N)$	n-particle correlation function
h	Planck's constant
$h(\mathbf{r}_1, \mathbf{r}_2)$	Total correlation function
$\hat{h}(k)$	Fourier transform of the total correlation function
$\mathscr{H}$	Hamiltonian
k_B	Boltzmann's constant
L	Center-to-center distance in polyatomic molecular fluids
m	(a) Mass of a particle
	(b) Number of components in a mixture
n	(a) Exponent of a potential (in particular, exponent of the SS potential)
	(b) Number of beads in a chain molecule
	(c) Number density of electrons in a metal
$\mathbf{n}$	Unit vector
N	Number of particles
$\mathbf{p}_i$	Momentum of particle i
P	Pressure
q_i	Generalized coordinates
Q	Canonical partition function
$\mathcal{Q}$	Effective mass of the heat reservoir
r	Distance

$\mathbf{r}_i$	Position vector of particle i
r_{ij}	Distance between the centers of particles i and j
r_m	Position of the minimum of the potential
R	(a) Diameter ratio in binary hard-sphere mixtures
	(b) $1/4\pi$ times the mean curvature integral of a hard body
	(c) A particular distance or radius
$S(k)$	Static structure factor
$t(r)$	Tail function
T	Absolute temperature
T_c	Critical temperature
T^*	$= k_B T/\varepsilon$, reduced temperature
$u(r), u(\mathbf{r}_i, \mathbf{r}_j)$	Pair potential
U	Internal energy
$\mathcal{U}_N \equiv \mathcal{U}(\mathbf{r}_1, \mathbf{r}_2, \dots, \mathbf{r}_N)$	Total potential energy
$\mathbf{v}_i$	Velocity of a particle
v	Molecular volume
v^e	Excluded volume
V	Volume
$\mathscr{V}(\mathbf{r})$	External potential
x_i	Mole fraction of component i
$y(r)$	Cavity distribution function
z	(a) Charge number of a particle
	(b) Fugacity or activity
Z	$= PV/Nk_B T$, compressibility factor

Greek Letters

α	(a) Excess volume relative to the close packing volume
	(b) Parameter coupling the reference and perturbation potentials
	(c) $= RS/3v$ nonsphericity parameter
	(d) A critical exponent
β	(a) $= 1/k_B T$
	(b) A critical exponent
ϵ	Dielectric constant
ε	Energy parameter of the potential
γ	(a) Length-to-breadth ratio of hard spherocylinders
	(b) Parameter in the Gaussian approximation for the density profile of a solid
	(c) A critical exponent
$\gamma(r)$	Indirect correlation function
Γ	$= \rho\sigma^3(\varepsilon/k_B T)^{3/n}$, coupling parameter in soft-sphere systems
η	(a) $= \rho v$, packing fraction for or hard-body systems
	(b) A critical exponent

ϕ	Fraction of occupied sites in a lattice
$\Phi[\rho(\mathbf{r})]$	Dimensionless free energy density
κ	Inverse length parameter in the HCY potential
κ_D	Debye screening parameter
κ_T	Isothermal compressibility
λ	(a) Potential range, in units of the diameter σ, of the SW and SSH potentials
	(b) Wavelength
Λ	Thermal wavelength
μ	(a) Chemical potential
	(b) Dipolar moment
$\Theta(x)$	Heaviside step function
Ω	(a) Set of angles (or solid angle) defining the orientation
	(b) Grand potential
ρ	$=N/V$, number density
ρ^*	$=\rho\sigma^3$, reduced density
ρ_c	Critical density
$\rho^{(1)}(\mathbf{r}) \equiv \rho(\mathbf{r})$	Single-particle density or density profile
$\rho^{(n)}(\mathbf{r}_1, \mathbf{r}_2, \ldots, \mathbf{r}_n)$	n-particle density distribution
$\rho(q_i, p_i; V)$	Probability density of a given configuration in the canonical ensemble
σ	Distance parameter of the potential. In particular, diameter of a hard sphere
σ_{ij}	Distance of closest approach between two spheres with diameters σ_i and σ_j
τ	Dimensionless parameter in the SHS potential
$\Psi(r)$	Potential of mean force
ξ	Correlation length
Ξ	Grand canonical partition function
ω	Bond angle

Subscripts and Superscripts

id Ideal gas
E   Excess (over-ideal gas)
c   (a) Configurational
     (b) Critical

Abbreviations

AO	Asakura–Oosawa (depletion potential)
bcc	Body-centered cubic
BACK	Boublík–Alder–Chen–Kreglewski
BHS	Bonded hard-sphere
BMCSL	Boublík–Mansoori–Carnahan–Starling–Leland
BPGG	Ballone–Pastore–Galli–Gazzillo
CHS	Charged hard spheres

COR	Chain-of-rotators (equation of state)
DCF	Direct correlation function
DFT	Density functional theory
DH	Duh–Haymet
DLVO	Derjaguin–Landau–Verwey–Overbeek (potential)
ELA	Effective liquid approximation
EMA	Extended modified weighted density approximation
EOS	Equation of state
EXP	Exponential approximation
fcc	Face-centered cubic
FH	Flory–Huggins
FHS	Fused hard spheres
FMSA	First-order mean spherical approximation
FMT	Fundamental measure theory
FSS	Finite size scaling
GC	Group contribution
GELA	Generalized effective liquid approximation
GF	Generalized Flory
GFA	Generating function approach
GFD	Generalized Flory-dimer
GFH	Generalized Flory–Huggins
GMSA	Generalized mean spherical approximation
GvdW	Generalized van der Waals
HCB	Hard convex body
HCY	Hard-core Yukawa
HMD	Homonuclear diatomics
HNC	Hypernetted chain
HRT	Hierarchical reference theory
HS	Hard-sphere
HSE	Hard-sphere expansion
HTD	Heteronuclear diatomics
HTE	High-temperature expansion
HWDA	Hybrid weighted density approximation
IET	Integral equation theory
ISPT	Improved scaled particle theory
lc	Local compressibility
LDA	Local density approximation
LEXP	Linearized exponential approximation
LFHS	Linear fused hard spheres
LJ	Lennard–Jones
LNST	Linear nonsymmetric triatomics
LST	Linear symmetric triatomics
LTHS	Linear tangent hard spheres
mc	Macroscopic compressibility
MC	Monte Carlo
MD	Molecular dynamics

MDA	Mean density approximation
MELA	Modified effective liquid approximation
MF	Mean field
MFA	Mean field approximation
ML	Malijevský–Labík
MRJ	Maximally random jammed state
MS	Martynov–Sarkisov
MSA	Mean spherical approximation
MTPT	Modified thermodynamic perturbation theory (modified Wertheim theory)
MWDA	Modified weighted density approximation
NAHS	Nonadditive hard spheres
OCP	One-component plasma
OCT	Optimized cluster theory
OER	Oblate ellipsoid of revolution
OMWDA	Optimized modified weighted density approximation
ORPA	Optimized random phase approximation
OSC	Oblate spherocylinder
OZ	Ornstein–Zernike
PER	Prolate ellipsoid of revolution
PHNC	Perturbative hypernetted chain theory
PHSC	Perturbed hard-sphere-chain (theory)
pI	Isoelectric point
PLJC	Perturbed Lennard–Jones chain (equation of state)
PM	Primitive model
PRISM	Polymer reference interaction site model
PSC	Prolate spherocylinder
PWDA	Perturbation weighted density approximation
PY	Percus–Yevick
RCP	Random close packing
RDF	Radial distribution function
RFA	Rational function approximation
RG	Renormalization group
RHNC	Reference hypernetted chain
RIS	Rotational isomeric state (model)
RISM	Reference interaction site model
RPA	Random phase approximation
RPM	Restricted primitive model
RPY	Reference Percus–Yevick
RY	Rogers–Young
SA	Scaling approximation
SAFT	Statistical associating fluid theory
sc	Simple cubic
SCOZA	Self-consistent Ornstein–Zernike approximation
SGA	Square-gradient approximation
SHS	Sticky hard-sphere (or adhesive hard-sphere)

SMSA	Soft MSA
SPT	Scaled particle theory
SPWDA	Simplified perturbation weighted density approximation
SS	Soft-sphere
SSH	Square-shoulder
SW	Square-well
SWDA	Simple weighted density approximation
TL	Tang–Lu
TPI	Test particle insertion method
TPT	Thermodynamic perturbation theory
TSMC	Temperature-scaling Monte Carlo
vdW	van der Waals
vdW1	van der Waals one-fluid model
VM	Verlet modified
WDA	Weighted density approximation
ZH	Zerah–Hansen
ZSEP	Zero-separation theorem-based closure

1 Introduction

1.1 AGGREGATION STATES OF MATTER

Everybody is familiar with the three classical aggregation states of matter: gas, liquid, and solid, and everybody knows some of the characteristics that differentiate each other. Gases are easily compressible and do not have a fixed shape nor volume, so that they tend to occupy the whole volume of the container; liquids are hardly compressible and do not have a fixed shape, so that they are easily deformable, but have a well-defined volume which only varies slightly with temperature and pressure; solids are hardly compressible and deformable and, similarly to liquids, their volume is nearly constant with temperature and pressure. A fourth aggregation state of matter, less commonly known, that can be added to the three former is the plasma, in which the particles are strongly ionized.

Most people also know that the aggregation states are connected to each other. Thus, cooling a gas will result in its condensation to a liquid and, upon further cooling, the system will transform into a solid. On the opposite side, upon strong heating of a gas, a plasma state can be finally reached. One can intuitively infer that the finite size of the particles determines to a great extent the differences in the properties of the aggregation states. In a dilute gas, the average distances between the centers of the neighboring particles are much greater than the characteristic size of the particles, which explains the low density and the high compressibility of the gas. In a liquid, these two lengths are the same order of magnitude, whence the high density and low compressibility of a liquid; however, the fact that a liquid is easily deformable implies that the constituent particles can move more or less freely throughout the whole volume occupied by the liquid. In a solid, the particles are packed together so closely that not only the system has a high density and low compressibility but also the particles are essentially fixed in certain positions that cannot abandon in a significant amount. However, the finite size of the particles alone cannot explain everything. Thus, the existence of a liquid requires the presence of attractive forces between the molecules, whereas this is not the case of a solid. On the other hand, the properties of a plasma are mainly related to the presence of Coulombian interactions between the charged particles. Therefore, the free volume available for the movement of the particles cannot completely explain the properties of the aggregation states, so that other kinds of interactions need to be considered. All of these interactions are described by the intermolecular forces, which depend on the nature of the system under consideration.

The four cited aggregation states do not exhaust the classification of the states of matter. As one knows from elementary thermodynamics, even simple monocomponent systems, that is to say monocomponent systems whose particles interact by means of spherically symmetric forces, may present different phases in the same aggregation state. Such a system will have a single gaseous phase but, depending on the nature of the intermolecular forces, several liquid phases may be possible and this situation may be responsible for the anomalous properties in certain liquids (see Solana 2008, and references therein). The liquid can be undercooled below the triple point and, eventually, it may become a glassy solid. Glassy solids are not in equilibrium states, as they tend to relax toward crystalline solids. However, the relaxation time may be enormous as compared to any realistic measurement time, and so glasses can often be regarded as being in equilibrium. In any case, the properties of a glassy state may depend not only on temperature and density, but also on the way the glassy state was achieved. A simple system may also have different crystalline phases, depending on temperature and density.

For molecular systems, the variety of phases may increase. Thus, a system consisting of elongated molecules may present a number of fluid phases, such as nematic, in which particles are oriented to more or less extent along a given direction; smectic, in which molecules, in addition to having nematic ordering, are distributed in layers; or columnar, in which the molecules are grouped in columns, among others. A system consisting of disk-shaped molecules may also present columnar phases as well as discotic phases; in the latter, the flattened surfaces are oriented more or less parallel to a given surface. All of these phases are intermediate between the isotropic fluid and a crystalline solid.

For multicomponent fluids new phases may appear. Even the simplest binary systems may exhibit demixing in the fluid phase, with phases enriched in one of the components, and a rich variety of solid phases (see Barrio and Solana 2008, and references therein).

The complex phase behavior that may exhibit a given system makes difficult its theoretical treatment. Fortunately, there are many theoretical schemes that apply to different phases provided we introduce the appropriate information. The most obvious cases are the gaseous and liquid phases of simple monocomponent systems, for which many theories apply without significant changes. Other examples are certain perturbation theories, which are the main subject of interest here, that can be applied to monodisperse fluids as well as to fluid mixtures and crystalline solids. It seems enriching to discuss the way similar theoretical frameworks apply to different kinds of systems and phases. For this reason, this book, although for obvious reasons, devotes a considerable space to simple fluids, goes beyond them and deals also with crystalline solids, molecular fluids, and mixtures.

1.2 NATURE OF THE INTERMOLECULAR FORCES

The structure and thermodynamic properties of a system at equilibrium are determined by the intermolecular forces. It is well known that gases consisting of neutral atoms or molecules are more compressible at low densities than an ideal gas at the same temperature and density, and the contrary occurs at high densities. From this fact we can infer that intermolecular forces are attractive at large distances and repulsive at short distances. Much more information on intermolecular forces can be extracted from scattering experiments. Intermolecular forces, or the corresponding intermolecular potentials, can be obtained from *ab initio* calculations based on quantum mechanics, although these calculations are complex, tedious, and computationally demanding.

A detailed account on the nature of intermolecular forces and their determination can be found in classical literature about this matter (Hirschfelder et al. 1954; Hirschfelder 1967). We will give only a brief survey on this subject here.

Consider two neutral atoms placed not too close to each other. The movement of the electrons around the nucleus of one of the atoms leads to instantaneous fluctuations in the charge distribution giving rise to instantaneous dipoles. The electric field created by these dipoles induces instantaneous dipoles in the other atom. The interaction between the instantaneous dipoles of the two atoms results in an attractive force acting on them, the London dispersion force (London 1930a) so called because the oscillating dipoles are the cause of the scattering of light. It can be proved that the potential of the dispersive forces varies as a power series of the inverse of the distance r separating the centers of the atoms in the form

$$u_{lr}(r) = -\frac{c_6}{r^6} - \frac{c_8}{r^8} - \frac{c_{10}}{r^{10}} - \cdots . \tag{1.1}$$

The explicit expression of coefficient c_8 can be obtained from quantum mechanics. Approximate expressions are also available. One of them, derived by London (1930b), is

$$c_6 = \frac{3}{2}\alpha_1\alpha_2\frac{\bar{E}_1\bar{E}_2}{\bar{E}_1 + \bar{E}_2}, \tag{1.2}$$

where
   α_i is the polarizability of atom i
   $\bar{E}_i$ is an average energy approximately equal to the ionization energy of atom i

These quantities can be determined from experiment. Higher order coefficients are much more difficult to obtain from *ab initio* calculations. Fortunately, the first term in expansion (1.1) is dominant at long distances.

When the two atoms become close enough to each other, the corresponding electron clouds overlap and a strong repulsion arises, mainly due to the Pauli exclusion principle. Obviously, chemical bonding may also take place in case the atoms have incomplete electronic orbitals, but we do not consider that situation here. The short-range repulsive force also can be obtained from *ab initio* calculations, but this involves a much higher degree of complexity than that of the long-range attractive force. The former varies with distance much more quickly than the latter. Theoretical calculations, as well as scattering experiments, provide evidence of an approximately exponential decay of the short-range potential with distance, thus $u_{sr}(r) \propto e^{-ar}$, although it is often approximated by a power function of the inverse of the distance r.

At some intermediate distance, the long-range attractive and the short-range repulsive forces become balanced and this allows for the existence of the liquid state below the critical point. However, it is to be noted that many particle interactions may become important in systems with moderate to high densities. For a system consisting of N particles placed at positions $\mathbf{r}_1, \mathbf{r}_2, \ldots, \mathbf{r}_N$, the total potential energy can be expanded in the form

$$\mathcal{U}_N \equiv \mathcal{U}(\mathbf{r}_1, \mathbf{r}_2, \ldots, \mathbf{r}_N) = \sum_{i=1}^{N-1} \sum_{j=i+1}^{N} u(\mathbf{r}_i, \mathbf{r}_j) + \sum_{i=1}^{N-2} \sum_{j=i+1}^{N-1} \sum_{k=j+1}^{N} u(\mathbf{r}_i, \mathbf{r}_j, \mathbf{r}_k) + \cdots,$$

(1.3)

where the sums' limits have been chosen so as to avoid counting each term several times. The first term of the right-hand side is the contribution due to pair interactions, the next one is that due to triplet interactions, and so on. Neglecting all the contributions except the first one on the right-hand side, it is equivalent to considering that the interaction energy is pairwise additive. This is a frequently used approximation, although for high density systems higher order interactions, especially three-body effects, may have a nonnegligible contribution. The effect of many-body interactions can be determined from computer simulation, and the result used to determine an effective pair potential which will depend on density in addition to distance. The long-range contribution to the three-body interactions due to the induced dipole–dipole–dipole interactions between three atoms was determined by Axilrod and Teller (1943) with the result

$$u_{\mathrm{AT}}(\mathbf{r}_i, \mathbf{r}_j, \mathbf{r}_k) = C \frac{\cos \gamma_i \cos \gamma_j \cos \gamma_k + 1}{r_{ij}^3 r_{jk}^3 r_{ki}^3},$$

(1.4)

where
$r_{ij} = |\mathbf{r}_i - \mathbf{r}_j|$
γ_i is the angle formed by the sides r_{ij} and r_{ki} of the triangle defined by the positions
    $\mathbf{r}_i, \mathbf{r}_j,$ and $\mathbf{r}_k$ of the three atoms, and similar definitions apply to the remaining
    distances and angles
C is a constant which depends on the polarizabilities and ionization energies of
    the three atoms

Depending on the values of the involved angles, u_{AT} can be a positive or a negative quantity.

Up to this point, we have considered interactions between atoms. When the interacting particles are polyatomic molecules, the interaction energies will depend on the relative orientations of the molecules. A useful and frequently used approach, based on the assumption of pairwise additivity, consists in considering that the potential energy for the interaction between two polyatomic molecules is the sum of all site–site interactions between atoms belonging to different molecules.

Thus far, we have considered only neutral molecules for which the dispersion forces arise from instantaneous anisotropies in the charge distributions. Other kinds of interactions take place when particles have permanent charges or permanent dipoles or multipoles. According to Coulomb's law, the potential energy between two ions, whose centers are a distance r apart, having charges z_1 and z_2 in units of the absolute value e of the charge of the electron, is

$$u(r) = \frac{z_1 z_2 e^2}{r}.$$ (1.5)

On the other hand, the potential energy of interaction between two permanent dipoles with dipolar moments μ_1 and μ_2, whose centers are a distance r apart, depends on their relative orientations (θ, φ). In a low-density dipolar fluid at temperature T, there are two competing effects: the mutual interaction that tends to align the dipoles along the same direction and the thermal motion of the molecules that tends to produce a random orientation of the dipoles. At a given finite temperature, the distribution of orientations is determined by the corresponding Boltzmann factor $\exp[-u(r, \theta, \varphi)/k_B T]$, where k_B is the Boltzmann constant. At high temperatures, a weighted average of the dipole–dipole interaction, using the Boltzmann distribution, gives (Hirschfelder et al. 1954)

$$u(r) = -\frac{2}{3} \frac{\mu_1^2 \mu_2^2}{k_B T} \frac{1}{r^6},$$ (1.6)

so that the dipole–dipole potential is temperature dependent.

The interaction of two particles with different charge characteristics may lead to new kinds of interactions, such as ion-induced dipole, ion–dipole, or dipole-induced dipole. A very special case of polar interactions is the hydrogen bonding, occurring in certain molecular liquids, like water, in which hydrogen is covalently bonded to a strongly electronegative atom, like oxygen. Then, a positive charge appears at the hydrogen site and a negative charge at the site of the electronegative atom, resulting in a strong tendency of the hydrogen atom to attach to a negatively charged site belonging to another molecule. Intramolecular hydrogen bonding can also occur in certain molecules, like DNA, RNA, and proteins.

We will end this section mentioning the associative interactions, consisting in the aggregation of particles to form greater complexes. Hydrogen bonding is frequently the cause of association, but other kinds of interactions may also be responsible for such behavior. The strength of the associative interactions is intermediate between dispersive interactions and chemical bonding, and the energy involved is of the order of the thermal energy, so that association is governed by temperature. In solutions of associating particles with more than one bonding site, association often leads to polydisperse mixtures, and the degree of polydispersity depends on temperature.

1.3 SIMPLE POTENTIAL MODELS

As discussed in Section 1.2, the *ab initio* determination of the intermolecular potentials is a complex task, even for simple fluids, and leads to numerical results that are unsuitable for theoretical calculations. For this reason, one often resorts to suitable analytical functions depending on a number of parameters that, for a particular system, are determined from the fitting of *ab initio* potentials or from available experimental data for some thermodynamic property. Restricting ourselves to neutral nonpolar molecules interacting with each other by means of spherically symmetric pair potentials $u(r)$, the general features of the potential are as follows: it is positive for $r < \sigma$, with a strongly negative slope, and negative for $r > \sigma$, so that it is null for $r = \sigma$; it reaches its minimum value $-\varepsilon$ at $r = r_m$, and decays toward zero as $r \to \infty$. Therefore, the intermolecular force $F(r) = -du(r)/dr$ is strongly positive, that is strongly repulsive, for $r < r_m$, and negative, that is attractive, for $r > r_m$, in agreement with the considerations in Section 1.2. The parameter σ gives an estimation of the effective diameter of the molecules, although it is to be noted that the actual effective diameter is slightly temperature dependent.

A potential model reproducing qualitatively these features is the *Lennard–Jones* (LJ) potential, shown in Figure 1.1a, whose mathematical form is

$$u(r) = 4\varepsilon \left[\left(\frac{\sigma}{r} \right)^{12} - \left(\frac{\sigma}{r} \right)^{6} \right]. \tag{1.7}$$

The minimum of the potential occurs at $r_m = 2^{1/6}\sigma$. This potential model is suitable for certain monatomic fluids, such as the noble gases, and even for some fluids with nearly spherically shaped molecules such as methane. For polyatomic molecular fluids, the LJ potential is used frequently as a model for the site–site interactions. The LJ potential is a particular case of the *Mie* potential, which obeys Equation 1.7, but replacing the exponents 12 and 6 by m and n, respectively, and the coefficient 4 with $[n/(n - m)]\,(n/m)^{m/(n-m)}$.

Still simpler potential models are often used for particular applications. The most simple potential model is the *hard-sphere* (HS) potential

$$u(r) = \begin{cases} \infty, & r < \sigma \\ 0, & r > \sigma \end{cases}, \tag{1.8}$$

shown in Figure 1.1b. This potential approaches those for real nonpolar molecules with spherically symmetric potentials at very high temperatures. As attractive forces are absent, liquid phase is not possible in the HS system. Instead, crystalline solid phases are possible due to entropic forces, which arise from the competition between the decrease of entropy associated with ordering and the increase of entropy due to the increase in the average free volume per molecule available in the ordered crystal, as compared with the disordered fluid, with predominance of the latter effect. The HS system plays an essential role in most of the perturbation theories for fluids with spherically symmetric potentials, as we will see in Chapters 4 through 7 and 9.

A potential more appropriate for modeling the interactions in simple systems at high temperatures is the *soft-sphere* (SS) potential

$$u(r) = \varepsilon \left(\frac{\sigma}{r} \right)^{n}, \tag{1.9}$$

with n being a positive number. Its shape for $n = 12$, which mimics the repulsive part of the LJ potential, is shown in Figure 1.1c. Other values of n are of interest, such as the case $n = 1$ that corresponds to Coulombian repulsive interactions.

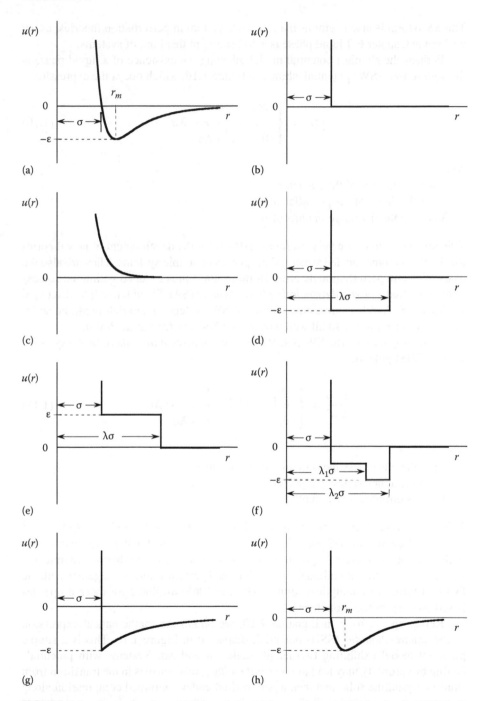

FIGURE 1.1 Some simple potential models: (a) Lennard–Jones; (b) hard sphere; (c) soft sphere; (d) square well; (e) square shoulder; (f) double square-well; (g) hard-core Yukawa; (h) hard-core Lennard–Jones.

The SS system is also useful as the reference system in perturbation theories, as we will see in Chapter 6. Liquid phase is also lacking in this kind of systems.

Perhaps the simplest potential model allowing the existence of a liquid phase is the *square-well* (SW) potential, shown in Figure 1.1d, which obeys the expression

$$u(r) = \begin{cases} \infty, & r < \sigma \\ -\varepsilon, & \sigma < r < \lambda\sigma \,, \\ 0, & r > \lambda\sigma \end{cases} \qquad (1.10)$$

where

σ is the diameter of the particles
$-\varepsilon$ is the depth of the potential well
λ is the potential range in units of σ

This kind of systems are very useful as testbeds for the development of new theories for the thermodynamic and structural properties of simple systems. They are also the basis of certain perturbation theories for molecular fluids grouped within the context of the so-called *statistical associating fluid theories* (SAFT), which will be discussed in Chapter 8. Another interesting feature of SW systems is the rich phase behavior they present for not too small well widths (see Serrano-Illán et al. 2006).

A limiting case of the SW potential is the *adhesive hard-sphere* or *sticky hard-sphere* (SHS) potential

$$\frac{u(r)}{k_B T} = \begin{cases} \infty, & r < \sigma \\ \ln\left[\frac{12\tau(\lambda-1)}{\lambda}\right], & \sigma < r < \lambda\sigma \,, \\ 0, & r > \lambda\sigma \end{cases} \qquad (1.11)$$

where

τ is a dimensionless measure of the temperature
τ^{-1} is the adhesiveness parameter
$\lambda - 1$ is infinitesimally small

This potential reduces to the SW potential for $\lambda \to 1$, $\varepsilon \to \infty$, and $\tau = \frac{1}{12}(\lambda - 1)^{-1}$ $e^{-\varepsilon/k_B T}$ finite, and to the HS potential for $T \to \infty$. The interest of the SHS fluid resides in the fact that it is able to reproduce some of the features of the thermodynamic and structural properties of colloidal suspensions and protein solutions, together with the fact that there is an analytical solution (Baxter 1968) available for these properties based on integral equation theory.

Replacing $-\varepsilon$ by ε in Equation 1.10, we obtain the mathematical expression of the *square-shoulder* (SSH) potential, displayed in Figure 1.1e. This is a simple potential model exhibiting two length scales: σ and $\lambda\sigma$. Systems with potentials having this property may lead to *reentrant melting*; this consists in the transition from fluid to crystalline solid and then again to fluid under continued compression along isotherms.

Figure 1.1f shows a combination of two SW with different well width and well depth. Combining two SW or a SW with a SSH may lead to anomalous behavior. Everybody is more or less familiar with the anomalous temperature-dependence of the density of water around 4°C, but there are also anomalies in other thermodynamic properties such as the thermal expansion coefficient, the isothermal compressibility, and the constant-pressure heat capacity in water supercooled below 0°C (see Mishima and Stanley 1998, and references therein). This anomalous behavior has been attributed to the existence of two liquid phases, a *low-density liquid* (LDL) and a *high-density liquid* (HDL), which would be related to the *low-density amorphous* (LDA) phase and the *high-density amorphous* (HDA) phase, respectively, found experimentally. Mishima and Stanley (1998) explain the existence of two liquid phases in water in terms of an interaction potential consisting of two subwells: an inner subwell, relatively shallow and wide, and an outer subwell, deeper and narrower. The potential model shown in Figure 1.1f roughly reproduces that shape. Theory and computer simulation show that models consisting either SW+SW or SSH+SW may exhibit liquid–liquid transition. There is also experimental evidence of the existence of liquid–liquid phase transitions for real liquids other than water.

Another potential model widely used is the attractive *hard-core Yukawa* (HCY) potential, whose mathematical definition is

$$u(r) = \begin{cases} \infty, & r < \sigma \\ -\varepsilon\sigma\dfrac{e^{-\kappa(r-\sigma)}}{r}, & r > \sigma \end{cases}, \qquad (1.12)$$

where σ is again the diameter of the HS forming the core and parameter κ determines the decaying rate of the attractive tail of the potential. The shape of this potential for $\kappa = 2$ is shown in Figure 1.1g. Replacing $-\varepsilon$ by ε in Equation 1.12, the repulsive HCY potential results. HCY potentials are used frequently for modeling the interaction between colloidal particles. One interesting feature is that certain integral equation theories lead to analytical solutions for this potential model. We will return to this point in Chapter 4.

A potential model closely resembling the HCY potential is the *Sutherland* (or generalized Sutherland) potential

$$u(r) = \begin{cases} \infty, & r < \sigma \\ -\varepsilon\left(\dfrac{\sigma}{r}\right)^n, & r > \sigma \end{cases}, \qquad (1.13)$$

in which n is a positive quantity which plays a similar role as parameter κ in the HCY potential.

We will end this short account of simple potential models for interactions between equal molecules by mentioning the *hard-core Lennard–Jones* (HCLJ) potential, defined in the form

$$u(r) = \begin{cases} \infty, & r < R \\ 4\varepsilon\left[\left(\dfrac{\sigma}{r}\right)^{12} - \left(\dfrac{\sigma}{r}\right)^6\right], & r > R \end{cases}, \qquad (1.14)$$

so that it consists of a hard spherical core with diameter R plus a tail in the form of the LJ potential (Equation 1.7). Therefore, in contrast with the LJ potential, in the HCLJ potential the diameter of the particles is constant independently of the temperature. Its shape for $R = \sigma$ is displayed in Figure 1.1h.

The preceding potential models can be readily extended to interactions between unequal molecules in mixtures. However, there are some particular kinds of potential models for mixtures that deserve a special mention.

The simplest potential model for mixtures, like for single-component fluids, is the HS potential

$$u_{ij}(r) = \begin{cases} \infty, & r < \sigma_{ij} \\ 0, & r > \sigma_{ij} \end{cases}. \tag{1.15}$$

Here σ_{ij} is the distance of closest approach between the centers of the spheres of species i and j with diameters σ_i and σ_j, respectively. HS mixtures play a similar role in perturbation theory for mixtures as the pure HS system for single-component fluids.

For mixtures of *charged hard spheres* (CHS) embedded in a dielectric continuum of dielectric constant ϵ, two popular models that combine a hard spherical core with a tail of the form of Equation 1.5 are the *restricted primitive model* (RPM) of electrolytes

$$u_{ij}(r) = \begin{cases} \infty, & r < \sigma \\ \dfrac{z_i z_j e^2}{\epsilon r}, & r > \sigma \end{cases}, \tag{1.16}$$

for spheres of equal diameter, and the *primitive model* (PM) of electrolytes

$$u_{ij}(r) = \begin{cases} \infty, & r < \sigma_{ij} \\ \dfrac{z_i z_j e^2}{\epsilon r}, & r > \sigma_{ij} \end{cases}, \tag{1.17}$$

for spheres of different diameter.

REFERENCES

Axilrod, B. M. and E. Teller. 1943. Interaction of the van der Waals type between three atoms. *J. Chem. Phys.* 11:299.

Barrio, C. and J. R. Solana. 2008. Binary mixtures of additive hard spheres. Simulations and theories. In *Theory and Simulation of Hard-Sphere Fluids and Related Systems*, ed., A. Mulero, *Lect. Notes Phys.* 753:133–182. Berlin,Germany: Springer-Verlag.

Baxter, R. J. 1968. Percus-Yevick equation for hard spheres with surface adhesion. *J. Chem. Phys.* 49:2770.

Hirschfelder, J. O., Ed. 1967. Intermolecular forces. *Adv. Chem. Phys.* 12.

Hirschfelder, J. O., C. F. Curtiss, and R. B. Bird. 1954. *Molecular Theory of Gases and Liquids*. New York: John Wiley & Sons.

London, F. 1930a. Theory and system of molecular forces. *Z. Phys.* 63:245.

London, F. 1930b. Properties and applications of molecular forces. *Z. Phys. Chem.* 11:222.

Mishima, O. and H. E. Stanley. 1998. The relationship between liquid, supercooled and glassy water. *Nature* 396:329.

Serrano-Illán, J., G. Navascués, and E. Velasco. 2006. Noncompact crystalline solids in the square-well potential. *Phys. Rev. E* 73:011110.

Solana, J. R. 2008. Thermodynamic properties of double square-well fluids: Computer simulations and theory. *J. Chem. Phys.* 129:244502.

2 Some Basics on Statistical Mechanics

In this chapter, we will introduce a number of basic concepts and expressions which will be useful throughout the book. These include the virial theorem, the correlation functions, the pressure, energy, and compressibility equations for monocomponent as well as for multicomponent systems, and the static structure factor. All these matters can be found in many textbooks on statistical mechanics, but it is convenient to review them here for further reference.

2.1 VIRIAL THEOREM AND THE EQUATION OF STATE

Let us consider a three-dimensional system consisting of N molecules with mass m interacting with each other by means of a spherically symmetric potential $u(r)$. Let us define the following function of the position vectors $\mathbf{r}_i$ of the molecules

$$\mathcal{M} = \frac{1}{2} \sum_{i=1}^{N} m\mathbf{r}_i^2. \tag{2.1}$$

The time derivative $\dot{\mathcal{M}}$ is

$$\dot{\mathcal{M}} = \sum_{i=1}^{N} m\mathbf{r}_i \cdot \dot{\mathbf{r}}_i. \tag{2.2}$$

Deriving again we obtain

$$\ddot{\mathcal{M}} = \sum_{i=1}^{N} m\dot{\mathbf{r}}_i^2 + \sum_{i=1}^{N} m\mathbf{r}_i \cdot \ddot{\mathbf{r}}_i. \tag{2.3}$$

The time average of the latter function will be

$$\left\langle \ddot{\mathcal{M}} \right\rangle = \frac{1}{t} \int_0^t \ddot{\mathcal{M}} dt = \frac{1}{t} \left(\dot{\mathcal{M}}_t - \dot{\mathcal{M}}_0 \right). \tag{2.4}$$

On the other hand, from Equation 2.3 we have

$$\left\langle \ddot{\mathcal{M}} \right\rangle = \left\langle \sum_{i=1}^{N} m\dot{\mathbf{r}}_i^2 \right\rangle + \left\langle \sum_{i=1}^{N} m\mathbf{r}_i \cdot \ddot{\mathbf{r}}_i \right\rangle. \tag{2.5}$$

13

As $\mathbf{r}_i$ and $\dot{\mathbf{r}}_i$ are finite quantities, so must be $\dot{\mathcal{M}}_t$ and $\dot{\mathcal{M}}_0$ as well as their difference. Therefore, in the limit $t \to \infty$, the right-hand side of Equation 2.4, and hence that of Equation 2.5, must be zero. Now introducing the Clausius' virial function

$$\mathcal{V} = -\frac{1}{2}\left\langle \sum_{i=1}^{N} m\mathbf{r}_i \cdot \ddot{\mathbf{r}}_i \right\rangle = -\frac{1}{2}\left\langle \sum_{i=1}^{N} \mathbf{r}_i \cdot \mathbf{F}_i \right\rangle, \tag{2.6}$$

into Equation 2.5, the aforementioned condition leads to

$$\mathcal{V} = \frac{1}{2}\left\langle \sum_{i=1}^{N} m\dot{\mathbf{r}}_i^2 \right\rangle = E, \tag{2.7}$$

where E is the total averaged kinetic energy of the system. This is the mathematical expression of the *virial theorem*. However, when expressed in this way, the virial theorem is of little use to us. Next, we will obtain an alternative expression which will allow us to obtain the EOS.

To this end, consider the ideal gas as a particular case. In this system, there are no intermolecular forces but only collisions of the molecules with the walls of the container, so that

$$\mathcal{V}_{id} = E = \frac{3}{2}Nk_BT = \frac{3}{2}PV, \tag{2.8}$$

where we have made use of the principle of equipartition of energy and of the EOS of the ideal gas. Therefore, the contribution to the virial due to the collisions of the molecules with the walls is $(3/2)PV$.

Now consider a real gas whose particles interact with each other by means of a pairwise additive potential $u(r)$ depending only on the relative distance r between the interacting particles. In this case, the force $\mathbf{F}_i$ acting on the particle i consists of two contributions, one of them is the sum of the interaction forces $\mathbf{F}_{ij}$ of the particle i with each of the remaining particles j, and the other is the force $\mathbf{F}_{iw}$ that the walls exert on the particle, that is,

$$\mathbf{F}_i = \sum_{j \neq i} \mathbf{F}_{ij} + \mathbf{F}_{iw}. \tag{2.9}$$

The second of these forces contributes to the virial by an amount $(3/2)PV$, as we have seen. On the other hand, the force that particle j exerts on particle i due to their interaction potential can be put in the form

$$\mathbf{F}_{ij} = -u'\left(r_{ij}\right)\mathbf{n}_{ij}, \tag{2.10}$$

where

$u'(r_{ij}) = du(r)/dr|_{r=r_{ij}}$ is the modulus of the intermolecular force between molecules i and j

$\mathbf{n}_{ij} = \mathbf{r}_{ij}/r_{ij}$ is the unit vector in the direction joining the centers of the two molecules directed from j to i

From Equations 2.7 through 2.10, one obtains

$$\frac{3}{2}Nk_BT = \frac{3}{2}PV - \frac{1}{2}\left\langle \sum_i \sum_{j\neq i} \mathbf{r}_i \cdot \mathbf{F}_{ij} \right\rangle = \frac{3}{2}PV + \frac{1}{2}\left\langle \sum_i \sum_{j\neq i} \mathbf{r}_i \cdot \mathbf{n}_{ij}\, u'\left(r_{ij}\right) \right\rangle.$$

(2.11)

In the last term, each pair of molecules appears twice, one as $\mathbf{r}_i \cdot \mathbf{n}_{ij}u'(r_{ij})$ and the other as $\mathbf{r}_j \cdot \mathbf{n}_{ji}u'(r_{ji})$. Taking into account that $u'(r_{ji}) = u'(r_{ij})$ and $\mathbf{n}_{ji} = -\mathbf{n}_{ij}$, we can rearrange the two terms into a single term of the form $(\mathbf{r}_i - \mathbf{r}_j) \cdot \mathbf{n}_{ij}u'(r_{ij})$, and dividing the sum by 2, to avoid counting each pair twice, we finally obtain

$$PV = Nk_BT - \frac{1}{6}\left\langle \sum_i \sum_{j\neq i} r_{ij}\, u'\left(r_{ij}\right) \right\rangle.$$

(2.12)

This equation, which is a more usual form of the virial theorem, is also known as *virial equation* or *pressure equation*. According to the preceding equation, knowing the form of the interaction potential $u(r)$ and carrying out the indicated average, we can obtain the EOS. However, Equation 2.12 is not suitable for theoretical calculations; a more appropriate expression will be given in Section 2.3.

In case of angle-dependent pair potentials, the average in Equation 2.12 must be performed not only on the positions but also on the orientations of the molecules. In case of site–site interactions between polyatomic molecules, for each pair ij in Equation 2.12 we must perform an additional summation over all site–site interactions. Equation 2.12 is also applicable to mixtures.

2.2 DISTRIBUTION FUNCTIONS

Let us again consider a real monocomponent gas consisting of N particles interacting with each other by means of a spherically symmetric potential. The probability density in the canonical ensemble is

$$\rho\left(\mathbf{r}_1, \ldots, \mathbf{r}_N, \mathbf{p}_1, \ldots, \mathbf{p}_N\right) = \frac{1}{N!h^{3N}}\frac{e^{-\beta\mathcal{H}}}{Q},$$

(2.13)

where $\beta = 1/k_BT$, $\mathcal{H}$ is the Hamiltonian of the system

$$\mathcal{H} = \sum_{i=1}^{N} \frac{\mathbf{p}_i^2}{2m} + \mathcal{U}\left(\mathbf{r}_1, \ldots, \mathbf{r}_N\right),$$

(2.14)

in which $\mathbf{p}_i$ are the momenta of the particles, and

$$Q = \frac{1}{N!h^{3N}} \int \cdots \int e^{-\beta\mathcal{H}}\, d\mathbf{r}_1 \ldots d\mathbf{r}_N d\mathbf{p}_1 \ldots d\mathbf{p}_N$$

(2.15)

is the canonical partition function which, integrating over momenta, can be put in the form

$$Q = \frac{\Lambda^{-3N}}{N!} Q_c, \tag{2.16}$$

in which

$$\Lambda = \left(\frac{2\pi m k_B T}{h^2}\right)^{-\frac{1}{2}} \tag{2.17}$$

is the thermal wavelength and

$$Q_c = \int_V \ldots \int_V e^{-\beta \mathcal{U}(\mathbf{r}_1, \ldots, \mathbf{r}_N)} d\mathbf{r}_1 \ldots d\mathbf{r}_N \tag{2.18}$$

is the configurational integral of the system with volume V.

The probability density for a particular configuration $\mathbf{r}_1, \ldots, \mathbf{r}_N$, independently of the momenta of the particles, is obtained from integrating Equation 2.13 over momenta, with the result

$$\rho(\mathbf{r}_1, \ldots, \mathbf{r}_N) = \frac{e^{-\beta \mathcal{U}(\mathbf{r}_1, \ldots, \mathbf{r}_N)}}{Q_c}. \tag{2.19}$$

The *n-particle density distribution function*, or simply *n-particle density*, gives the probability density of having a particle at position $\mathbf{r}_1$, another particle at $\mathbf{r}_2, \ldots$, another at $\mathbf{r}_n$. It is obtained from Equation 2.19 by integration for the positions of the remaining $N - n$ particles, that is,

$$\rho^{(n)}(\mathbf{r}_1, \ldots, \mathbf{r}_n) = \frac{N!}{(N-n)!} \frac{1}{Q_c} \int_V \ldots \int_V e^{-\beta \mathcal{U}(\mathbf{r}_1, \ldots, \mathbf{r}_N)} d\mathbf{r}_{n+1} \ldots d\mathbf{r}_N, \tag{2.20}$$

which fulfills the normalization condition

$$\int_V \ldots \int_V \rho^{(n)}(\mathbf{r}_1, \mathbf{r}_2, \ldots, \mathbf{r}_n) d\mathbf{r}_1 \ldots d\mathbf{r}_n = \frac{N!}{(N-n)!}. \tag{2.21}$$

The factor $N!/(N-n)!$ in Equation 2.20 arises from the fact that we have N possible choices to place a particle at position $\mathbf{r}_1$, $N-1$ possible choices to place a second particle at position $\mathbf{r}_2, \ldots$, and then $N-n$ possible choices to place a particle at position $\mathbf{r}_n$. Obviously, for "a particle at position $\mathbf{r}_i$" we mean "a particle within a volume element around $\mathbf{r}_i$."

In particular, for $n = 1$ we have the *single-particle density distribution*, or simply *single-particle density* or *density profile*

$$\rho^{(1)}(\mathbf{r}_1) = \frac{N}{Q_c} \int_V \ldots \int_V e^{-\beta \mathcal{U}(\mathbf{r}_1, \ldots, \mathbf{r}_N)} d\mathbf{r}_2 \ldots d\mathbf{r}_N, \tag{2.22}$$

which gives the averaged local particle density. For an homogeneous and isotropic system, the translational invariance requires that the local density must be equal in any volume element to the mean number density $\rho = N/V$, that is,

$$\rho^{(1)}(\mathbf{r}_i) = \rho. \tag{2.23}$$

It is easy to see that the single-particle density can be expressed in the form

$$\rho^{(1)}(\mathbf{r}) = \left\langle \sum_{i=1}^{N} \delta(\mathbf{r} - \mathbf{r}_i) \right\rangle, \tag{2.24}$$

where $\delta(x)$ is the Dirac delta function and the angular brackets mean an average over all configurations.

For a perfect crystalline solid at absolute zero, all the particles are placed exactly at their equilibrium positions. For $T > 0$ K, the particles in the crystalline solid can move around their equilibrium positions in the lattice and, correspondingly, the single-particle density distribution shows a number of peaks centered at the equilibrium positions.

In a similar way, from Equation 2.20, for $n = 2$ we will have

$$\rho^{(2)}(\mathbf{r}_1, \mathbf{r}_2) = \frac{N(N-1)}{Q_c} \int_V \ldots \int_V e^{-\beta\mathcal{U}(\mathbf{r}_1\ldots\mathbf{r}_N)} d\mathbf{r}_3 \ldots d\mathbf{r}_N, \tag{2.25}$$

which is the *pair density distribution*, or simply the *pair density*, which gives the probability density of having a particle at position $\mathbf{r}_2$ on condition that there is another particle at position $\mathbf{r}_1$.

The *n-particle correlation function* is defined in the form

$$g(\mathbf{r}_1, \ldots, \mathbf{r}_n) = \frac{\rho^{(n)}(\mathbf{r}_1, \ldots, \mathbf{r}_n)}{\rho^{(1)}(\mathbf{r}_1) \ldots \rho^{(n)}(\mathbf{r}_n)}. \tag{2.26}$$

For $n = 2$ we have the *pair correlation function*

$$g(\mathbf{r}_1, \mathbf{r}_2) = \frac{\rho^{(2)}(\mathbf{r}_1, \mathbf{r}_2)}{\rho^{(1)}(\mathbf{r}_1)\rho^{(1)}(\mathbf{r}_2)}. \tag{2.27}$$

The *total correlation function* is defined as

$$h(\mathbf{r}_1, \mathbf{r}_2) = g(\mathbf{r}_1, \mathbf{r}_2) - 1. \tag{2.28}$$

In a homogeneous and isotropic system with particles interacting with each other by means of a spherically symmetric pair potential, $g(\mathbf{r}_1, \mathbf{r}_2)$ depends only on the relative distance r between the two particles and is called RDF and denoted as $g(r)$

$$g(r) = \frac{\rho^{(2)}(r)}{\rho^2}, \tag{2.29}$$

and, in this case, the total correlation function is

$$h(r) = g(r) - 1. \tag{2.30}$$

We can write

$$g(r) = e^{-\beta \Psi(r)}, \tag{2.31}$$

where $\Psi(r)$ is termed *potential of mean force* between two particles separated a distance r, because from Equation 2.25 it is clear that

$$\mathbf{f}(\mathbf{r}_1) = -\frac{\partial \Psi(r)}{\partial \mathbf{r}_1} = -\frac{\int_V \cdots \int_V e^{-\beta \mathcal{U}(\mathbf{r}_1 \ldots \mathbf{r}_N)} \frac{\partial \mathcal{U}_N}{\partial \mathbf{r}_1} d\mathbf{r}_3 \ldots d\mathbf{r}_N}{\int_V \cdots \int_V e^{-\beta \mathcal{U}(\mathbf{r}_1 \ldots \mathbf{r}_N)} d\mathbf{r}_3 \ldots d\mathbf{r}_N} \tag{2.32}$$

is the mean force acting on a particle at position $\mathbf{r}_1$ given that there is another particle at position $\mathbf{r}_2$, with $r = |\mathbf{r}_2 - \mathbf{r}_1|$, averaged over the positions of the remaining $N - 2$ particles.

The physical meaning of the RDF is easily understood if we take into account that the number of particles whose centers lie at a distance between r and $r + dr$ of the center of a given particle, taken as a reference, is

$$dN(r, r + dr) = \rho g(r) 4\pi r^2 dr. \tag{2.33}$$

The RDF gives an idea about the structure of the system. It indicates the influence that a particle has on the location of the remaining particles of the system. In an ideal gas, the potential energy is zero and, therefore, Equation 2.25 reduces to

$$\rho_{id}^{(2)}(r) = \frac{N(N-1)V^{N-2}}{V^N} \approx \rho^2, \tag{2.34}$$

and, according to Equation 2.29, the RDF is $g(r) = 1$, independently of the distance r, which indicates that the ideal gas is structureless. In a real fluid, the RDF presents an oscillatory character, with the successive maxima and minima progressively attenuated with distance, so that it holds $\lim_{r\to\infty} g(r) = 1$. This is the reason why it is said that a fluid has short-range order. The lower is the density of the fluid, the greater is the rate of attenuation of the successive peaks and the distance between their maxima, and in the low density limit,

$$\lim_{\rho\to 0} g(r) = e^{-\beta u(r)}. \tag{2.35}$$

The behavior of the RDF for a real fluid is illustrated in Figure 2.1 for a LJ system at several reduced densities $\rho^* = \rho \sigma^3$, where $\rho = N/V$ is the number density, and reduced temperatures $T^* = k_B T/\varepsilon$. This potential model is suitable for argon, among other simple systems, with $\sigma = 3.405$ Å and $\varepsilon/k_B = 119.8$ K. For reference, at the triple point $T_t^* = 0.694$, $\rho_{tl}^* = 0.845$, and $\rho_{ts}^* = 0.961$ (Mastny and de Pablo 2007), where subscripts l and s refer to the liquid and solid, respectively, and at the critical

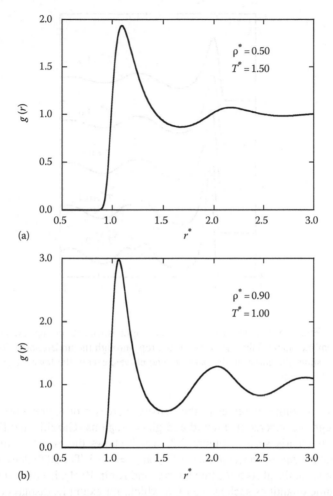

FIGURE 2.1 RDF of an LJ fluid at two different states: (a) a slightly supercritical temperature and moderate density: (b) a state in the liquid phase close to the freezing line.

point $T_c^* = 1.312$ and $\rho_c^* = 0.316$ (Potoff and Panagiotopoulos 1998). Figure 2.1a shows the RDF at a slightly supercritical temperature and moderate density; for short reduced distances $r^* = r/\sigma$, the RDF rapidly decays to zero due to the sharp increase of the potential function (1.7) at these distances (see Figure 1.1a), and for reduced distances beyond the first maximum, the successive maxima rapidly attenuate and the RDF tends to the ideal gas value. At $T^* = 1.0$ and $\rho^* = 0.90$, a state corresponding to the liquid phase close to the freezing line, the maxima appear more marked and displaced to shorter distances (Figure 2.1b).

The bottom curve in Figure 2.2 shows the RDF in a glassy state. It is remarkable the splitting of the second peak of the RDF. This is a characteristic feature of the RDF at high densities in the region of undercooled liquid and glassy solid. O'Malley and Snook (2005) attribute this fact to the development of precursor structures to crystallization. This splitting first manifests as a flattening or a shoulder in the second

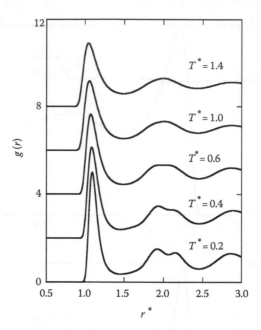

FIGURE 2.2   Evolution of the RDF of an LJ system upon isochoric cooling at reduced density $\rho^* = 0.95$, from the stable fluid, the curve on the top, through the undercooled liquid region, up to a glassy state, the curve on the bottom. The curves, except the lowest one, have been shifted upward for clarity.

peak and the splitting progresses as the density increases or temperature decreases further through the supercooled liquid and glassy regions (Gazzillo and Della Valle 1993). This is clearly seen in Figure 2.2, which shows the evolution of the RDF with temperature during an isochoric cooling at $\rho^* = 0.95$. The RDF does not exhibit marked changes at the glass transition (Barrat and Klein 1991), in contrast with other thermodynamic quantities such as the EOS which, for example, displays changes in the slope of the isobaric density versus temperature curve.

The splitting of the second peak of the RDF was observed by Finney (1970) in dense random packings of HS, with subpeaks at reduced distances $r^* \simeq \sqrt{3}$ and 2.0. These distances approximately correspond to the positions of the third and fourth nearest neighbors in a close packed *face-centered cubic* (fcc) HS lattice, but the lack of the peak at $r^* = \sqrt{2}$, corresponding to the second nearest neighbors in the fcc lattice, discards the existence of local crystalline order. The splitting is also present in the *inherent structure* of fluids (Stillinger and Weber 1984, 1985, Stillinger and LaViolette 1985, 1986, LaViolette and Stillinger 1986). According to these authors, the short-range order in liquids consists of thermal vibrations superimposed to inherent structures consisting in amorphous particle packings. The inherent structure, which is temperature independent, emerges when the fluid, starting from a particular state, is quenched to potential energy minima removing the thermal motion.

For systems with very short-ranged interaction potentials, a splitting of the second peak may arise at extremely low temperatures even at low densities (Babu

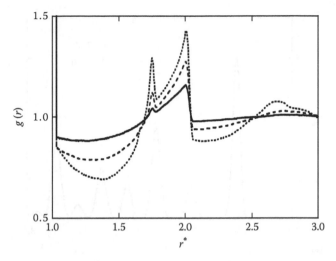

FIGURE 2.3   RDF for a SW fluid with $\lambda = 1.03$ at $T^* = 0.35$, and reduced densities $\rho^* = 0.20$ (continuous curve), $\rho^* = 0.60$ (dashed curve), and $\rho^* = 0.90$ (dotted curve).

et al. 2009, Zhou and Solana 2009) indicating the presence of clustering. This is illustrated in Figure 2.3 for a system with a SW potential (1.10) with $\lambda = 1.03$ at the reduced temperature $T^* = 0.35$ and several densities. For reference, the critical temperature for this value of λ is $T_c^* = 1.31$ (Largo et al. 2008), although it is to be noted that for $\lambda \lesssim 1.25$ the liquid phase is metastable. Figure 2.3 shows that, for all densities, the two subpeaks appear at positions $r^* \simeq \sqrt{3}$ and 2.0, as in dense random packings of HS.

In a crystalline solid, the pair correlation function (2.27) not only depends on distance, but also on direction, so that we cannot talk about an RDF strictly speaking, but we can define an angle-averaged RDF. The structure of peaks of the latter, which depends on the kind of lattice, manifests at any distance and for this reason it is said that a crystalline solid has long-range order. This is illustrated in Figure 2.4 for an LJ crystalline solid in the fcc lattice at $T^* = 0.60$ and two reduced densities. In the high-density solid (Figure 2.4a) the peaks are well defined; there are little or no overlapping between neighboring peaks. As density decreases, the overlapping and merging between neighboring peaks increases and, at densities close to melting (Figure 2.4b), the overlapping is considerable, and some peaks have disappeared and the crystalline structure is close to dissolve giving rise to a fluid structure. Provided that there is no overlapping between neighboring peaks, the integration of Equation 2.33 for a given peak gives the number of particles corresponding to that peak. Thus, the result for the first, second, etc., peaks is the number of nearest neighbors, or coordination number, second nearest neighbors, etc., of a particle in the lattice. These numbers, like the structure itself, are characteristics of the lattice considered (see Hirschfelder et al. 1954 the number of neighbors and their distances for several crystalline structures).

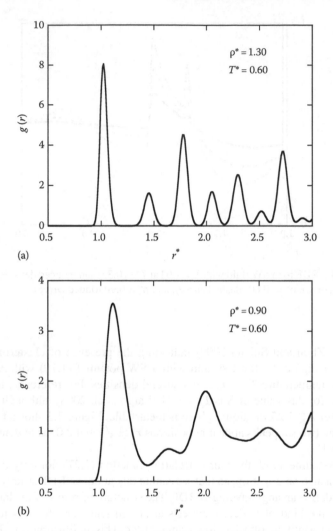

(a)

(b)

FIGURE 2.4 Angle-averaged RDF for an LJ crystalline solid with fcc structure at $T^* = 0.60$ and several densities: (a) a density near close packing; (b) a density near melting.

2.3 THERMODYNAMIC PROPERTIES IN TERMS OF THE RADIAL DISTRIBUTION FUNCTION

Using Equation 2.33, the virial theorem (Equation 2.12) can be expressed in a more convenient way in terms of the RDF in the form

$$Z = \frac{PV}{Nk_BT} = 1 - \frac{2}{3}\pi\frac{\rho}{k_BT}\int_0^\infty g(r)\frac{du(r)}{dr}r^3dr, \qquad (2.36)$$

which is a more usual form of the *pressure* or *virial equation*.

The compressibility factor can be expanded in power series of the density, the *virial expansion*,

$$Z = 1 + \sum_{n=2}^{\infty} B_n(T) \rho^{n-1}, \tag{2.37}$$

where the virial coefficients $B_n(T)$ depend only on temperature and are related to the pair potential. In particular, the second virial coefficient of a fluid with spherically symmetric pair potential can be easily obtained from the virial equation (2.36) introducing the low density expression (2.35) of the RDF, yielding

$$B_2 = -\frac{2\pi}{3}\beta \int_0^{\infty} \frac{du(r)}{dr} e^{-\beta u(r)} r^3 dr = -2\pi \int_0^{\infty} f(r) r^2 dr, \tag{2.38}$$

or

$$B_2 = -\frac{1}{2} \int_V f(r_{12}) d\mathbf{r}_{12}, \tag{2.39}$$

where $f(r) = e^{-\beta u(r)} - 1$ is the Mayer function. The derivation of the expressions for the third and higher order virial coefficients are more involved and the details may be found in most textbooks on statistical mechanics. The result for the third virial coefficient is

$$B_3 = -\frac{1}{3} \int_V \int_V f(r_{12}) f(r_{13}) f(r_{23}) d\mathbf{r}_{12} d\mathbf{r}_{13}. \tag{2.40}$$

The second virial coefficient can be analytically determined for many simple potentials. Analytical expressions for the third and fourth virial coefficients have been obtained for some simple potential models. For more complex potentials as well as for higher order virial coefficients, the calculation of the virial coefficients involve numerical calculations with complexity increasing with the order of the virial coefficient.

For a fluid whose particles interact with each other by means of a discontinuous potential, as is the case of several of the potentials displayed in Figure 1.1, the RDF presents discontinuous jumps in each of the discontinuities of the potential. In such situations, it is useful to make use of the *cavity distribution function* $y(r)$ defined in the form

$$y(r) = g(r) e^{\beta u(r)}. \tag{2.41}$$

This function has the property of being continuous even when the RDF is discontinuous (Percus 1964) and is related to the excess chemical potential μ^E over that of the ideal gas by (Hoover and Poirier 1962)

$$\mu^E = k_B T \ln y(r). \tag{2.42}$$

Using the cavity distribution function, for the particular case of a fluid with an HS potential (1.8), the virial equation (2.36) reduces to

$$Z = 1 + \frac{2}{3}\pi\rho\sigma^3 g(\sigma), \tag{2.43}$$

where $g(\sigma)$ is the value of the RDF $g(r)$ at contact distance $r = \sigma$.

Proceeding in a similar way with a fluid with the SW potential (1.10), one obtains

$$Z = 1 + \frac{2}{3}\pi\rho\sigma^3 \left\{ g(\sigma) - \lambda^3 \left[g\left(\lambda\sigma^-\right) - g\left(\lambda\sigma^+\right) \right] \right\}, \tag{2.44}$$

where $g(\lambda\sigma^-)$ and $g(\lambda\sigma^+)$ are the values of the RDF at the limit $r = \lambda\sigma$ from the left and from the right, respectively.

Finally, for potentials consisting of an HS core plus a continuous tail, the virial equation (2.36) leads to

$$Z = 1 + \frac{2}{3}\pi\rho \left[\sigma^3 g(\sigma) - \frac{1}{k_B T} \int\limits_{\sigma}^{\infty} \frac{du(r)}{dr} g(r) r^3 dr \right]. \tag{2.45}$$

The RDF allows us also to obtain the configurational energy by means of the *energy equation*, which is easily derived from Equation 2.33 in the form

$$U^E = 2\pi N\rho \int\limits_{0}^{\infty} g(r) u(r) r^2 dr, \tag{2.46}$$

where U^E is the excess energy over that of an ideal gas at the same temperature, that is to say, the configurational energy.

In the grand canonical ensemble, the number of particles is not fixed. The probability density takes the form

$$\rho(\mathbf{r}_1, \dots, \mathbf{r}_n, \mathbf{p}_1, \dots, \mathbf{p}_N, N) = \frac{z^N}{N! h^{3N}} \frac{e^{-\beta \mathcal{H}}}{\Xi}, \tag{2.47}$$

where $z = e^{\beta\mu}$ is the *fugacity* or *activity*, μ is the chemical potential, and

$$\Xi = \sum_{N=0}^{\infty} \frac{z^N}{N! h^{3N}} \int \dots \int e^{-\beta \mathcal{H}} d\mathbf{r}_1 \dots d\mathbf{r}_n d\mathbf{p}_1 \dots d\mathbf{p}_N \tag{2.48}$$

is the grand canonical partition function. Comparing this expression with Equation 2.15, and taking into account Equation 2.16, one easily arrives at

$$\Xi = \sum_{N=0}^{\infty} z^N Q(N) = \sum_{N=0}^{\infty} \frac{z^N \Lambda^{-3N}}{N!} Q_c(N), \tag{2.49}$$

where

    $Q(N)$ is the canonical partition function for an N-particle system

    $Q_c(N)$ is the corresponding configurational integral

The n-particle density in the grand canonical ensemble is defined in the form

$$\rho^{(n)}(\mathbf{r}_1,\ldots,\mathbf{r}_n) = \frac{\sum_{N=n}^{\infty} \frac{z^N \Lambda^{-3N}}{N!} Q_c(N) \rho_N^{(n)}(\mathbf{r}_1,\ldots,\mathbf{r}_n)}{\sum_{N=0}^{\infty} \frac{z^N \Lambda^{-3N}}{N!} Q_c(N)}, \tag{2.50}$$

where $\rho_N^{(n)}(\mathbf{r}_1,\ldots,\mathbf{r}_n)$ is the canonical n-particle density, Equation 2.19, for an N-particle system. Now, the normalization condition is

$$\int_V \ldots \int_V \rho^{(n)}(\mathbf{r}_1,\ldots,\mathbf{r}_n)d\mathbf{r}_1 \ldots d\mathbf{r}_n = \left\langle \frac{N!}{(N-n)!} \right\rangle. \tag{2.51}$$

As a particular case, for $n = 2$ we have

$$\int_V \int_V \rho^{(2)}(\mathbf{r}_1,\mathbf{r}_2)d\mathbf{r}_1 d\mathbf{r}_2 = \langle N^2 \rangle - \langle N \rangle. \tag{2.52}$$

On the other hand, from any textbook on statistical mechanics (see, e.g., McQuarrie 1976) we know that the fluctuations in the particle number in the grand canonical ensemble are related to the isothermal compressibility κ_T in the form

$$\sigma_N^2 = \frac{\langle N \rangle^2 k_B T \kappa_T}{V} \tag{2.53}$$

where $\sigma_N^2 = \langle N^2 \rangle - \langle N \rangle^2$. Combining Equations 2.52 and 2.53, after some algebra we easily arrive at

$$1 + \rho \int [g(r) - 1]d\mathbf{r} = \rho k_B T \kappa_T, \tag{2.54}$$

which is the *compressibility equation*, an alternative route to obtain the EOS. Note that this equation does not involve the assumption of pairwise additivity of the intermolecular forces.

    Equation 2.36 is readily extended to m-component mixtures. Thus,

$$Z = 1 - \frac{2}{3}\pi \frac{\rho}{k_B T} \sum_{i=1}^{m} \sum_{j=1}^{m} x_i x_j \int_0^{\infty} g_{ij}(r) \frac{du_{ij}(r)}{dr} r^3 dr, \tag{2.55}$$

where

    x_i and x_j are the mole fractions of species i and j in the mixture

    $u_{ij}(r)$ the pair potential for the interactions between the two kinds of molecules

    $g_{ij}(r)$ is the corresponding RDF

As a particular case, for HS mixtures the preceding equation reduces to

$$Z = 1 + \frac{2}{3}\pi\rho \sum_{i=1}^{m} \sum_{j=1}^{m} x_i x_j \sigma_{ij}^3 g_{ij}\left(\sigma_{ij}\right), \tag{2.56}$$

where σ_{ij} is the distance of closest approach between the centers of spheres i and j. In a similar way, the energy equation (2.46) for mixtures takes the form

$$U^E = 2\pi N\rho \sum_{i=1}^{m} \sum_{j=1}^{m} x_i x_j \int_0^\infty g_{ij}\left(r\right) u_{ij}\left(r\right) r^2 dr. \tag{2.57}$$

Finally, in the case of a pure fluid with an anisotropic and axially symmetric pair potential $u(r_{12}, \Omega_1, \Omega_2)$, where r_{12} is the distance between the centers of any two molecules 1 and 2 and the angles Ω_1 and Ω_2 define the orientations of these molecules, the pressure and energy equations include integration over angles and become

$$Z = 1 - \frac{1}{6(4\pi)^2}\frac{\rho}{k_B T} \iiint g\left(r_{12}, \Omega_1, \Omega_2\right) \frac{\partial u\left(r_{12}, \Omega_1, \Omega_2\right)}{\partial r_{12}} r_{12} d\mathbf{r}_{12} d\Omega_1 d\Omega_2$$

$$= 1 - \frac{1}{6}\frac{\rho}{k_B T} \int \left\langle g\left(r_{12}, \Omega_1, \Omega_2\right) \frac{\partial u\left(r_{12}, \Omega_1, \Omega_2\right)}{\partial r_{12}} \right\rangle r_{12} d\mathbf{r}_{12}, \tag{2.58}$$

$$U^E = \frac{1}{2(4\pi)^2} N\rho \iiint g\left(r_{12}, \Omega_1, \Omega_2\right) u\left(r_{12}, \Omega_1, \Omega_2\right) d\mathbf{r}_{12} d\Omega_1 d\Omega_2$$

$$= \frac{1}{2} N\rho \int \left\langle g\left(r_{12}, \Omega_1, \Omega_2\right) u\left(r_{12}, \Omega_1, \Omega_2\right) \right\rangle d\mathbf{r}_{12}, \tag{2.59}$$

respectively, where the angular brackets mean unweighted averages over orientations.

2.4   STATIC STRUCTURE FACTOR

For an isotropic system, the RDF $g(r)$ is related to the Fourier transform of the so-called *static structure factor* $S(k)$ in the form

$$g\left(r\right) = 1 + \frac{1}{2\pi^2 r\rho} \int_0^\infty \left[S\left(k\right) - 1\right] \sin\left(kr\right) k dk. \tag{2.60}$$

The importance of this relationship arises in the fact that the structure factor can be experimentally measured by means of diffraction techniques. As a matter of fact, the structure factor $S(k)$ for an N-particle system is the ratio of the intensity of the radiation of wavelength λ scattered by the system in a direction forming an angle θ with the incident beam, with $k = (4\pi/\lambda)\sin(\theta/2)$, to the intensity scattered by a set of N independent particles. Conversely, the inverse Fourier transform of the RDF allows us to obtain the structure factor

$$S(k) = 1 + \frac{4\pi\rho}{k} \int_0^\infty [g(r) - 1] \sin(kr) \, r \, dr. \tag{2.61}$$

For nonisotropic systems, Equations 2.60 and 2.61 read

$$g(\mathbf{r}) = 1 + \frac{1}{\rho(2\pi)^3} \int [S(\mathbf{k}) - 1] e^{i\mathbf{k}\cdot\mathbf{r}} d\mathbf{k} \tag{2.62}$$

and

$$S(\mathbf{k}) = 1 + \rho \int [g(\mathbf{r}) - 1] e^{-i\mathbf{k}\cdot\mathbf{r}} d\mathbf{r}, \tag{2.63}$$

respectively.

Figure 2.5 shows the structure factor for an LJ fluid for $\rho^* = 0.50$ and $T^* = 1.50$, a state relatively close to the critical point, displaying the characteristic damped oscillations. The $k = 0$ value is given by the compressibility equation (2.54)

$$S(0) = \rho k_B T \kappa_T. \tag{2.64}$$

The precise behavior of the static structure factor depends on the state of the system and the nature of the intermolecular forces. For an LJ fluid at high densities, the structure factor is very similar to that corresponding to a fluid with an HS potential (1.8), which indicates that in such circumstances the repulsive forces are dominant

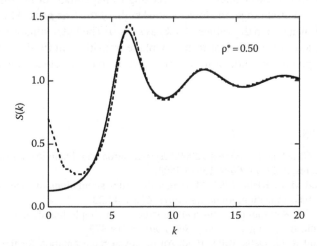

FIGURE 2.5 Structure factor for an LJ fluid (dashed curve) for $\rho^* = 0.50$ and $T^* = 1.50$, the same state as that considered in Figure 2.1a for the RDF, compared with the static structure factor of a HS fluid (continuous curve) with the same reduced density.

(Verlet 1968, Vliegenthart et al. 1999); at low k the structure factor decays to values much lower than 1 and for high enough densities $S(0) \to 0$, reflecting the fact that the fluid becomes nearly incompressible. The main peak is placed at $k_0 = 2\pi/r_0$, where r_0 is slightly lower than σ (Verlet 1968), which means that the position of the first peak is determined by the effective size of the particles. At moderate densities, the shape of the structure factor of the LJ fluid departs from that of the HS fluid, as shown in Figure 2.5. For the latter, in particular, as k is lowered from the value k_0 corresponding to the first peak, the structure factor continuously decreases toward the $k = 0$ value. In contrast, for the LJ fluid, $S(k)$ first decreases with decreasing k below k_0 and then rises again as $k \to 0$; this is because the attractive forces, which become important at moderate to low densities, give rise to an increase in the compressibility of the LJ fluid as compared to that of the HS fluid. The strong rising of $S(k)$ for low k values exhibited by the LJ fluid in Figure 2.5 reveals the presence of large wavelength fluctuations (Vliegenthart et al. 1999), which are related to local density fluctuations that arise when the system approaches a critical point or a binodal. For $k \gtrsim 3$, the structure factor of the LJ fluid continues to be quite well reproduced by that of the HS fluid; this is a reflect of the fact that the oscillations of $S(k)$ arise from the repulsive part of the potential (Verlet 1968).

The Hansen–Verlet freezing criterion (Hansen and Verlet 1969) states that the freezing of an LJ system, at least in the temperature range from the triple point to twice the critical temperature, takes place when the height of the first peak of the radial structure factor is $S(k_0) = 2.85$, as in the HS system. Hansen and Schiff (1973) showed that the criterion was fulfilled, at least in an approximate way, by systems with SS potentials of the form (1.9) with exponents ranging from $n = 1$, the Coulomb potential for which $S(k_0) = 2.55$, to $n = \infty$, the HS potential for which $S(k_0) = 2.85$ as stated earlier. However, for systems that, in addition to the repulsive forces, have shorter-ranged attractive interactions than those in the LJ potential, the values of $S(k_0)$ may be much lower than 2.85 at freezing (Vliegenthart et al. 1999), and the shorter is the potential range, the lower is the height of the first peak. This behavior is accompanied by a rise in the values of $S(k)$ as $k \to 0$. The latter situation may also be present, even at low densities far from any phase transition, in fluids with extremely short-ranged potentials, indicating the presence of particle clustering (Malijevský et al. 2006).

REFERENCES

Babu, S., J.-C. Gimel, and T. Nicolai. 2009. Crystallization and dynamical arrest of attractive hard spheres. *J. Chem. Phys.* 130:064504.

Barrat, J.-L. and M. L. Klein. 1991. Molecular dynamics simulations of supercooled liquids near the glass transition. *Annu. Rev. Phys. Chem.* 42:23.

Finney, J. L. 1970. Random packings and the structure of simple liquids I. The geometry of random close packing. *Proc. Roy. Soc. Lond. A* 319:479.

Gazzillo, D. and R. G. Della Valle. 1993. An improved representation for the high-density structure of Lennard-Jones systems: From liquid toward glass. *J. Chem. Phys.* 99:6915.

Hansen, J. P. and D. Schiff. 1973. Influence of interatomic repulsion on the structure of liquids at melting. *Mol. Phys.* 25:1281.

Hansen, J. P. and L. Verlet. 1969. Phase transitions of the Lennard-Jones system. *Phys. Rev.* 184:151.

Hirschfelder, J. O., C. F. Curtiss, and R. B. Bird. 1954. *Molecular Theory of Gases and Liquids.* New York: John Wiley & Sons.

Hoover, W. G. and J. C. Poirier. 1962. Determination of virial coefficients from the potential of mean force. *J. Chem. Phys.* 37:1041.

Largo, J., M. A. Miller, and F. Sciortino. 2008. The vanishing limit of the square-well fluid: The adhesive hard-sphere model as a reference system. *J. Chem. Phys.* 128:134513.

LaViolette, R. A. and F. H. Stillinger. 1986. Thermal disruption of the inherent structure of simple liquids. *J. Chem. Phys.* 85:6027.

Malijevský, A., S. B. Yuste, and A. Santos. 2006. How Òsticky Ó are short-range square-well fluids? *J. Chem. Phys.* 125:074507.

Mastny, E. A. and J. J. de Pablo. 2007. Melting line of the Lennard-Jones system, infinite size, and full potential. *J. Chem. Phys.* 127:104504.

McQuarrie, D. A. 1976. *Statistical Mechanics.* New York: Harper and Row.

O'Malley, B. and I. Snook. 2005. Structure of hard-sphere fluid and precursor structures to crystallization. *J. Chem. Phys.* 123:054511.

Percus, K. 1964. The pair distribution function in classical statistical mechanics. In *The Equilibrium Theory of Classical Fluids*, ed. H. L. Frisch and J. L. Lebowitz, II, pp. 33–170. New York: Benjamin.

Potoff, J. J. and A. Z. Panagiotopoulos. 1998. Critical point and phase behavior of the pure fluid and a Lennard-Jones mixture. *J. Chem. Phys.* 109:10914.

Stillinger, F. H. and R. A. LaViolette. 1985. Sensitivity of liquid-state inherent structure to details of intermolecular forces. *J. Chem. Phys.* 83:6413.

Stillinger, F. H. and R. A. LaViolette. 1986. Local order in quenched states of simple atomic substances. *Phys. Rev. B* 34:5136.

Stillinger, F. H. and T. A. Weber. 1984. Inherent pair correlation in simple liquids. *J. Chem. Phys.* 80:4434.

Stillinger, F. H. and T. A. Weber. 1985. Inherent structure theory of liquids in the hard-sphere limit. *J. Chem. Phys.* 83:4767.

Verlet, V. 1968. Computer "experiments" on classical fluids. II. Equilibrium correlation functions. *Phys. Rev.* 165:201.

Vliegenthart, G. A., J. F. M. Lodge, and H. N. W. Lekkerkerker. 1999. Strong weak and metastable liquids structural and dynamical aspects of the liquid state. *Physica A* 263:378.

Zhou, S. and J. R. Solana. 2009. Low temperature behavior of thermodynamic perturbation theory. *Phys. Chem. Chem. Phys.* 11:11528.

3 Overview of Computer Simulation Methods

In this chapter, we will give an overview of computer simulation methods. This must not be intended as an exhaustive course on the matter qualifying the inexperienced reader to carry out their own computer simulations, for which there are available excellent and more specialized books. The aim is rather to provide to those readers unfamiliar with the subject, and not particularly interested in it, a guide to make easier the understanding of the foundations of the most common techniques used in computer simulation of the thermodynamic and structural properties of systems, addressing the interested readers to the appropriate literature. We will restrict ourselves to the simulation of equilibrium properties, as this book is restricted to these.

3.1 COMPUTER SIMULATIONS

The ultimate objective of any statistical theory of the thermodynamic and structural properties is to reliably predict these properties for real systems. To do so, we must know accurately the corresponding interaction potential. However, as mentioned in Section 1.2, this is often a very complicated task even for simple molecules. Many-body effects and state dependence of the interaction potential increase the complexity of the problem. For this reason, one often resorts to simple potential functions, as quoted in Section 1.3, resembling more or less closely the actual intermolecular potential of a given substance. When comparing the theoretical results, obtained using the approximate potential, with experiment, surely discrepancies will appear, and we will be unable to discriminate which part of the discrepancy will be due to the inaccuracy of the theory and which part will be due to the inaccuracy of the potential function used. In the development of a theory, it is desirable to separate the question of the accuracy of the potential from that of the theory itself. To this end, we can fix in advance a suitable potential function, but then the problem that arises is that there will not be a real substance with particles interacting exactly through that potential, so that we will not have experimental data to compare with the theory.

To overcome this problem, one can resort to computer simulation. In a computer we can simulate a system consisting of particles interacting through a given potential function and measure its thermodynamic and structural properties. But we can go farther than that. Computer experiments allow us to explore thermodynamic states that are unattainable in real experiments, or follow the evolution with time of the system or a part of it at a microscopic scale, thus providing us new insight into certain properties of the system, or obtain effective pair potentials incorporating many-body effects, as mentioned in Section 1.2.

Two techniques are used to perform computer simulations of the thermodynamic and structural properties of the substances: *Monte Carlo* (MC) and *molecular dynamics* (MD). We will not give here a detailed account of these techniques, as there are excellent books devoted to the subject (Allen and Tildesley 1987, Haile 1992, Frenkel and Smit 2002, Rapaport 2004). We will only give a brief survey on this matter with the aim to provide to the readers unfamiliar with it enough information so as to gain insight into the basics of these techniques and to refer the interested readers to the appropriate literature.

3.2 MONTE CARLO METHOD

The MC method was initially developed by von Neumann, Ulam, and Metropolis little after the end of the Second World War in order to study the diffusion of neutrons in a fissionable material (see Metropolis 1987 for a historical recreation of the birth of the method). The name refers to the most famous casino in the world, because of the use of random numbers in this method. It is to be noted that experiments based on the sampling of random variables were being used since much earlier times. The original contribution of these researchers was, on the one hand, that they thought that certain problems in physics can be solved as a probabilistic problem using a random sampling procedure and, on the other hand, that they realized the usefulness of the recently born electronic computers to carry out these calculations.

Soon after its first implementation, the MC method found applications in many fields of physics, and in particular in statistical mechanics, where one might say that it gave rise to a true revolution together with the MD method developed nearly at the same time. Thus, Metropolis et al. (1953) carried out the first MC simulation to obtain the thermodynamic properties of a simple model fluid, and this pioneering work was followed by many others from an increasing number of researchers and dealing with systems with increasing complexity, in parallel with the increase of the speed and the availability of computers.

3.2.1 MONTE CARLO METHOD IN THE CANONICAL ENSEMBLE

The MC method allows us to sample the configuration space of a system and, therefore, to measure the configurational contribution to the equilibrium thermodynamic properties. The essence of the Metropolis et al. (1953) procedure is as follows. Let us consider a thermodynamic system consisting of N particles in equilibrium at fixed temperature T and volume V, so that the appropriate ensemble is the canonical one. The probability density distribution in this ensemble is given by Equation 2.13, which we rewrite in the form

$$\rho\left(q_i, p_i\right) = \frac{e^{-\beta \mathcal{H}\left(q_i, p_i\right)}}{\int e^{-\beta \mathcal{H}\left(q_i, p_i\right)} dq_i \, dp_i}, \tag{3.1}$$

where, q_i symbolizes the set of generalized position coordinates defining a particular configuration of the system, for example, the position coordinates of the particles or angles defining the orientations of the molecules, p_i symbolizes the set of conjugate momenta, $\mathcal{H}(q_i, p_i)$ is the Hamiltonian of the system, and the integral is obviously a multidimensional integral.

In the canonical ensemble, thermodynamic quantities such as the energy or the pressure fluctuate in a given equilibrium state. Let us consider a quantity $A(q_i, p_i)$ depending on the coordinates and momenta of the particles; its average $\langle A \rangle$ is obtained in the form

$$\langle A \rangle = \int A(q_i, p_i) \, \rho(q_i, p_i) \, dq_i \, dp_i. \tag{3.2}$$

In practice, the integration over momenta in general is easy to carry out, even analytically, as seen in Section 2.2, and will cancel out due to the denominator in (3.1). Therefore, we need only average over configurations, that is,

$$\langle A \rangle = \int A(q_i) \, \rho(q_i) \, dq_i, \tag{3.3}$$

where

$$\rho(q_i) = \frac{e^{-\beta \mathcal{U}_N(q_i)}}{\int e^{-\beta \mathcal{U}_N(q_i)} dq_i} = \frac{e^{-\beta \mathcal{U}_N(q_i)}}{Q_c} \tag{3.4}$$

is the probability density distribution in the configuration space, $\mathcal{U}_N(q_i)$ is the total potential energy of the system in the configuration q_i, and Q_c is the configurational integral introduced in Section 2.2. The probability density that the system, starting from configuration q_i, evolves to a new configuration q_i' is

$$\frac{\rho(q_i')}{\rho(q_i)} = e^{-\beta [\mathcal{U}_N(q_i') - \mathcal{U}_N(q_i)]} = e^{-\beta \Delta \mathcal{U}_N}. \tag{3.5}$$

The MC simulation proceeds as follows. Starting from a given configuration, defined by the set of coordinates q_i of the N particles, we choose particles, one at a time either randomly or sequentially, and attempt to displace them at a new position selected randomly. The displacement of one particle will cause a change $\Delta \mathcal{U}_N$ in the configuration energy of the system. If $\Delta \mathcal{U}_N < 0$, the displacement is accepted; otherwise, it is accepted with a probability given by the corresponding Boltzmann factor $e^{-\beta \Delta \mathcal{U}_N}$. We repeat the trial move with a new particle and so on. After a number of trials, say N, we measure the thermodynamic properties and the structure of the resulting configuration. We repeat the procedure a great number n of times and average the results. An estimation of the statistical uncertainty of the averaged values can be obtained from the standard deviation. Note that the averages obtained are the canonical averages, as we actually sample the configuration space according to the canonical distribution.

3.2.2 Monte Carlo Method in the Isothermal–Isobaric Ensemble

The MC method can be applied in other ensembles. The microcanonical ensemble, in which N, V, and the total energy U of the system are fixed in an equilibrium thermodynamic state, is rarely used in MC simulation. Instead, the isothermal–isobaric ensemble, in which N, T, and the pressure P are fixed at equilibrium, is one of the most frequently used. The probability density distribution in this ensemble is

$$\rho\left(q_i, p_i; V\right) = \frac{e^{-\beta PV} e^{-\mathcal{H}(q_i, p_i)}}{\int_0^\infty e^{-\beta PV} dV \int e^{-\mathcal{H}(q_i, p_i)} dq_i dp_i}. \tag{3.6}$$

Carrying out the integration over momenta, we obtain the probability density of a configuration determined by a set of coordinates q_i and a volume V, in the form

$$\rho\left(q_i; V\right) = \frac{e^{-\beta PV} e^{-\beta \mathcal{U}_N(q_i)}}{\int_0^\infty e^{-\beta PV} dV \int e^{-\beta \mathcal{U}_N(q_i)} dq_i}, \tag{3.7}$$

from which we can obtain the average of a quantity $A(q_i; V)$ as

$$\langle A \rangle = \int_0^\infty dV \int A\left(q_i; V\right) \rho\left(q_i; V\right) dq_i, \tag{3.8}$$

so that we must sample not only the configuration space, but also the fluctuating volume of the system.

To establish the way of sampling, let us assume that the volume of the system is $V = L^3$. For simplicity we will assume that all q_i are position coordinates. Then, performing the variable change $q_i^* = q_i/L$, from Equations 3.7 and 3.8 we have

$$\langle A \rangle = \frac{\int_0^\infty e^{-\beta PV} V^N dV \int A\left(q_i^*; V\right) e^{-\beta \mathcal{U}_N(q_i^*)} dq_i^*}{\int_0^\infty e^{-\beta PV} V^N dV \int e^{-\beta \mathcal{U}_N(q_i^*)} dq_i^*}, \tag{3.9}$$

which can be rewritten in the form

$$\langle A \rangle = \frac{\int A\left(q_i^*; V\right) e^{-\beta PV - \beta \mathcal{U}_N(q_i^*) + N \ln V} dV \, dq_i^*}{\int e^{-\beta PV - \beta \mathcal{U}_N(q_i^*) + N \ln V} dV \, dq_i^*}, \tag{3.10}$$

which is analogous to the expression corresponding to the canonical ensemble, except for the fact that now, instead of the Boltzmann factor $e^{-\beta \mathcal{U}_N(q_i)}$, we have $e^{-\beta PV - \beta \mathcal{U}_N(q_i^*) + N \ln V}$. Denoting $E = -PV - \mathcal{U}_N\left(q_i^*\right) + N\beta^{-1} \ln V$, the MC procedure is as follows. Starting from a configuration determined by the set of coordinates q_i and volume V, we attempt either a change in the position of a particle or a volume change. If this change leads to a decrease in the energy, that is, $\Delta E < 0$, the change is accepted; otherwise, the change is accepted with probability $e^{-\beta \Delta E}$. In practice, a change in volume is attempted after a number of, for example, N, trial particle moves. The thermodynamic averages and the error estimates are determined in a similar way as in the canonical ensemble.

3.2.3 Monte Carlo Method in the Grand Canonical Ensemble

In the grand canonical ensemble, the variables held constant in equilibrium are T, V, and the chemical potential μ, whereas other quantities fluctuate. The probability density in this ensemble is given by Equation 2.47, which we rewrite in the form

$$\rho(q_i, p_i; N) = \frac{\frac{z^N}{N!h^{3N}} e^{-\mathcal{H}(q_i, p_i; N)/kT}}{\sum_{N=0}^{\infty} \frac{z^N}{N!h^{3N}} \int e^{-\mathcal{H}(q_i, p_i; N)/kT} dq_i\, dp_i}. \tag{3.11}$$

The ensemble average of a quantity $A(q_i, p_i; N)$ is

$$\langle A \rangle = \sum_{N=0}^{\infty} z^N \int dp_i \int A(q_i, p_i; N)\, \rho(q_i, p_i; N)\, dq_i. \tag{3.12}$$

If the quantity A depends only on the number and positions of the particles, we can get rid of the integration over momenta, and performing the same variable change as in the NPT ensemble, we obtain

$$\langle A \rangle = \frac{\sum_{N=0}^{\infty} \frac{z^N}{N!} \frac{V^N}{\Lambda^{3N}} \int A(q_i^*; N)\, e^{-\beta\, \mathcal{U}_N(q_i^*)} dq_i^*}{\sum_{N=0}^{\infty} \frac{z^N}{N!} \frac{V^N}{\Lambda^{3N}} \int e^{-\beta\, \mathcal{U}_N(q_i^*)} dq_i^*}, \tag{3.13}$$

where $\Lambda = \left(2\pi mkT/h^2\right)^{-1/2}$ is the thermal wavelength. The preceding expression can be written in the alternative way

$$\langle A \rangle = \frac{\sum_{N=0}^{\infty} \int A(q_i^*; N)\, e^{\beta\, \mu N - N \ln \Lambda^3 - \ln N! - \beta\, \mathcal{U}_N(q_i^*) + N \ln V}\, dq_i^*}{\sum_{N=0}^{\infty} \int e^{\beta\, \mu N - N \ln \Lambda^3 - \ln N! - \beta\, \mathcal{U}_N(q_i^*) + N \ln V}\, dq_i^*}, \tag{3.14}$$

which is again analogous to the expression corresponding to the canonical ensemble, but for the fact that instead of the Boltzmann factor $e^{-\beta \mathcal{U}_N(q_i)}$ we have $e^{\beta \mu N - N \ln \Lambda^3 - \ln N! - \beta \mathcal{U}_N(q_i^*) + N \ln V}$. Then, denoting $E = \beta \mu N - N \ln \Lambda^3 - \ln N! - \beta\, \mathcal{U}_N q_i^* + N \ln V$, the MC sampling proceeds in a similar way as in the isothermal–isobaric ensemble. In this case, we start from a configuration determined by the set of coordinates q_i and number of particles N, and then we attempt either a change in the position of a particle or the insertion/deletion of a particle. If this change leads to a decrease in the energy, that is, $\Delta E < 0$, the change is accepted; otherwise, the change is accepted with probability $e^{-\beta \Delta E}$. The thermodynamic averages and the error estimates are determined in a similar way as in other ensembles.

3.2.4 Configurational-Bias Monte Carlo

For chain molecular fluids at high densities, particle moves as in conventional MC procedures become inefficient because, in addition to translational and rotational trial moves, trial changes of the chain conformation are needed, and the latter are particularly inefficient. Configurational-bias MC methods have been devised to tackle this problem (Siepmann 1990, Frenkel et al. 1992, de Pablo et al. 1992, 1993, Siepmann and Frenkel 1992). These procedures are ultimately based on a much earlier method (Rosenbluth and Rosenbluth 1955) developed to sample configurations of molecular chains from MC simulation considering the problem as a self-avoiding walk on a lattice. To this end, a chain consisting of M equal segments was constructed on a cubic lattice by placing the start of the first segment at a random position in the lattice, then one of the six neighboring positions is chosen at random to place the other end of the segment and the start of a new segment, next a new neighboring position is selected randomly, with the restriction that must not be occupied by the end of a segment, and the end of the second segment is placed in that position, and so on until the M segments are placed. Each configuration so obtained is weighted by its probability of occurrence, which is determined as follows. If m segments have been placed, and at the end of this chain there are n positions available to place the end of the segment $m+1$, with $n \leq 5$ in the cubic lattice except for the end of the first segment for which $n = 6$, then the probability of occurrence of any of the n possible configurations of the $m+1$ chain is $W_{m+1} = (n/5)W_m$, with $W_1 = 6$. In a more general situation, the segments in the molecule that would represent the bonds in a chain molecule may have discrete or continuously varying bond angles and torsional angles, and the probability of insertion of a new segment of the chain, the center of an atom, will depend on the corresponding Boltzmann factor. The configurational-bias MC consists in choosing a neighboring position with probability depending on the corresponding Boltzmann weight, instead of equal probability, so that the most probable configurations, those most largely contributing to the corresponding averages, are sampled preferently. Changes in the configuration of a molecule in the fluid will be performed by trying to add one or more segments to one of the ends of the molecules and, in case of success, deleting an equal number of segments in the opposite end (*reptation*). Therefore, the MC simulation of a chain molecular fluid will consist of conventional trial translational and rotational moves of the molecules and trial changes of the molecular configurations by a reptation procedure.

3.3   MOLECULAR DYNAMICS METHOD

MD method, so called because it allows to solve the Newton equations of motion for a set of particles, was developed by Alder and Wainwright (1957). In contrast with the MC method, which is stochastic in nature, the MD method is deterministic. Instead of sampling microstates from a given ensemble, as does MC, the MD methods follows the temporal evolution of the system. This allows the latter to be applied to both equilibrium and transport properties, whereas MC is restricted nearly exclusively to equilibrium properties.

In MD, the average of a quantity A is obtained from measurements performed a number of times at equal time intervals. To this end, it is necessary to numerically integrate the equations of motion. One might think that this would be the cause of a severe problem, as the molecular motion is chaotic in nature. This means that, starting from different, but extremely close, initial conditions, the phase space trajectories of the system will diverge with time. Even starting from exactly the same conditions, the small errors involved in the numerical calculations will result in a trajectory diverging from the one that would have resulted in an exact calculation. Fortunately this has little importance, as we are interested in statistical averages and not in the exact trajectory.

3.3.1 ALGORITHMS FOR SOLVING THE EQUATIONS OF MOTION

In the general case, for a simple system with continuous potential, the problem to solve is the integration of the second-order differential equation

$$\sum_{j \neq i} \mathbf{F}_{ij} = m\ddot{\mathbf{r}}_i, \tag{3.15}$$

which gives the force acting on particle i due to its interaction with the remaining particles. Or, equivalently, we can solve the two first-order differential equations

$$\mathbf{p}_i = m\dot{\mathbf{r}}_i \tag{3.16}$$

and

$$\sum_{j \neq i} \mathbf{F}_{ij} = \dot{\mathbf{p}}_i. \tag{3.17}$$

Such integration must be performed numerically, for which we need to use an appropriate algorithm. One of the most commonly used is the *Verlet algorithm* (Verlet 1967), based on the integration of Equation 3.15 to obtain the position of the particle at time $t + \Delta t$ in terms of its position at time $t - \Delta t$ by means of the expression

$$\mathbf{r}_i(t + \Delta t) = -\mathbf{r}_i(t - \Delta t) + 2\mathbf{r}_i(t) + \frac{1}{m}\sum_{j \neq i} \mathbf{F}_{ij}(t)(\Delta t)^2 + \mathcal{O}(\Delta t)^4, \tag{3.18}$$

which is obtained from the Taylor series expansions of $\mathbf{r}_i(t + \Delta t)$ and $\mathbf{r}_i(t - \Delta t)$ around $\mathbf{r}_i(t)$ and adding the two results. If, instead, we subtract these results, we easily arrive at

$$\mathbf{v}_i(t) = \frac{\mathbf{r}_i(t + \Delta t) - \mathbf{r}_i(t - \Delta t)}{2\Delta t} + \mathcal{O}(\Delta t)^2, \tag{3.19}$$

which allows us to calculate the velocity.

The Verlet algorithm, in spite of being excellent in many aspects, suffers from some drawbacks. On the one hand, as the position at time $t + \Delta t$ depends on the position at time $t - \Delta t$, this poses a problem to start the simulation, that is, at $t = 0$. On the other hand, as the velocity is obtained from the positions at times $t + \Delta t$ and $t - \Delta t$, whose magnitudes may be large but close to each other, the involved error may be considerable. Further, in Equation 3.18, $\mathbf{r}_i(t + \Delta t)$ is obtained from the addition of terms $\mathcal{O}(\Delta t)^2$ and $\mathcal{O}(\Delta t)^0$, which may involve truncation errors (Allen and Tildesley 1987).

These drawbacks can be overcome by using the *velocity form of the Verlet algorithm* (Swope et al. 1982), which transforms Equations 3.18 and 3.19 into

$$\mathbf{r}_i(t + \Delta t) = \mathbf{r}_i(t) + \mathbf{v}_i(t)\,\Delta t + \frac{1}{2m}\sum_{j\neq i}\mathbf{F}_{ij}(t)\,(\Delta t)^2 \tag{3.20}$$

and

$$\mathbf{v}_i(t + \Delta t) = \mathbf{v}_i(t) + \frac{1}{2m}\sum_{j\neq i}\left[\mathbf{F}_{ij}(t + \Delta t) + \mathbf{F}_{ij}(t)\right]\Delta t, \tag{3.21}$$

respectively.

In the *leap–frog algorithm* (Hockney and Eastwood 1981), which can be derived from the Verlet algorithm, the velocities are calculated at half timesteps, so that

$$\mathbf{r}_i(t + \Delta t) = \mathbf{r}_i(t) + \dot{\mathbf{r}}_i\!\left(t + \frac{\Delta t}{2}\right)\Delta t = \mathbf{r}_i(t) + \mathbf{v}_i\!\left(t + \frac{\Delta t}{2}\right)\Delta t \tag{3.22}$$

and

$$\mathbf{v}_i\!\left(t + \frac{\Delta t}{2}\right) = \mathbf{v}_i\!\left(t - \frac{\Delta t}{2}\right) + \frac{1}{2m}\sum_{j\neq i}\mathbf{F}_{ij}(t)\Delta t. \tag{3.23}$$

Therefore, to calculate the positions at time $t + \Delta t$ we need the velocities at time $t + \Delta t/2$, and the new positions are used to obtain the velocities in the next half timestep, so that each of these quantities is alternatively "leaping" over the other, whence the name of the algorithm. The velocities at time t can be calculated approximately from the relationship

$$\mathbf{v}_i(t) = \frac{1}{2}\left[\mathbf{v}_i\!\left(t + \frac{\Delta t}{2}\right) + \mathbf{v}_i\!\left(t - \frac{\Delta t}{2}\right)\right]. \tag{3.24}$$

Verlet's and other related algorithms are the most frequently used, as they combine easy implementation, low requirement of computational resources, good energy conservation properties, which allows to use relatively long timesteps thus saving computational time, etc., so that for a great variety of applications they are the most appropriate. However, sometimes it is necessary to use higher order algorithms, although in general their characteristics are not so advantageous as those of the former. These higher order algorithms belong to the class of the *predictor–corrector methods*. A predictor–corrector algorithm of order n in essence consists in using the positions of the particles and their first n derivatives at time t to predict the positions and their first n derivatives at time $t + \Delta t$ from a Taylor series expansion around t. In this stage, the equations of motion are not used, so that the predictions will not be accurate. Next, using the equations of motion, the forces and the corresponding accelerations are obtained and the latter are used to correct the predicted values of the accelerations, the positions, and the remaining $n - 1$ derivatives. To do so, the differences between the values of the accelerations calculated from the equations of motion and those predicted, multiplied by suitable coefficients different for each of the quantities, are added to the predicted values to obtain the corrected ones. Gear (1971) reported the sets of coefficients for different values of n and for different orders of the differential equation to be solved; the resulting scheme is referred to as the *Gear prediction–correction algorithm*. To improve the precision of the results, the prediction–correction steps can be iterated until convergence is achieved. However, the iteration is very time consuming, so that it is desirable to reduce the iterations to a minimum reducing, if necessary, the timestep Δt.

Systems with step potentials, such as the HS (1.8) and the SW (1.10) potentials, require a different procedure. In the HS system, particles move along straight lines between collisions. When two particles collide they change their velocities according to the laws of elastic collisions. If two particles i and j with diameter σ, which at $t = 0$ are placed at position $\mathbf{r}_i$ and $\mathbf{r}_j$ and with velocities $\mathbf{v}_i$ and $\mathbf{v}_j$, will collide in the future, the time t_{ij} required to do so must fulfill the condition

$$\left| \mathbf{r}_{ij} + \mathbf{v}_{ij} t_{ij} \right| = \sigma, \tag{3.25}$$

where $\mathbf{r}_{ij} = \mathbf{r}_i - \mathbf{r}_j$ and $\mathbf{v}_{ij} = \mathbf{v}_i - \mathbf{v}_j$. The solution of the quadratic equation resulting from Equation 3.25 is

$$t_{ij} = \frac{-\mathbf{r}_{ij} \cdot \mathbf{v}_{ij} \pm \sqrt{\left(\mathbf{r}_{ij} \cdot \mathbf{v}_{ij}\right)^2 - v_{ij}^2 \left(r_{ij}^2 - \sigma^2\right)}}{v_{ij}^2}. \tag{3.26}$$

The existence of real solutions requires that $\left(\mathbf{r}_{ij} \cdot \mathbf{v}_{ij}\right)^2 - v_{ij}^2 \left(r_{ij}^2 - \sigma^2\right) > 0$ and the condition $t_{ij} > 0$ requires that $\mathbf{r}_{ij} \cdot \mathbf{v}_{ij} < 0$. Then, from the two solutions, the lowest one, that is, that corresponding to the minus sign, is the solution sought.

Therefore, the problem reduces to the calculation of the time $t_{min} = \min(t_{ij})$ to the next collision. Then, each particle of the system is displaced for a time t_{min} with its own velocity so that the two particles that collide will come in contact, that is, their centers will be at a distance σ apart. The collision will change the velocities of the two particles in the form

$$\Delta \mathbf{v}_i = -\Delta \mathbf{v}_j = -\frac{\mathbf{r}_{ij} \cdot \mathbf{v}_{ij}}{\sigma^2} \mathbf{r}_{ij}, \tag{3.27}$$

and the process is repeated for the next collision.

The procedure was extended to SW systems by Alder and Wainwright (1959). In this case, we have two kinds of collisions: repulsive collisions and attractive collisions. In the latter case, particles will leave the potential well if the combined kinetic energy exceeds the attractive potential barrier and will bounce otherwise. Other possible situations are, the two particles entering the mutual potential well or not interacting at all. All these situations can be described by an expression like Equation 3.26 by replacing σ with d, where $d = \sigma$ for repulsive collisions and $d = \lambda\sigma$ for attractive collisions. Further details can be found in Alder and Wainwright (1959).

3.3.2 MOLECULAR DYNAMICS METHOD IN THE CANONICAL ENSEMBLE

In standard MD simulations, the total energy U is kept constant, so that the simulations are performed in the microcanonical (NVU) ensemble. However, MD simulations can be performed in other ensembles, such as the canonical, isothermal–isobaric, and grand canonical ensembles. One of the most frequently used procedures to perform MD simulations in the canonical ensemble is the so-called *Andersen thermostat* (Andersen 1980), in which, from time to time, a particle is selected at random and its velocity is replaced by a new velocity randomly chosen from the Maxwell distribution corresponding to the desired temperature. Then, the energy of the system changes and the system samples the phase space corresponding to the new energy until a new particle experiences a change in its velocity on the basis of the Maxwell distribution. This, in practice, is equivalent to the system interacting with a heat bath. As the method involves an stochastic process, the Andersen thermostat is not purely an MD method, but rather a hybrid of MD and MC methods.

Perhaps the most widely used procedure in canonical MD is the so-called *Nosé–Hoover thermostat* (Nosé 1984a,b, Hoover 1985). The procedure devised by Nosé (1984a) consists in introducing an additional degree of freedom s, with conjugate momentum p_s, which acts as a heat reservoir with which the system can exchange heat, thus giving rise to fluctuations in the total energy of the system. The momenta $\mathbf{p}_i$ of the particles are transformed to $\mathbf{p}'_i$, now considered as the true momenta, in the form $\mathbf{p}'_i = \mathbf{p}_i/s$, which is equivalent to scaling the simulation timestep Δt to $\Delta t' = \Delta t/s$, now considered as the true timestep. A potential energy $(f+1)k_BT \ln s$, where f is the number of degrees of freedom of the actual system and T is the fixed temperature, and a kinetic energy $\frac{1}{2}Q\dot{s}^2 = p_s^2/2Q$ are associated with s, whose conjugate momentum is p_s, so that the Hamiltonian of the extended system is

$$\mathcal{H} = \sum_i \frac{\mathbf{p}_i^2}{2ms^2} + \mathcal{U}_N(\mathbf{r}_1, \ldots, \mathbf{r}_N) + \frac{p_s^2}{2Q} + (f+1)k_BT \ln s. \tag{3.28}$$

It was shown (Nosé 1984a) that with such a particular choice of the potential energy the system truly samples the canonical ensemble, while the extended system evolves in the microcanonical ensemble since the total energy is preserved. The kinetic energy term is introduced to allow to apply the equations of motion to s, with parameter Q, which acts as an effective mass of the heat reservoir, controlling the size of the fluctuations in the energy of the actual system.

From the Hamiltonian (3.28), the canonical equations $\partial H/\partial q_i = -\dot{p}_i$ and $\partial H/\partial p_i = \dot{q}_i$ give

$$\dot{\mathbf{r}}_i = \frac{\mathbf{p}_i}{ms^2}\,\mathbf{r}, \quad \dot{\mathbf{p}}_i = \sum_{j \neq i} \mathbf{F}_{ij},$$

$$\dot{s} = \frac{p_s}{Q}, \qquad \dot{p}_s = \sum_i \frac{\mathbf{p}_i^2}{ms^3} - (f+1)k_B\frac{T}{s},$$

$$\tag{3.29}$$

where $\mathbf{F}_{ij}$ is the force acting on particle i due to the interaction with particle j, Equation 2.10, and we are assuming pairwise additive interactions. $\mathbf{r}_i$, $\mathbf{p}_i$, and t are *virtual* variables defined for the extended system, whereas the *real* variables of the actual system are $\mathbf{r}'_i = \mathbf{r}_i$, $\mathbf{p}'_i = \mathbf{p}_i/s$, and $t' = \int_0^t dt/s$. The equations of motion (3.29) can be expressed in terms of the real variables, but the resulting equations are no longer canonical nor the transformed equation 3.28 is a Hamiltonian (Nosé 1984b).

On the other hand, with the change $dt' = dt/s$, Equations 3.29 transform into (Hoover 1985)

$$\dot{\mathbf{r}}_i = \frac{\mathbf{p}_i}{ms}, \quad \dot{\mathbf{p}}_i = s\sum_{j\neq i}\mathbf{F}_{ij},$$

$$\dot{s} = \frac{sp_s}{Q}, \quad \dot{p}_s = \sum_i \frac{\mathbf{p}_i^2}{ms^2} - (f+1)k_BT, \tag{3.30}$$

where the dot over a variable means a derivative with respect to t', instead of t. We can now eliminate the variable s from the equations of motion by performing the time derivative of the first equation in (3.30) and combining the result with the remaining equations, with the result

$$\ddot{\mathbf{r}}_i = \frac{\dot{\mathbf{p}}_i}{ms} - \frac{\mathbf{p}_i}{ms}\frac{\dot{s}}{s} = \frac{1}{m}\sum_{j\neq i}\mathbf{F}_{ij} - \frac{\dot{\mathbf{r}}_ip_s}{Q} = \frac{1}{m}\sum_{j\neq i}\mathbf{F}_{ij} - \zeta_T\dot{\mathbf{r}}_i, \tag{3.31}$$

where $\zeta_T = \dot{s}/s = p_s/Q$ acts as a friction coefficient with changes with time in the form

$$\dot{\zeta}_T = \frac{\dot{p}_s}{Q} = \frac{1}{Q}\left[\sum_i \frac{\mathbf{p}_i^2}{ms^2} - (f+1)k_BT\right] = \frac{1}{Q}\left[\sum_i m\dot{\mathbf{r}}_i^2 - (f+1)k_BT\right]. \tag{3.32}$$

Replacing $f+1$ by f, Hoover (1985) showed that Equations 3.31 and 3.32 lead the system to sample canonically the phase space, while avoiding the timescaling involved in the Nosé scheme. Defining an instantaneous kinetic temperature $T(t)$ as

$$T(t) = \frac{1}{fk_B}\sum_i \frac{\mathbf{p}_i^2}{ms^2} = \frac{1}{fk_B}\sum_i m\dot{\mathbf{r}}_i^2, \tag{3.33}$$

Equation 3.32 transforms into

$$\dot{\zeta}_T = \frac{1}{\tau_T^2}\left[\frac{T(t)}{T} - 1\right], \tag{3.34}$$

where $\tau_T = (Q/fk_BT)^{1/2}$ has the dimensions of time. Then, the Nosé–Hoover algorithm can be implemented, within the scheme of the leap–frog algorithm, as follows (Smith et al. 2009):

$$\zeta_T\left(t + \frac{1}{2}\Delta t\right) = \zeta_T\left(t - \frac{1}{2}\Delta t\right) + \frac{\Delta t}{\tau_T^2}\left[\frac{T(t)}{T} - 1\right], \tag{3.35}$$

$$\zeta_T(t) = \frac{1}{2}\left[\zeta_T\left(t + \frac{1}{2}\Delta t\right) + \zeta_T\left(t - \frac{1}{2}\Delta t\right)\right], \tag{3.36}$$

$$\mathbf{v}_i\left(t + \frac{1}{2}\Delta t\right) = \mathbf{v}_i\left(t - \frac{1}{2}\Delta t\right) + \Delta t\left[\frac{1}{m}\sum_{j\neq i}\mathbf{F}_{ij} - \zeta_T\mathbf{v}_i(t)\right], \tag{3.37}$$

$$\mathbf{v}_i(t) = \frac{1}{2}\left[\mathbf{v}_i\left(t + \frac{1}{2}\Delta t\right) + \mathbf{v}_i\left(t - \frac{1}{2}\Delta t\right)\right], \tag{3.38}$$

$$\mathbf{r}_i(t + \Delta t) = \mathbf{r}_i(t) + \Delta t\,\mathbf{v}_i\left(t + \frac{1}{2}\Delta t\right), \tag{3.39}$$

with $\mathbf{v}_i \equiv \dot{\mathbf{r}}_i$. The calculation of both $T(t)$ and $\mathbf{v}_i(t)$ requires the previous knowledge of $\mathbf{v}_i(t)$, so that an iteration procedure must be used.

3.3.3 MOLECULAR DYNAMICS METHOD IN THE ISOTHERMAL–ISOBARIC ENSEMBLE

Andersen (1980) also devised a procedure to perform MD simulations at constant pressure. The procedure is based on introducing a new variable V, the volume of the system that now fluctuates, with conjugate momentum p_V, and associated kinetic energy $\frac{1}{2}\mathcal{M}\dot{V}^2 = p_V^2/2\mathcal{M}$ and potential energy PV, where $\mathcal{M}$, with dimensions of mass × volume$^{-4/3}$, is the "mass" associated with a hypothetical piston forcing the system to compress or expand isotropically, and P is the fixed pressure. Combining this procedure with the Andersen thermostat, MD simulations in the isothermal–isobaric ensemble can be performed.

The Andersen procedure for MD simulations at constant pressure was also combined with the Nosé thermostat to perform MD simulations in the NPT ensemble (Nosé 1984a,b). To this end, assuming that the volume of the system is a cubic cell with edge L, so that $V = L^3$, the position vector of the particles are expressed in reduced units as

$$\mathbf{x}_i = \frac{\mathbf{r}_i}{V^{1/3}}. \tag{3.40}$$

The Hamiltonian of this extended system is

$$\mathscr{H} = \sum_i \frac{\mathbf{p}_i^2}{2ms^2V^{2/3}} + \mathcal{U}\left(V^{1/3}\mathbf{x}_1, \ldots, V^{1/3}\mathbf{x}_N\right) + \frac{p_s^2}{2Q} + (f+1)\,k_BT\ln s + \frac{p_V^2}{2\mathcal{M}} + PV, \tag{3.41}$$

and the equations of motion are (Nosé 1984b)

$$\dot{\mathbf{x}}_i = \frac{\mathbf{p}_i}{ms^2 V^{2/3}}, \quad \dot{\mathbf{p}}_i = V^{1/3} \sum_{j \neq i} \mathbf{F}_{ij},$$

$$\dot{s} = \frac{p_s}{Q}, \qquad \dot{p}_s = \sum_i \frac{\mathbf{p}_i^2}{ms^3 V^{2/3}} - (f + 1)k_B \frac{T}{s}, \qquad (3.42)$$

$$\dot{V} = \frac{p_V}{\mathcal{M}}, \qquad \dot{p}_V = \frac{1}{3} \sum_i \frac{\mathbf{p}_i^2}{ms^2 V^{5/3}} + \frac{1}{3} \sum_{j \neq i} \mathbf{r}'_i \cdot \frac{\mathbf{F}_{ij}}{V^{2/3}} - P.$$

Introducing the scaled variables $\mathbf{x}_i = \mathbf{r}_i/V^{1/3}$, $\mathbf{p}'_i = \mathbf{p}_i/sV^{1/3}$, and $t' = \int_0^t dt/s$, which are the real variables for the system, the Hoover formulation for the isothermal–isobaric equations of motion is (Hoover 1985)

$$\dot{\mathbf{x}}_i = \frac{\mathbf{p}'_i}{mV^{1/3}}, \qquad\qquad \dot{\mathbf{p}}'_i = \sum_{j \neq i} \mathbf{F}_{ij} - (\zeta_P + \zeta_T)\mathbf{p}'_i,$$

$$\zeta_T = \frac{\left[\sum_i \mathbf{p}_i^2/m - fk_B T\right]}{Q}, \qquad\qquad (3.43)$$

$$\zeta_P = \frac{\dot{V}}{3V}, \qquad\qquad \dot{\zeta}_P = [P(t) - P]V/\tau_P^2 k_B T,$$

where
   τ_P is a relaxation time
   $P(t)$ is the instantaneous pressure obtained from the virial theorem (2.12) by
      replacing the temperature T with the instantaneous kinetic temperature $T(t)$
      and suppressing the averages

However, it was shown that with the equations of motion (3.43), the system does not sample exactly the *NPT* ensemble (Melchionna et al. 1993). To remedy this, the first of these equations was changed to

$$\dot{\mathbf{r}}_i = \frac{\mathbf{p}'_i}{m} + \zeta_P(\mathbf{r}_i - \mathbf{R}_0), \qquad (3.44)$$

where $\mathbf{R}_0$ is the center of mass of the system and $\mathbf{r}'_i = \mathbf{r}_i$.
   The Nosé–Hoover algorithm for MD simulations in the *NPT* ensemble, with the Melchionna et al. (1993) modification, can be implemented in a similar way as for the Nosé–Hoover thermostat, namely,

$$\zeta_T\left(t + \frac{1}{2}\Delta t\right) = \zeta_T\left(t - \frac{1}{2}\Delta t\right) + \frac{\Delta t}{\tau_T^2}\left[\frac{T(t)}{T} - 1\right], \qquad (3.45)$$

$$\zeta_T(t) = \frac{1}{2}\left[\zeta_T\left(t + \frac{1}{2}\Delta t\right) + \zeta_T\left(t - \frac{1}{2}\Delta t\right)\right],\tag{3.46}$$

$$\zeta_P\left(t + \frac{1}{2}\Delta t\right) = \zeta_P\left(t - \frac{1}{2}\Delta t\right) + \frac{\Delta t V(t)}{f k_B T \tau_P^2}[P(T) - P],\tag{3.47}$$

$$\zeta_P(t) = \frac{1}{2}\left[\zeta_P\left(t + \frac{1}{2}\Delta t\right) + \zeta_P\left(t - \frac{1}{2}\Delta t\right)\right],\tag{3.48}$$

$$\mathbf{v}_i\left(t + \frac{1}{2}\Delta t\right) = \mathbf{v}_i\left(t - \frac{1}{2}\Delta t\right) + \Delta t\left[\frac{1}{m}\sum_{j \neq i}\mathbf{F}_{ij} - (\zeta_T + \zeta_P)\,\mathbf{v}_i(t)\right],\tag{3.49}$$

$$\mathbf{v}_i(t) = \frac{1}{2}\left[\mathbf{v}_i\left(t + \frac{1}{2}\Delta t\right) + \mathbf{v}_i\left(t - \frac{1}{2}\Delta t\right)\right],\tag{3.50}$$

$$\mathbf{r}_i(t + \Delta t) = \mathbf{r}_i(t) + \Delta t\left\{\mathbf{v}_i\left(t + \frac{1}{2}\Delta t\right) + \zeta_P\left(t + \frac{1}{2}\Delta t\right)\left[\mathbf{r}_i\left(t + \frac{1}{2}\Delta t\right) - \mathbf{R}_0\right]\right\},\tag{3.51}$$

$$\mathbf{r}_i\left(t + \frac{1}{2}\Delta t\right) = \frac{1}{2}[\mathbf{r}_i(t) + \mathbf{r}_i(t + \Delta t)].\tag{3.52}$$

As in the case of the algorithm for the Nosé–Hoover thermostat, this algorithm requires an iteration procedure.

3.3.4 MOLECULAR DYNAMICS METHOD IN THE GRAND CANONICAL ENSEMBLE

Pettitt and coworkers (Çağin and Pettitt 1991a,b, Lynch and Pettitt 1997) developed a procedure to perform MD simulations in the grand canonical ensemble which was latter modified by Eslami and Müller-Plathe (2007). The procedure is based on coupling the real system to heat and particle reservoirs. In the approach of Eslami and Müller-Plathe (2007) the number of particles of the system is considered as a continuously varying dynamical variable, so that in a given instant the system has N real particles plus one additional particle "scaled" by a factor ν between 0 and 1, so that for $\nu = 0$ the system has N identical particles, for $\nu = 1$ the system has $N + 1$ identical particles, and for values of ν between these two, the system has N identical particles plus one fractional particle. The Hamiltonian of this extended system is

$$\mathcal{H} = \sum_{i=1}^{N}\frac{\mathbf{p}_i^2}{2ms^2} + \frac{p_s^2}{2Q} + \frac{\mathbf{p}_f^2}{2m_f s^2} + \frac{p_\nu^2}{2W} + \mathcal{U}_N + \mathcal{U}_s + \mathcal{U}_f + \mathcal{U}_\nu,\tag{3.53}$$

where $p_f^2/2m_f s^2$ is the kinetic energy of the fractional particle, with mass $m_f = \nu^{1/3}m$, $p_\nu^2/2W$ is the kinetic energy associated with the particle reservoir, $\mathcal{U}_N$ is the potential energy of the N-particle system, $\mathcal{U}_s = (f + 1)k_B T \ln s$, with $f = 3(N + 1)$, is the potential energy associated with the heath reservoir, $\mathcal{U}_f = \sum_i u_{if}$ is the potential energy for the interaction between the fractional particle and the remaining N particles of the system, in which u_{if} is the interaction potential between particle i and the fractional particle, and $\mathcal{U}_v$ is the potential energy associated with the particle reservoir, which is given by the excess Helmholtz free energy of the extended system with respect to an ideal gas mixture consisting of N equal particles plus one fractional particle, that is,

$$\mathcal{U}_v = F^E = N\mu + \mu_f + Nk_BT\ln\left[\left(\frac{2\pi mk_BT}{h^2}\right)^{3/2}\frac{V}{N}\right]$$

$$+ k_BT\ln\left[\left(\frac{2\pi v^{1/3}mk_BT}{h^2}\right)^{3/2}V\right], \tag{3.54}$$

where
   μ is the chemical potential of the ordinary particles
   μ_f is that of the fractional particle

From Hamiltonian (3.53), the equations of motion are

$$m\ddot{\mathbf{r}}_i = \frac{1}{s^2}\left(\mathbf{F}_{if} + \sum_{j\neq i}\mathbf{F}_{ij}\right) - 2m\frac{\dot{s}}{s}\dot{\mathbf{r}}_i, \tag{3.55}$$

$$v^{1/3}m\ddot{\mathbf{r}}_f = \frac{1}{s^2}\mathbf{F}_{if} - 2v^{1/3}m\frac{\dot{s}}{s}\dot{\mathbf{r}}_f - \frac{1}{3}m\frac{\dot{v}}{v^{2/3}}\dot{\mathbf{r}}_f, \tag{3.56}$$

$$Q\ddot{s} = \sum_{i=1}^{N}ms\dot{\mathbf{r}}_i^2 + v^{1/3}ms\dot{\mathbf{r}}_f^2 - \frac{1}{s}(f+1)k_BT, \tag{3.57}$$

$$W\ddot{v} = -\sum_{i=1}^{N}\frac{\partial u_{if}}{\partial v} + \mu - k_BT\ln\left[\left(\frac{2\pi mk_BT}{h^2}\right)^{3/2}\frac{N}{V}\right]$$

$$+ \frac{1}{6}mv^{-2/3}s^2\dot{\mathbf{r}}_f^2 - \frac{1}{2}\frac{k_BT}{v}, \tag{3.58}$$

where
   $\mathbf{F}_{if}$ is the force that the fractional particle f exerts on particle i
   $\mathbf{F}_{ij}$ is the force that particle j, other than f and i, exerts on particle i

These equations can be solved (Lynch and Pettitt 1997, Eslami and Müller-Plathe 2007) using an algorithm first developed for MD simulations in the *NPT* ensemble by Fox and Andersen (1984).

   Although we have restricted the discussion in this section to simple atomic systems, the procedures can be extended to deal with polyatomic molecules.

3.4　SOME TECHNICAL DETAILS

3.4.1　NUMBER OF PARTICLES

If we are dealing with noncrystalline systems, the number of particles to be used in the simulation is arbitrary, without other limitations that the fact that a too small number of particles will give inaccurate estimates of the thermodynamic properties and a too great number of particles will result in an enormous increase in computational time without necessarily increasing significantly the accuracy.

In contrast, in the simulation of crystalline solids, the number of particles to be used is regulated by the crystalline structure, because otherwise a defective solid will result. Thus, for example, the face center cubic (fcc) structure is determined by four particles placed at positions $(0, 0, 0)$, $(0, a/2, a/2)$, $(a/2, 0, a/2)$, and $(a/2, a/2, 0)$. Repeating this structure in each direction we will have a perfect fcc crystal. Therefore, the number of particles is determined by the condition $N = 4n^3$. This gives, for example, 108, 256, 500, 864, 1372, 2048,... , particles for $n = 3, 4, 5, 6, 7, 8,....$ The *body centered cubic* (bcc) structure is built from two particles placed at positions $(0, 0, 0)$ and $(a/2, a/2, a/2)$, so that $N = 2n^3$, and the *simple cubic* (sc) structure can be constructed from a single particle placed at position $(0, 0, 0)$, so that $N = n^3$.

3.4.2　PERIODIC BOUNDARY CONDITIONS

Unavoidably, the number of particles in the simulation box is limited. In most cases, only a few hundreds of particles are used, in some cases a few thousands, and rarely more. In such small systems the surface effects would be very important, strongly influencing the thermodynamic and structural properties. To avoid this problem, periodic boundary conditions are imposed to the system. This means that the simulation box is repeated in each direction an infinite number of times, as illustrated in Figure 3.1, which results in an infinite system. Each time a particle is moved, each of its images is moved in the same way.

In spite of the use of periodic boundary conditions, certain problems remain. Thus, for example, the limited size of the simulation box prevents the existence of long wavelength fluctuations in a fluid near the critical point, in contrast with the situation in a true infinite system.

3.4.3　MINIMUM IMAGE CONVENTION

In the interaction of a particle i with another particle j, one considers the replica of the latter that is nearest to particle i, as shown in Figure 3.1.

3.4.4　CUTOFF DISTANCE OF A CONTINUOUS POTENTIAL

For potentials with continuous tails, we ought to consider interactions up to infinite distances. Obviously, this is impossible and, moreover, meaningless, because for enough large distances the potential tail becomes extremely small, so that its effect can be neglected. In such cases, a cutoff distance r_c is introduced beyond which the potential is set to zero. The choice of such distance is somewhat arbitrary, although

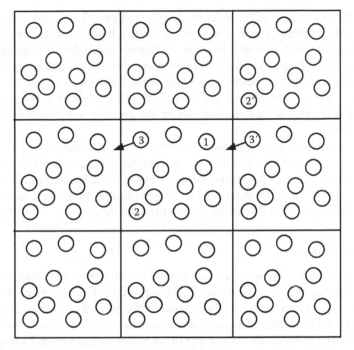

FIGURE 3.1 Illustration, for a two-dimensional system, of the periodic boundary conditions and of the minimum image convention. The central box is the simulation cell and the remaining boxes are identical copies of the former. In the interaction between particles 1 and 2, the image 2′ of particle 2 is considered because it is nearer than the particle 2 itself. When a particle 3 exits the simulation box, its image 3′ enters the opposite side.

a reasonable value must take into account the decay rate of the potential tail. For the LJ potential, the usual choice is $r_c = 2.5\sigma$.

The truncation of the potential requires the introduction of tail corrections in the thermodynamic properties to account for the effect of the truncation. These corrections will be explained in Section 3.5. In MD simulations, however, the truncation of the potential leads to an unsuitable discontinuity in the force. One way to overcome this drawback is to truncate and shift the potential so that it becomes zero at $r = r_c$. Then, the truncated and shifted potential $u^{\mathrm{TS}}(r)$ will be

$$u^{\mathrm{TS}}(r) = \begin{cases} u(r) - u(r_c), & r \le r_c \\ 0, & r > r_c \end{cases} \tag{3.59}$$

3.4.5 SETTING THE INITIAL STATE

In the simulation of a crystalline solid, the initial configuration would be a perfect crystalline lattice with the appropriate structure. In the case of a fluid, any initial configuration would be suitable. For convenience, we can start from any crystalline structure, because subsequently the system will evolve spontaneously toward a fluid if this is the more stable phase under the fixed conditions. To accelerate the melting of the initial crystalline structure, we can start with a high temperature and/or a low density state and then gradually increase the density and reduce the temperature to achieve the desired state. An alternative procedure, which leads efficiently to melting even for temperatures well below the melting point, has been also reported (McBride et al. 2005). If we are to perform simulations for different states in the same phase, it may be preferable to take the last one of the previous simulation as the initial configuration.

In MD simulations, we also need to set the initial velocities of the particles. We can assign random initial velocities to the particles, because in the course of the simulation the velocities will spontaneously evolve toward a Maxwellian distribution. We can approach the instantaneous temperature of the system, as defined by Equation 3.33, to a desired value by scaling the velocities.

3.4.6 EQUILIBRATION

Obviously, the system in the initial state, established in the way just described, will be far from thermodynamic equilibrium. Therefore, it must be equilibrated before performing measurements of the thermodynamic and structural properties. The number of MC steps or MD timesteps required by the system to reach equilibrium depends on how far from equilibrium the system was in the initial state as well as on temperature and density. During the equilibration process, any measured thermodynamic quantity will show a tendency toward increasing or decreasing. Once equilibrium is achieved, the thermodynamic properties will show oscillations around an average value.

3.4.7 SETTING THE LENGTH OF THE STEPS

Once the particles are placed in the initial configuration, one has to establish the length of the steps, that is, the maximum displacement Δ, either positive or negative, allowed for each coordinate of a particle in MC simulations or the timestep Δt in MD simulations.

In MC simulations, once the value of Δ is determined ,we will choose particles at random, one at a time, and then generate random displacements Δx, Δy, and Δz for the coordinates of the center of the particle by selecting three random values between $-\Delta$ and $+\Delta$. Obviously, in the case of molecular systems we would also generate random changes in the angles defining the orientation of the molecules. In any case, if the value of Δ is too small, most of the attempted displacements will be accepted, because the changes in energy surely will be small and, in case of particles with a hard core, the probability of overlapping will also be small, but the successive configurations will be highly correlated, so that the system would be inefficiently sampling the phase space and then larger runs would be necessary to accurately obtain the thermodynamic properties. If, instead, Δ is too big, although the correlation between successive configurations will be low, most of the attempted displacements will be rejected and so, again, we would be inefficiently sampling the phase space. It is commonly agreed that an acceptance ratio of 30%–50% is reasonable, although of course at low densities the acceptance ratio may be much higher.

The situation is much the same in MD simulations, but for the fact that now there is no rejection of the moves. However, a new problem arises, namely, that for large timesteps the conservation of the energy may be poor, depending on the algorithm used. Therefore, for a given algorithm we must determine the optimal choice for the timestep Δt.

3.4.8 SAVING COMPUTATIONAL TIME

The most time-consuming part of the simulation is usually the calculation of the interactions between all pairs of particles in the system. Strictly speaking, once we have fixed a cutoff radius r_c, we only need to calculate the interactions between pairs of particles separated at a distance smaller than r_c, but to do so we would still need to calculate the distances between all pairs of particles, which would take a time of the order of N^2 for a N-particle system. To save computational time, two procedures commonly used are the Verlet list and the cell list.

In the *Verlet list*, a second radius $r_0 > r_c$ is fixed in such a way that $r_c - r_0$ is the maximum distance that the centers of any pair of particles can approach each other in a given number of steps, either MC steps or MD timesteps. Then, a list of neighbors situated at a distance $r < r_0$ of each particle is associated to that particle. When we calculate the interaction of a particle i with the remaining particles, we only have to determine which particles in the list of particle i are at a distance $r < r_c$ and then calculate the corresponding interactions. If the number of neighbors within a distance r_0 of particle i is n, the number of distances to calculate is of the order of nN, with $n \ll N$ for large N, instead of N^2. Every time that a particle enters or leaves the sphere of radius r_0 of particle i, the associated list must be updated.

In the *cell list*, the simulation box, assumed cubic, is divided into cubic boxes with edge r_c. Then a list of particles within each of the small boxes is associated to it. To calculate the interactions of a particle i we only need to consider those particles in its box and in the nearest neighboring boxes that are a distance $r < r_c$ apart from particle i. The number of distances to be calculated is of the order of $n'N$, where again $n' \ll N$ for large N.

3.4.9 DEALING WITH LONG-RANGE INTERACTIONS

The use of periodic boundary conditions introduces some degree of ordering in an otherwise isotropic fluid, which represents a problem when we are dealing with long-ranged interactions. Moreover, in this case, the truncation of the potential is not feasible because the involved error, even introducing tail corrections in the thermo-dynamic properties, will be considerable. In particular, in the case of interactions between charged particles, the truncation may introduce anomalous distribution of charges within the sphere of radius r_c (Allen and Tildesley 1987). Minimum image convention is not useful when the effective range of the interaction is greater than the size of the box. A better approach to the problem will consist in considering the inter-action of each particle not only with the nearest image of the others, but also with all the images of the particles in all the boxes. In this case, the total interaction energy of the system, assumed pairwise additivity and spherically symmetric potential, will be

$$\mathcal{U}(\mathbf{r}_1,\ldots,\mathbf{r}_N) = \frac{1}{2}\sum_{\mathbf{n}}{}'\sum_{i=1}^{N}\sum_{j=1}^{N} u\left(|\mathbf{r}_i + \mathbf{n}L - \mathbf{r}_j|\right), \tag{3.60}$$

where $\mathbf{n}$ is an integer vector, L is the edge of the box, the prime in the first sum means that the interaction of particle i with itself is not counted for $\mathbf{n} = 0$, and the factor $1/2$ is introduced to avoid counting twice the interaction of each pair ij.

As the sum in Equation 3.60 is poorly convergent for long-ranged potentials, the convergence must be improved in some way. One of the most widely used procedures for systems with charged particles as well as dipolar systems is the *Ewald sum*, initially devised to calculate the Madelung energy of ionic solids, which basically consists in rearranging the terms in the sum by grouping them in nearly spherical layers around the central box. A simple introduction to these procedures can be found in Allen and Tildesley (1987), Frenkel and Smit (2002) and Rapaport (2004).

3.5 THERMODYNAMIC AND STRUCTURAL PROPERTIES FROM COMPUTER SIMULATION

The procedure used to obtain the radial distribution function (RDF) from computer simulation is based on Equation 2.33. To this end, the intermolecular distances are divided into small intervals of width Δr. The number N_i of pairs separated a distance within the interval $(r_i - \Delta r/2, r_i + \Delta r/2)$, $i = 1, 2, 3, \ldots$, is averaged from measurements performed for a number of independent configurations. If $\langle N_i \rangle$ is such average then, provided that Δr is small enough, we will have

$$g(r_i) = \frac{\langle N_i \rangle}{4\pi\rho r_i^2 \Delta r}. \tag{3.61}$$

In computer simulations in the *NVT* ensemble for a system with continuos potential, the EOS is obtained from Equation 2.36, by replacing the integral with a sum using the RDF obtained in the form just described. In case of potentials with a hard spherical core plus a continuous tail, we must also calculate the contact value $g(\sigma)$, according to Equation 2.45, by extrapolating the RDF to $r = \sigma$, which involves some uncertainty especially for very short-ranged potentials at high densities. For potentials with more discontinuities, we must perform additional extrapolations, as is the case of the SW potential (1.10) for which, according to Equation 2.44, we have to extrapolate to $r = \lambda\sigma^-$ and $r = \lambda\sigma^+$ in addition to $r = \sigma$.

In computer simulations in the *NPT* ensemble, the compressibility factor Z can be calculated directly from its definition as $Z = P\langle V \rangle / Nk_BT$, where $\langle V \rangle$ is the average volume determined from a large number of different configurations.

In the microcanonical ensemble, the EOS can be obtained in the same way as in the canonical ensemble, because in the thermodynamic limit, $N \to \infty$ and $V \to \infty$ with $\rho = N/V$ being finite, all the ensembles provide the same results. This is not exactly true in computer simulations, because of the limited number of particles used. On the other hand, the temperature is not fixed in this ensemble, so that it must be calculated from averaging the instantaneous temperature as defined in Equation 3.33.

In any case, we must correct for the truncation of the potential at r_c. This may be achieved by approximating the RDF beyond the cutoff distance by that corresponding to the ideal gas, namely, $g(r) = 1$ for $r > r_c$. Then, the pressure equation (2.36) gives for the correction term Z_c the expression

$$Z_c = -\frac{2}{3}\pi \frac{\rho}{k_B T} \int_{r_c}^{\infty} \frac{du(r)}{dr} r^3 dr. \tag{3.62}$$

Obviously, this correction is not suitable for long-ranged potentials.

The excess energy in computer simulations can be obtained from the energy equation (2.46), replacing the integral with a sum and using the RDF obtained in the form (3.61). Again, a long-range correction U_c must be added to the result. From Equation 2.46, with $g(r) = 1$ for $r > r_c$, the correction term is

$$U_c = 2\pi N \rho \int_{r_c}^{\infty} u(r) r^2 dr. \tag{3.63}$$

In case of simulations performed with a cut and shifted potential, if one wishes to obtain the energy corresponding to the system with the full, unshifted potential, apart from the long-range correction, we must introduce an additional correction U_s to account for the effect of the shifting on the energy. This correction can be approximated by

$$U_s = 2\pi N \rho u(r_c) \int_0^{r_c} g(r) r^2 dr. \tag{3.64}$$

The constant-volume heat capacity C_V of the system can be obtained from the fluctuations of the energy U in the canonical ensemble, as is well known from statistical mechanics (see, e.g., McQuarrie 1976), in the form

$$C_V = \frac{\sigma_U^2}{k_B T^2}, \tag{3.65}$$

where $\sigma_U^2 = \langle U^2 \rangle - \langle U \rangle^2$.

In the microcanonical ensemble, the total energy is constant, but the kinetic and potential energies of the system both fluctuate and the constant-volume heat capacity can be obtained from the fluctuations of any of these two quantities (Lebowitz et al. 1967).

A number of second-order properties, those obtained from second-order derivatives of a thermodynamic potential, can be easily obtained from simulations in the isothermal–isobaric ensemble. Thus, the constant-pressure heat capacity C_P of the system can be obtained from the fluctuations in the enthalpy $H = U + PV$ in this ensemble, according to the expression

$$C_P = \frac{\sigma_H^2}{k_B T^2},$$ (3.66)

where $\sigma_H^2 = \langle H^2 \rangle - \langle H \rangle^2$. In MC simulations, as only the configuration space is sampled, and not the momentum space, we can only compute the configurational contribution to the constant-pressure heat capacity, that is, the constant-pressure excess heat capacity C_P^E with respect to that of the ideal gas, which is given by (Lagache et al. 2001)

$$C_P^E = \frac{1}{k_B T^2} \left(\langle U^E \widehat{H} \rangle - \langle U^E \rangle \langle \widehat{H} \rangle \right) + \frac{P}{k_B T^2} \left(\langle V \widehat{H} \rangle - \langle V \rangle \langle \widehat{H} \rangle \right) - N k_B,$$ (3.67)

where $\widehat{H} = U^E + PV$.

The isothermal compressibility is related to the fluctuations in volume in the *NPT* ensemble by means of the relationship (McQuarrie 1976)

$$\kappa_T = \frac{1}{k_B T} \frac{\sigma_V^2}{\langle V \rangle}.$$ (3.68)

with $\sigma_V^2 = \langle V^2 \rangle - \langle V \rangle^2$.

The isobaric thermal expansivity α_P is other quantity that can be easily obtained from simulations in the *NPT* ensemble in the form

$$\alpha_P = \frac{1}{\langle V \rangle k_B T^2} \left(\langle VH \rangle - \langle V \rangle \langle H \rangle \right),$$ (3.69)

and for MC simulations (Lagache et al. 2001)

$$\alpha_P = \frac{1}{\langle V \rangle k_B T^2} \left(\langle V \widehat{H} \rangle - \langle V \rangle \langle \widehat{H} \rangle \right).$$ (3.70)

Other second-order thermodynamic properties, apart from those quoted here, can be obtained from the microcanonical, canonical, and grand canonical ensembles (see, e.g., Allen and Tildesley 1987), but the procedures outlined here are perhaps the most frequently used in computer simulations.

Another thermodynamic quantity of great interest is the chemical potential which, in the canonical ensemble, is given by $\mu = (\partial F / \partial N)_{N,V,T}$, where $F = -k_B T \ln Q$ is the Helmholtz free energy and Q is the canonical partition function (2.16). The chemical potential can be separated into two contributions: One of them is the chemical potential of an ideal gas with the same density and temperature as the system under consideration

$$\mu_{id} = k_B T \ln \left(\rho \Lambda^3 \right),$$ (3.71)

and the other arises from the interactions and gives the excess chemical potential μ^E over that of the ideal gas. The latter can be expressed in the form

$$\mu^E = -k_B T \ln \langle e^{-\Delta\mathcal{U}/k_B T} \rangle, \tag{3.72}$$

where $\Delta\mathcal{U}$ is the change in the potential energy of the system when an additional particle, the test particle, is introduced and the angular brackets mean an average over all configurations of the N-particle system and, for each configuration, over all possible positions of the additional particle. This is the basis of the *Widom test particle insertion* (TPI) *method* (Widom 1963), one of the most frequently used procedures to obtain from simulation the excess chemical potential of fluids. The method can be used in combination with either MC or MD simulation and has been extended to the isothermal–isobaric (Shing and Chung 1987) and microcanonical (Frenkel 1986, Lustig 1994) ensembles. The latter is particularly interesting, as the microcanonical ensemble is the most straightforward to perform MD simulations, as seen in Section 3.3.

The Widom method becomes inefficient in fluids at high densities, especially for hard-core potentials. Therefore, several procedures have been devised to improve the efficiency in the calculation of the chemical potential for high-density fluids. One of them is the *umbrella sampling* (Torrie and Valleau 1974, 1977), which uses a suitable reference system, with known Helmholtz free energy, to calculate the difference between the free energy of the system of interest and that of the reference system. It is based on using a biased sampling that favors those configurations, largely contributing to the chemical potential, that are not efficiently sampled in a conventional sampling according to the Boltzmann distribution, with the bias removed in the final result. In particular, if the reference system is the actual N-particle system, and the "system of interest" is the $(N + 1)$-particle system, the excess chemical potential is given by (Ding and Valleau 1993)

$$\mu^E = -k_B T \ln \left[\frac{\langle \exp\left(-\beta\mathcal{U}_{N+1}\right)/\pi \rangle_\pi}{\langle \exp\left(-\beta\mathcal{U}_N\right)/\pi \rangle_\pi} \right]. \tag{3.73}$$

In this expression, subscript π in the angular brackets means that the average is performed by sampling on a distribution $\pi(\mathbf{r}_1, \ldots, \mathbf{r}_{N+1}) \propto w(\Delta\mathcal{U}) \exp\left(-\beta\mathcal{U}_N\right)$, in which $\Delta\mathcal{U} = \mathcal{U}_{N+1} - \mathcal{U}_N$ and $w(\Delta\mathcal{U})$ is a function chosen to adequately cover the range of values of $\Delta\mathcal{U}$ relevant to the accurate calculation of μ^E. This requires that the sampling distribution must cover the configurations relevant to the $(N + 1)$-particle system as well as those relevant to the N-particle system. These two regions of the configuration are nearly nonoverlapping at high densities, in contrast to the situation at low densities (Ding and Valleau 1993).

The *acceptance ratio method*, developed by Bennett (1976), also uses a reference system, with potential energy $\mathcal{U}_0$ and known free energy, to determine the free energy difference $\Delta F = F_1 - F_0$ between the actual system, with potential energy $\mathcal{U}_1$, and the reference system, through the expression

$$\Delta F = -k_B T \ln \left[\frac{Q_c^{(0)}}{Q_c^{(1)}} \right] = -k_B T \ln \left[\frac{\langle w \exp\left(-\mathcal{U}_0/k_B T\right) \rangle_1}{\langle w \exp\left(-\mathcal{U}_1/k_B T\right) \rangle_0} \right], \qquad (3.74)$$

where
    w is a suitable weighting function
    $Q_c^{(0)}$ and $Q_c^{(1)}$ are the configurational integrals of systems 0 and 1, respectively
    angular brackets mean averages
    the subscripts in angular brackets indicate the system in whose configurational
        space are performed the averages

The chemical potential can be obtained from the thermodynamic relationship $\beta\mu = \beta F/N + Z$.

A procedure related to the previous one is the *overlap distribution method* of Bennett (1976). Again, two systems with potential energies $\mathcal{U}_0$ and $\mathcal{U}_1$ are considered. From simulations in the two systems, we can obtain two probability densities $\rho_1(\Delta\mathcal{U})$ and $\rho_0(\Delta\mathcal{U})$, with $\Delta\mathcal{U} = \mathcal{U}_1 - \mathcal{U}_0$, defining the distributions of the energy difference $\Delta\mathcal{U}$ in systems 0 and 1, respectively. Then, the free energy difference is obtained from the expression

$$\Delta F = \Delta\mathcal{U} + k_B T \ln \left[\frac{\rho_1(\Delta\mathcal{U})}{\rho_0(\Delta\mathcal{U})} \right]. \qquad (3.75)$$

To apply this procedure it is necessary that the ratio $\rho_1(\Delta\mathcal{U})/\rho_0(\Delta\mathcal{U})$ be finite, that is, the two distributions must overlap, whence the name. When the two distributions do not overlap, we can consider a number of intermediate systems, with potential energies between $\mathcal{U}_0$ and $\mathcal{U}_1$, and perform the calculation in several intermediate steps.

Another proposed procedure consists in the combination of the Widom and inverse Widom methods, the latter consisting in the fictitious removal of a particle of the system (Shing and Gubbins 1982). Often denoted as the *f-g method*, it can be obtained as a particular case of the overlapping distribution method (Frenkel 1986).

A useful method consists in the gradual switching on of the interaction between a particle and the remaining particles of the system (Mon and Griffiths 1985, Shing and Chung 1987). The procedure is based on an exact thermodynamic relationship (Kirkwood 1935) which can be given in the form

$$F = F_0 + \int_0^1 \langle \mathcal{U} - \mathcal{U}_0 \rangle_\alpha \, d\alpha, \qquad (3.76)$$

in which $\mathcal{U}$ is the potential energy of the actual system, $\mathcal{U}_0$ that of a reference system, eventually the ideal gas, and subscript α in the angular brackets means that the average is performed on a system with potential energy $\mathcal{U} = \mathcal{U}_0 + \alpha(\mathcal{U} - \mathcal{U}_0)$.

In a more general way, considering an arbitrary dependence of $\mathcal{U}$ on α, it is easy to show that

$$F = F_0 + \int_0^1 \left\langle \frac{\partial \mathcal{U}}{\partial \alpha} \right\rangle_\alpha d\alpha. \tag{3.77}$$

In particular, for real and reference systems with spherically symmetric pair potentials $u(r)$ and $u_0(r)$, respectively, Equation 3.76 leads to

$$\mu = \mu_0 + 4\pi\rho \int_0^1 d\alpha \int_V g(r, \alpha) [u(r) - u_0(r)] \, r^2 dr, \tag{3.78}$$

where $g(r, \alpha)$ is the RDF of the system with potential $u_\alpha(r) = u_0(r) + \alpha[u(r) - u_0(r)]$.

An accurate method to obtain the chemical potential of fluids with hard-core potentials, suitable even for extremely high densities in the metastable fluid region beyond the freezing density, is the so-called *scaled particle Monte Carlo* (Labík and Smith 1994, Labík et al. 1995, 1999, Barosova et al. 1996), which essentially consists in trial insertions of smaller copies of the actual particles of the fluid followed by trial scalings of the inserted particles together with trial increases of the core of randomly selected particles of the fluid. The changes in the chemical potentials in these two kinds of scalings are used to extrapolate to the actual sizes of the particles of the fluid to obtain the chemical potential of the latter.

A particularly useful method for mixtures is the *difference method* (Sindzingre et al. 1987), in which the difference between the excess chemical potentials of two components A and B in the mixture is given by

$$\mu_A^E - \mu_B^E = -k_B T \ln \left\langle e^{-\Delta\mathcal{U}^{A+B^-}/k_B T} \right\rangle_{N_A, N_B} = -k_B T \ln \left\langle e^{-\Delta\mathcal{U}^{B+A^-}/k_B T} \right\rangle_{N_A, N_B}, \tag{3.79}$$

where $\Delta\mathcal{U}^{A+B^-}$ is the change in the potential energy when a particle of species A is replaced by a particle of species B and the opposite change corresponds to $\Delta\mathcal{U}^{B+A^-}$.

In the grand canonical ensemble, μ, V, and T are fixed, but the number of particles N fluctuates. As a result of the simulation we will obtain the average number density $\langle\rho\rangle$ and, therefore, the chemical potential as a function of temperature and average density.

Configurational-bias MC techniques are useful to obtain the chemical potential for chain molecular fluids (Frenkel et al. 1991, de Pablo et al. 1992). To this end, trial insertions of the molecule are performed gradually, segment by segment, in the form outlined in Section 3.2.4. The excess chemical potential can be obtained from the expression (de Pablo et al. 1992)

$$\mu^E = -k_B T \ln \left\langle W e^{-\Delta\mathcal{U}/k_B T} \right\rangle, \tag{3.80}$$

where $\Delta\mathcal{U}$ is the change in energy associated to the insertion of the test molecule with a particular configuration in a given position, W is the probability of that particular molecular configuration, and the average is performed over all possible positions and all possible molecular configurations.

Most of the aforementioned methods are unsuitable to obtain the chemical potential of a crystalline solid. However, procedures based on the coupling between the actual potential energy of the system and that of a reference system, similar to those in Equations 3.76 and 3.77, are useful (Frenkel and Ladd 1984, Frenkel and Smit 2002). In this case, the reference system is the Einstein crystal, whose thermodynamic properties are known, with the same structure as the solid under consideration.

The chemical potential of either fluids or solids can be obtained indirectly from integration of the compressibility factor obtained from simulation along a reversible path, using the exact relationship

$$\beta\mu = \beta\mu_0 + \int_{\rho_0}^{\rho} (Z - Z_0) \frac{d\rho'}{\rho'} + Z - Z_0. \tag{3.81}$$

To this end, we must know the chemical potential μ_0 and the compressibility factor Z_0 in a reference state. For a fluid, the usual reference state is the ideal gas limit, and so $\rho_0 = 0$, $Z_0 = 1$, and we can obtain in this way the excess chemical potential μ^E. Obviously, for a solid, the ideal gas is not appropriate as a reference.

Another indirect procedure to obtain the chemical potential is based on the thermodynamic relationship

$$\beta\mu = \beta\mu_0 + \frac{1}{N} \int_0^{\beta} (U - U_0) \, d\beta' + Z - Z_0, \tag{3.82}$$

where
   U is the internal energy
   subscript 0 refers to the reference system corresponding to the limit $\beta = 0$, or
      equivalently $T \rightarrow \infty$, of the actual system

In the case of a system with a hard-core potential model, the reference system will be the one consisting of hard particles with the same shape as the hard cores of the particles in the actual system. Then, Z_0 and μ_0 can be obtained either from computer simulation or from one of the accurate analytical equations of state available for these systems, some of which will be cited in Chapters 5 and 8. Again, if the reference system is the ideal gas, $Z_0 = 1$, $U - U_0 = U^E$, and $\mu - \mu_0 = \mu^E$. Equation 3.82 can also be used for crystalline solids by replacing the lower limit of integration with ∞, that is to say, $T = 0$. In this case, the appropriate reference state is the Einstein crystal.

As in other thermodynamic properties, the chemical potential must be corrected in case of truncation of the potential at a distance r_c. This correction is of the form $\mu_c = 2U_c/N$, where U_c is given by Equation 3.63.

Obviously, all the corrections to the thermodynamic properties for the truncation of the potential, in the forms explained in this section, are reasonable for fluids, as for them $g(r) \rightarrow 1$ for large r. In the case of solids, such approximation can hardly be justified.

3.6 COMPUTER SIMULATION OF PHASE EQUILIBRIA

Simulation of phase coexistence in first-order phase transitions using a simulating cell with a reduced number of particles is handicapped by the considerable amount of energy required to create an interface, so that the simulated system will be in one or another phase, but not simultaneously in the two phases. Even for large systems, direct simulation of coexisting phases separated by an interface is challenging because of difficulties in equilibration and surface effects. Therefore, other procedures are more commonly used. In first-order phase transitions, the phase coexistence is characterized by equality of temperature, pressure, and chemical potential in the two phases, so that, in principle, any procedure allowing to obtain these properties in the two phases will be suitable for this purpose. In addition, a number of simulation techniques have been developed specially to obtain coexistence properties. Some of these methods will be briefly described here.

3.6.1 TEMPERATURE- AND DENSITY-SCALING MONTE CARLO

In the umbrella sampling procedure to obtain the Helmholtz free energy, the reference system can be the actual system at a different temperature T_0 for which the free energy F_0 is known. This is the basis of the *temperature-scaling Monte Carlo* (TSMC) procedure of Torrie and Valleau (1977) to obtain, from a single sampling, the free energy of the system over a range of temperatures, for which the configuration space is adequately sampled by the probability distribution $\pi(\mathbf{r}_1, \ldots, \mathbf{r}_N)$, according to the expression (Graham and Valleau 1990)

$$\frac{F}{Nk_BT} = \frac{F_0}{Nk_BT_0} - \ln\left[\frac{\langle \exp\left(-\mathcal{U}_N/k_BT\right)/\pi\rangle_\pi}{\langle \exp\left(-\mathcal{U}_N/k_BT_0\right)/\pi\rangle_\pi}\right]. \tag{3.83}$$

One can choose the probability density π so that its dependence on the configuration is captured through a dependence on the potential energy, that is, $\pi(\mathbf{r}_1, \ldots, \mathbf{r}_N) = \pi(\mathcal{U}_N)$, and, to cover a wide range of temperatures, $\pi(\mathcal{U}_N)$ must be enough uniform for a wide range of potential energies (Graham and Valleau 1990).

The TSMC method is not particularly suitable for the study of coexistence properties, because one needs the free energy surface as a function of density and temperature for the two phases. To this end, the free energy for a range of densities along an isotherm, taken as a reference, is to be determined by means of any appropriate simulation procedure. Then, using TSMC, the free energy along different isochores for the two phases is obtained. The results are fitted to a suitable function and the coexistence properties are determined from the usual conditions of equal temperature, pressure, and chemical potential.

The *density-scaling MC* technique proposed by Valleau (1991a) is based on similar grounds. Now, the difference in the excess Helmholtz free energy of a system at two different densities along the same isotherm is given by

$$\Delta F^E = -k_BT \ln\left\{\frac{\langle \exp\left[-\mathcal{U}\left(\mathbf{x}_1, \ldots, \mathbf{x}_N, V\right)/k_BT\right]/\pi\rangle_\pi}{\langle \exp\left[-\mathcal{U}\left(\mathbf{x}_1, \ldots, \mathbf{x}_N, V_0\right)/k_BT\right]/\pi\rangle_\pi}\right\}, \tag{3.84}$$

where $\mathbf{x}_i = \mathbf{r}_i/V^{1/3}$ for the system in the state of interest and $\mathbf{x}_i = \mathbf{r}_i/V_0^{1/3}$ for the system in the reference state. In this way, the free energy difference is determined for the desired range of densities along a given isotherm, eventually performing several runs to cover different, but overlapping, density ranges. The coexistence densities for that isotherm are then determined from the common tangent method combined with the condition of equality of the chemical potentials (Valleau 1991b). Although the common tangent method, in principle, would allow to obtain the two coexisting densities, as those for which there are a common tangent in the free energy versus density plot, the small curvature at low densities prevents an accurate estimation of the coexisting vapor density, whence the need of using the chemical potential too.

A logical extension of the two scalings just outlined is the *temperature- and density-scaling Monte Carlo* (TDSMC), or simply thermodynamic scaling MC, (Orkoulas and Panagiotopoulos 1996, Brilliantov and Valleau 1998), which allows to sample states with different temperatures and densities in a single run. The expression for the free energy difference in this case is

$$
\frac{F(N,V,T)}{Nk_BT} = \frac{F_0(N,V_0,T_0)}{Nk_BT_0} - \ln \left\{ \frac{\langle \exp\left[-\mathcal{U}(\mathbf{x}_1,\ldots,\mathbf{x}_N;V)/k_BT\right]/\pi \rangle_\pi}{\langle \exp\left[-\mathcal{U}(\mathbf{x}_1,\ldots,\mathbf{x}_N;V_0)/k_BT_0\right]/\pi \rangle_\pi} \right\}.
$$
(3.85)

A suitable form of the sampling distribution is

$$
\pi(x_1,\ldots,x_N) = \sum_j \sum_k w_{jk} \exp\left[\frac{-\mathcal{U}(\mathbf{x}_1,\ldots,\mathbf{x}_N;V_k)}{k_BT_j} \right],
$$
(3.86)

considering a number of temperatures T_j and densities ρ_k adequately covering the corresponding ranges of interest, and weighting factors w_{ij} chosen to guarantee a uniform sampling in these ranges. This can be achieved taking $w_{ij} \propto \exp[F^E(N,T_j,V_k/k_BT_j)]$, but as F^E is not known in advance, one can use an approximate theoretical estimate and an iterative process (Brilliantov and Valleau 1998).

3.6.2 NPT AND NVT PLUS TEST PARTICLE INSERTION METHODS

In the *NPT* plus TPI method, introduced by Möller and Fischer (1990), the chemical potential of the liquid and vapor phases is determined along isotherms as a function of the pressure by means of the Widom TPI method, although other procedures to determine the chemical potential can also be used. Then, the intersection point of the two branches for each isotherm determines the coexistence pressure and, from the latter, the coexistence densities. The pressure dependence of the chemical potential along isotherms in the neighborhood of the coexistence curve can be approximated by a Taylor series in terms of the pressure truncated at first order, so that two simulated states close to coexistence for each phase suffice. The method has been extended by Boda et al. (1995) by using an expansion of the chemical potential in terms of both pressure and inverse temperature (or β) up to third order in both quantities. The terms in the expansion can be determined from fluctuations at two state points near coexistence for the two branches. This allows us to obtain the coexistence properties within a certain temperature interval, without the need of performing simulations for each temperature within the interval. These procedures have also been extended to mixtures (Vrabec et al. 1995, Kronome et al. 2000).

The *NPT + TPI* method becomes inaccurate near the critical point because of the large volume fluctuations. Better accuracy in this situation seems to be achieved from simulation in the canonical ensemble combined with TPI (*NVT + TPI*) proposed by Okumura and Yonezawa (2001, 2002). This method proceeds in a similar way as the *NPT + TPI*, but for the fact that now the chemical potential is determined as a function of volume along isotherms for the two phases and the intersection of the two branches gives the coexistence pressure.

3.6.3 Gibbs Ensemble Simulation

In the Gibbs ensemble simulation (Panagiotopoulos 1987, Panagiotopoulos et al. 1988), the system, with fixed volume, temperature, and number of particles, is divided into two subsystems, one for liquid and another for vapor, each of them in a separate simulation cell. Particles are allowed to move within each cell and to transfer from one cell to another, and the volume of the cells can fluctuate, in order to allow them satisfy the conditions of equality of temperature, pressure, and chemical potential. Particle transfer is performed by means of the TPI method, which requires a proper formulation for the Gibbs ensemble (Smit and Frenkel 1989) that reduces to the expression (3.72) for the *NVT* ensemble only for large particle number and negligible density fluctuations. Because of the use of the TPI method, the Gibbs ensemble simulation will become inefficient for liquids at high densities, especially in case of very asymmetric molecules. For chain molecular fluids, the efficiency is much improved by combining configurational-bias MC with the Gibbs ensemble simulation (Laso et al. 1992, Mooij et al. 1992, Smit et al. 1995). For subcritical temperatures close to the critical point, the interfacial tension between liquid and vapor decreases, so that each of the two subsystems will not be longer in a single phase, but rather an average density will be observed (Smit et al. 1989). Therefore, it is difficult to obtain reliable estimates of the coexistence densities near the critical point. In spite of this, quite accurate values of the critical parameters can be obtained from extrapolation of subcritical data using appropriate scalings laws. The procedure has also been extended to mixtures by Lopes and Tildesley (1997).

The Gibbs ensemble simulation can be combined with thermodynamics scaling MC (Kiyohara et al. 1996), resulting in a more efficient method than any of the two techniques separately when one is interested in obtaining the coexistence properties at many different coexistence points.

Although the Gibbs ensemble simulation was originally implemented within the context of the MC method, the procedure has also been extended to MD by Palmer and Lo (1994).

3.6.4 Gibbs–Duhem Integration

The Gibbs–Duhem integration method, proposed by Kofke (1993a,b), is based on the Clausius–Clapeyron equation, known from elementary thermodynamics, which gives the slope of the pressure versus temperature curve along the phase coexistence in a first-order transition. For a monocomponent system, it takes the form

$$\frac{dP}{dT} = \frac{\Delta h}{T \Delta v},$$ (3.87)

or, equivalently,

$$\frac{d \ln P}{dT} = \frac{\Delta h}{PT \Delta v} = f(T, P).$$ (3.88)

In these expressions, Δh and Δv are the changes in molar enthalpy and molar volume, respectively, in the transition at temperature T and pressure P. Equation 3.88 can be integrated to give the temperature dependence of the pressure along the coexistence curve, provided that we know the pressure and temperature for a coexistence point. The Clausius–Clapeyron equation can be readily derived from the Gibbs–Duhem relationship

$$-SdT + VdP - Nd\mu = 0,$$ (3.89)

whence the name of this technique.

The function $f(T, P)$ in Equation 3.88 is determined from NPT simulations for the two phases. The integration is conveniently performed using a predictor–corrector method (see Kofke 1993b, for further details). The coexistence data for the reference point can be obtained from other simulation technique for phase equilibria, such as the Gibbs ensemble simulation, or from a suitable theoretical procedure.

One advantage of the Gibbs–Duhem integration with respect to other methods for phase coexistence is that it does not use particle insertion, which may be difficult in certain circumstances as seen earlier. Another advantage is that it can be applied to solid–fluid coexistence (Hagen et al. 1993, Silva et al. 2001) and even to solid–liquid–vapor, triple point, coexistence (Agrawal et al. 1994). The procedure has been extended to mixtures (Mehta and Kofke 1994, Bolhuis and Kofke 1996), polymers (Escobedo and de Pablo 1997a,b), rigid chain molecular fluids (Galindo et al. 2004), isotropic-nematic and isotropic-discotic phase coexistence (Camp et al. 1996, Camp and Allen 1997), and phase coexistence in water models, including vapor-liquid (Vega et al. 2006), melting curves of plastic crystal phases (Aragonés and Vega 2009), and the phase diagram at negative pressures (Conde et al. 2009), among other applications, which proves the versatility of the method.

3.6.5 Histogram-Reweighting Grand Canonical Monte Carlo Simulation

The liquid–vapor coexistence properties can also be obtained by grand canonical MC simulations. In this ensemble, temperature, volume, and chemical potential are fixed in an equilibrium state, as stated earlier, whereas energy and pressure fluctuate.

From Equations 2.47 and 2.48, the probability density $\rho(N, U)$ that the system has N particles and total energy U can be written as follows:

$$\rho(N, U) = \frac{D(N, U) e^{\beta(\mu N - U)}}{\Xi}, \tag{3.90}$$

where $D(N, U)$ is the density of states (number of microstates per unit energy interval in a system with N particles and energy U). Then, for T, V, and μ fixed, we can go further by integrating over energy for each value of N, thus obtaining the probability $p(N)$ that the system has N particles. In computer simulation, this probability is recorded as a histogram obtained from averaging over many independent samples (Wilding 2001). In a liquid–vapor equilibrium at a given temperature, volume, and chemical potential, the probability versus density thus obtained will show two peaks, as illustrated in Figure 3.2, corresponding to the liquid and vapor phases. As the system will be exploring the two phases with equal probability, the area under the two peaks must be equal. As a matter of fact, the integration of the numerator on the right-hand side of Equation 3.90 over particle number and energy gives the grand canonical partition function which, as is well known from statistical mechanics, is related to the pressure by means of the relationship $PV = k_B T \ln \Xi$. Therefore, the condition of equal area under the two peaks of $p(N)$ ensures that the pressure is the same in the two phases and, as the temperature and chemical potential are also equal, this completes the fulfillment of the equilibrium conditions. Then, to determine the liquid–vapor coexistence properties, we would adjust the temperature and

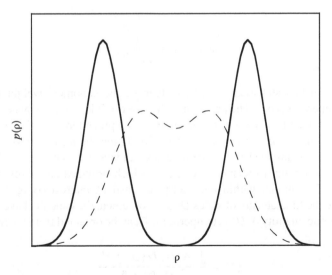

FIGURE 3.2 Illustration of the probability versus density distribution in a liquid–vapor equilibrium at a slightly subcritical temperature, dashed curve, and at a lower temperature, continuous curve. The two peaks of each curve correspond to the vapor and liquid at coexistence. The height between the peaks and the valley is an indication of the energetic barrier between the two phases.

chemical potential until the equal area rule is accomplished (Panagiotopoulos 2000, Wilding 2001).

However, there is a problem with the procedure just outlined. There is an energy barrier between the two phases, which is reflected in the height between the peaks and the valley. In the neighborhood of the critical point, the probability distribution $p(N)$ between the two peaks is very flat (see Figure 3.2), so that the two phases are efficiently sampled, but as the temperature is lowered the energy barrier rapidly increases in magnitude, thus preventing the efficient sampling of the two phases. The problem can be overcome by using the *multicanonical sampling* (Berg and Neuhaus 1991, 1992, Berg 1992), which is based on similar grounds as the umbrella sampling and the thermodynamic scaling MC procedures, previously mentioned, as it also uses a biased sampling by means of an appropriate probability distribution, instead of the Boltzmann distribution, which ideally should be flat in the region of interest. In the canonical ensemble, the probability density $\rho(U)$ that the system has energy U is

$$\rho(U) = \frac{D(U)\,e^{-\beta U}}{Q} \propto D(U)\,w_c, \qquad (3.91)$$

with $w_c = e^{-\beta U}$. The multicanonical probability density ρ_{muc} is chosen in such a way that

$$\rho_{muc} \propto D(U)\,w_{muc} = \text{constant}, \qquad (3.92)$$

that is,

$$w_{muc}(U) \propto \frac{1}{D(U)}. \qquad (3.93)$$

Sampling with this statistical weight, instead of the canonical weight $w_c = e^{-\beta U}$, avoids any energy barrier. The problem is that we don't know in advance the density of states $D(U)$ and so we must use an iterative procedure, starting with a trial statistical weight w_0 and determining from simulation the histogram of energies h_0 that it produces, using the latter to obtain a new statistical weight as $w_1 = h_0/w_0$, and so on until we obtain a statistical weight w_n which provides an energy probability distribution flat enough within the energy interval of interest (Berg 1998, Janke 1998). Once the final density of states $D(U)$ is obtained, the canonical average of any thermodynamic quantity $A(U)$ at temperature T can be obtained from the relationship

$$\langle A \rangle = \frac{\int_U A(U)\,D(U)\,e^{-\beta U}}{\int_U D(U)\,e^{-\beta U}}. \qquad (3.94)$$

Note that from a single multicanonical simulation providing a density of states accurate for some range of energies, we can obtain canonical averages for a range of temperatures, which may be considered as equivalent to performing a great number of canonical simulations for different temperatures within that temperature range, whence the term "multicanonical."

The use of the multicanonical method is not restricted to the framework of first-order phase transitions and, in particular, it is useful to overcome problems with the loss of ergodicity in low temperature simulations. This technique is not limited only to MC simulations, as the procedure has also been used in combination with MD simulations (Hansmann et al. 1996, Jang et al. 2002, Shimizu 2004, 2005).

In multicanonical sampling in the grand canonical ensemble to determine the liquid–vapor coexistence, the statistical weight $w_{muc}(U)$ of Equation 3.93 can be replaced by a statistical weight $w(N)$ depending on the particle number N (Wilding 2001). If the weighted probability obtained in the course of the simulation is $p_w(N)$, the unweighted probability $p(N)$ can be recovered by reweighting as $p(N) = p_w(N)/w(N)$. Once we have obtained the probability $p_0(N)$, or the probability density $\rho_0(N, U)$ for T_0 and μ_0, we can estimate the probabilities $p(N)$ and $\rho(N, U)$ for other close values T and μ by means of the *histogram-reweighting* technique (Ferrenberg and Swendsen 1988, Wilding 1995) as, from Equation 3.90, the ratio of the two probability densities is

$$\frac{\rho(N, U)}{\rho_0(N, U)} \propto e^{(\beta\mu - \beta_0\mu_0)N - (\beta - \beta_0)U}. \tag{3.95}$$

Details on the practical procedure to obtain the liquid–vapor coexistence properties by combining histogram reweighting and multicanonical sampling with MC simulations in the grand canonical ensemble have been reported by Wilding (2001).

The precise calculation of the critical point parameters is more cumbersome. Near the critical point, fluctuations occur at all length scales and, as the critical point is approached, the correlation length and other properties tend to diverge, whereas still other properties tend to zero, according to power laws involving *critical exponents*. The properties of the system become dependent on the system size, which is particularly important in computer simulations due to the limited size of the simulation box, and *universal*. The latter means that the properties of many different systems, belonging to the same *universality class*, exhibit the same functional dependence, and with the same critical exponents, on certain quantities, the *scaling fields*, measuring the departure of the state of the system with respect to the critical point. This leads to a universal dependence of the critical properties on the system size. Therefore, to determine the critical point properties from simulation it is necessary to carry out runs with different system sizes and then extrapolate the results to the infinite system according to the appropriate scaling laws. The *finite size scaling* (FSS) theory provides expressions for these scaling laws. Ordinary fluids belong to the same universality class as the Ising model, and so the well-known scaling laws and critical exponents for this system can be used to obtain the critical properties of fluids (Bruce and Wilding 1992, Wilding and Bruce 1992, Wilding 1995, 1997a).

Extensions to mixtures of the histogram-reweighting grand canonical MC method are also available (Wilding 1997b, Potoff and Panagiotopoulos 1998, Wilding et al. 1998).

3.6.6 PSEUDOENSEMBLES, EXPANDED ENSEMBLES, AND MORE

Several of the simulation techniques for phase equilibria described in this section require trial insertion of particles, as are the cases of the *NPT* plus TPI, Gibbs ensemble procedure, and grand canonical MC methods. In addition, the first of these methods also requires trial volume changes. Trial insertions become inefficient for high-density fluids and for large molecules, particularly in mixtures of large and small molecules, and trial volume changes become inefficient for large systems. To overcome these difficulties, a number of procedures have been proposed, some of which will be briefly described here.

In the *pseudo-grand canonical simulation* method (Mehta and Kofke 1995), the density changes are performed by replacing the trial insertion–deletion of particles by volume fluctuations, whereas the chemical potential is still determined from the TPI method. In the *pseudo-Gibbs ensemble simulation* technique (Camp and Allen 1996), the particle exchange between the two simulation boxes is replaced by density changes in them by rescaling their volumes. An improved version of the pseudo-grand canonical method together with a *pseudo-NPT ensemble simulation* procedure was developed by Escobedo and de Pablo (1997a). The pseudoensemble technique has also been extended to phase equilibria in multicomponent mixtures (Escobedo 1998).

The *semigrand ensemble* method (Kofke and Glandt 1988) was devised to improve the efficiency in the simulation of phase equilibria of multicomponent mixtures, in which the trial exchange of particles between the different phases may be inefficient, as stated earler. In this procedure, the chemical potential of only one of the species of the mixture at one state needs to be obtained from usual procedures, whereas equilibration is achieved by changing randomly chosen particles from an species to another. The procedure can be implemented at constant pressure or at constant volume, or within the Gibbs ensemble scheme (Kofke and Glandt 1988, de Miguel et al. 1995). However, when the molecules involved in the trial change from one species to another are very disparate in size, the probability of acceptance may be extremely small and the method will not work.

The *expanded ensemble* method (Lyubartsev et al. 1992) may be regarded as a generalization of the temperature-scaling technique. In the *temperature expanded canonical ensemble*, M canonical configurational partition functions for a given system are considered, each of them corresponding to a different temperature T, or equivalently β. Then the configurational integral Q_c of the expanded ensemble is constructed in the form

$$Q_c = \sum_{i=1}^{M} w_i Q_{ci}, \qquad (3.96)$$

where
   Q_{ci} is the configurational integral of a system with temperature $T_i = k_B/\beta_i$
   w_i are weighting factors introduced to ensure an adequate sampling (ideally with
      equal probability) of all β_i

The term "expanded" refers to the fact that we are sampling M different temperatures or, equivalently, M different canonical ensembles. The probability p_i that the system is in a configuration corresponding to the ensemble with temperature T_i is

$$p_i = \frac{w_i Q_{ci}}{Q_c}, \tag{3.97}$$

so that the relative probability of two states corresponding to T_i and T_j is

$$\frac{p_i}{p_j} = \frac{w_i Q_{ci}}{w_j Q_{cj}}. \tag{3.98}$$

Recording the M probabilities for the M different temperatures in the course of a run, we can obtain the change in the free energy between any two temperatures within the range considered.

Up to this point, the expanded ensemble formalism is closely related to the temperature-scaling procedure, discussed in Section 3.6, and to the so-called *simulated tempering* method (Marinari and Parisi 1992). However, instead of considering β as the parameter varying from one ensemble to another within the expanded ensemble, we can consider another parameter, for example, the number of particles (expanded grand canonical ensemble) or a parameter accounting for the gradual insertion or deletion of a large molecule in the system. This may be achieved in different ways as, for example, the gradual switching on or off of the interaction between the tagged particles or by gradually inserting or deleting segments of a large polyatomic molecule. In this case, we can consider a coupling parameter α_i, $i = 1, \ldots, M$, accounting for M different states of partial coupling or insertion of the tagged particle, with subscripts 1 and M corresponding to the null insertion and full insertion of the particle, respectively. Then, the partition function in the expanded grand canonical ensemble can be expressed in the form (Escobedo and de Pablo, 1996)

$$\Xi_e = \sum_{N=0}^{\infty} \sum_{i=1}^{M} e^{\beta \mu N} w_i Q\left(N, \alpha_i, V, T\right), \tag{3.99}$$

where w_i is a weighting factor chosen to ensure an equal probability of all states α_i. Denoting $Q_i = Q(N, \alpha_i, V, T)$, the probability that the system is in the state i will be

$$p_i = \frac{e^{\beta \mu N} w_i Q_i}{\Xi_e}, \tag{3.100}$$

and the relative probabilities of two states i and j

$$\frac{p_i}{p_j} = \frac{w_i Q_i}{w_j Q_j}. \tag{3.101}$$

If all probabilities are equal, $p_i/p_j = 1$, and

$$\ln\left(\frac{w_j}{w_i}\right) = -\ln\left(\frac{Q_j}{Q_i}\right) = \beta\left(F_j - F_i\right) \propto \beta\left(\mu_j^E - \mu_i^E\right). \qquad (3.102)$$

This suggests weighting factors of the form $w_i = \exp(\beta\mu_i^E) = \exp(a_i\beta\mu^E)$, where μ^E is the excess chemical potential of the N-particle system and a_i are coefficients such that $a_1 = 0$, $a_M = 1$ and intermediate values can be obtained from a previous simulation.

The expanded ensemble method can also be applied to simulations in the Gibbs and other ensembles (Escobedo and de Pablo 1996, 1997c, Chang and Sandler 2003).

REFERENCES

Agrawal, R., M. Mehta, and D. A. Kofke. 1994. Efficient evaluation of three-phase coexistence lines. *Int. J. Thermophys.* 15:1073.

Alder, B. J., and T. E. Wainwright. 1957. Phase transition for a hard sphere system. *J. Chem. Phys.* 27:1208.

Alder, B. J., and T. E. Wainwright. 1959. Studies in molecular dynamics. I. General method. *J. Chem. Phys.* 31:459.

Allen, M. P., and D. J. Tildesley. 1987. *Computer Simulation of Liquids*. Oxford,U.K.: Clarendon Press.

Andersen, H. C. 1980. Molecular dynamics simulations at constant pressure and/or temperature. *J. Chem. Phys.* 72:2384.

Aragonés, J. L., and C. Vega. 2009. Plastic crystal phases of simple water models. *J. Chem. Phys.* 130:244504.

Barosová, M., M. Malijevský, S. Labík, and W. R. Smith. 1996. Computer simulation of the chemical potentials of binary hard-sphere mixtures. *Mol. Phys.* 87:423.

Bennett, C. H. 1976. Efficient estimation of free energy differences from Monte Carlo data. *J. Comput. Phys.* 22:245.

Berg, B. 1992. The multicanonical ensemble: A new approach to computer simulations. *Int. J. Mod. Phys. B* 3:1083.

Berg, B. A. 1998. Algorithmic aspects of multicanonical simulations. *Nucl. Phys. B* 63A-C:982.

Berg, B. A., and T. Neuhaus. 1991. Multicanonical algorithms for first order phase transitions. *Phys. Lett. B* 267:249.

Berg, B. A., and T. Neuhaus. 1992. Multicanonical ensemble: A new approach to simulate first-order phase transitions. *Phys. Rev. Lett.* 68:9.

Boda, D., J. Liszi, and I. Szalai. 1995. An extension of the NPT plus test particle method for the determination of the vapour-liquid equilibria of pure fluids. *Chem. Phys. Lett.* 235:140.

Bolhuis, P. G., and D. A. Kofke. 1996. Monte Carlo study of freezing of polydisperse hard spheres. *Phys. Rev. E* 54:634.

Brilliantov, N. V., and J. P. Valleau. 1998. Thermodynamic scaling Monte Carlo study of the liquid-gas transition in the square-well fluid. *J. Chem. Phys.* 108:1115.

Bruce, A. D., and N. B. Wilding. 1992. Scaling fields and universality of the liquid-gas critical point. *Phys. Rev. Lett.* 68:193.

Çagin, T., and B. M. Pettitt. 1991a. Grand molecular dynamics: A method for open systems. *Mol. Simul.* 6:5.

Çagin, T., and B. M. Pettitt. 1991b. Molecular dynamics with a variable number of molecules. *Mol. Phys.* 72:169.

Camp, P. J., and M. P. Allen. 1996. Phase coexistence in a pseudo Gibbs ensemble. *Mol. Phys.* 88:1459.

Camp, P. J., and M. P. Allen. 1997. Phase diagram of the hard biaxial ellipsoid fluid. *J. Chem. Phys.* 106:6681.

Camp, P. J., C. P. Mason, M. P. Allen, A. A. Khare, and D. A. Kofke. 1996. The isotropic-nematic phase transition in uniaxial hard ellipsoid fluids: Coexistence data and the approach to the Onsager limit. *J. Chem. Phys.* 105:2837.

Chang, J., and S. I. Sandler. 2003. Determination of liquid-solid transition using histogram reweighting method and expanded ensemble simulations. *J. Chem. Phys.* 118:8390.

Conde, M. M., C. Vega, G. A. Tribello, and B. Slater. 2009. The phase diagram of water at negative pressures: Virtual ices. *J. Chem. Phys.* 131:034510.

Ding, K., and J. P. Valleau. 1993. Umbrella-sampling realization of "Widom" chemical potential estimation. *J. Chem. Phys.* 98:3306.

Escobedo, F. A. 1998. Novel pseudoensembles for simulation of multicomponent phase equilibria. *J. Chem. Phys.* 108:8761.

Escobedo, F. A., and J. J. de Pablo. 1996. Expanded grand canonical and Gibbs ensemble Monte Carlo simulation of polymers. *J. Chem. Phys.* 105:4391.

Escobedo, F. A., and J. J. de Pablo. 1997a. Pseudo-ensemble simulations and Gibbs-Duhem integrations for polymers. *J. Chem. Phys.* 106:2911.

Escobedo, F. A., and J. J. de Pablo. 1997b. Gibbs-Duhem integration in lattice systems. *Europhys. Lett.* 40:111.

Escobedo, F. A., and J. J. de Pablo. 1997c. Monte Carlo simulation of athermal mesogenic chains: Pure systems, mixtures, and constrained environments. *J. Chem. Phys.* 106:9858.

Eslami, H., and F. Müller-Plathe. 2007. Molecular dynamics simulation in the grand canonical ensemble. *J. Comput. Chem.* 28:1763.

Ferrenberg, A. M., and R. H. Swendsen. 1988. New Monte Carlo technique for studying phase transitions. *Phys. Rev. Lett.* 61:2635.

Fox, J. R., and H. C. Andersen. 1984. Molecular dynamics simulations of a supercooled monatomic liquid and glass. *J. Phys. Chem.* 88:4019.

Frenkel, D. 1986. Free energy computations and first order phase transitions. In *Molecular Dynamics Simulations of Statistical Mechanics Systems, Proceedings of the 97th International "Enrico Fermi" School of Physics*, eds. G. Ciccotti and W. G. Hoover, pp.151–188. Amsterdam,the Netherlands: North-Holland.

Frenkel, D., and A. J. C. Ladd. 1984. New Monte Carlo method to compute the free energy of arbitrary solids. Application to the fcc and hcp phases of hard spheres. *J. Chem. Phys.* 81:3188.

Frenkel, D., G. C. A. M. Mooij, and B. Smit. 1991. Novel scheme to study structural and thermal properties of continuously deformable molecules. *J. Phys.: Condens. Matter* 3:3053.

Frenkel, D., and B. Smit. 2002. *Understanding Molecular Simulation*, 2nd edn. San Diego,CA: Academic Press.

Galindo, A., C. Vega, E. Sanz, L. G. MacDowell, E. de Miguel, and F. J. Blas. 2004. Computer simulation study of the global phase behavior of linear rigid Lennard-Jones chain molecules: Comparison with flexible models. *J. Chem. Phys.* 120:3957.

Gear, C. W. 1971. *Numerical Initial Value Problems in Ordinary Differential Equations.* Englewood Cliffs, NJ: Prentice-Hall.

Graham, I. S., and J. P. Valleau. 1990. A Monte Carlo study of the coexistence region of the restricted primitive model. *J. Phys. Chem.* 94:7894.

Hagen, M. H. J., E. J. Meijer, G. C. A. M. Mooij, D. Frenkel, and H. N. W. Lekkerkerker. 1993. Does C_{60} have a liquid phase? *Nature* 365:425.

Haile, J. M. 1992. *Molecular Dynamics Simulation*. New York: John Wiley & Sons.

Hansmann, U. H. E, Y. Okamoto, and F. Eisenmenger. 1996. Molecular dynamics, Langevin and hybrid Monte Carlo simulations in a multicanonical ensemble. *Chem. Phys. Lett.* 259:321.

Hockney, R. W., and J. W. Eastwood. 1981. *Computer Simulation Using Particles*. New York: McGraw-Hill.

Hoover, W. G. 1985. Canonical dynamics: Equilibrium phase-space distributions. *Phys. Rev. A* 31:1695.

Jang, S., Y. Pak, and S. Shin. 2002. Multicanonical ensemble with Nosé-Hoover molecular dynamics simulation. *J. Chem. Phys.* 116:4782.

Janke, W. 1998. Multicanonical Monte Carlo simulations. *Physica A* 254:164.

Kirkwood, J. G. 1935. Statistical mechanics of fluid mixtures. *J. Chem. Phys.* 3:300.

Kiyohara, K., T. Spyriouni, K. E. Gubbins, and A. Z. Panagiotopoulos. 1996. Thermodynamic scaling Gibbs ensemble Monte Carlo: A new method for determination of phase coexistence properties of fluids. *Mol. Phys.* 89:965.

Kofke, D. 1993a. Gibbs-Duhem integration: A new method for direct evaluation of phase coexistence by molecular simulation. *Mol. Phys.* 78:1331.

Kofke, D. 1993b. Direct evaluation of phase coexistence by molecular simulation via integration along the saturation line. *J. Chem. Phys.* 98:4149

Kofke, D., and E. D. Glandt. 1988. Monte Carlo simulation of multicomponent equilibria in a semigrand canonical ensemble. *Mol. Phys.* 64:1105.

Kronome, G., I. Szalai, M. Wendland, and J. Fischer. 2000. Extension of the NPT + test particle method for the calculation of phase equilibria of nitrogen + ethane. *J. Mol. Liq.* 85:237.

Labík, S., V. Jirásek, M. Malijevský, and W. R. Smith. 1995. Computer simulation of the chemical potentials of fused hard sphere diatomic fluids. *Chem. Phys. Lett.* 247:227.

Labík, S., M. Malijevský, R. Kao, W. R. Smith, and F. del Río. 1999. The SP-MC computer simulation method for calculating the chemical potential of the square-well fluid. *Mol. Phys.* 96:849.

Labík, S., and W. R. Smith. 1994. Scaled particle theory and the efficient calculation of the chemical potential of hard spheres in the NVT ensemble. *Mol. Simul.* 12:23.

Lagache, M., P. Ungerer, A. Boutin, and A. H. Fuchs. 2001. Prediction of thermodynamic derivative properties of fluids by Monte Carlo simulation. *Phys. Chem. Chem. Phys.* 3:4333.

Laso, M., J. J. de Pablo, and U. W. Suter. 1992. Simulation of phase equilibria for chain molecules. *J. Chem. Phys.* 97:2817.

Lebowitz, J. L., J. K. Percus, and L. Verlet. 1967. Ensemble dependence of fluctuations with application to machine computations. *Phys. Rev.* 153:250.

Lopes, J. N. C., and D. J. Tildesley. 1997. Multiphase equilibria using the Gibbs ensemble Monte Carlo method. *Mol. Phys.* 92:187.

Lustig, R. 1994. Statistical thermodynamics in the classical molecular dynamics ensemble. II. Application to computer simulation. *J. Chem. Phys.* 100:3060.

Lynch, G. C., and B. M. Pettitt. 1997. Grand canonical ensemble molecular dynamics simulations: Reformulation of extended system dynamics approaches. *J. Chem. Phys.* 107:8594.

Lyubartsev, A. P., A. A. Martsinovski, S. V. Shevkunov, and P. N. Vorontsov-Velyaminov. 1992. New approach to Monte Carlo calculation of the free energy: Method of expanded ensembles. *J. Chem. Phys.* 96:1776.

Marinari, E., and G. Parisi. 1992. Simulated tempering: A new Monte Carlo scheme. *Europhys. Lett.* 19:451.

McBride C., C. Vega, and E. Sanz. 2005. Non-Markovian melting: A novel procedure to generate initial liquid like phases for small molecules for use in computer simulation studies. *Comput. Phys. Commun.* 170:137.

McQuarrie, D. A. 1976. *Statistical Mechanics*. New York: Harper and Row.

Mehta, M., and D. A. Kofke. 1994. Coexistence diagrams of mixtures by molecular simulation. *Chem. Eng. Sci.* 49:2633.

Mehta, M., and D. A. Kofke. 1995. Molecular simulation in a pseudo grand canonical ensemble. *Mol. Phys.* 86:139.

Melchionna, S., G. Ciccotti, and B. L. Holian. 1993. Hoover *NPT* dynamics for systems varying in shape and size. *Mol. Phys.* 78:533.

Metropolis, N. 1987. The beginning of the Monte Carlo method. *Los Alamos Sci.* 15:125.

Metropolis, N., A. W. Rosenbluth, M. N. Rosenbluth, A. H. Teller, and E. Teller. 1953. Equation of state calculations by fast computing machines. *J. Chem. Phys.* 21:1087.

de Miguel, E., E. Martín del Río, and M. M. Telo da Gama. 1995. Liquid-liquid phase equilibria of symmetrical mixtures by simulation in the semigrand canonical ensemble. *J. Chem. Phys.* 103:6188.

Möller, D., and J. Fischer. 1990. Vapour liquid equilibrium of a pure fluid from test particle method in combination with *NPT* molecular dynamics simulation. *Mol. Phys.* 69:463.

Mon, K. K., and R. B. Griffiths. 1985. Chemical potential by gradual insertion of a particle in Monte Carlo simulation. *Phys. Rev. A* 31:956.

Mooij, G. C. A. M., D. Frenkel, and B. Smit. 1992. Direct simulation of phase equilibria of chain molecules. *J. Phys: Condens. Matter* 4:L255-L259.

Nosé, S. 1984a. A molecular dynamics method for simulations in the canonical ensemble. *Mol. Phys.* 52:255.

Nosé, S. 1984b. A unified formulation of the constant temperature molecular dynamics methods. *J. Chem. Phys.* 81:511.

Okumura, H., and F. Yonezawa. 2001. Method for liquid-vapor coexistence curves by test-particle insertions in the canonical ensemble. *J. Non-Cryst. Solids* 293–295:715.

Okumura, H., and F. Yonezawa. 2002. Precise determination of the liquid-vapor critical point by the *NVT* plus test particle method *J. Non-Cryst. Solids* 312-314:256.

Orkoulas, G., and A. Z. Panagiotopoulos. 1996. Phase diagram of the two-dimensional Coulomb gas: A thermodynamic scaling Monte Carlo study. *J. Chem. Phys.* 104:7205.

de Pablo, J. J., M. Laso, J. I. Siepmann, and U. W. Suter. 1993. Continuum-configurational-bias Monte Carlo simulations of long-chain alkanes. *Mol. Phys.* 80:55.

de Pablo, J. J., M. Laso, and U. W. Suter. 1992. Estimation of the chemical potential of chain molecules by simulation. *J. Chem. Phys.* 96:6157.

Palmer, B. J., and C. Lo. 1994. Molecular dynamics implementation of the Gibbs ensemble calculation. *J. Chem. Phys.* 101:10899.

Panagiotopoulos, A. Z. 1987. Direct determination of phase coexistence properties of fluids by Monte Carlo simulation in a new ensemble. *Mol. Phys.* 61:813.

Panagiotopoulos, A. Z. 2000. Monte Carlo methods for phase equilibria of fluids. *J. Phys.: Condens. Matter* 12:R25.

Panagiotopoulos, A. Z., N. Quirke, M. Stapleton, and D. J. Tildesley. 1988. Phase equilibria by simulation in the Gibbs ensemble. Alternative derivation, generalization and application to mixture and membrane equilibria. *Mol. Phys.* 63:527.

Potoff, J. J., and A. Z. Panagiotopoulos. 1998. Critical point and phase behavior of the pure fluid and a Lennard-Jones mixture. *J. Chem. Phys.* 109:10914.

Rapaport, D. C. 2004. *The Art of Molecular Dynamics Simulation*. 2nd edn. Cambridge, MA: Cambridge University Press.

Rosenbluth, M. N., and A. W. Rosenbluth. 1955. Monte Carlo calculation of the average extension of molecular chains. *J. Chem. Phys.* 23:356.

Shimizu, H. 2004. Estimation of the density of states by multicanonical molecular dynamics simulation. *Phys. Rev. E* 70:056704.

Shimizu, H. 2005. Measure of accuracy for multicanonical molecular-dynamics simulation. *J. Chem. Phys.* 123:104106.

Shing, K. S., and S. T. Chung. 1987. Computer simulation methods for the calculation of solubility in supercritical extraction systems. *J. Phys. Chem.* 91:1674.

Shing, K. S., and K. E. Gubbins. 1982. The chemical potential in dense fluids and fluid mixtures via computer simulation. *Mol. Phys.* 46:1109.

Siepmann, J. I. 1990. A method for the direct calculation of chemical potentials for dense chain systems. *Mol. Phys.* 70:1145.

Siepmann, J. I., and D. Frenkel. 1992. Configurational bias Monte Carlo: A new sampling scheme for flexible chains. *Mol. Phys.* 75:59.

Silva, F. M. S., R. P. S. Fartaria, Land F. F. M. Freitas. 2001. The starting state in simulations of the fluid-solid coexistence by Gibbs-Duhem integration. *Comput. Phys. Commun.* 141:403.

Sindzingre, P., G. Ciccotti, C. Massobrio, and D. Frenkel. 1987. Partial enthalpies and related quantities in mixtures from computer simulation. *Chem. Phys. Lett.* 136:35.

Smit, B., and D. Frenkel. 1989. Calculation of the chemical potential in the Gibbs ensemble. *Mol. Phys.* 68:951.

Smit, B., S. Karaborni, and J. I. Siepmann. 1995. Computer simulations of vapor-liquid phase equilibria of n-alkanes. *J. Chem. Phys.* 102:2126.

Smit, B., Ph. de Smedt, and D. Frenkel. 1989. Computer simulations in the Gibbs ensemble. *Mol. Phys.* 68:931.

Smith, W., T. R. Forrester, and I. T. Todorov. 2009. The DL_POLY_2 user manual. v. 2.20. STFC Daresbury Laboratory, Cheshire, U.K. http://www.cse.scitech.ac.uk/ccg/software/DL_POLY/ (accessed on December 14, 2009).

Swope, W. C., H. C. Andersen, P. H. Berens, and K. R. Wilson. 1982. A computer simulation method for the calculation of equilibrium constants for the formation of physical clusters of molecules: Application to small water clusters. *J. Chem. Phys.* 76:637.

Torrie, G. M., and J. P. Valleau. 1974. Monte Carlo free energy estimates using non-Boltzmann sampling: Application to the sub-critical Lennard-Jones fluid. *Chem. Phys. Lett.* 28:578.

Torrie, G. M., and J. P. Valleau. 1977. Nonphysical sampling distributions in Monte Carlo free energy estimation: umbrella sampling. *J. Comput. Phys.* 13:187.

Valleau, J. P. 1991a. Density-scaling: A new Monte Carlo technique in statistical mechanics. *J. Comput. Phys.* 96:193.

Valleau, J. P. 1991b. The Coulombic phase transition: Density-scaling Monte Carlo. *J. Chem. Phys.* 95:584.

Vega, C., J. L. F. Abascal, and I. Nezbeda. 2006. Vapor-liquid equilibria from the triple point up to the critical point for the new generation of TIP4P-like models: TIP4P/Ew, TIP4P/2005, and TIP4P/ice. *J. Chem. Phys.* 125:034503.

Verlet, V. 1967. Computer "experiments" on classical fluids. I. Thermodynamical properties of Lennard-Jones molecules. *Phys. Rev.* 159:98.

Vrabec, J., A. Loffi, Land J. Fischer. 1995. Vapour liquid equilibria of Lennard-Jones model mixtures from the *NPT* plus test particle method. *Fluid Phase Equilib.* 112:173.

Widom, B. 1963. Some topics in the theory of fluids. *J. Chem. Phys.* 39:2808.

Wilding, N. B. 1995. Critical-point coexistence-curve properties of the Lennard-Jones fluid: A finite-size scaling study. *Phys. Rev. E* 52:602.

Wilding, N. B. 1997a. Simulation studies of fluid critical behaviour. *J. Phys.: Condens. Matter* 9:585.

Wilding, N. B. 1997b. Critical end point behavior in a binary fluid mixture. *Phys. Rev. E* 55:6624.

Wilding, N. B. 2001. Computer simulation of fluid phase transitions. *Am. J. Phys.* 69:1147.

Wilding, N. B., and A. D. Bruce. 1992. Density fluctuations and field mixing in the critical fluid. *J. Phys.: Condens Matter* 4:3087.

Wilding, N. B., F. Schmid, and P. Nielaba. 1998. Liquid-vapor phase behavior of a symmetrical binary fluid mixture. *Phys. Rev. E* 58:2201.

4 Integral Equation Theories

This chapter summarizes the most widely used closure conditions to solve the *Ornstein–Zernike* (OZ) equation. Several procedures proposed to achieve thermodynamic consistency between the different routes for obtaining the thermodynamic properties are also discussed. Special attention is devoted to integral equation perturbation theories. This chapter ends analyzing the results for the thermodynamic and structural properties obtained from several *integral equation theories* (IETs) as compared with simulation data.

4.1 ORNSTEIN–ZERNIKE EQUATION

An important equation relating several correlation functions and which is the basis of the IETs is the OZ relationship (Ornstein and Zernike 1914) which, for a homogeneous and isotropic fluid, is expressed in the form

$$c(r) = h(r) - \rho \int c(\mathbf{r}')h(|\mathbf{r} - \mathbf{r}'|)d\mathbf{r}', \tag{4.1}$$

where $c(r)$ is the *direct correlation function* (DCF) defined by means of this expression. This function has the property (Percus and Yevick 1964)

$$c(r) \to -\beta u(r), \quad \text{if} \quad r \to \infty. \tag{4.2}$$

Alternatively, the OZ equation can be expressed in the following way:

$$h(r) = c(r) + \gamma(r), \tag{4.3}$$

which indicates that the total correlation function $h(r)$ consists of two contributions: the DCF between two particles $c(r)$ and the *indirect correlation function* $\gamma(r)$ between them because of their interactions with the remaining particles. From Equation 4.1 we have

$$\gamma(r) = \rho \int c(\mathbf{r}')\mathbf{h}(|\mathbf{r} - \mathbf{r}'|)d\mathbf{r}'. \tag{4.4}$$

Still another form of the OZ equation results from taking Fourier transforms on both sides of Equation 4.1, with the result

$$\hat{h}(k) = \frac{\hat{c}(k)}{1 - \rho\hat{c}(k)}, \tag{4.5}$$

where $\hat{c}(k)$ and $\hat{h}(k)$ are the Fourier transforms of the functions $c(r)$ and $h(r)$, respectively.

The static structure factor can also be expressed in terms of the Fourier transform of the DCF

$$S(k) = \frac{1}{1 - \rho\hat{c}(k)}. \tag{4.6}$$

As a consequence, the compressibility equation (2.54) can be written in terms of $\hat{c}(0)$, the Fourier transform of $c(r)$ for $k = 0$, as follows:

$$\rho\hat{c}(0) = 1 - \frac{\kappa_T^{id}}{\kappa_T}, \tag{4.7}$$

where κ_T^{id} is the isothermal compressibility of an ideal gas.

Baxter (1968) showed that, when $c(r) = 0$ for $r > R$, the total correlation function and DCF, coupled by means of the OZ equation, can be decoupled into a pair of equations that involve a third function $q(r)$ as follows:

$$rc(r) = -q'(r) + 2\pi\rho \int_r^R q'(t)q(t-r)\,dt, \quad 0 < r < R, \tag{4.8}$$

$$rh(r) = -q'(r) + 2\pi\rho \int_0^R h(|r-t|)q(t)(r-t)\,dt, \quad r > 0, \tag{4.9}$$

where $q'(r) = dq(r)/dr$ and $q(r) = 0$ for $r > R$ by definition.

4.2 CLOSURE CONDITIONS

IETs are based on solving the integral equation (4.1), for which we need an additional equation, the *closure condition*. The exact closure condition is (Van Leeuwen et al. 1959)

$$c(r) = h(r) - \ln y(r) + B(r), \tag{4.10}$$

where $B(r)$ is the so-called *bridge function*. Unfortunately, this function is not known exactly, although it can be obtained from simulation with great accuracy. The preceding expression can be put in the alternative, and also exact, form

$$c(r) = f(r)y(r) + t(r), \tag{4.11}$$

where
   $f(r) = e^{-\beta u(r)} - 1$ is the *Mayer function*
   $t(r)$ is the so-called *tail function*

For $B(r)$, or equivalently for $t(r)$ or for the closure relation, several analytical approximations have been proposed. Some of them are described in the following.

4.2.1 SIMPLE CLOSURES

Hypernetted chain (HNC) approximation, developed independently by several authors (Van Leeuwen et al. 1959, Meeron 1960, Morita 1960, Rushbrooke 1960, Verlet 1960)

$$B(r) = 0, \tag{4.12}$$

$$c(r) = h(r) - \ln y(r). \tag{4.13}$$

Percus–Yevick (PY) *approximation* (Percus and Yevick 1958)

$$B(r) = \ln y(r) - y(r) + 1, \tag{4.14}$$

$$c(r) = f(r)y(r), \tag{4.15}$$

which corresponds to setting $t(r) = 0$ in the exact relationship (4.11). An interesting feature of this closure is that it provides analytical solutions for the EOS, the DCF, and the radial distribution function (RDF) for certain simple potential models, especially for the HS fluid as well as for HS fluid mixtures as we will see in Chapters 5 and 7.

Mean spherical approximation (MSA) (Lebowitz and Percus 1966) for fluids with a spherical hard core with diameter σ

$$c(r) = -\beta u(r), \quad \text{for} \quad r > \sigma. \tag{4.16}$$

For the particular case of a HCY potential (1.12), the MSA provides an explicit expression for the RDF (Høye and Blum 1977).

The MSA has been extended to potentials with a soft core (SMSA) by introducing an effective HS diameter (Rosenfeld and Ashcroft 1979a), by introducing an effective diameter σ and using the PY closure for $r < \sigma$ and the MSA closure for $r > \sigma$ (Blum and Narten 1972, Narten et al. 1974), or by splitting the potential into short-range $u_0(r)$ and long-range $u_1(r)$ contributions (Madden and Rice 1980) according to Weeks et al. (1971) prescription (see Section 6.4 for details).

Generalized mean spherical approximation (GMSA), first developed for the HS fluid (Waisman 1973) for which the DCF is taken to be

$$c(r) = K\sigma \frac{e^{z(r-\sigma)}}{r}, \quad \text{for} \quad r > \sigma, \tag{4.17}$$

where K and z are parameters that are determined by means of appropriate conditions. This approximation arises from the MSA equation (4.16) by simply taking for the DCF of HS a fictitious potential of the form of the Yukawa potential (1.12). However, the denomination "generalized mean spherical approximation" was proposed later (Høye et al. 1974) for a more general approach for potential models with a spherical hard core in which the DCF beyond the core is approximated by a sum of terms, one of which is of the Yukawa form and the remaining account for the actual potential of the fluid considered, namely,

$$c(r) = K\sigma \frac{e^{z(r-\sigma)}}{r} - \beta u(r), \quad \text{for} \quad r > \sigma, \tag{4.18}$$

The GMSA approach has been applied also to an arbitrary number of Yukawa tails (Høye and Blum 1977). In particular, the two Yukawa case can be solved in an essentially analytical way (Høye et al. 1976, Høye and Stell 1984a,b).

Martynov–Sarkisov (MS) *approximation* (Martynov and Sarkisov 1983)

$$B(r) = -\frac{1}{2}[\ln y(r)]^2, \tag{4.19}$$

$$c(r) = h(r) - \ln y(r) - \frac{1}{2}[\ln y(r)]^2. \tag{4.20}$$

Alternatively, combining Equations 4.3, 4.10, and 4.19, we can write for the MS bridge function

$$B(r) = -1 - \gamma(r) + [1 + 2\gamma(r)]^{1/2}. \tag{4.21}$$

Further refinements based on similar grounds as this closure have been reported by several authors (Martynov et al. 1999, Sarkisov 2001, 2002). Ballone et al. (1986) generalized the MS expression (4.21) for the bridge function in the form

$$B(r) = -1 - \gamma(r) + [1 + s\gamma(r)]^{1/s}, \tag{4.22}$$

(*Ballone–Pastore–Galli–Gazzillo* (BPGG) *approximation*) where s is a parameter to be determined from some optimizing condition.

Verlet modified (VM) *approximation*, based on the approximate bridge function

$$B(r) = -\frac{a[\gamma(r)]^2}{2[1 + b\gamma(r)]}, \tag{4.23}$$

was first proposed by Verlet (1980) for the HS fluid on a semiempirical basis, with $a = 1$ and $b = 4/5$, and later modified and applied to other fluids by allowing parameters a and b to be density and eventually temperature dependent (Labík et al. 1991a,b, Choudhury and Ghosh 2002, 2003). Lomba and Lee (1996) introduced an additional modification to Equation 4.23 by replacing $\gamma(r)$ by $\gamma^*(r) = \gamma(r) - \beta u_1(r)$, where $u_1(r)$ is the long-range part of the potential as given in Equation 6.69. Another modification to Equation 4.23 has been proposed by Zhou and Solana (2009), and Zhou et al. (2003a,b) proposed a closure that reduces to the VM closure as a limiting case. A generalization of Equation 4.23 in the form

$$B(r) = -\frac{c[\gamma(r)]^2}{2}\left[1 - \frac{a'\gamma(r)}{1 + b'\gamma(r)}\right] \tag{4.24}$$

was also proposed by Verlet (1981).

Zero-separation theorem-based closure (ZSEP) (Lee 1995) is based on the generalized Verlet closure, Equation 4.24, but replacing $\gamma(r)$ with $\gamma^*(r) = \gamma(r)+(\rho/2)f(r)$ for HS (Lee 1995, Fernaud et al. 2000) and $\gamma^*(r) = \gamma(r) - \beta u_1(r)$, as before, for SS potentials (Lee et al. 1996). Parameters a, b, and c are determined from the *zero separation theorems*, which are exact relationships that must fulfill the functions $y(r)$, $\gamma(r)$, and $B(r)$ for $r=0$ (Hoover and Poirier 1962, Meeron and Siegert 1968, Grundke and Henderson 1972, Barboy and Tenne 1976, Zhou and Stell 1988).

For hard-core potential models, the aforementioned closures are supplemented with the so-called *core condition*, namely, $h(r) =-1$ for $r < \sigma$, which is an exact condition. On the other hand, although the preceding description of simple closures has been focused on single-component fluids, the closures can be readily extended to mixtures.

4.2.2 THERMODYNAMICALLY SELF-CONSISTENT INTEGRAL EQUATION THEORIES

A drawback of many IETs, as is the case of those resulting from most of the closures cited in this section, is the lack of *thermodynamic consistency*. This means that different routes used to obtain the thermodynamic properties, as are the virial (2.36), energy (2.46), and compressibility (2.54) equations, provide results that are at variance with each other. The problem can be overcome, at least partially, by introducing some kind of parameter which is determined from the condition that two of the routes just mentioned, generally the pressure and compressibility equations, give the same result. This is the situation of the GMSA where there are two undetermined parameters K and z in the Yukawa tail that can be used to this end. Similar considerations apply to the BPGG, VM, and ZSEP closures, as all of them have at least one undetermined parameter that can be used to force thermodynamic consistency. In particular, the Lomba and Lee (1996) version of the VM closure belongs to this class of approximations, but in this case, together with the consistency between the pressure and compressibility equations, the exact thermodynamic relationship $(\partial p/\partial \mu)_T = \rho$ is used, because the first condition may be satisfied by more than one pair of values of a and b in Equation 4.23 and so the second condition is used to chose a particular solution. Three new closures in which the consistency is enhanced are described in the following.

Self-consistent Ornstein–Zernike approximation (SCOZA): Høye and Stell (1977) termed in this way a class of closures for potentials with hard spherical cores based on a modification of the MSA by introducing one or more state-dependent parameters which are determined from self-consistency conditions. In its simplest form, the SCOZA closure is (Høye and Stell 1984b)

$$c(r) = -a(\rho, \beta)u(r), \quad \text{for} \quad r > \sigma, \tag{4.25}$$

where $a(\rho, \beta)$ is the parameter to be determined from some consistency condition. A more general version of the SCOZA closure is (Pini et al. 1998)

$$c(r) = c_{HS}(r) - a(\rho, \beta)u(r), \quad \text{for} \quad r > \sigma, \tag{4.26}$$

where $c_{HS}(r)$ is the HS DCF, for which one can choose a Yukawa form, as in the GMSA equation (4.17), or an expression known in advance by any other means.

SCOZA can be extended to soft-core potentials by exploiting the fact that many simple continuous potentials can be approximated by linear combinations of Yukawa tails (Kahl et al. 2002, Schöll-Paschinger and Kahl 2003), by introducing an effective hard-core diameter (Høye and Reiner 2006), by removing the condition $r > \sigma$ in Equation 4.25 (Raineri et al. 2004, Mladek et al. 2006), or by replacing $c_{HS}(r)$ in Equation 4.26 with a suitable DCF for the soft core (Raineri et al. 2004).

Bridge function expansion closure (Martynov and Vompe 1993, Vompe and Martynov 1994): In this approximation, the bridge function is expressed as an expansion of the form

$$B(r) = \sum_{i=1}^{\infty} a_i \varphi_i(r),$$ (4.27)

where

$\varphi_i(r)$ are suitable functions, depending on density and temperature

a_i are temperature- and density-dependent parameters that are determined by imposing full consistency

This is achieved by relating the pressure and compressibility equations through the definition of isothermal compressibility

$$\kappa_T = \frac{1}{\rho}\left(\frac{\partial \rho}{\partial P}\right)_T,$$ (4.28)

and pressure and energy equations through the exact thermodynamic relationship

$$\left(\frac{\partial U}{\partial V}\right)_T = T\left(\frac{\partial P}{\partial T}\right)_V - P.$$ (4.29)

This leads to two consistency conditions depending on $c(r)$, $g(r)$, and the temperature and density derivatives of the latter function, which can be obtained from the OZ equation and their derivatives. The two consistency conditions in turn can be derived, thus giving rise to an infinite set of conditions.

In addition, in some cases two different closures are mixed by means of a parameter that is determined from a consistency condition. Two well-known theories of this kind are described in the following.

Rogers–Young (RY) *approximation* (Rogers and Young 1984): Combines the HNC and PY closures to give the RDF

$$g(r) = \exp\left[-\beta u(r)\right]\left[1 + \frac{\exp\left[\gamma(r)\,\varphi(r)\right] - 1}{\varphi(r)}\right],$$ (4.30)

with $\varphi(r) = 1 - \exp(-ar)$, so that $0 \le \varphi(r) \le 1$, $\varphi(0) = 0$, and $\varphi(\infty) = 1$. It is easy to see that Equation 4.30 reduces to the PY result for $\varphi = 0$ and to the HNC result for $\varphi = 1$. Parameter a is determined from the condition of consistency between the pressure and compressibility equations.

Zerah–Hansen (ZH or HMSA) *approximation* (Zerah and Hansen 1986), a modification of the RY IET in the form

$$g(r) = \exp\left[-\beta u_0(r)\right]\left[1 + \frac{\exp\left[\gamma^*(r)\,\varphi(r)\right] - 1}{\varphi(r)}\right], \qquad (4.31)$$

where $\varphi(r)$ has the same meaning as in the RY expression (Equation 4.30) and is determined from the same consistency condition, and $u_0(r)$ and $\gamma^*(r)$ have the same meaning as earlier. It is easy to see that the RDF of Equation 4.31 interpolates between the HNC and SMSA closures and for this reason was called HMSA.

4.2.3 MOLECULAR FLUIDS

Several of the above-mentioned simple closures have been applied to molecular fluids. Among them are the PY (Chen and Steele 1971, Perera et al. 1987), the HNC (Perera et al. 1987), the VM (Labík et al. 1991a,b), and the self-consistent RY (Singh et al. 1996) approximations. In addition, general frameworks for extending IETs to nonspherical molecular fluids have also been reported (Lado 1982a, Fries and Patey, 1985, 1986, Perera et al. 1987, 1992).

However, a more widely used approach to study polyatomic molecular fluids from IET is the *reference interaction site model* (RISM) theory (Chandler and Andersen 1972; see also Chandler et al. 1982 for a more rigorous formulation of the theory). It is based on a generalization of the OZ equation for simple fluids to molecular fluids in which the pair interaction between two molecules i and j can be considered as the sum of all site–site interactions between a site α of molecule i and a site γ of molecule j, that is,

$$u\left(\mathbf{r}_i, \mathbf{r}_j\right) = \sum_{\alpha=1}^{n}\sum_{\gamma=1}^{n} u_{\alpha\gamma}\left(\left|\mathbf{r}_i^{\alpha} - \mathbf{r}_j^{\gamma}\right|\right), \qquad (4.32)$$

where n is the number of sites in each molecule, supposed to be equal. The OZ equation (4.1) can be expressed in the form

$$h(r) = c(r) + \rho c(r) * h(r). \qquad (4.33)$$

Here, the symbol "$*$" means a convolution that, for two functions $a(r)$ and $b(r)$, is defined as

$$a(r) * b(r) = \int a\left(\left|\mathbf{r} - \mathbf{r}'\right|\right) b\left(\mathbf{r}'\right) d\mathbf{r}' = \int b\left(\left|\mathbf{r} - \mathbf{r}'\right|\right) a\left(\mathbf{r}'\right) d\mathbf{r}' = b(r) * a(r). \qquad (4.34)$$

The generalization of the OZ equation for molecular fluids with site–site interactions is (Ladanyi and Chandler 1975)

$$\mathbf{h}(r) = \mathbf{w}(r) * \mathbf{c}(r) * \mathbf{w}(r) + \rho\mathbf{w}(r) * \mathbf{c}(r) * \mathbf{h}(r). \qquad (4.35)$$

In this expression, $\mathbf{h}(r)$ is a $(n \times n)$ matrix whose elements are $h_{\alpha\gamma}(r)$, the site–site total correlation functions between site α in one molecule and site γ in another molecule, $\mathbf{c}(r)$ is the matrix whose elements are the site–site DCFS $c_{\alpha\gamma}(r)$, and $\mathbf{w(r)}$ is a matrix whose elements are

$$w_{\alpha\gamma}(r) = \delta_{\alpha\gamma}\delta\mathbf{r}_{\alpha\gamma} + s_{\alpha\gamma}^{(2)}(r). \qquad (4.36)$$

In the latter expression, $\delta(x)$ is the Dirac delta function and $s_{\alpha\gamma}^{(2)}(r)$ is the intramolecular site–site correlation function between sites α and γ belonging to the same molecule. Equation 4.35 is supplemented with a suitable closure for atomic fluids, usually the PY or HNC closure. The theory has also been extended to multicomponent molecular fluids (Lombardero and Enciso 1981).

The RISM theory becomes intractable for flexible polymer melts because of the huge number of nonlinear coupled integral equations involved in Equation 4.35. On the basis of the RISM scheme, Curro and Schweizer developed a simplified procedure which renders the problem tractable (Curro and Schweizer 1987a,b, Schweizer and Curro 1988a,b). Their approach is known as the *polymer reference interaction site model* (PRISM).

On the other hand, it is to be noted that the extension of the pressure equation (2.36) to chain molecular fluids is not straightforward, because of the presence of internal degrees of freedom related to bond vibration, rotation, and bending, with the associated intramolecular potentials in addition to the intermolecular potential. Honnell et al. (1987) derived the expressions of the pressure equation for chain molecular fluids interacting with site–site potentials with different degrees of flexibility of the molecules.

4.2.4 ALGORITHMS FOR NUMERICAL SOLUTION OF INTEGRAL EQUATION THEORIES

Although for some potential models certain IET can be solved analytically, in most cases one must resort to a numerical procedure. The first techniques used were based on iteration (the Picard method) starting from an initial estimate of one of the functions involved in the OZ equation and solving the equation to obtain a new estimate and so on until convergence is achieved (see Watts 1973, for an early review on these procedures). However, this method is quite inefficient because the convergence of the iteration is slow, especially at high densities. A better procedure is based on the Newton–Raphson method for solving nonlinear equations. As it is well known, for a single nonlinear equation of the form $f(x) = 0$, an approximation of order $n + 1$ to the solution is obtained from the approximation of order n as

$$x_{n+1} = x_n - \frac{f'(x_n)}{f(x_n)}, \qquad (4.37)$$

where $f'(x_n)$ is the derivative of $f(x)$ at $x = x_n$, so that starting from an initial guess x_0 we can obtain successive improved estimates until the result converges to the desired accuracy. The problem now is that, as we have to solve the OZ equation in a huge number M of points, covering the whole range of distances of interest, we will have a number M of simultaneous equations whose solution involves a square matrix of dimension M. A much more efficient method, consisting in a combination of the Picard and Newton–Raphson methods, was developed by Gillan (1979). To this end, the indirect correlation function $\gamma(r)$ is divided into coarse and fine contributions, namely,

$$\gamma_i = \sum_{j=1}^{m} a_j P_{ij} + \Delta \gamma_i, \qquad (4.38)$$

where $i = 1, \ldots, M$, P_{ij} are basis functions, and $m \ll M$.

The first term of the right-hand side of the preceding expression is the coarse contribution and the second one is the fine contribution. Coefficients a_i in the coarse contribution are determined from the Newton–Raphson method, which now involves a small number of equations, and the fine contribution is determined using the Picard method, which now converges rapidly as $\Delta \gamma(r)$ is small. The method was extended by Monson (1982) to the RISM theory.

Still more efficient than the method developed by Gillan is the one proposed by Labík et al. (1985) for systems with spherically symmetrical pair potentials. It is also based on a combination of the Picard and Newton–Raphson methods together with the expansion of the function $\Gamma(r) = r[h(r) - c(r)]$ in terms of suitable basis functions. Further improvements on these methods have been developed by Kinoshita and Harada (1988) and by Lomba (1989).

4.3   UNIVERSALITY OF THE BRIDGE FUNCTION

As stated earlier, if the bridge function $B(r)$ were known exactly for a particular fluid, we could obtain the exact $g(r)$ for that fluid by solving the OZ equation. Conversely, from the simulation data of $g(r)$, exact to within the accuracy of the simulations, we could obtain the corresponding exact $B(r)$ from the OZ equation and the exact closure (Equation 4.10). Obviously, this would be of little practical interest, because we would need to obtain the bridge function for each fluid of interest from the simulation data of the RDF for the same fluid and then use the OZ equation with the exact closure (Equation 4.10) to obtain the RDF. But if we knew the RDF of the fluid from simulation, there would not be need of using IET to obtain that function! Fortunately, there are two facts that may simplify the problem. On the one hand, the RDF of simple fluids is quite insensitive to the precise form of the bridge function (Malijevský and Labík 1987, Kolafa et al. 2002). On the other hand, the Rosenfeld and Ashcroft (1979b) hypothesis on the universality of the bridge function states that "the bridge functions constitute the same family of curves, irrespective of the assumed pair potential." Under these assumptions we would only need to obtain from simulation the function $B(r)$ for a given fluid, representative of all simple fluids, and then use it to obtain the RDF of any other simple fluid from the OZ equation (4.1) and the exact closure (Equation 4.10).

In high density fluids with steeply repulsive potentials at short distances, the structure is mainly determined by the repulsive forces, as mentioned in Section 2.4, and it is commonly agreed that the HS fluid provides a quite good description of the structure in such circumstances. Precisely this is the basis of most of the proposed perturbation theories for fluids. Therefore, one may reasonably expect that the bridge function for the HS fluid will be a good representative of the bridge function of simple fluids. The HS bridge function was obtained by Malijevský and Labík (1987) from simulation data for the RDF and the results were parameterized as a simple function of density for reduced distances $r^* \geq 1$. The reliability of the *Malijevský and Labík* (ML) parameterization at high densities was confirmed by Kolafa et al. (2002) from comparison with their extremely accurate values of $B(r)$ obtained from simulation. Gazzillo and Della Valle (1993) proposed a procedure to obtain the HS bridge function for $0 \leq r^* \leq 1$ based on the use of the exact relationship (Equation 4.10), together with the Groot et al. (1987) expression for $c(r)$ and the Henderson and Grundke (1975) expression for $\ln y(r)$.

The procedure to obtain the bridge function from simulation data has also been applied to other simple fluids. Thus, numerical values have been reported for the LJ (Llano-Restrepo and Chapman 1992) and SS (Llano-Restrepo and Chapman 1994) fluids. Duh and Haymet (1995) showed that upon splitting the LJ potential into two suitably chosen density-dependent parts, a reference part $u_0(r, \rho)$ and a perturbation part $u_1(r, \rho)$, the curves of $B(r)$ for different densities obtained from simulation lie approximately on a single curve when $B(r)$ is plotted versus $\gamma^*(r) = \gamma(r) - \beta u_1(r)$. On this basis, they proposed the approximate closure

$$B\left(\gamma^*\right) = -\frac{\gamma^{*2}}{2}\left(1 + \frac{5\gamma^* + 11}{7\gamma^* + 9}\gamma^*\right)^{-1}. \tag{4.39}$$

This lends support to the so-called *unique functionality assumption* (Lee 1992), according to which the bridge function for a given fluid can be expressed as a unique function of the indirect correlation function, that is, $B = B(\gamma)$, provided that we replace γ with γ^* and optimize the choice of perturbation part $u_1(r)$ of the potential.

At this point, it is worthwhile to test whether the universality principle is fulfilled by the LJ fluid, by comparing the data from Llano-Restrepo and Chapman (1992) for the LJ $B(r)$ with the results for the HS $B(r)$ derived from the parameterization of Malijevský and Labík (1987) for $r^* \geq 1$ and from the procedure of Gazzillo and Della Valle (1993) for $0 \leq r^* \leq 1$. To this end, we need a way to obtain an effective HS diameter d for the LJ fluid as a function of density and temperature. In Section 4.4, and especially in Chapter 6, we will see several ways of obtaining effective diameters for molecules with soft repulsive cores. Here we will use a parameterization reported by Gazzillo and Della Valle (1993). The results, displayed in Figure 4.1, show that the LJ fluid conforms fairly well the universality condition, at least for moderate to high densities

However, the situation seems not to be the same for other fluids. Thus, for the *one-component plasma* (OCP), a fluid consisting of particles interacting through a Coulomb pair potential of the form (1.5) and immersed in a uniform compensating background, it has been found that the bridge function is well reproduced by the bridge function of the HS fluid at short range, but at long range small deviations

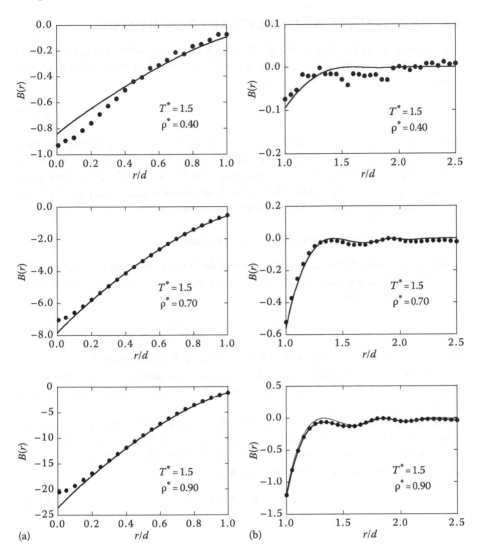

FIGURE 4.1 Bridge function $B(r)$ for the LJ fluid at reduced temperature T^* and several reduced densities $\rho^* = \rho\sigma^3$, where σ is the length parameter of the LJ potential, as a function of the reduced distance r/d. The points are the data obtained from simulation by Llano-Restrepo and Chapman (1992) and the curves are the results for the HS bridge function obtained from the procedure of Gazzillo and Della Valle (1993) for $0 \leq r^* \leq 1$ (a) and from the parameterization of Malijevský and Labík (1987) for $r^* \geq 1$ (b).

appear (Poll et al. 1988). More marked differences with respect to the HS $B(r)$ at medium and long range have been reported for the bridge function obtained from simulation for some models of liquid metals (Lomba et al. 1992a). In spite of this, the universality ansatz is a useful approximation which has led to a class of IETs for simple fluids that will be presented in the following section.

4.4 INTEGRAL EQUATION PERTURBATION THEORIES

We will include in this category these IETs, like perturbation theories in general, based on splitting the potential $u(r)$ into reference $u_0(r)$ and perturbation $u_1(r)$ contributions, namely,

$$u(r) = u_0(r) + u_1(r), \tag{4.40}$$

and using the structural and thermodynamic properties of a system with potential $u_0(r)$, the reference system, to obtain the corresponding properties of the actual system. The usefulness of these procedures lies on the fact that the properties of the reference system are often known beforehand, usually with great accuracy. Next, we will describe the foundations of a number of these approaches.

4.4.1 CORRECTED INTEGRAL EQUATION THEORIES

Madden and Fitts (1975) proposed to correct the RDF $g^{IE}(r)$ obtained from an IET for a fluid with a potential of the form of Equation 4.40 to obtain a more accurate result $g(r)$ in the form

$$g(r) \approx g_0(r) + g^{IE}(r) - g_0^{IE}(r), \tag{4.41}$$

where g_0^{IE} and $g_0(r)$ are the RDF of the reference fluid obtained from the same IET and from an independent procedure, for example, from computer simulation, respectively. Provided that $g_0(r)$ is more accurate than g_0^{IE}, this will lead to a more accurate estimate of $g(r)$.

Smith and Henderson (1978) denoted *hybrid theories* to this kind of approximations as they lie between the ordinary and *corrected integral equations*, proposed by them, based on splitting the total function and DCF into the contributions from the reference system and the perturbation, namely,

$$h(r) = h_0(r) + h_1(r), \tag{4.42}$$

and

$$c(r) = c_0(r) + c_1(r), \tag{4.43}$$

respectively. If the total $h_0(r)$ and direct $c_0(r)$ correlation functions for the reference system are known, and an approximate relation between $h_1(r)$ and $c_1(r)$ is introduced, we can solve the OZ equation (4.1) to obtain the latter quantities and, from Equations 4.42 and 4.43, the correlation functions $h(r)$ and $c(r)$ for the actual system.

4.4.2 OPTIMIZED RANDOM PHASE APPROXIMATION

For a system for which the intermolecular potential $u(r)$ can be split in the form of Equation 4.40, we can approximate the DCF by

$$c(r) = c_0(r) - \beta u_1(r), \tag{4.44}$$

Which defines the so-called *random phase approximation* (RPA) (Andersen et al. 1972), where $c_0(r)$ is the DCF of the reference system with potential $u_0(r)$. If the reference potential is the HS one and the corresponding PY approximation (Equation 4.15) is taken for $c_0(r)$, then Equation 4.44 reduces to the MSA Equation 4.16.

A useful feature of the RPA is that the static structure factors of the actual and reference systems, $S(k)$ and $S_0(k)$, respectively, are related in a simple way (Andersen and Chandler 1971)

$$S(k) = \frac{S_0(k)}{1 + S_0(k)\,\rho\beta\hat{u}_1(k)}, \tag{4.45}$$

where $\hat{u}_1(k)$ is the Fourier transform of $u_1(r)$.

In the case where $u_0(r) = u_{HS}(r)$, the HS potential, and so $u(r) = u_1(r)$ for $r > \sigma$, where σ is the HS diameter, Equation 4.44 yields the correct asymptotic bahavior (4.2). In this case, we can solve the OZ equation with the MSA closure for $c_1(r) = -\beta u_1(r)$ and then obtain $g(r)$ if we know $g_{HS}(r)$ from an independent source.

Also, for potentials with an HS core, introducing a *renormalized potential* $C(r)$ such that

$$C(r) = g_{RPA}(r) - g_{HS}(r), \tag{4.46}$$

the *exponential* (EXP) *approximation* (Andersen and Chandler 1972) reads

$$g(r) = g_{HS}(r)\exp[C(r)]. \tag{4.47}$$

This result is readily obtained by applying the exact closure (Equation 4.10) to the actual and reference HS fluids, introducing the approximation $B(r) \approx B_{HS}(r)$, and taking into account Equations 4.44 and 4.46. For $\rho \to 0$, $C(r) \to -\beta u_1(r)$ (Andersen et al. 1976) and so the EXP approximation gives the right second virial coefficient. The *linearized exponential* (LEXP) *approximation* (Verlet and Weis 1974) is obtained by expanding to first-order $\exp[C(r)]$ in Equation 4.47, namely,

$$g(r) = g_{HS}(r)[1 + C(r)]. \tag{4.48}$$

In contrast with the EXP approximation, for potentials with an HS core the RPA will not satisfy, in general, the condition $g(r) = 0$ for $r < \sigma$. To correct for this unphysical behavior one can supplement the RPA equation (4.44) with the condition

$$g(r) = 0, \quad r < \sigma, \tag{4.49}$$

or, equivalently,

$$C(r) = 0, \quad r < \sigma, \tag{4.50}$$

which, together with Equation 4.44, defines the *optimized random phase approximation* (ORPA) developed by Andersen and Chandler (1971). As the definition of the perturbation potential within the core is somewhat arbitrary, the fulfillment of condition (4.49) is achieved by determining a suitable perturbation potential within the core.

If $u_0(r)$ is continuous but harshly repulsive, we can replace the reference potential with the HS potential for $r < d$ and take $u(r) = u_1(r)$ for $r > d$, where d is some effective HS diameter, and rewrite Equation 4.44 as

$$c(r) = c_{HS}(r) - \beta u_1(r), \quad \text{for} \quad r > d. \tag{4.51}$$

For fluids with narrow attractive potentials, approximation (4.51) becomes inaccurate, and so the ORPA is not suitable for these fluids. Pini et al. (2002) have proposed to replace closure (4.51) with

$$c(r) = c_{HS}(r) + f_1(r), \quad \text{for} \quad r > d, \tag{4.52}$$

where $f_1(r) = \exp[-\beta u_1(r)] - 1$ is the Mayer function for the perturbation potential $u_1(r)$. Approximation (4.52) together with condition (4.49) defines the *nonlinear ORPA*, or *NLORPA*. We will return to the ORPA and other related approximations in Section 6.5.

4.4.3 REFERENCE INTEGRAL EQUATION THEORIES

Starting again from the splitting of the potential in the form of Equation 4.40, one can apply the exact closure (Equation 4.10) to the actual potential $u(r)$ and to the reference potential $u_0(r)$. Then, one easily arrives at

$$g(r) = \exp\left[-\beta u(r) + \gamma(r) + B(r)\right] = g_0(r) \exp\left[-\beta \Delta u(r) + \Delta \gamma(r) + \Delta B(r)\right], \tag{4.53}$$

where
$$\Delta u(r) = u(r) - u_0(r)$$
$$\Delta \gamma(r) = \gamma(r) - \gamma_0(r)$$
$$\Delta B(r) = B(r) - B_0(r)$$

One can use a known $B_0(r)$ for the reference fluid and some approximation for $\Delta B(r)$. This gives rise to a kind of theories which were denoted by Larsen (1978) *reference IETs*, although this approach was proposed earlier by Lado (1964, 1973).

One of the most simple and successful of these theories is the *reference hypernetted chain* (RHNC) theory, proposed by Lado (1964, 1973), consisting in setting $\Delta B(r) = 0$ in Equation 4.53 or, equivalently, $B(r) = B_0(r)$. In many cases, the reference system is taken to be the HS fluid, for which the bridge function is accurately known, as seen in Section 4.3. The fact that the universality of the bridge function holds approximately for simple fluids implies that for them $\Delta B(r) = 0$ is a good approximation, whence the success of the theory.

The RHNC theory* requires a procedure to determine the parameters of the reference system. This can be achieved by imposing thermodynamic consistency.

---

* Sometimes this theory is referred to as modified HNC or MHNC. However, under the latter denomination are often included different theories based on replacing the exact (unknown) $B(r)$ with some approximate one in Equation 4.10, not necessarily related to any particular reference fluid, which avoids the need of determining the parameters of the reference fluid. With this in mind, the RHNC theories might be considered as a subset of the MHNC class of theories.

However, a better procedure (Lado 1982b, Lado et al. 1983), based on minimizing the free energy, uses the constraint

$$\int d\mathbf{r} \left[g\left(r\right) - g_0\left(r\right) \right] \delta B_0\left(r\right) = 0, \tag{4.54}$$

which, for a reference potential $u_0(r)$ depending on a length parameter σ and an energy parameter ε, gives rise to the conditions

$$\int d\mathbf{r} \left[g\left(r\right) - g_0\left(r\right) \right] \sigma \frac{\partial B_0\left(r\right)}{\partial \sigma} = 0 \tag{4.55}$$

and

$$\int d\mathbf{r} \left[g\left(r\right) - g_0\left(r\right) \right] \varepsilon \frac{\partial B_0\left(r\right)}{\partial \varepsilon} = 0. \tag{4.56}$$

The RHNC theory has been extended to diatomic molecular fluids using different choices for the reference system and the corresponding bridge function. Thus, for example, Lado (1982c) uses an equivalent HS fluid as the reference system taking for the bridge function accurate parameterizations. Lado (1988) takes a hard diatomic molecular fluid within the PY approximation as the reference fluid. Lomba et al. (1992b) and Lombardero et al. (1996) also use the hard diatomic as the reference fluid solved by means of the OZ equation with the exact closure (Equation 4.10), taking for $B_0(r)$ the VM closure for molecular fluids.

In a similar way, the application of the RHNC theory to mixtures can be performed in several ways. For mixtures of spherical molecules, one can use an HS mixture as the reference system (Enciso et al. 1987). In case the mixture contains nonspherical molecules, an equivalent HS fluid mixture can be used as reference (Anta et al. 1997). Asymmetric binary mixtures of HS can be treated as an effective one-component fluid (Clément-Cottuz et al. 2000). The repulsive part of the interaction potential is well suited for the reference system in liquid alloys within the framework of the RHNC (Mori et al. 1991), and so on. We will address perturbation theories for fluid mixtures in more detail in Chapter 7.

An integral equation perturbation theory closely related to the RHNC is the so-called *perturbative hypernetted chain* (PHNC) theory, proposed by Kang and Ree (1995a), which uses density-dependent reference and perturbation potentials[*] and, as in the RHNC, approximates the bridge function $B(r)$ of the actual system with that of the reference system $B_0(r)$. The latter is determined by solving the OZ equation using the MS or BPGG approximations.

Another reference IET that is sometimes used is the *reference Percus–Yevick* (RPY) theory (Lado 1964), which is based on replacing the tail function $t(r)$ in the exact closure (4.11) with $t_0(r)$, the tail function corresponding to the reference system.

---

[*] We will return to the Kang and Ree splitting of the potential in Chapter 6 in a different context (see Equations 6.75 and 6.76).

4.4.4 PERTURBATIVE SOLUTIONS OF THE PERCUS–YEVICK AND MEAN SPHERICAL APPROXIMATIONS

Tang and Lu (1993, 1997a) developed a procedure based on perturbation theory to solve the OZ equation in the framework of the PY or MSA approximations for potentials consisting of a spherical hard core plus an arbitrary tail. To this end, they expand the total $h(r)$ and direct $c(r)$ correlation functions in terms of a suitable parameter ε around those for the HS system, taken as a reference, in the form

$$h(r) = h_0(r) + \varepsilon h_1(r) + \varepsilon^2 h_2(r) + \cdots, \qquad (4.57)$$

$$c(r) = c_0(r) + \varepsilon c_1(r) + \varepsilon^2 c_2(r) + \cdots, \qquad (4.58)$$

where subscripts 0 and i refer to the reference HS fluid and to the ith perturbative contribution, respectively, so that for $\varepsilon = 0$ the correlation functions of the HS reference fluid are recovered. Introducing Equations 4.57 and 4.58 into the PY closure equation (4.15) or the MSA closure equation (4.16), the OZ equation can be solved, yielding the RDF in the form of a series expansion, namely

$$g(r) = \sum_{i=0}^{\infty} g_i(r). \qquad (4.59)$$

The procedure devised by *Tang and Lu* (TL) allows to obtain analytical expressions for the first-order term $g_1(r)$ for a number of hard-core potential models in the mean spherical approximation (*first-order MSA* or FMSA) (Tang and Lu 1997a, Díez et al. 2007). Another useful feature of this theory is that it also provides analytical expressions for the first-order DCF $c_1(r)$ for some hard-core fluids (Tang 2003, 2007, Hlushak et al. 2009). A general extension of the TL theory to fluid mixtures of particles with hard-core potentials was reported also by Tang and Lu (1995). We will address the TL solution for the RDF of the HS fluid with some detail in Chapter 5.

On the other hand, Henderson et al. (1995) reported an analytical expression, derived on the basis of the MSA, for the free energy of HCY fluids in the form of an inverse temperature expansion up to fifth order. The series was later re-summed to obtain an approximate, though very accurate, expression for the free energy in a closed form (Duh and Mier-y-Terán 1997) and the procedure was also extended to mixtures of HCY fluids (Mier-y-Terán et al. 1998).

4.5 SOME RESULTS FROM INTEGRAL EQUATION THEORIES FOR SELECTED POTENTIAL MODELS

In the preceding sections we reviewed a number of the most commonly used IETs for fluids. Some of these theories have a sound theoretical basis, as are the cases of the HNC and PY theories, and therefore they allow for systematic improvements as are the RHNC and RPY theories, respectively. Other theories are semiempirical in nature, and so improvements must be introduced in a more heuristic way.

We will now analyze the performance of a number of IETs as compared with simulation data for several of the most commonly used potential models. This will include the SW, HCY, SS, and LJ potentials for simple fluids, as well as some simple models of molecular fluids. The results for the HS fluid obtained from several IETs will be analyzed with some detail in Chapter 5.

In general, one may expect that the HNC theory will be accurate at low densities, because in that situation the bridge function is small, as seen in Figure 4.1. On the other hand, introducing the HNC approximation $B(r) = 0$ into the exact closure (Equation 4.10), for $r \to \infty$, and so $g(r) \to 1$, one readily obtains the result (4.2). A similar result is arrived at from the PY equation (4.15) for $c(r)$. It can be shown that both, the PY and HNC theories, yield the right low-density expansion of the cavity distribution function $y(r)$, or the RDF $g(r)$, up to the first-order term in the density, and, therefore, the right second and third virial coefficients of the EOS (see, e.g., Barker and Henderson 1976). In contrast, the MSA approximation fails to provide the right second virial coefficient. In a similar way, the EXP approximation yields the exact second virial coefficient, whereas the RPA and the ORPA do not.

4.5.1 Square-Well Fluids

For the SW fluid, the PY and MS closures yield poor results at moderate temperatures for the EOS either using the compressibility or the virial route (Sarkisov et al. 1993). The PY results obtained from the compressibility (virial) equations lie below (over) the simulation data, whereas the contrary occurs in the MS theory, although in the latter the difference between the two routes is generally smaller. Instead, the performance of both theories for the excess energy is quite good.

In contrast, the MSA theory yields excellent results for the EOS from the energy route, as well as for the excess energy, as seen in Figure 4.2. The FMSA also provides reasonable results for the thermodynamic and structural properties at moderate to high densities, temperatures, and potential widths (Tang and Lu 1994a, Largo et al. 2005).

The HNC theory provides quite good agreement with simulations for moderate temperatures and well widths, and still better agreement is achieved with the RHNC theory for wide ranges of temperatures, densities, and potential widths (Gil-Villegas et al. 1995, Giacometti et al. 2009). This is a reflect of the excellent accuracy in the predicted values of the RDF from the RHNC theory, as illustrated in Figure 4.3.

A common drawback of many IETs for SW fluids is their poor performance in predicting the coexistence properties in the neighborhood of the critical region. A noteworthy exception is the SCOZA, which gives quite satisfactory results for the critical parameters. This is shown in Figure 4.4 for the reduced critical temperature T_c^*. Similar accuracy is achieved for the critical pressure and somewhat lower for the critical density, although it is to be noted that the simulation data for these two quantities suffer from greater uncertainty, and the results reported by different authors differ appreciably with each other, which prevents a better assessment of the quality of the theoretical predictions.

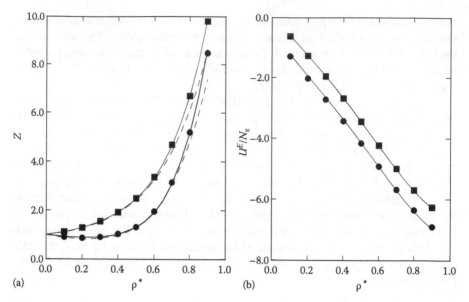

FIGURE 4.2 (a) Compressibility factor Z and (b) excess energy U^E for the SW fluid with $\lambda = 1.5$. Points: simulation data for $T^* = 2$ (circles) and 5 (squares), from Largo and Solana (2003). Curves: MSA results from Smith et al. (1977); the continuous and dashed curves in the left correspond to the energy and virial routes, respectively. The curve and simulation data for the excess energy at $T^* = 2$ have been displaced downward by 0.5 for clarity.

4.5.2 HARD-CORE YUKAWA FLUIDS

For HCY fluids, the analytical expression from Henderson et al. (1995) for the inverse temperature expansion of the MSA provides excellent agreement of the EOS with simulation data for moderate values of the inverse range parameter κ and supercritical temperatures, as shown in Figure 4.5 for $\kappa = 1.8$ and 4. The agreement is also very satisfactory for the excess energy in the first case, whereas for the shorter-ranged case $\kappa = 4$ some deviation can be seen at low temperatures. The results from the FMSA and those from the re-summed series by Duh and Mier-y-Terán (1997) are indistinguishable at the scale of Figure 4.5 from those of the inverse temperature expansion of the MSA.

Caccamo and Pellicane (2002) compared the predictions of the RHNC and HMSA theories with simulation for $\kappa = 1.8$ and 7, with the result that both can provide excellent accuracy for the excess energy and the EOS, but this may depend strongly on the consistency criterion used, especially for subcritical temperatures.

On the other hand, Caccamo et al. (1999) analyzed the performance of RHNC, GMSA, and SCOZA theories, among others, for the same values of κ. Their results reveal that all three theories are of similar accuracy in predicting the excess energy and the pressure. The SCOZA yields quite poor results for the contact value $g(\sigma)$ of the RDF at low temperatures, but in contrast provides better results for the isothermal compressibility. The accuracy of the SCOZA in predicting the liquid–vapor coexistence even for relatively large values of the inverse range parameter κ is clearly

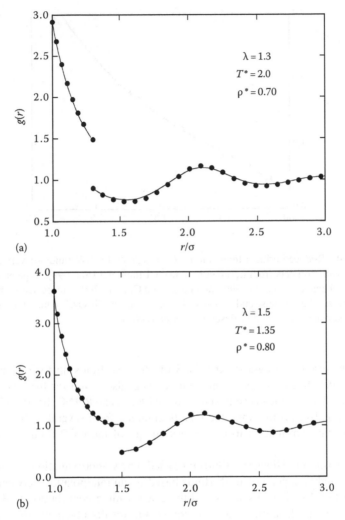

FIGURE 4.3 (a and b) Two examples of the RHNC predictions for the RDF of SW fluids. Points: simulation data from Largo et al. (2005). Curves: RHNC results from Gil-Villegas et al. (1995).

appreciated in Figure 4.6. This includes the accurate prediction of the critical point temperature and density (Reiner and Høye 2008).

4.5.3 SOFT-SPHERE FLUIDS

The thermodynamic properties of the SS fluids depend on a single-coupling parameter $\Gamma = \rho\sigma^3(\varepsilon/k_B T)^{3/n}$. The extreme case $n = 1$ is related to the OCP, consisting in point charges q in an uniform neutralizing background, for which the coupling parameter is $\Gamma = (q^2/k_B T)(4\pi\rho/3)^{1/3}$.

Rogers and Young (1984) solved the OZ equation with the RY closure (Equation 4.30) for SS fluids with $n = 12, 9, 6$, and 4, and for the OCP. The results in general

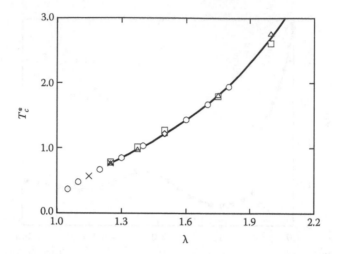

FIGURE 4.4 Reduced critical temperature $T_c^* = k_B T_c / \varepsilon$ for SW fluids of variable width. Points are simulation data from Largo et al. (2008), Elliott and Hu (1999), Vega et al. (1992), Orkoulas and Panagiotopoulos (1999), and Pagan and Gunton (2005), respectively, for circles, squares, triangles, diamonds, and crosses. The curve is the SCOZA data (interpolated for displaying purposes) from Schöll-Paschinger et al. (2005).

agreed with simulation data for the EOS of all these fluids up to freezing within a few parts per thousand. Excellent accuracy was also found for the RDF close to freezing. Similar accuracy is achieved with the PHNC and RHNC theories (Kang and Ree 1995a,b). To illustrate this, the results obtained from several of these theories for the thermal contribution to the excess energy of the OCP fluid are shown in Figure 4.7.

Kambayashi and Hiwatari (1990) proposed a new approximation for the bridge function of highly supercooled SS fluids based on an interpolation between the PY bridge function for short distances and the exact leading contribution to the bridge function for large distances. They compared the results obtained from the HNC modified in this way with those from the RY and RHNC approximations for supercooled SS fluids with $n = 6$ and 12, and concluded that for temperatures above freezing all three theories yield similar results, whereas for supercooled temperatures the modified HNC yields better agreement with simulations. In particular, the latter theory correctly predicts the splitting of the second peak of the RDF for strong supercooling.

Llano-Restrepo and Chapman (1994) reported an expression correlating their simulation data for the bridge function of the SS fluid with $n = 12$ for use in the context of the RHNC and used this approach for SS and LJ fluids.

4.5.4 LENNARD–JONES FLUID

Much of the research on IETs for simple fluids has focused on the LJ fluid, as a prototype of this kind of fluids, and so a considerable amount of results

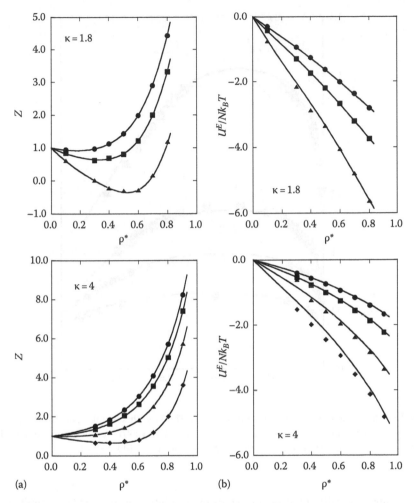

FIGURE 4.5 EOS (a) and excess energy (b) for two examples of HCY fluids. Curves: analytical expressions derived by Henderson et al. (1995) of the inverse temperature expansion of the MSA. Points: simulation data for $T^* = 2$ (circles), 1.5 (squares), 1 (triangles), and 0.7 (diamonds) from Shukla (2000), for $\kappa = 1.8$, and from Garnett et al. (1999), for $\kappa = 4$.

from different theories are available in the literature, some of which will be analyzed here.

The RDF that results from the PY theory for the LJ fluid is only in moderate agreement with simulation data, particularly in the neighborhood of the first maximum (Madden and Fitts 1974). The EOS obtained from the virial (compressibility) route considerably overestimates (underestimates) the simulation data at moderate to high densities (Mandel et al. 1970). In contrast, the excess energy is accurately predicted. Better accuracy for the RDF is achieved from the HNC

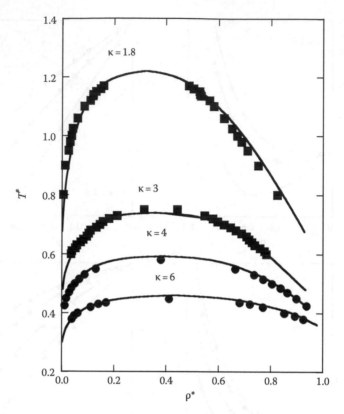

FIGURE 4.6 Liquid–vapor equilibrium for HCY fluids with several values of the inverse range parameter κ. Points: simulation data from Duda et al. (2007) (circles) and from Shukla (2000) (squares). Curves: SCOZA results from Pini and Stell (2002).

theory, although still some deviation from simulation is present, especially near the first maximum (Madden and Fitts 1974).

More satisfactory results for the RDF at low temperatures are obtained from the MSA, in spite of which this theory provides only moderate accuracy for the energy and the EOS obtained from the virial route at high densities and the accuracy worsens at high temperatures (Madden and Rice 1980). The FMSA predictions for the RDF are poorer than those from the MSA (Tang and Lu 1994b, 1997b) but the results for the energy and the EOS are in fairly good agreement with simulations.

The modified MS closure, developed by Martynov et al. (1999), was shown to yield excellent accuracy for the RDF at all densities and temperatures, and the results for the energy and the EOS were fairly good as well (Martynov et al. 1999, Sarkisov 2001). Similar considerations apply to the VM closure (Choudhury and Ghosh 2002).

The performance of the ZSEP closure for the LJ fluid was analyzed by Lee et al. (1996). They concluded that this procedure leads to accurate values of the RDF and the thermodynamic properties over a wide range of densities and temperatures.

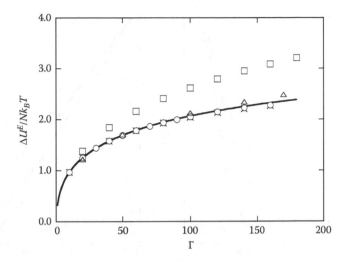

FIGURE 4.7 Thermal contribution to the excess energy for the OCP fluid from several theories. The curve corresponds to the simulation data as parameterized by Slattery et al. (1982). Squares: HNC, from Ng (1974). Circles: RHNC, from Lado et al. (1983) and from Kang and Ree (1995b). Triangles: RY approximation, from Rogers and Young (1984). Crosses: PHNC results from Kang and Ree (1995b) hardly distinguishable from those of the RHNC at the scale of the figure.

Lombardero et al. (1992) applied the RHNC theory with the VM closure for the HS reference system to an LJ fluid with different temperatures and densities. They found excellent agreement with simulation data for the RDF, the EOS and the energy. Gazzillo and Della Valle (1993) used the Malijevský and Labík (1987) parameterization for $B_0(r)$ in combination with the RHNC. Apart from obtaining excellent accuracy for the thermodynamic and structural properties for wide ranges of densities and temperatures, they showed that the RHNC-ML IET is able to reproduce the splitting of the second peak of the RDF in the neighborhood of the glass transition. Figure 4.8 displays some results for the EOS and the excess energy obtained from the RHNC-ML, modified MS, and VM closures.

Concerning the liquid–vapor coexistence, Duh and Haymet (1995) reported values for the coexistence densities obtained from the MSA theory as well as from their own bridge function approach equation (4.39). Comparison with simulation data reveal that the MSA theory yields too low (high) values of the coexistence densities for the vapor (liquid) phase, whereas those from the *Duh-Haymet* (DH) closure are quite accurate Figure 4.9. Sarkisov (2001) also found very good agreement with simulations from the Martynov et al. (1999) modified MS closure. Still more accurate are the results from a modified DH closure proposed by Duh and Henderson (1996) and similar accuracy was achieved by Lomba (1989) with the RHNC theory using the ML bridge function. To illustrate this, the latter three theories are compared with simulation data in Figure 4.9 where one can see that the modified DH and the RHNC-ML closures are in excellent agreement with simulations except perhaps in the neighborhood of the critical point.

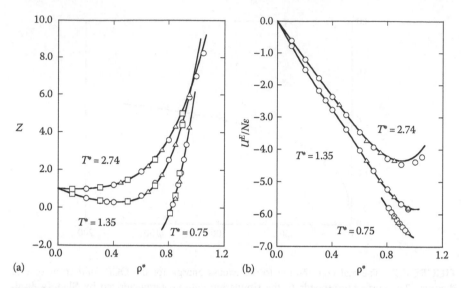

FIGURE 4.8   (a) Compressibility factor Z and (b) excess energy U^E for the LJ fluid at three reduced temperatures $T^* = 0.75$ (subcritical), $T^* = 1.35$ (slightly supercritical), and $T^* = 2.74$ (supercritical). The curves are the results from the Johnson et al. (1993) fitting of the simulation data. Circles: Martynov et al. (1999) modified MS closure (from Sarkisov 2001). Squares: VM closure (Choudhury and Ghosh 2002). Triangles: RHNC with the ML bridge function (Gazzillo and Della Valle 1993).

4.5.5 Nonspherical Potentials

One of the most simple models of fluids with nonspherical interactions is the hard homonuclear diatomic fluid, whose molecules consist of two fused HS with diameter σ and reduced center-to-center distance $L^* = L/\sigma$. Lado (1982c) applied to this kind of fluids the RHNC theory with a spherically symmetric bridge function for the reference system, obtaining fairly good results for the EOS for low values of L^* but quite unsatisfactory for larger values. Labík et al. (1991a) compared the performance of the VM, PY, and RHNC theories, using for the latter a nonspherical bridge function previously developed by Labík et al. (1990), in predicting the structure and the EOS of this kind of fluids with L^* ranging from 0.2 to 1.0. They found excellent agreement with simulation data from the VM theory, quite good results for the EOS and not so good for the structural properties from the RHNC theory, and quite poor results from the PY theory. Munaò et al. (2009a,b) use the RISM formalism with the PY, HNC, and a thermodynamically consistent RHNC with HS bridge function. The results obtained for the EOS are in general quite disappointing, except for values of L^* close to 1.0, when using the PY-virial route and the RHNC closures. More satisfactory are the results for the site–site RDF resulting from the RHNC closure.

Labík et al. (1991b) applied the VM and PY closures to hard *heteronuclear diatomics* (HTD) (molecules formed by fused HS with different diameters). They found excellent agreement with simulation data for the structural properties obtained from the VM closure at all densities, whereas the PY closure was satisfactory only

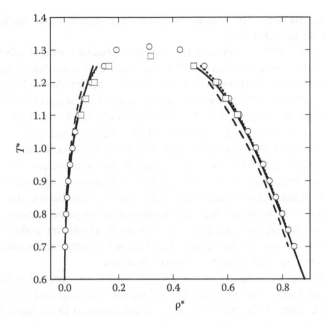

FIGURE 4.9 Liquid–vapor equilibrium for the LJ fluid. Points: simulation data from Lofti et al. (1992) (circles) and from Panagiotopoulos (1994) (squares). Continuous curves: Duh and Henderson (1996) modification of the DH closure (Equation 4.39). Dashed curves: Martynov et al. (1999) modified MS closure, from Sarkisov (2001). Dotted curves: RHNC with the ML bridge function, from Lomba (1989).

for low densities. The results for the EOS obtained from the virial route were also quite accurate using the first of these closures and much worse using the second of them. A remarkable feature also found was the fact that the inconsistency between the virial and compressibility routes was small for the VM closure, in contrast with the PY one.

The RHNC theory was extended to two-center LJ fluids, using a hard diatomics as the reference fluid, by Lado (1988), with the reference fluid solved in the PY approximation, and by Lombardero et al. (1992), using the VM closure for the reference fluid. The RHNC-VM was later extended to heteronuclear two-center LJ fluids by Lombardero et al. (1994). The results from this theory for the thermodynamic and structural properties of these fluids are in excellent agreement with simulations. Jirásek et al. (2000) carried out an exhaustive analysis of the performance of the RHNC-VM, RHNC with HS bridge function, PY, and HNC closures for homo- and heteronuclear SW diatomic fluids. The RHNC-VM theory proved to be more accurate for the excess energy and much more accurate for the compressibility factor than all the others. The results for the structure obtained from the RHNC-VM theory were also excellent. Similar accuracy was achieved with this theory for polar heteronuclear diatomic fluids (Lombardero et al. 1996). Martín et al. (1995) compared the predictions of the RHNC-VM with the RISM-PY and RISM-HNC for the atomic structure factors and site–site distribution functions of diatomic molecular fluids with

LJ site–site interactions. The results showed a clear superiority of the first of these approaches over the other two.

Yetiraj et al. (1990) compared the structure obtained from PRISM with the PY closure with simulation data for fluids composed of chain molecules modeled as pearl-necklaces of freely jointed HS of diameter σ. For the $g(r)$ obtained from averaging the intermolecular site–site distribution function $g_{ij}(r)$ for n-mers with $n = 4$, 8, and 16, they found that the theory predicts values too high for reduced distances $r^* < 1.5$ and quite accurate for larger distances. The predictions for the compressibility factor were also quite poor. Much better results for the structure were obtained by combining PRISM with a Yukawa closure with the two free parameters of the latter determined from forcing the theory to reproduce a prescribed EOS and the simulation data for the contact values of the site–site distribution function (Yetiraj and Hall 1990). This approach was later improved by the same authors (Yetiraj and Hall 1992) who developed a procedure to theoretically obtain the contact values $g(\sigma)$ of the averaged site–site intermolecular distribution function, thus avoiding the need of obtaining these quantities from complex simulations.

In the original formulation of the PRISM theory, an expression for $w(r)$, the average intramolecular site–site correlation function, was required as input together with a closure relation. To this end, $w(r)$ was determined in an approximate way by assuming that the intermolecular and intramolecular correlation functions are independent of each other, which is not the case in general. Several procedures to determine the intra- and intermolecular structures in a self-consistent way have been proposed (Schweizer et al 1992, Grayce and Schweizer 1994).

REFERENCES

Andersen, H. C., and D. Chandler. 1971. Mode expansion in equilibrium statistical mechanics. III. Optimized convergence and application to ionic solution theory. *J. Chem. Phys.* 55:1497.

Andersen, H. C., and D. Chandler. 1972. Optimized cluster expansions for classical fluids. I. General theory and variational formulation of the mean spherical model and hard-sphere Percus-Yevick equations. *J. Chem. Phys.* 57:1918.

Andersen, H. C., D. Chandler, and J. D. Weeks. 1972. Roles of repulsive and attractive forces in liquids: The optimized random phase approximation. *J. Chem. Phys.* 56:3812.

Andersen, H. C., D. Chandler, and J. D. Weeks. 1976. Roles of repulsive and attractive forces in liquids: The equilibrium theory of classical fluids. *Adv. Chem. Phys.* 34:105.

Anta, J. A., E. Lomba, M. Alvarez, C. Martín, and M. Lombardero. 1997. Integral equation approaches to mixtures of atomic and molecular fluids. *J. Chem. Phys.* 106:2712.

Ballone, P., G. Pastore, G. Galli, and D. Gazzillo. 1986. Additive and non-additive hard sphere mixtures. Monte Carlo simulation and integral equation results. *Mol. Phys.* 59:275.

Barboy, B., and R. Tenne. 1976. Distribution functions at zero separation and an equation of state for hard-core particles with a finite interaction tail. *Mol. Phys.* 31:1749.

Barker, J. A., and D. Henderson. 1976. What is liquid? Understanding the states of matter. *Rev. Mod. Phys.* 48:587.

Baxter, R. J. 1968. Ornstein-Zernike relation for a disordered fluid. *Aust. J. Phys.* 21:563.

Blum, L., and A. H. Narten. 1972. Mean spherical model for the structure of liquid metals. *J. Chem. Phys.* 56:5197.

Caccamo, C., and G. Pellicane. 2002. Microscopic theories of model macromolecular fluids and fullerenes: The role of thermodynamic consistency. *J. Chem. Phys.* 117:5072.

Caccamo, C., G. Pellicane, D. Costa, D. Pini, and G. Stell. 1999. Thermodynamically self-consistent theories of fluids interacting through short-range forces. *Phys. Rev. E* 60:5533.

Chandler, D., and H. C. Andersen. 1972. Optimized cluster expansions for classical fluids. II. Theory of molecular Liquids. *J. Chem. Phys.* 57:1930.

Chandler, D., R. Silbey, and B. M. Ladanyi. 1982. New and proper integral equations for site-site equilibrium correlations in molecular fluids. *Mol. Phys.* 46:1335.

Chen, Y., and W. A. Steele. 1971. Statistical mechanics of linear molecules. VI. Solutions of the Percus-Yevick integral equation for a hard-core model. *J. Chem. Phys.* 54:703.

Choudhury, N., and S. K. Ghosh. 2002. Integral equation theory of Lennard-Jones fluids: A modified Verlet bridge function approach. *J. Chem. Phys.* 116:8517.

Choudhury, N., and S. K. Ghosh. 2003. Integral equation theory of penetrable sphere fluids: A modified Verlet bridge function approach. *J. Chem. Phys.* 119:4827.

Clément-Cottuz, J., S. Amokrane, and C. Regnaut. 2000. Phase diagram of highly asymmetric binary mixtures: A study of the role of attractive forces from the effective one-component approach. *Phys. Rev. E* 61:1692.

Curro, J. G., and K. S. Schweizer. 1987a. Theory of polymer melts: An integral equation approach. *Macromolecules.* 20:1928.

Curro, J. G., and K. S . Schweizer. 1987b. Equilibrium theory of polymer liquids: Linear chains. *J. Chem. Phys.* 87:1842.

Díez, A., J. Largo, and J. R. Solana. 2007. Structure and thermodynamic properties of Sutherland fluids from computer simulation and the Tang-Lu integral equation theory. *Fluid Phase Equilib.* 253:67.

Duda, Y., A. Romero-Martínez, and P. Orea. 2007. Phase diagram and surface tension of the hard-core attractive Yukawa model of variable range: Monte Carlo simulations. *J. Chem. Phys.* 126:224510.

Duh, D.-M., and A. D. J. Haymet. 1995. Integral equation theory for uncharged liquids: The Lennard-Jones fluid and the bridge function. *J. Chem. Phys.* 103:2625.

Duh, D.-M., and D. Henderson. 1996. Integral equation theory for Lennard-Jones fluids: The bridge function and applications to pure fluids and mixtures. *J. Chem. Phys.* 104:6742.

Duh, D.-M., and L. Mier-y-Terán. 1997. An analytical equation of state for the hard-core Yukawa fluid. *Mol. Phys.* 90:373.

Elliott, J. R., and L. Hu. 1999. Vapor-liquid equilibria of square-well spheres. *J. Chem. Phys.* 110:3043.

Enciso, E., F. Lado, M. Lombardero, J. L. F. Abascal, and S. Lago. 1987. Extension of the optimized RHNC equation to multicomponent liquids. *J. Chem. Phys.* 87:2249.

Fernaud, M. J., E. Lomba, and L. L. Lee. 2000. A self-consistent integral equation study of the structure and thermodynamics of the penetrable sphere fluid. *J. Chem. Phys.* 112:810.

Fries, P. H., and G. N. Patey. 1985. The solution of the hypernetted-chain approximation for fluids of nonspherical particles. A general method with application to dipolar hard spheres. *J. Chem. Phys.* 82:429.

Fries, P. H., and G. N. Patey. 1986. The solution of the Percus-Yevick approximation for fluids with angle-dependent pair interactions. A general method with results for dipolar hard spheres. *J. Chem. Phys.* 85:7307.

Garnett, E., L. Mier-y-Terán, and F. del Río. 1999. On the hard core Yukawa fluid of variable range: Monte Carlo simulations and test of the MSA equation of state. *Mol. Phys.* 97:597.

Gazzillo, D., and R. G. Della Valle. 1993. An improved representation for the high-density structure of Lennard-Jones systems: From liquid toward glass. *J. Chem. Phys.* 99:6915.

Giacometti, A., G. Pastore, and F. Lado. 2009. Liquid-vapor coexistence in square-well fluids: An RHNC study. *Mol. Phys.* 107:555.

Gillan, M. J. 1979. A new method of solving the liquid structure integral equations. *Mol. Phys.* 38:1781.

Gil-Villegas, A., C. Vega, F. del Río, and A. Malijevský. 1995. Structure of variable-width square-well fluids from the reference hypernetted chain theory. *Mol. Phys.* 86:857.

Grayce, C. J., and K. S. Schweizer. 1994. Solvation potentials for macromolecules. *J. Chem. Phys.* 100:6846.

Groot, R. D., J. P. van der Eerden, and N. M. Faber. 1987. The direct correlation function in hard sphere fluids. *J. Chem. Phys.* 87:2263.

Grundke, E.W., and D. Henderson. 1972. Distribution functions of multi-component fluid mixtures of hard spheres. *Mol. Phys.* 24:269.

Henderson, D., L. Blum, and J. P. Noworyta. 1995. Inverse temperature expansion of some parameters arising from the solution of the mean spherical approximation integral equation for a Yukawa fluid. *J. Chem. Phys.* 102:4973.

Henderson, D., and E. W. Grundke. 1975. Direct correlation function: Hard sphere fluid. *J. Chem. Phys.* 63:601.

Hlushak, S., A. Trokhymchuk, and S. Sokolowski. 2009. Direct correlation function of the square-well fluid with attractive well width up to two particle diameters. *J. Chem. Phys.* 130:234511.

Honnell, K. G., C. K. Hall, and R. Dickman. 1987. On the pressure equation for chain molecules. *J. Chem. Phys.* 87:664.

Hoover, W. G., and J. C. Poirier. 1962. Determination of virial coefficients from the potential of mean force. *J. Chem. Phys.* 37:1041.

Høye, J. S., and L. Blum. 1977. Solution of the Yukawa closure of the Ornstein-Zernike equation. *J. Stat. Phys.* 16:399.

Høye, J. S., J. L. Lebowitz, and G. Stell. 1974. Generalized mean spherical approximations for polar and ionic fluids. *J. Chem. Phys.* 61:3253.

Høye, J. S., and A. Reiner. 2006. Self-consistent Ornstein-Zernike approximation for molecules with soft cores. *J. Chem. Phys.* 125:104503.

Høye, J. S., and G. Stell. 1977. New self-consistent approximations for ionic and polar fluids. *J. Chem. Phys.* 67:524.

Høye, J. S., and G. Stell. 1984a. Ornstein-Zernike equation for a two-Yukawa $c(r)$ with core condition. II. Further analytic explication and simplification. *Mol. Phys.* 52:1057.

Høye, J. S., and G. Stell. 1984b. Ornstein-Zernike equation for a two-Yukawa $c(r)$ with core condition. III. A self-consistent approximation for a pair potential with hard core and Yukawa tail. *Mol. Phys.* 52:1071.

Høye, J. S., G. Stell, and E. Waisman. 1976. Ornstein-Zernike equation for a two-Yukawa $c(r)$ with core condition. *Mol. Phys.* 32:209.

Jirásek, V., S. Labík, A. Malijevský, and M. Lísal. 2000. An integral equation and Monte Carlo study of homo- and hetero-nuclear square-well diatomic fluids. *Mol. Phys.* 98:2033.

Johnson, J. K., J. A. Zollweg, and K. E. Gubbins. 1993. The Lennard-Jones equation of state revisited. *Mol. Phys.* 78:591.

Kahl, S., E. Schöll-Paschinger, and G. Stell. 2002. Phase transitions and critical behaviour of simple fluids and their mixtures. *J. Phys.: Condens. Matter* 14:9153.

Kambayashi, S., and Y. Hiwatari. 1990. Improved integral equation for highly supercooled liquids: Numerical tests for soft-sphere fluids. *Phys. Rev. A* 41:1990.

Kang, H. S., and F. H. Ree. 1995a. New integral equation for simple fluids. *J. Chem. Phys.* 103:3629.

Kang, H. S., and F. H. Ree. 1995b. Applications of the perturbative hypernetted-chain equation to the one-component plasma and the one-component charged hard-sphere systems. *J. Chem. Phys.* 103:9370.

Kinoshita, M., and M. Harada. 1988. Numerical solution of the HNC equation for ionic systems. *Mol. Phys.* 65:599.

Kolafa, J., S. Labík, and A. Malijevský. 2002. The bridge function of hard spheres by direct inversion of computer simulation data. *Mol. Phys.* 100:2629.

Labík, S., A. Malijevský, and W. R. Smith. 1991a. An accurate integral equation for molecular fluids. I. Hard homonuclear diatomics. *Mol. Phys.* 73:87.

Labík, S., A. Malijevský, and W. R. Smith. 1991b. An accurate integral equation for molecular fluids. Part II. Hard heteronuclear diatomics. *Mol. Phys.* 73:495.

Labík, S., A. Malijevský, and P. Voňka. 1985. A rapidly convergent method of solving the OZ equation. *Mol. Phys.* 56:709.

Labík, S., W. R. Smith, R. Pospísil, and A. Malijevský. 1990. Non-spherical bridge function theory of molecular fluids. I. The hard-dumbbell fluid. *Mol. Phys.* 69:649.

Ladanyi, B. M., and D. Chandler. 1975. New type of cluster theory for molecular fluids: Interaction site cluster expansion. *J. Chem. Phys.* 62:4308.

Lado, F. 1964. Perturbation correction to the radial distribution function. *Phys. Rev. A* 135:1013.

Lado, F. 1973. Perturbation correction to the free energy and structure of simple fluids. *Phys. Rev. A* 8:2548.

Lado, F. 1982a. Integral equations for fluids of linear molecules I. General formulation. *Mol. Phys.* 47:283.

Lado, F. 1982b. A local thermodynamic criterion for the reference-hypernetted chain equation. *Phys. Lett. A* 89:196.

Lado, F. 1982c. Integral equations for fluids of linear molecules. II. Hard dumbell solutions. *Mol. Phys.* 47:299.

Lado, F. 1988. Reference-hypernetted chain equation with anisotropic bridge function for fluids of diatomic molecules. *J. Chem. Phys.* 88:1950.

Lado, F., S. M. Foiles, and N. W. Ashcroft. 1983. Solution of the reference-hypernetted-chain equation with minimized free energy. *Phys. Rev. A* 28:2374.

Largo, J., M. A. Miller, and F. Sciortino. 2008. The vanishing limit of the square-well fluid: The adhesive hard-sphere model as a reference system. *J. Chem. Phys.* 128:134513.

Largo, J., and J. R. Solana. 2003. Generalized van der Waals theory for the thermodynamic properties of square-well fluids. *Phys. Rev. E* 67:066112.

Largo, J., J. R. Solana, S. B. Yuste, and A. Santos. 2005. Pair correlation function of short-ranged square-well fluids. *J. Chem. Phys.* 122:084510.

Larsen, B. 1978. Studies in statistical mechanics of Coulombic systems. III. Numerical solutions of the HNC and RHNC equations for the restricted primitive model. *J. Chem. Phys.* 68:4511.

Lebowitz, J. L., and J. K. Percus. 1966. Mean spherical model for lattice gases with extended hard cores and continuum fluids. *Phys. Rev.* 144:251.

Lee, L.L. 1992. Chemical potentials based on the molecular distribution functions. An exact diagrammatical representation and the star function. *J. Chem. Phys.* 97:8606.

Lee, L. L. 1995. An accurate integral equation theory for hard spheres: Role of the zero-separation theorems in the closure relation. *J. Chem. Phys.* 103:9388.

Lee, L. L., D. Ghonasgi, and E. Lomba. 1996. The fluid structures for soft-sphere potentials via the zero-separation theorems on molecular distribution functions. *J. Chem. Phys.* 104:8058.

Llano-Restrepo, M., and W. G. Chapman. 1992. Bridge function and cavity correlation function for the Lennard-Jones fluid from simulation. *J. Chem. Phys.* 97:2046.

Llano-Restrepo, M., and W. G. Chapman. 1994. Bridge function and cavity correlation function for the soft sphere fluid from simulation: Implications on closure relations. *J. Chem. Phys.* 100:5139.

Lofti, A., J. Vrabec, and J. Fischer. 1992. Vapour liquid equilibria for the Lennard-Jones fluid from the NpT plus test particle method. *Mol. Phys.* 76:1319.

Lomba, E. 1989. An efficient procedure for solving the reference hypernetted chain equation (RHNC) for simple fluids. Illustrative results with application to phase coexistence for a Lennard-Jones fluid. *Mol. Phys.* 68:87.

Lomba, E., M. Alvarez, G. Stell, and J. A. Anta. 1992a. Bridge functions for models of liquid metals. *J. Chem. Phys.* 97:4349.

Lomba, E., and L. L. Lee. 1996. Consistency conditions for the integral equation of liquid structures. *Int. J. Thermophys.* 17:663.

Lomba, E., C. Martín, M. Lombardero, and J. A. Anta. 1992b. On the use of semiphenomenological closures in integral equations for classical fluids *J. Chem. Phys.* 96:6132.

Lombardero, M., and E. Enciso. 1981. A "RISM" theory for multicomponent molecular fluids. *J. Chem. Phys.* 74:1357.

Lombardero, M., C. Martín, and E. Lomba. 1992. A reference hypernetted-chain equation for soft potentials. Atomic and molecular Lennard-Jones systems. *J. Chem. Phys.* 97:2724.

Lombardero, M., C. Martín, and E. Lomba. 1994. Structure and thermodynamics of heteronuclear two-centre Lennard-Jones fluids from Monte Carlo simulation and a reference hypernetted chain equation. *Mol. Phys.* 81:1313.

Lombardero, M., C. Martín, E. Lomba, and F. Lado. 1996. Monte Carlo simulation and reference hypernetted chain equation results for structural, thermodynamic, and dielectric properties of polar heteronuclear diatomic fluids. *J. Chem. Phys.* 104:6710.

Madden, W. G., and D. D. Fitts. 1974. A re-examination of the HNC theory for the radial distribution function. *J. Chem. Phys.* 61:5475.

Madden, W. G., and D. D. Fitts. 1975. Integral-equation perturbation theory for the radial distribution function of simple fluids. *Mol. Phys.* 30:809.

Madden, W. G., and S. A. Rice. 1980. The mean spherical approximation and effective pair potential in liquids. *J. Chem. Phys.* 72:4208.

Malijevský, A., and S. Labík. 1987. The bridge function for hard spheres. *Mol. Phys.* 60:663.

Mandel, F., R. J. Bearman, and M. Y. Bearman. 1970. Numerical solutions of the Percus-Yevick equation for the Lennard-Jones (6-12) and hard-sphere potentials. *J. Chem. Phys.* 52:3315.

Martín, C., M. Lombardero, M. Alvarez, and E. Lomba. 1995. Atomic structure factors from a molecular integral equation theory: An application to homonuclear diatomic fluids. *J. Chem. Phys.* 102:2092.

Martynov, G. A., and G. N. Sarkisov. 1983. Exact equations and the theory of liquids. V. *Mol. Phys.* 49:1495.

Martynov, G. A., G. N. Sarkisov, and A. G. Vompe. 1999. New closure for the Ornstein-Zernike equation. *J. Chem. Phys.* 110:3961.

Martynov, G. A., and A. G. Vompe. 1993. Differential condition of thermodynamic consistency as a closure for the Ornstein-Zernike equation. *Phys. Rev. E* 47:1012.

Meeron, E. 1960. Nodal expansions. III. Exact integral equations for particle correlation functions. *J. Math. Phys.* 1:192.

Meeron. E., and A. J. F. Siegert. 1968. Statistical mechanics of hard-particle systems. *J. Chem. Phys.* 48:3139.

Mier-y-Terán, L., S. E. Quiñones-Cisneros, I. D. Núñez-Riboni, and E. Lemus-Fuentes. 1998. An analytical equation of state for the hard core Yukawa fluid; the electroneutral mixture. *Mol. Phys.* 95:179.

Mladek, B. M., D. Kahl, and M. Neumann. 2006. Thermodynamically self-consistent liquid state theories for systems with bounded potentials. *J. Chem. Phys.* 124:064503.

Monson, P. A. 1982. Numerical solution of the RISM equations for the site-site 12-6 potential. *Mol. Phys.* 47:435.

Mori, H., K. Hoshino, and M. Watabe. 1991. A new bridge function scheme in the modified hypernetted-chain approximation for liquid alloys. *J. Phys.: Condens. Matter* 3:9791.

Morita, T. 1960. Theory of classical fluids: Hyper-netted chain approximation. III. A new integral equation for the pair distribution function. *Prog. Theor. Phys.* 23:829.

Munaò, G., D. Costa, and C. Caccamo. 2009a. Thermodynamically consistent reference interaction site model theory of the tangent diatomic fluid. *Chem. Phys. Lett.* 470:240.

Munaò, G., D. Costa, and C. Caccamo. 2009b. Reference interaction site model investigation of homonuclear hard dumbbells under simple fluid theory closures: Comparison with Monte Carlo simulations. *J. Chem. Phys.* 130:144504.

Narten, A. H., L. Blum, and R. H. Fowler. 1974. Mean spherical model for the structure of Lennard-Jones fluids. *J. Chem. Phys.* 60:3378.

Ng, K.-C. 1974. Hypernetted chain solutions for the classical one-component plasma up to $\Gamma = 7000$. *J. Chem. Phys.* 61:2680.

Orkoulas, G., and A. Z. Panagiotopoulos. 1999. Phase behavior of the restricted primitive model and square-well fluids from Monte Carlo simulations in the grand canonical ensemble. *J. Chem. Phys.* 110:1581.

Ornstein, L.S., and F. Zernike. 1914. Accidental deviations of density and opalescence at the critical point of a single substance. *Proc. Acad. Sci. (Amsterdam)* 17:793.

Pagan, D. L., and J. D. Gunton. 2005. Phase behavior of short-range square-well model. *J. Chem. Phys.* 122:184515.

Panagiotopoulos, A. Z. 1994. Molecular simulation of phase coexistence: Finite-size effects and determination of critical parameters for two- and three-dimensional Lennard-Jones fluids. *Int. J. Thermophys.* 15:1057.

Percus, J. K., and G. J. Yevick. 1958. Analysis of classical statistical mechanics by means of collective coordinates. *Phys. Rev.* 110:1.

Percus, J. K., and G. J. Yevick. 1964. Hard-core insertion in the many-body problem. *Phys. Rev. B* 136:290.

Perera, A., P. G. Kusalik, and G. N. Patey. 1987. The solution of the hypernetted chain and Percus-Yevick approximations for fluids of hard nonspherical particles. Results for hard ellipsoids of revolution. *J. Chem. Phys.* 87:1295.

Perera, A., F. Sokolič, and M. Moreau. 1992. Fluids of linearly fused Lennard-Jones sites: Comparison between simulations and integral equation theories. *J. Chem. Phys.* 97:1969.

Pini, D., A. Parola, and L. Reatto. 2002. A Simple approximation for fluids with narrow attractive potentials. *Mol. Phys.* 100:1507.

Pini, D., and G. Stell. 2002. Globally accurate theory of structure and thermodynamics for soft-matter liquids. *Physica A* 306:270.

Pini, D., G. Stell, and N. B. Wilding. 1998. A liquid-state theory that remains successful in the critical region. *Mol. Phys.* 95:483.

Poll, P. D., N. W. Ashcroft, and H. E. DeWitt. 1988. One-component plasma bridge function. *Phys. Rev. A* 37:1672.

Raineri, F. O., G. Stell, and D. Ben-Amotz. 2004. Progress in thermodynamic perturbation theory and self-consistent Ornstein-Zernike approach relevant to structural-arrest problems. *J. Phys.: Condens. Matter.* 16:S4887.

Reiner, A., and J. S. Høye. 2008. Self-consistent Ornstein-Zernike approximation for the Yukawa fluid with improved direct correlation function. *J. Chem. Phys.* 128:114507.

Rogers, F. J., and D. A. Young. 1984. New, thermodynamically consistent, integral equation for simple fluids. *Phys. Rev. A* 30:999.

Rosenfeld, Y., and N. W. Ashcroft. 1979a. Mean-spherical model for soft potentials: The hard core revealed as a perturbation. *Phys. Rev. A* 20:2162.

Rosenfeld, Y., and N. W. Ashcroft. 1979b. Theory of simple classical fluids: Universality in the short-range structure. *Phys. Rev. A* 20:1208.

Rushbrooke, G. S. 1960. On the hyper-chain approximation in the theory of classical fluids. *Physica* 26:259.

Sarkisov, G. 2001. Approximate integral equation theory for classical fluids. *J. Chem. Phys.* 114:9496.

Sarkisov, G. 2002. Molecular distribution functions of stable, metastable and amorphous classical models. *Phys. Usp.* 45:597.

Sarkisov, G., D. Tikhonov, J. Malinsky, and Yu. Magarshak. 1993. Martynov-Sarkisov integral equation for the simple fluids. *J. Chem. Phys.* 99:3926.

Schöll-Paschinger, E., A. L. Benavides, and R. Castañeda-Priego. 2005. Vapor-liquid equilibrium and critical behavior of the square-well fluid of variable range: A theoretical study. *J. Chem. Phys.* 123:234513.

Schöll-Paschinger, E., and G. Kahl. 2003. Accurate determination of the phase diagram of model fullerenes. *Europhys. Lett.* 63:538.

Schweizer, K. S., and J. G. Curro. 1988a. Equation of state of polymer melts: General formulation of a microscopic integral equation theory. *J. Chem. Phys.* 89:3342.

Schweizer, K. S., and J. G. Curro. 1988b. Equation of state of polymer melts: Numerical results for athermal freely jointed chain fluids. *J. Chem. Phys.* 89:3350.

Schweizer, K. S., K. G. Honnell, and J. G. Curro. 1992. Reference interaction site model theory of polymeric liquids: Self-consistent formulation and non ideality effects in dense solutions and melts. *J. Chem. Phys.* 96:3211.

Shukla, K. P. 2000. Phase equilibria and thermodynamic properties of hard core Yukawa fluids of variable range from simulations and an analytical theory. *J. Chem. Phys.* 112:10358.

Singh, R. C., J. Ram, and Y. Singh. 1996. Thermodynamically self-consistent integral-equation theory for pair-correlation functions of a molecular fluid. *Phys. Rev. E* 54:977.

Slattery, W. L., G. D. Dolen, and H. E. DeWitt. 1982. N dependence in the classical one-component plasma Monte Carlo calculations. *Phys. Rev. A* 26:2255.

Smith, W. R., and D. Henderson. 1978. Some corrected integral equations and their results for the square-well fluid *J. Chem. Phys.* 69:319.

Smith, W. R., D. Henderson, and Y. Tago. 1977. Mean spherical approximation and optimized cluster theory for the square-well fluid. *J. Chem. Phys.* 67:5308.

Tang, Y. 2003. On the first-order mean spherical approximation. *J. Chem. Phys.* 118:4140.

Tang, Y. 2007. Direct correlation function for the square-well potential. *J. Chem. Phys.* 127:164504.

Tang, Y., and B. C.-Y. Lu. 1993. A new solution of the Ornstein-Zernike equation from the perturbation theory. *J. Chem. Phys.* 99:9828.

Tang, Y., and B.C.-Y. Lu. 1994a. An analytical analysis of the square-well fluid behaviors. *J. Chem. Phys.* 100:6665.

Tang, Y., and B.C.-Y. Lu. 1994b. First-order radial distribution functions based on the mean spherical approximation for square-well, Lennard-Jones, and Kihara fluids. *J. Chem. Phys.* 100:3079.

Tang, Y., and B. C.-Y. Lu. 1995. Analytical solution of the Ornstein-Zernike equation for mixtures. *Mol. Phys.* 84:89.

Tang, Y., and B. C.-Y. Lu. 1997a. Analytical representation of the radial distribution function for classical fluids. *Mol. Phys.* 90:215.

Tang, Y., and B.C.-Y. Lu. 1997b. Analytical description of the Lennard-Jones fluid and its application. *AIChE J.* 43:2215.

Van Leeuwen, J. M. J., J. Groeneveld, and J. De Boer. 1959. New method for the calculation of the pair correlation function. I. *Physica* 25:792.

Vega, L., E. de Miguel, L. F. Rull, G. Jackson, and I. A. McLure. 1992. Phase equilibria and critical behavior of square-well fluids of variable width by Gibbs ensemble Monte Carlo simulation. *J. Chem. Phys.* 96:2296.

Verlet, L. 1960. On the theory of classical fluids. *Nuovo Cimento* 18:77.

Verlet, L. 1980. Integral equations for classical fluids. I. The hard sphere case. *Mol. Phys.* 41:183.

Verlet, L. 1981. Integral equations for classical fluids. II. Hard spheres again. *Mol. Phys.* 42:1291.

Verlet, L., and J.-J. Weis. 1974. Perturbation theories for polar fluids. *Mol. Phys.* 28:665.

Vompe, A. G., and G. A. Martynov. 1994. The bridge function expansion and the self-consistency problem of the Ornstein-Zernike equation solution. *J. Chem. Phys.* 100:5249.

Waisman, E. 1973. The radial distribution function for a fluid of hard spheres at high densities. Mean spherical integral equation approach. *Mol. Phys.* 25:45.

Watts, R. O. 1973. Integral equation approximations in the theory of fluids. In *Statistical Mechanics*, Vol. 1, ed. K. Singer, 1–70. London, U.K.: The Chemical Society.

Weeks, J. D., D. Chandler, and H. C. Andersen. 1971. Role of repulsive forces in determining the equilibrium structure of simple liquids. *J. Chem. Phys.* 54:5237.

Yetiraj, A., and C. K. Hall. 1990. Local structure of fluids containing chain-like molecules: Polymer reference interaction site model with Yukawa closure. *J. Chem. Phys.* 93:5315.

Yetiraj, A., and C. K. Hall. 1992. Monte Carlo simulations and integral equation theory for microscopic correlations in polymeric fluids. *J. Chem. Phys.* 96:797.

Yetiraj, A., C. K. Hall, and K. G. Honnell. 1990. Site-site correlations in short chain fluids. *J. Chem. Phys.* 93:4453.

Zerah, G., and J.-P. Hansen. 1986. Self-consistent integral equations for fluid pair distribution functions: Another attempt. *J. Chem. Phys.* 84:2336.

Zhou, S., H. Chen, S. Ling, X. Xiang, and X. Zhang. 2003a. Statistical mechanics approach for uniform and non-uniform fluid with hard core and interaction tail. *Commun. Theor. Phys.* 39:331.

Zhou, S., H. Chen, and X. Zhang. 2003b. A new uniform phase bridge functional: Test and its application to non-uniform phase fluid. *Commun. Theor. Phys.* 39:231.

Zhou, S., and J. R. Solana. 2009. Inquiry into thermodynamic behavior of hard sphere plus repulsive barrier of finite height. *J. Chem. Phys.* 131:204503.

Zhou, Y., and G. Stell. 1988. The hard-sphere fluid: New exact results with applications. *J. Stat. Phys.* 52:1389.

5 Radial Distribution Function and Equation of State of the Hard-Sphere Fluid and Solid

The thermodynamic and structural properties of the HS system play an essential role in some integral equation perturbation theories, as addressed in Chapter 4, as well as in other kinds of perturbation theories that will be addressed in Chapter 6. For this reason, in this chapter, after describing the different phases of the HS system, we present a number of theories for the radial distribution function and the EOS of the HS fluid and solid. Concerning the fluid, we will start with the derivation of the *scaled particle theory* (SPT) which, although does not provide a way to obtain the radial distribution function, the EOS arising from this theory is related to that resulting from other theories and, moreover, its extension to hard-body molecular fluids and mixtures is the basis of certain perturbation theories for molecular fluids and mixtures. Concerning the radial distribution function, we will restrict ourselves to those theories leading to analytical expressions. This includes not only the PY theory, but also other less-known theories such as the *rational function approximation* (RFA) and the FMSA. The latter two theories, apart from being analytical and reasonably simple to use, are very accurate and free from empirical corrections. We will end this chapter describing simple theories for the EOS and the angle-averaged radial distribution function of the solid.

5.1 FLUID AND SOLID PHASES IN THE HARD-SPHERE SYSTEM

Computer simulations reveal that the HS system may exist as an isotropic fluid, metastable fluid, crystalline solid, either fcc or hexagonal close packed (hcp), and amorphous (glassy) solid. The HS system undergoes a first-order phase transition between a crystalline solid and the isotropic fluid at high densities. The coexistence densities for the fluid and fcc solid phases were established to be $\rho^* = 0.943$ and $\rho^* = 1.041$, respectively (Hoover and Ree 1968). Other crystalline phases are possible for the HS solid, such as the aforementioned hcp. The problem of determining from computer simulation which of the crystalline phases, fcc or hcp, is the most stable one is a long-standing one. Now it seems well established that the fcc is the most stable one at any density, although the difference in free energies is very small (Bruce et al. 2000). As a matter of fact, computer simulations performed with very large systems seem to indicate that in the thermodynamic limit the stable crystalline structure for the

HS system would be a mixture of fcc and random stacked hexagonal (rhcp) structures (Kendall et al. 2002).

Beyond the normal freezing density $\rho^* = 0.943$, the system may remain as a metastable fluid until finally crystallization is unavoidable. In extensive simulations with large systems performed by Kolafa et al. (2004a), it was found that at density $\rho^* = 1.03$ crystallization occurs very fast, although crystallization may take place at lower densities too. Therefore, the limiting density of the metastable fluid can be considered as $\rho^* \approx 1.03$. The HS fluid can undergo a transition to a glassy state only on condition that the metastable fluid is squeezed quickly to avoid spontaneous crystallization. The glassy state reached will depend on the starting state of the fluid and the quenching rate. On the basis of molecular dynamics(MD) simulations, the glass transition was located by Woodcock (1981) at $\rho^* \approx 1.09$ and a similar result was obtained later by Speedy (1998). However, the question about the existence and location of a glass transition is quite unclear. It has been argued (Rintoul and Torquato 1996a,b) that the existence of HS glasses as revealed by computer simulations is an effect of the small size of the systems used. These authors found that for systems with more than 2000 particles, crystallization occurs even for densities well above the assumed glass transition density, provided that the system is allowed to equilibrate for enough long time.

Continued compression of the fcc or hcp HS crystals will finally end at the *regular close packing density* $\rho^* = \sqrt{2}$, the maximum density allowed for these crystalline structures, corresponding to a *packing fraction* $\eta \equiv (\pi/6)\rho\sigma^3 = 0.74048$. In contrast, the limiting density for random packings of HS, that is, packings without local crystalline order, or *random close packing* (RCP) *density* also called *Bernal density*, was determined from measurements performed on ball bearings using mechanical devices by different authors pioneered by Scott (1960) and Bernal and Mason (1960), and accurately estimated to be $\rho^* = 1.2158$, or equivalently $\eta = 0.6366$ (Scott and Kilgour 1969, Finney 1970).

By computer generation of random packings of HS, jammed systems with diverse degrees of ordering may be obtained with packing fractions ranging from 0.52 to 0.74 (Kansal et al. 2002). Therefore, instead of the RCP density, the concept of *maximally random jammed* (MRJ) state was proposed by Torquato et al. (2000). The MRJ state is defined as that configuration, among all jammed HS systems, which maximizes disorder. The packing fraction corresponding to the MRJ state was estimated to be $\eta \approx 0.64$, in agreement with the Bernal density. A similar packing fraction was obtained by Truskett et al. (2000) as the limiting density for an HS glass in the limit of infinity compression rate.

5.2   SCALED PARTICLE THEORY

The starting point of the SPT (Reiss et al. 1959, Helfand et al. 1961) is the probability of insertion of a spherical particle of radius r into a fluid consisting of HS of radius $\sigma/2$. Such probability is given by

$$p_0(r) = e^{-W(r)/k_BT}, \tag{5.1}$$

where $W(r)$ is the work needed to introduce the particle of radius r. This work can be written as

$$W(r) = W(0) + 4\pi\rho k_B T \int_0^r G\left(r' + \frac{\sigma}{2}\right)\left(r' + \frac{\sigma}{2}\right)^2 dr', \qquad (5.2)$$

where $W(0)$ is the work needed to introduce a point particle ($r=0$) and the second term in the right-hand side is the additional work required to increase its size up to a radius r. The function $G(r + \sigma/2)$ is the radial distribution function (RDF) at contact distance between a test particle of radius r and an ordinary particle of the system, with radius $\sigma/2$. From Equation 5.2, we obtain

$$G\left(r + \frac{\sigma}{2}\right) = \frac{1}{4\pi(r + \sigma/2)^2 \rho k_B T} \frac{\partial W(r)}{\partial r}. \qquad (5.3)$$

To insert a test particle of radius r with its center placed at any arbitrary position of the fluid, it is needed that no ordinary particle of the fluid, with radius $\sigma/2$, has its center within a sphere of radius $r+\sigma/2$ centered at that position. For $r \leq 0$, drawing a hypothetical sphere of radius $r+\sigma/2$ centered at an arbitrary point of the fluid, inside it can be located at most the center of a particle. The probability of this occurrence is $p_1(r) = \rho(4/3)\pi(r + \sigma/2)^3$. Therefore, the probability that not inside the sphere is the center of any particle is $p_0(r) = 1 - p_1(r) = 1 - \rho(4/3)\pi(r + \sigma/2)^3$, so that, from Equation 5.1

$$W(r) = -k_B T \ln\left[1 - \rho\left(\frac{4}{3}\right)\pi\left(r + \frac{\sigma}{2}\right)^3\right], \quad r \leq 0, \qquad (5.4)$$

and from Equation 5.3

$$G\left(r + \frac{\sigma}{2}\right) = \frac{1}{1 - \rho(4/3)\pi(r + \sigma/2)^3}, \quad r \leq 0. \qquad (5.5)$$

For $r > 0$, $G(r + \sigma/2)$ is not known exactly, although obviously it holds that $G(\sigma) = g(\sigma)$, where $g(\sigma)$ is the RDF at contact distance between the particles of the system. Moreover, it is also true that $G(\infty) = Z$ (Reiss et al. 1959), where $Z = PV/Nk_B T$ is the compressibility factor, so that, making use of the virial equation for the HS potential, Equation 2.43, one obtains

$$G(\infty) = 1 + 4\eta G(\sigma). \qquad (5.6)$$

For $r \geq 0$, the SPT takes for $G(r)$ a polynomial of the form

$$G(r) = 1 + a_0 + \sum_{i=1}^{2} a_i \left(\frac{\sigma}{r + \sigma/2}\right)^i, \quad r \geq 0. \qquad (5.7)$$

The coefficients a_i are determined from the condition of Equation 5.6, together with the conditions that for $r = 0$ Equations 5.5 and 5.7 as well as their first derivatives must be equal. Note that, in particular, according to Equations 5.6 and 5.7, it holds that $Z = 1 + a_0$. Then, one readily obtains

$$a_0 = -1 + \frac{1 + \eta + \eta^2}{(1 - \eta)^3}, \quad a_1 = -\frac{3}{2} \frac{\eta + \eta^2}{(1 - \eta)^3}, \quad a_2 = \frac{3}{4} \frac{\eta^2}{(1 - \eta)^3}, \tag{5.8}$$

and so

$$g(\sigma) = \frac{1}{1 - \eta} + \frac{3}{2} \frac{\eta}{(1 - \eta)^2} + \frac{3}{4} \frac{\eta^2}{(1 - \eta)^3} \tag{5.9}$$

and

$$Z = \frac{1 + \eta + \eta^2}{(1 - \eta)^3}. \tag{5.10}$$

The same results would have been achieved by considering that for large r the work needed to grow the point particle up to a radius r is mainly due to the work of expansion against a constant pressure P, namely, $P(4/3)\pi r^3$, whereas for r close to zero $W(r)$ may be approximated by its Taylor series expansion up to second order (Lebowitz et al. 1965), thus

$$W(r) = W(0) + W'(0) r + \frac{1}{2} W''(0) r^2 + P\frac{4}{3}\pi r^3, \quad r \geq 0. \tag{5.11}$$

The coefficients in this expansion may be obtained from Equation 5.4 and the condition of continuity of $W(r)$ and its first two derivatives at $r = 0$. Then, from Equation 5.3 for $r = \sigma/2$ we will obtain for $g(\sigma)$ the same expression as in Equation 5.9.

The results from the EOS (5.10) are compared in Figure 5.1 with the extremely accurate simulation data reported by Kolafa et al. (2004a), which extends up to high densities close to the glass transition. It is seen that the SPT equation yields values too high in the high-density region.

5.3   SOLUTION OF THE PERCUS–YEVICK EQUATION

Introducing the PY approximation (4.15) into the OZ equation (4.1), we obtain

$$y(r) = 1 + \rho \int f(r') y(r') \left\{ [1 + f(|\mathbf{r} - \mathbf{r}'|)] y(|\mathbf{r} - \mathbf{r}'|) - 1 \right\} d\mathbf{r}', \tag{5.12}$$

which is the PY integral equation (Percus and Yevick 1958).

This equation was solved for an HS fluid independently by Wertheim (1963) and Thiele (1963). Later, several other authors obtained different analytical expressions for the RDF of the HS fluid on the basis of the PY theory (Throop and Bearman 1965, Smith and Henderson 1970, Chang and Sandler 1994). Of course, all of these

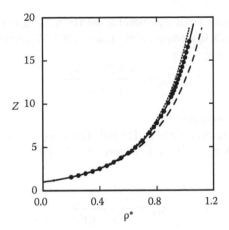

FIGURE 5.1 Compressibility factor Z for the HS fluid as a function of the reduced density ρ^*. Points: simulation data from Kolafa et al. (2004a). Dotted curve: SPT equation (5.10), which coincides with the PY compressibility equation (5.17). Dashed curve: PY virial equation (5.16).

solutions are equivalent and provide the same results. This fluid satisfies the following relationships between the RDF $g(r)$, the DCF $c(r)$ and the cavity distribution function $y(r)$ (Wertheim 1963)

$$\left.\begin{array}{ll} g\,(r) = 0, & r < \sigma \\ g\,(r) = y\,(r), & r > \sigma \\ c\,(r) = -y\,(r), & r < \sigma \\ c\,(r) = 0, & r > \sigma \end{array}\right\} . \qquad (5.13)$$

Analyzing the series expansion of Equation 5.12, Wertheim (1963) concludes that the DCF must have the form of a third-degree polynomial in $x = r/\sigma$. To determine the four coefficients of the polynomial, Wertheim takes into account that from Equation 5.12 it follows that $y(r)$ and its first two derivatives are continuous at $r = \sigma$ and, moreover, it holds that $y(0) = 1 + 24\eta K$, where K is given by

$$K = \int_0^1 y\,(x)\,x^2 dx . \qquad (5.14)$$

These four conditions allow us to obtain the DCF $c(x)$ as follows:

$$c\,(x) = -\frac{(1+2\eta)^2}{(1-\eta)^4} + \frac{6\eta\left(1+\eta/2\right)^2}{(1-\eta)^4}\,x - \frac{\eta\,(1+2\eta)^2}{2\,(1-\eta)^4}\,x^3. \qquad (5.15)$$

From this expression, the static structure factor $S(k)$ of the HS fluid within the PY approximation can be easily obtained through Equation 4.6. Introducing the result of Equation 5.15 into the virial equation for the HS fluid (2.43), taking into account also the relationships (5.13), one obtains (Thiele 1963, Wertheim 1963)

$$Z_{PY}^v = \frac{1 + 2\eta + 3\eta^2}{(1-\eta)^2}, \tag{5.16}$$

which is the *PY virial equation*.

Alternatively, using expression (5.15) and taking into account the compressibility equation in the form of (4.7), one obtains (Thiele 1963) the *PY compressibility equation*

$$Z_{PY}^c = \frac{1 + \eta + \eta^2}{(1-\eta)^3}, \tag{5.17}$$

which is the same as the SPT equation (5.10).

The same result of Equation 5.15, and hence the PY compressibility equation (5.17), can be obtained from Equations 4.8 and 4.9, taking into account that for HS $R = \sigma$ together with the first and last conditions of (5.13). Following Baxter (1968), denoting

$$a = 1 - 2\pi\rho \int_0^\sigma q(t)\, dt, \quad b = 2\pi\rho \int_0^\sigma q(t)\, t\, dt, \tag{5.18}$$

and integrating Equation 4.9 with respect to r, one readily obtains

$$q(r) = \frac{a}{2}\left(r^2 - \sigma^2\right) + b(r - \sigma), \quad 0 < r < \sigma. \tag{5.19}$$

Then, solving the system (5.18) with $q(r)$ given by Equation 5.19 yields

$$a = \frac{1 + 2\eta}{(1-\eta)^2}, \quad b = -\frac{3}{2}\frac{\sigma\eta}{(1-\eta)^2}. \tag{5.20}$$

Finally, solving Equation 4.8 with $q(r)$ given by Equations 5.19 and 5.20 yields the result of Equation 5.15 for the DCF in the PY approximation.

When the results from the two PY equations of state are compared with the simulation data, one finds that the first of them yields too low and the second too high values (see Figure 5.1). However, combining the two weighted by 1/3 and 2/3 respectively, one obtains

$$Z_{CS} = \frac{2}{3}Z_{PY}^c + \frac{1}{3}Z_{PY}^v = \frac{1 + \eta + \eta^2 - \eta^3}{(1-\eta)^3}, \tag{5.21}$$

which is the celebrated Carnahan–Starling (CS) EOS (Carnahan and Starling 1969), which provides excellent accuracy, as shown in Figure 5.1. The same result can be arrived at by multiplying the last term of the SPT equation (5.9) for $g(\sigma)$ by a factor $2/3$ and applying the virial equation (2.43) for the HS fluid (Boublík 1970).

To obtain the RDF, Wertheim (1963) takes the Laplace transform of Equation 5.12. Denoting $G(t)$ to the Laplace transform of $xg(x)$, that is to say,

$$G(t) = \mathscr{L}[xg(x)] = \int_1^\infty xy(x)\, e^{-tx}dx, \tag{5.22}$$

and taking into account the result of Equation 5.15, together with the conditions (5.13), and simplifying one obtains

$$G(t) = \frac{tL(t)}{12\eta L(t) + S(t)\, e^t}, \tag{5.23}$$

so that $xg(x)$ is given by the inverse Laplace transform of the preceding expression, namely,

$$xg(x) = \mathscr{L}^{-1}\left[\frac{tL(t)}{12\eta L(t) + S(t)\, e^t}\right], \tag{5.24}$$

where

$$S(t) = (1-\eta)^2 t^3 + 6\eta(1-\eta)t^2 + 18\eta^2 t - 12\eta(1+2\eta), \tag{5.25}$$

$$L(t) = \left(1 + \frac{1}{2}\eta\right)t + 1 + 2\eta. \tag{5.26}$$

In order to obtain an explicit expression of $g(x)$, Wertheim expands the denominator of Equation 5.24, from which the integrals involved in the calculation of the inverse Laplace transform of Equation 5.24 can be determined from the residue theorem. The result is of the form (Wertheim 1963, Smith and Henderson 1970)

$$g(x) = \sum_{n=1}^\infty \Theta(x-n)g_n(x), \tag{5.27}$$

where $\Theta(x)$ is the Heaviside step function.

Wertheim also provides an analytical expression for $xg(x)$ in the first coordination shell, corresponding to $1 < x < 2$, as well as a general method to obtain the expressions corresponding to any other coordination shell. Later, Throop and Bearman (1965) reported the values of $g_n(x)$ for the first three coordination shells, $1 \leq x \leq 4$ or $n \leq 3$, obtained from numerical calculation of the inverse Laplace transform, for $0.1 \leq \rho^* \leq 1.1$ with step 0.1. Smith and Henderson (1970) reported analytical expressions of $xg_n(x)$ for the first four coordination shells, that is, for $1 \leq x \leq 5$ or $n \leq 4$. More recently, Chang and Sandler (1994) developed an alternative procedure, though equivalent to the preceding ones, to obtain an analytical solution for $g(x)$ in terms of real functions for the first two coordination shells on the basis of

Equations 4.9 and 5.19. For the reasons that will become clear later, we will follow here a formalism analogous to that used by Bravo Yuste and Santos (1991). To this end, we will rewrite expression (5.23) for $G(t)$ in the form

$$G(t) = t \frac{F(t) e^{-t}}{1 + 12\eta F(t) e^{-t}}, \tag{5.28}$$

where

$$F(t) = -\frac{1}{12\eta} \frac{L(t)}{S(t)}, \tag{5.29}$$

in which

$$L(t) = 1 + L_1 t, \quad S(t) = 1 + S_1 t + S_2 t^2 + S_3 t^3, \tag{5.30}$$

with

$$L_1 = \frac{1 + \frac{1}{2}}{1 + 2\eta} \eta, \quad S_1 = -\frac{3}{2} \frac{\eta}{1 + 2\eta}, \quad S_2 = -\frac{1}{2} \frac{1 - \eta}{1 + 2\eta}, \quad S_3 = -\frac{(1 - \eta)^2}{12\eta (1 + 2\eta)}. \tag{5.31}$$

Then, proceeding as mentioned earlier we obtain a result of the form of Equation 5.27, which we will rewrite in the notation of Bravo Yuste and Santos (1991) as

$$x g(x) = \sum_{n=1}^{\infty} (-12\eta)^{n-1} f_{n-1}(x) \Theta(x - n), \tag{5.32}$$

where the functions $f_n(x)$ are defined from the condition

$$\mathscr{L}[f_{n-1}(x)] = t [F(t)]^n. \tag{5.33}$$

The first four of these functions are

$$f_0(x) = \sum_{i=1}^{3} a_i e^{t_i(x-1)}, \quad f_1(x) = \sum_{i=1}^{3} b_i e^{t_i(x-2)}, \tag{5.34}$$

$$f_2(x) = \sum_{i=1}^{3} c_i e^{t_i(x-3)}, \quad f_3(x) = \sum_{i=1}^{3} d_i e^{t_i(x-4)}, \tag{5.35}$$

where

$$a_i = -\frac{1}{12\eta}\frac{(1+L_1 t_i)t_i}{S_1 + 2S_2 t_i + 3S_3 t_i^2},\tag{5.36}$$

$$b_i = \frac{a_i}{t_i}\left\{\left[1 + t_i(x-2)\right]\frac{a_i}{t_i} + 2t_i\sum_{j\neq i}\frac{\frac{a_j}{t_j}}{t_i - t_j}\right\},\tag{5.37}$$

$$c_i = \left(\frac{a_i}{t_i}\right)^3\left[(x-3) + \frac{1}{2}t_i(x-3)^2\right]$$

$$+ 3\left(\frac{a_i}{t_i}\right)^2\sum_{j\neq i}\frac{a_j}{t_j}\frac{-t_j\left[1 + t_i(x-3)\right] + t_i^2(x-3)}{(t_i - t_j)^2}$$

$$+ 3\frac{a_i}{t_i}\sum_{j\neq i}\left(\frac{a_j}{t_j}\right)^2\frac{t_i}{(t_i - t_j)^2} + 6\frac{a_1 a_2 a_3}{t_1 t_2 t_3}\frac{t_i}{(t_i - t_j)(t_i - t_k)},\tag{5.38}$$

$$d_i = \frac{1}{2}\left(\frac{a_i}{t_i}\right)^4\left[(x-4)^2 + \frac{1}{3}t_i(x-4)^3\right]$$

$$+ 2\left(\frac{a_i}{t_i}\right)^3\sum_{j\neq i}\frac{\frac{a_j}{t_j}}{(t_i - t_j)^3}\left\{-2t_j\left[-1 + t_i(x-4) + t_i^2(x-4)^2\right]\right.$$

$$\left.+ t_j^2\left[2(x-4) + t_i(x-4)^2\right]\right\}$$

$$+ 6\left(\frac{a_i}{t_i}\right)^2\sum_{j\neq i}\frac{\left(\frac{a_j}{t_j}\right)^2}{(t_i - t_j)^3}\left[-t_i - t_j + t_i^2(x-4) + t_i t_j(x-4)\right] + 4a_i\sum_{j\neq i}\frac{\left(\frac{a_j}{t_j}\right)^3}{(t_i - t_j)^3}$$

$$+ \frac{12\left(\frac{a_i}{t_i}\right)^2\frac{a_j}{t_j}\frac{a_k}{t_k}}{(t_i - t_j)^2(t_i - t_k)^2}\left\{t_j t_k - t_i^2\left[1 + t_j(x-4) + t_k(x-4)\right]\right.$$

$$\left.+ t_i^3(x-4) + t_i t_j t_k(x-4)\right\}$$

$$+ 12a_i\sum_{j\neq i}\frac{\left(\frac{a_j}{t_j}\right)^2\frac{a_k}{t_k}}{(t_i - t_j)^2(t_i - t_k)},\tag{5.39}$$

with $i,j = 1,2,3$, and

$$t_1 = A + \frac{B}{C} + \frac{C}{3S_3},\tag{5.40}$$

$$t_2 = A - \frac{1+\sqrt{3}\,i}{2}\frac{B}{C} - \frac{1-\sqrt{3}\,i}{6}\frac{C}{S_3},\tag{5.41}$$

$$t_3 = A - \frac{1-\sqrt{3}\,i}{2}\frac{B}{C} - \frac{1+\sqrt{3}\,i}{6}\frac{C}{S_3},\tag{5.42}$$

where $\mathbf{i} = \sqrt{(-1)}$, and

$$A = -\frac{S_2}{3S_3}, \quad B = \frac{D}{3S_3}, \quad C = \frac{1}{\sqrt[3]{2}}\left[E + \sqrt{-4D^3 + E^2}\right]^{1/3},$$

$$D = S_2^2 - 3S_1S_3, \quad E = -2S_2^3 + 9S_1S_2S_3 - 27S_3^2. \tag{5.43}$$

As occurs with the EOS, the PY solution for the RDF of the HS fluid suffers from some inaccuracies with respect to simulation data, especially in the region close to contact where the predicted values are considerably low at high densities, as Figure 5.2 shows. Verlet and Weis (1972) developed an empirical modification, which has been widely used, to correct these defects. However, we will not give here more details on the subject, as there are other analytical expressions available, which are theoretically derived, very accurate, and with affordable complexity, as we will see in the following sections.

5.4 RATIONAL FUNCTION APPROXIMATION

Bravo Yuste and Santos (1991) (see also Bravo Yuste et al. 1996, López de Haro et al. 2008) developed the so-called RFA to obtain the RDF of the HS fluid. The starting point is the expression of the RDF as a series expansion in terms of the packing fraction η. The first two terms in this series are known exactly from a long time (de Boer 1949, Hirschfelder et al. 1954) in the first coordination shell ($1 \le x \le 2$). From them, to first order in the density

$$g(x) = \Theta(x-1) + \Theta(x-1)\Theta(2-x)\left(8 - 6x + \frac{1}{2}x^3\right)\eta + \mathcal{O}(\eta^2). \tag{5.44}$$

Then, the Laplace transform of $xg(x)$ is

$$G(t) = t[F_0(t) + F_1(t)\eta]e^{-t} - 12\eta t[F_0(t)]^2 e^{-2t} + \mathcal{O}(\eta^2), \tag{5.45}$$

with

$$F_0(t) = t^{-2} + t^{-3}, \tag{5.46}$$

$$F_1(t) = \frac{5}{2}t^{-2} - 2t^{-3} - 6t^{-4} + 12t^{-5} + 12t^{-6}. \tag{5.47}$$

Assuming that the series (5.45) continues indefinitely we will obtain for $G(t)$ an expression of the form

$$G(t) = \sum_{i=1}^{\infty}(-12\eta)^{i-1}t[F(t)]^i e^{-it} = t\frac{F(t)e^{-t}}{1 + 12\eta F(t)e^{-t}}, \tag{5.48}$$

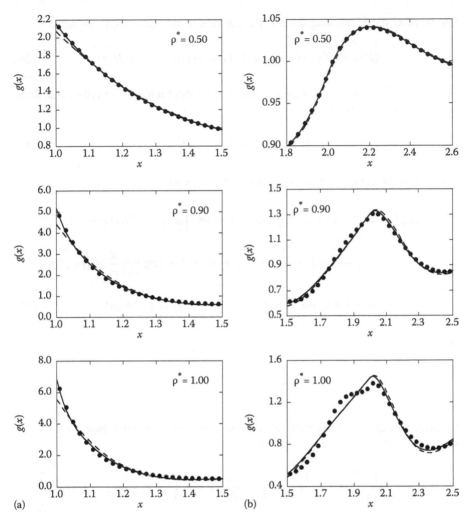

FIGURE 5.2 Radial distribution function of the HS fluid as a function of the reduced distance $x = r/\sigma$ for three different reduced densities $\rho^* = \rho\sigma^3$ in the region near contact (a) and in the region around the second maximum (b). The points are the simulation data from Kolafa et al. (2004b), the dashed curves are the results from the PY theory, and the continuous curves are the results from the RFA and FMSA theories (indistinguishable from each other at the scale of the figure). Note the shoulder in the simulation data for the second peak at $\rho^* = 1.10$, close to the glass transition.

whose inverse Laplace transform gives an expression of the form of Equation 5.32. Solving Equation 5.48 for $F(t)$ yields

$$F(t) = e^t \frac{G(t)}{t - 12\eta G(t)}. \tag{5.49}$$

On the other hand, denoting by $H(t)$ the Laplace transform of $xh(x)$, we have

$$G(t) = \mathscr{L}[xg(x)] = \mathscr{L}\{x[1 + h(x)]\} = t^{-2} + H(t). \tag{5.50}$$

Moreover, as $h(x)$ vanishes for large x, this requires that for low t the Laplace transform of $xh(x)$ must be of the form

$$H(t) = H_0 + H_1 t + O(t^2). \tag{5.51}$$

Then, from the compressibility equation (2.54), one has

$$\rho k_B T \kappa_T = 1 + 4\pi\rho \int r^2 h(r)\, dr = 1 + 4\pi\rho \lim_{t\to 0} \int r^2 e^{-tr} h(r)\, dr$$

$$= 1 - 4\pi\rho \lim_{t\to 0} \frac{d}{dt} \int e^{-tr} r h(r)\, dr = 1 - 4\pi\rho \lim_{t\to 0} \frac{d}{dt} \mathscr{L}[rh(r)]$$

$$= 1 - 4\pi\rho \lim_{t\to 0} \frac{d}{dt}\left[H_0 + H_1 t + O(t^2)\right] = 1 - 4\pi\rho H_1, \tag{5.52}$$

that is,

$$H_1 = \frac{1 + \rho k_B T \kappa_T}{24\eta}. \tag{5.53}$$

Introducing Equation 5.50, with (5.51), into Equation 5.49 and expanding in power series of t yields

$$F(t) = -\frac{1}{12\eta}\left[1 + t + \frac{1}{2}t^2 + \left(\frac{1}{6} + \frac{1}{12\eta}\right)t^3 + \left(\frac{1}{24} + \frac{1}{12\eta}\right)t^4 + O(t^5)\right]. \tag{5.54}$$

On the other hand, from Equation 5.32 it is obvious that $f_0(1) = g(1^+)$. Therefore, taking into account that $\mathscr{L}[g(1^+)] = g(1^+)/t$ and $\mathscr{L}[f_0(1)] = \lim_{t\to\infty} tF(t)$, for large values of t the function $F(t)$ must satisfy

$$\lim_{t\to\infty} t^2 F(t) = f_0(1) = g(1^+), \tag{5.55}$$

that is to say,

$$\lim_{t\to\infty} F(t) \sim t^{-2}. \tag{5.56}$$

In this way, the function $F(t)$, given by Equation 5.49, must satisfy the conditions (5.46), (5.47), (5.54), and (5.56).

As the functional form of $F(t)$ one can choose a Padé approximant of the form $F(t) = P_n(t)/P_m(t)$, where $P_n(t)$ and $P_m(t)$ are polynomials in t with degrees n and m, respectively. The coefficients of the polynomials are determined from the condition that the expansion of $F(t)$ in power series of t must coincide with Equation 5.54. This requires that $n + m \geq 4$ and condition (5.56) implies that $m = n + 2$. The simplest approximant fulfilling these conditions is that for $n = 1$ and $m = 3$, in which case

$$F(t) = F^{PY}(t) = -\frac{1}{12\eta}\frac{1 + L_1^{PY}t}{1 + S_1^{PY}t + S_2^{PY}t^2 + S_3^{PY}t^3}, \tag{5.57}$$

which is of the form of Equation 5.29. In fact, determining the coefficients in the aforementioned way, we obtain the results (5.31) corresponding to the PY solution, whence the notation used in Equation 5.57. Although conditions (5.46) and (5.47) have not been imposed, it is easy to prove that expression (Equation 5.57) of $F(t)$ fulfills them.

The next order approximant for $F(t)$, which corresponds to $n = 2$ and $m = 4$, reads

$$F(t) = F^{GMSA}(t) = -\frac{1}{12\eta}\frac{1 + L_1 t + L_2 t^2}{1 + S_1 t + S_2 t^2 + S_3 t^3 + S_4 t^4}. \tag{5.58}$$

Inserting this in Equation 5.48, the resulting expression of $G(t)$ is of the form that from Henderson and Blum (1976) and Høye and Blum (1977) corresponds to the GMSA, whence the notation used, although these authors did not yield explicit expressions of the coefficients L_i and S_i. They are again determined from the condition that the expansion of Equation 5.58 in power series of t must reproduce expansion (5.54). This gives (Bravo Yuste and Santos 1991)

$$L_1 = L_1^{PY} + \frac{12\eta}{1 + 2\eta}\left(\frac{1}{2}L_2 - S_4\right), \tag{5.59}$$

$$S_1 = S_1^{PY} + \frac{12\eta}{1 + 2\eta}\left(\frac{1}{2}L_2 - S_4\right), \tag{5.60}$$

$$S_2 = S_2^{PY} + \frac{12\eta}{1 + 2\eta}\left(\frac{1 - 4\eta}{12\eta}L_2 + S_4\right), \tag{5.61}$$

$$S_3 = S_3^{PY} - \frac{12\eta}{1 + 2\eta}\left(\frac{1 - \eta}{12\eta}L_2 + \frac{1}{2}S_4\right). \tag{5.62}$$

Coefficients L_2 and S_4 are determined from condition (5.55), with the result

$$g(1^+) = -\frac{1}{12\eta}\frac{L_2}{S_4}, \tag{5.63}$$

or equivalently

$$L_2 = -3(Z - 1)S_4. \tag{5.64}$$

Inserting Equations 5.59 through 5.62 into Equation 5.58 and performing the series expansion we can obtain the terms of order five and six in the expansion as a function of L_2 and S_4. The difference between these two terms depends of L_2, S_4, and $H_1(t)$. The latter quantity is related to the isothermal compressibility κ_T through Equation 5.53 so that, once the EOS Z, and thus κ_T, is fixed, this gives another relationship between L_2 and S_4, and upon substitution in Equation 5.64, Bravo Yuste et al. (1996) finally obtain

$$S_4 = \frac{1 - \eta}{36\eta(Z - 1/3)}\left\{1 - \left[1 + \frac{Z - 1/3}{Z - Z^{PY}}\left(\frac{\kappa_T}{\kappa_T^{PY}} - 1\right)\right]^{1/2}\right\}. \tag{5.65}$$

For Z we can use the CS equation (5.21) or any other suitable EOS for the HS fluid. With this, all the coefficients in the polynomials of Equation 5.58 are completely specified. Then, introducing the result into Equation 5.48, and performing the inverse Laplace transform of the resulting $G(t)$, one obtains the RDF $g(x)$.

An analytical expression of $g(x)$ for the first coordination shell in terms of real functions, derived on the basis of the RFA approach, may be found in Largo (2003) and Largo and Solana (2003). More recently, an alternative, but equivalent, expression for the first coordination shell, as well as another for the second coordination shell, corresponding to the first two terms of Equation 5.32, has been derived (Díez 2009). The expressions for f_0 and f_1 are

$$f_0(x) = \sum_{i=1}^{4} a_i e^{t_i(x-1)}, \tag{5.66}$$

$$f_1(x) = \sum_{i=1}^{4} b_i e^{t_i(x-2)}, \tag{5.67}$$

with

$$a_i = -\frac{1}{12\eta}\frac{(1 + L_1 t_i + L_2 t_i^2)t_i}{S_1 + 2S_2 t_i + 3S_3 t_i^2 + 4S_4 t_i^3}, \tag{5.68}$$

$$b_i = \frac{a_i}{t_i}\left\{\left[1 + t_i(x - 2)\right]\frac{a_i}{t_i} + 2t_i \sum_{j \neq i}\frac{a_j/t_j}{t_i - t_j}\right\}, \tag{5.69}$$

where, with the notation of Largo and Solana (2003), but for small changes,

$$t_1 = -\frac{S_3}{4S_4} + y_p - y_+, \quad t_2 = -\frac{S_3}{4S_4} + y_p + y_+,$$

$$t_3 = -\frac{S_3}{4S_4} - y_p - y_-, \quad t_4 = -\frac{S_3}{4S_4} - y_p + y_-, \tag{5.70}$$

with

$$y_p = -\frac{1}{2}\sqrt{\frac{S_3^2}{4S_4^2} - \frac{2S_2}{3S_4} + y_r + \frac{C}{3S_4}}, \tag{5.71}$$

$$y_r = \frac{S_2^2 - 3S_1S_3 + 12S_4}{3S_4 C}, \tag{5.72}$$

$$y_\pm = \frac{1}{2}\sqrt{\frac{S_3^2}{2S_4^2} - \frac{4S_2}{3S_4} - y_r - \frac{C}{3S_4} \pm \frac{-\frac{S_3^3}{S_4^3} + \frac{4S_2S_3}{S_4^2} - \frac{8S_1}{S_4}}{8y_p}}, \tag{5.73}$$

$$C = \frac{1}{\sqrt[3]{2}}\left[E + \sqrt{-4(S_2^2 - 3S_1S_3 + 12S_4)^3 + E^2}\right]^{1/3}, \tag{5.74}$$

$$E = 2S_2^3 - 9S_1 S_2 S_3 + 27 S_3^2 + 27S_1^2 S_4 - 72S_2 S_4. \tag{5.75}$$

With the preceding expressions, together with Equations 5.59 through 5.62, 5.64, and 5.65, f_0 and f_1 are completely specified and so too are the first two shells of the RDF (Equation 5.24) of the HS fluid. The performance of the HS RDF obtained in the way just described is excellent, as shown in Figure 5.2, except in the region close to the second peak at densities close to the glass transition.

5.5   FIRST-ORDER MEAN SPHERICAL APPROXIMATION

The FMSA, cited in Section 4.4, was applied by Tang and Lu (1995) to the HS fluid. In this case, the Yukawa tail is used to correct the PY closure, so that the DCF is taken to be

$$c(x) = K\frac{e^{-z(x-1)}}{x}. \tag{5.76}$$

The FMSA then yields the RDF as the sum of a zero-order contribution $g_0(x)$, which is the PY solution, and a first-order contribution $g_1(x)$. The Laplace transform of $xg(x)$ is then

$$G(t) = G_0(t) + G_1(t), \tag{5.77}$$

where $G_0(t)$ and $G_1(t)$, the Laplace transforms of $xg_0(x)$ and $xg_1(x)$, respectively, are

$$G_0(t) = \frac{L(t) e^{-t}}{(1 - \eta)^2 Q_0(t) t^2},\tag{5.78}$$

$$G_1(t) = \frac{Ke^{-t}}{Q_0(t) Q_0(z) (t + z)},\tag{5.79}$$

with

$$Q_0(t) = \frac{S(t) + 12\eta L(t) e^{-t}}{(1 - \eta)^2 t^3},\tag{5.80}$$

and a similar expression for $Q(z)$ by replacing t with z. In the latter expression $S(t)$ and $L(t)$ are given by Equations. 5.25 and 5.26, respectively.

The parameters K and z of the Yukawa tail were determined by the Henderson and Blum (1976) procedure based on the expression of the RDF at contact and the compressibility equation in the form of Equation 4.7. To this end, Tang and Lu considered the CS EOS for the HS fluid. Then, the first of these conditions yields

$$g(1) = \frac{1 + \eta/2}{(1 - \eta)^2} + \frac{K}{Q_0^2(z)}.\tag{5.81}$$

On the other hand, the Fourier transform of the DCF $\hat{c}(k)$ is related to that of the total correlation function $\hat{h}(k)$ through the Fourier transform of the OZ equation (4.5), whose first-order contribution can be written as (Tang and Lu 1995)

$$\rho\hat{c}_1(k) = \rho\hat{h}_1(k) \left[1 - \rho\hat{c}_0(k)\right]^2.\tag{5.82}$$

Introducing this expression, with $k = 0$, into the compressibility equation (4.7), taking into account that, in the first-order approximation, $\hat{c}(k) = \hat{c}_0(k) + \hat{c}_1(k)$, and taking for $\hat{c}_0(0)$ the PY result derived from the compressibility equation, one obtains

$$\frac{1}{k_B T}\left(\frac{\partial P}{\partial \rho}\right)_T = \frac{(1 + 2\eta)^2}{(1 - \eta)^4} - \frac{24\eta(1 + 2\eta)[(1 - \eta)z + 1 + 2\eta]}{(1 - \eta)^4 Q_0^2(z) z^2}K.\tag{5.83}$$

Now, requiring consistency between the viral and compressibility routes leads to the following expressions for the parameters K and z

$$z = \frac{2(1 + 2\eta)}{(4 - \eta)(1 - \eta)^2}\left(3 + \sqrt{21 - 15\eta + 3\eta^2}\right),\tag{5.84}$$

$$K = \frac{\eta^2}{2(1 - \eta)^3}Q_0^2(z).\tag{5.85}$$

Introducing the latter result into the expression for $G_1(t)$, we have

$$G_1(t) = \frac{\eta^2}{2(1-\eta)^3} \frac{e^{-t}}{Q_0^2(t)(t+z)}. \qquad (5.86)$$

Then the Taylor series expansion of $G_1(t)$ is

$$G_1(t) = \frac{\eta^2(1-\eta)}{2} \sum_{n=0}^{\infty} (1+n)(-12\eta)^n \frac{t^6 L^n}{S^{n+2}(t+z)} e^{-(n+1)t}. \qquad (5.87)$$

Truncating the series at $n = 1$ yields

$$G_1(t) = \frac{\eta^2(1-\eta)}{2} \left[\frac{t^6}{S^2(t+z)} e^{-t} - 24\eta \frac{t^6 L}{S^3(t+z)} e^{-2t} \right]. \qquad (5.88)$$

The inverse Laplace transform of this expression can be performed following the same procedure as that used by Throop and Bearman (1965) to obtain the PY solution, based on the residue theorem. The final result is of the form (Tang and Lu 1995)

$$x g_{11}(x) = \frac{\eta^2(1-\eta) z^6}{2S^2(-z)} e^{-z(x-1)} + \sum_{i=1}^{3} [a_{10}(t_i) + a_{11}(t_i)(x-1)] e^{t_i(x-1)} \qquad (5.89)$$

$$x g_{21}(x) = -\frac{12\eta^3(1-\eta) z^6 L(-z)}{S^3(-z)} e^{-z(x-2)}$$

$$+ \sum_{i=1}^{3} [b_{10}(t_i) + b_{11}(t_i)(x-2) + b_{12}(t_i)(x-2)^2] e^{t_i(x-2)}, \qquad (5.90)$$

for the first-order perturbative contribution to the first and second shells of the RDF respectively. The t_i in Equations 5.89 and 5.90 have the same expressions as in the PY theory and the explicit expressions for the parameters a_{ij} and b_{ij} in these equations were reported by Tang and Lu (1995). Adding to the zero-order contributions for the first and second shells of the RDF, as given by the PY solution, the corresponding first-order contributions, as given by Equations 5.89 and 5.90, respectively, we obtain the complete solution for the HS RDF from the FMSA. The results from this theory are virtually identical to those from the RFA theory.

5.6 EQUATION OF STATE AND RADIAL DISTRIBUTION FUNCTION OF THE HARD-SPHERE SOLID

In a crystalline solid, the particles are confined within the cells or "cages" formed by their nearest neighbors. This is the basis of the cell, or *free volume theory*, according to which the configurational partition function of the solid is

$$Q_c = v_f^N, \qquad (5.91)$$

where v_f is the *free volume* per particle, which for the HS solid is simply the volume available for the movement of the center of a particle within its cell and can be approximated by (Hirschfelder et al. 1954)

$$v_f = b\left(r_1 - \sigma\right)^3, \tag{5.92}$$

where

  r_1 is the nearest-neighbor distance
  b is a parameter depending on the lattice structure

Equation 5.92 can be rewritten in terms of the close packing density ρ_0 as

$$v_f = \frac{8}{\rho_0}\left[\left(\frac{\rho_0}{\rho}\right)^{1/3} - 1\right]^3. \tag{5.93}$$

This expression can be applied to different crystalline structures provided we introduce the corresponding close packing density ρ_0. Thus, for example, for fcc and hcp lattices $\rho_0^* = \sqrt{2}$, for bcc $\rho_0^* = 3\sqrt{3}/4$, and for sc $\rho_0 = 1$ (Hirschfelder et al. 1954).

Then, the EOS is readily obtained as

$$Z \equiv \frac{PV}{Nk_BT} = \left[1 - \left(\frac{\rho}{\rho_0}\right)^{1/3}\right]^{-1}. \tag{5.94}$$

Alder et al. (1968) accurately determined the EOS of the fcc HS solid up to high densities from MD simulation, and assumed that obeys a series expansion of the form

$$Z = \frac{3}{\alpha} + C_0 + C_1\alpha + C_2\alpha^2 + C_3\alpha^3 + \cdots, \tag{5.95}$$

where $\alpha = V/V_0 - 1 \equiv \rho_0/\rho - 1$. The first term of the right-hand side of Equation 5.95 is the right asymptotic behavior for $\alpha \to 0$ (Salsburg and Wood 1962) also predicted by Equation 5.94. The next two coefficients in Equation 5.95 were determined by Alder et al. (1968) by fitting the simulation data, with the result $C_0 = 2.56 \pm 0.02$ and $C_1 = 0.56 \pm 0.08$ respectively. These two coefficients were subsequently calculated by Alder et al. (1971) directly by means of MD simulation of systems of dodecahedra for fcc and hcp configurations. This yielded $C_0 = 2.5658 \pm 0.0004$ and $C_1 = 0.52 \pm 0.01$ for fcc, and $C_0 = 2.5658 \pm 0.0002$ and $C_1 = 0.57 \pm 0.01$ for hcp.

Figure 5.3 compares the predictions from Equations 5.94 and 5.95 with the simulation data of Alder et al. (1968). One can see that the free volume theory equation (5.94) reproduces the correct behavior near close packing, but appreciably deviates from simulation data at lower densities. In contrast, Equation 5.95, truncated at the level of the first-order term in α and taking for C_0 and C_1 the values determined by Alder et al. (1971), is accurate up to densities near melting. For situations in which extreme accuracy at moderate to low densities is needed, Hall (1972) reported an EOS, based on the fitting of the simulation data of Alder et al. (1968), which has been widely used in perturbation theory of solids.

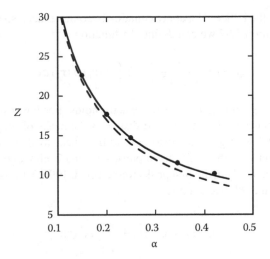

FIGURE 5.3 Compressibility factor Z for the HS solid as a function of $\alpha = V/V_0 - 1$. Points: simulation data from Alder et al. (1968). Dashed curve: free volume theory Equation 5.94. Continuous curve: Equation 5.95 with the parameters C_0 and C_1 determined by Alder et al. (1971).

With regard to the (angle-averaged) RDF of the solid, the most widely used approach is based on modeling the RDF of the solid as a sum of Gaussian-like functions, representing the successive peaks, whose parameters are determined by fitting the simulation data. Parameterizations obtained in this way are available for the fcc solid (Weis 1974, Kincaid and Weis 1977, Kang et al. 1986, Choi et al. 1991), as well as for the hcp solid (Jackson and Van Swol 1988, Choi et al. 1991).

A theoretical procedure, free from adjustable parameters, to obtain the RDF of the HS solid was developed by Rascón et al. (1996a,b,c). The starting point is the ansatz that the single-particle density $\rho^{(1)}(\mathbf{r}) \equiv \rho(\mathbf{r})$ is given by a sum of Gaussians of the form

$$\rho\left(\mathbf{r}\right) = \left(\frac{\gamma}{\pi}\right)^{3/2} \sum_i e^{-\gamma(\mathbf{r}-\mathbf{r}_i)^2}, \tag{5.96}$$

where $\mathbf{r}_i$ are the position vectors of the lattice sites.

On the other hand, the angle-averaged RDF is defined as

$$g\left(r\right) = \frac{1}{4\pi V \rho^2} \int d\Omega \int \rho\left(\mathbf{r}_1, \mathbf{r}_2\right) d\mathbf{r}_2, \tag{5.97}$$

where
$r = |\mathbf{r}_2 - \mathbf{r}_1|$, $\rho(\mathbf{r}_1, \mathbf{r}_2) \equiv \rho^{(2)}(\mathbf{r}_1, \mathbf{r}_2)$
Ω is the solid angle around the position $\mathbf{r}_1$

In the large distance limit $\rho(\mathbf{r}_1, \mathbf{r}_2) \rightarrow \rho(\mathbf{r}_1)\rho(\mathbf{r}_2)$, that is, the positions of the two particles are independent of each other and the sole correlation between them is due to the periodicity of the lattice. The approximation becomes exact also at close packing. Similarly to Equation 5.97 we can define the function

$$g_0(r) = \frac{1}{4\pi V \rho^2} \int d\Omega \int \rho(\mathbf{r}_1) \rho(\mathbf{r}_2) \, d\mathbf{r}_2. \tag{5.98}$$

In the close packing limit, $g(r) = g_0(r)$ exactly holds. For lower densities $g(r) \to g_0(r)$ at large distances, whereas at short distances the peaks of $g(r)$ appear slightly displaced toward the origin and distorted, with respect to those of $g_0(r)$, due to the correlations between the positions of particles 1 and 2, although in practice these effects are important only in the first peak (Rascón et al. 1996c). Introducing Equation 5.96 into Equation 5.98, one obtains

$$g_0(r) = g_0^{(0)}(r) + \sum_{i=1}^{\infty} g_0^{(i)}(r), \tag{5.99}$$

with

$$g_0^{(0)}(r) = \frac{1}{4\pi \rho} \left(\frac{\gamma}{2\pi}\right)^{1/2} 2\gamma e^{-(\gamma/2)r^2} \tag{5.100}$$

and

$$g_0^{(i)}(r) = \frac{1}{4\pi \rho} \left(\frac{\gamma}{2\pi}\right)^{1/2} n_i \frac{e^{-(\gamma/2)(r-r_i)^2} + e^{-(\gamma/2)(r+r_i)^2}}{r_i r}$$

$$\approx \frac{1}{4\pi \rho} \left(\frac{\gamma}{2\pi}\right)^{1/2} n_i \frac{e^{-(\gamma/2)(r-r_i)^2}}{r_i r}, \quad i \geq 1, \tag{5.101}$$

where n_i is the number of ith nearest neighbors corresponding to the lattice considered, that is, the number of lattice sites at distance r_i from a lattice site taken as a reference, and the last exponential has been neglected because it is many orders of magnitude lower than the first one (Rascón et al. 1996a).

Because of the similarity of $g(r)$ and $g_0(r)$ beyond the first peak, Rascón et al. (1996b) take for $g(r)$ a functional form like that for $g_0(r)$, that is,

$$g(r) = g^{(0)}(r) + \sum_{i=1}^{\infty} g^{(i)}(r), \tag{5.102}$$

and introduce the approximation $g^{(i)}(r) = g_0^{(i)}(r)$ for $i > 1$, whereas for $g^{(1)}(r)$ they propose the ansatz

$$g^{(1)}(r) = \frac{A e^{-\gamma_1 (r-r_1)^2 / 2}}{r}. \tag{5.103}$$

With these approximations, the functional form of the RDF is very much like that used in the aforementioned fittings of the simulation data. Instead, Rascón et al.

(1996b,c) determine the parameters by imposing appropriate conditions. The first one consists in approximating the average position of the nearest neighbors calculated from $g^{(1)}(r)$ with that calculated from $g_0^{(1)}(r)$. From Equation 2.33 this condition is equivalent to

$$\frac{\int_\sigma^\infty g^{(1)}(r) r^3 dr}{\int_\sigma^\infty g^{(1)}(r) r^2 dr} = \frac{\int_\sigma^\infty g_0^{(1)}(r) r^3 dr}{\int_\sigma^\infty g_0^{(1)}(r) r^2 dr}. \tag{5.104}$$

The second condition, which is exact, guarantees that the number of particles within the first peak is the coordination number of the lattice

$$4\pi\rho \int_\sigma^\infty g^{(1)}(r) r^2 dr = n_1. \tag{5.105}$$

The third condition, also exact, is the fulfillment of the virial equation (2.43) for a given EOS, as, for example, any of those cited in this section. These three conditions allow us to obtain parameters A, γ_1, and r_1 in $g^{(1)}(r)$. To complete the construction of $g(r)$ we need to determine γ. For Gaussian distributions, this parameter is related to the average square displacements of the particles from their lattice sites $\mathbf{r}_i$ as

$$\gamma = \frac{3}{2\langle(\Delta\mathbf{r})^2\rangle}, \tag{5.106}$$

where $\Delta\mathbf{r} = \mathbf{r} - \mathbf{r}_i$. Velasco et al. (1998, 1999) devised a procedure to approximately obtain γ by equating the volume of a sphere of radius $[\langle(\Delta\mathbf{r})^2\rangle]^{1/2}$ to the free volume per particle (5.93). This gives

$$\gamma = \frac{3}{2} \left(\frac{4\pi}{3v_f}\right)^{2/3}, \tag{5.107}$$

which provides a good estimate of γ for fcc solids as compared with the data derived from computer simulations (Velasco et al. 1998, 1999). This completes the procedure for obtaining the parameters involved in $g(r)$.

The procedure outlined can be applied not only to fcc or hcp crystals, but also to bcc and sc structures. It is to be noted that the latter two structures are unstable for the HS solid, an so we cannot assess the reliability of the theory, but the possibility of obtaining the EOS and the RDF of bcc or sc HS solids is not lack of interest, because they might be useful in perturbation theories for real solids with these crystalline structures (Velasco et al. 1998, Zhou 2007).

Figure 5.4 compares the values of the averaged RDF of the HS fcc solid as predicted by the procedure just described with the simulation data for three different densities. It is seen that the agreement is quite satisfactory.

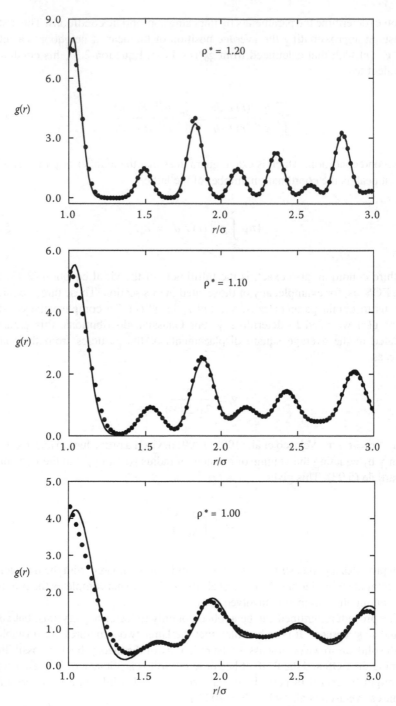

FIGURE 5.4   Angle-averaged radial distribution function of the HS fcc solid as a function of the reduced distance r/σ for three different reduced densities. Points are simulation data and curves are theoretical results.

REFERENCES

Alder, B. J., W. G. Hoover, and D. A. Young. 1968. Studies in molecular dynamics. V. High-density equation of state and entropy for hard disks and spheres. *J. Chem. Phys.* 49:3688.

Alder, B. J., D. A. Young, M. R. Mansigh, and Z. W. Salsburg. 1971. Hard sphere equation of state in the close-packed limit. *J. Comput. Phys.* 7:361.

Baxter, R. J. 1968. Ornstein-Zernike relation for a disordered fluid. *Aust. J. Phys.* 21:563.

Bernal, J. D., and J. Mason. 1960. Co-ordination of randomly packed spheres. *Nature* 188:910.

de Boer, J. 1949. Molecular distribution and equation of state of gases. *Rep. Prog. Phys.* 12:305.

Boublík, T. 1970. Hard-sphere equation of state. *J. Chem. Phys.* 53:471.

Bravo Yuste, S., M. López de Haro, and A. Santos. 1996. Structure of hard-sphere fluids. *Phys. Rev. E* 53:4820.

Bravo Yuste, S., and A. Santos. 1991. Radial distribution function for hard spheres. *Phys. Rev. A* 43:5418.

Bruce, A. D., A. N. Jackson, G. J. Ackland, and N. B. Wilding. 2000. Lattice-switch Monte Carlo method. *Phys. Rev. E* 61:906.

Carnahan, N. F., and K. E. Starling. 1969. Equation of state for nonattracting rigid spheres. *J. Chem. Phys.* 51:635.

Chang, J., and S. I. Sandler. 1994. A real representation for the structure of the hard-sphere fluid. *Mol. Phys.* 81:735.

Choi, Y., T. Ree, and F. H. Ree. 1991. Hard-sphere radial distribution functions for face-centered cubic and hexagonal close-packed phases: Representation and use in a solid-state perturbation theory. *J. Chem. Phys.* 95:7548.

Díez, A. 2009. Teoría y simulación de las propiedades termodinámicas de equilibrio de sistemas con potencial de esferas duras y cola atractiva de potencia inversa. PhD diss., Universidad de Cantabria.

Finney, J. L. 1970. Random packings and the structure of simple liquids. I. The geometry of random close packing. *Proc. R. Soc. A* 319:479.

Hall, K. H. 1972. Another hard sphere equation of state. *J. Chem. Phys.* 57:2252.

Helfand, E., H. L. Frisch, and J. L. Lebowitz. 1961. Theory of the two- and one-dimensional rigid sphere fluids. *J. Chem. Phys.* 34:1037.

Henderson, D., and L. Blum. 1976. Generalized mean spherical approximation for hard spheres. *Mol. Phys.* 32:1627.

Hirschfelder, J. O., C. F. Curtiss, and R. B. Bird. 1954. *Molecular Theory of Gases and Liquids*. New York: John Wiley & Sons.

Hoover, W. G., and F. H. Ree. 1968. Melting transition and communal entropy for hard spheres. *J. Chem. Phys.* 49:3609.

Høye, J. S., and L. Blum. 1977. Solution of the Yukawa closure of the Ornstein-Zernike equation. *J. Stat. Phys.* 16:399.

Jackson, G., and F. Van Swol. 1988. Perturbation theory of a model hcp solid. *Mol. Phys.* 65:161.

Kang, H. S., T. Ree, and F. H. Ree. 1986. A perturbation theory of classical solids. *J. Chem. Phys.* 84:4547.

Kansal, A. R., S. Torquato, and F. H. Stillinger. 2002. Diversity of order and densities in jammed hard-particle packings. *Phys. Rev. E* 66:041109.

Kendall, K., C. Stainton, F. van Swol, and L. V. Woodcock. 2002. Crystallization of spheres. *Int. J. Thermophys.* 23:175.

Kincaid, J. M., and J. J. Weis. 1977. Radial distribution function of a hard-sphere solid. *Mol. Phys.* 34:931.

Kolafa, J., S. Labík, and A. Malijevský. 2004a. Accurate equation of state of the hard sphere fluid in stable and metastable regions. *Phys. Chem. Chem. Phys.* 6:2335.

Kolafa, J., S. Labík, and A. Malijevský. 2004b. Radial distribution function of the hard sphere fluid. Poster presented at the *18th IUPAC International Conference on Chemical Thermodynamics*, Beijing, China. The simulation data are available at the URL: http://www.vscht.cz/fch/software/hsmd/ (accessed on June 18, 2010).

Largo, J. 2003. Teoría y simulación de las propiedades de equilibrio de fluidos de pozo cuadrado. PhD diss., Universidad de Cantabria.

Largo, J., and J. R. Solana. 2003. Theory and computer simulation of the first- and second-order perturbative contributions to the free energy of square-well fluids. *Mol. Simul.* 29:363.

Lebowitz, J. L., E. Helfand, and E. Praestgaard. 1965. Scaled particle theory of fluid mixtures. *J. Chem. Phys.* 43:774.

López de Haro, M., S. B. Yuste, and A. Santos. 2008. Alternative approaches to the equilibrium properties of hard-sphere liquids. In *Theory and Simulation of Hard-Sphere Fluids and Related Systems*, ed. A. Mulero, *Lect. Notes Phys.* 753:183–245. Berlin, Germany: Springer-Verlag.

Percus, J. K., and G. J. Yevick. 1958. Analysis of classical statistical mechanics by means of collective coordinates. *Phys. Rev.* 110:1.

Rascón, C., L. Mederos, and G. Navascués. 1996a. Equation of state of the hard-sphere crystal. *Phys. Rev. E* 53:5698.

Rascón, C., L. Mederos, and G. Navascués. 1996b. Theoretical approach to the correlations of a classical crystal. *Phys. Rev. E* 54:1261.

Rascón, C., L. Mederos, and G. Navascués. 1996c. Thermodynamic consistency of the hard-sphere solid distribution function. *J. Chem. Phys.* 105:10527.

Reiss, H., H. L. Frisch, and J. L. Lebowitz. 1959. Statistical mechanics of rigid spheres. *J. Chem. Phys.* 31:369.

Rintoul, M. D. and S. Torquato. 1996a. Computer simulations of dense hard-sphere systems. *J. Chem. Phys.* 105:9258.

Rintoul, M. D. and S. Torquato. 1996b. Metastability and Crystallization in hard-sphere systems. *Phys. Rev. Lett.* 77:4198.

Salsburg, Z. W., and W. W. Wood. 1962. Equation of state of classical hard spheres at high density. *J. Chem. Phys.* 37:798.

Scott, G. D. 1960. Packing of equal spheres. *Nature* 188:908.

Scott, G. D., and D. M. Kilgour. 1969. The density of random close packing of spheres. *J. Phys. D* 2:863.

Smith, W. R., and D. Henderson. 1970. Analytical representation of the Percus-Yevick hard-sphere radial distribution function. *Mol. Phys.* 19:411.

Speedy, R. J. 1998. The hard sphere glass transition. *Mol. Phys.* 95:169.

Tang, Y., and B. C.-Y. Lu. 1995. Improved expressions for the radial distribution function of hard spheres. *J. Chem. Phys.* 103:7463.

Thiele, E. 1963. Equation of state for hard spheres. *J. Chem. Phys.* 39:474.

Throop, G. J., and R. J. Bearman. 1965. Numerical solutions of the Percus-Yevick equation for the hard-sphere potential. *J. Chem. Phys.* 42:2408.

Torquato, S., T. M. Truskett, and P. G. Debenedetti. 2000. Is random close packing of spheres well defined? *Phys. Rev. Lett.* 84:2064.

Truskett, T. M., S. Torquato, and P. G. Debenedetti. 2000. Towards a quantification of disorder in materials: Distinguishing equilibrium and glassy sphere packings. *Phys. Rev. E* 62:993.

Velasco, E., L. Mederos, and G. Navascués. 1998. Phase diagram of colloidal systems. *Langmuir* 14:5652.

Velasco, E., L. Mederos, and G. Navascués. 1999. Analytical approach to the thermodynamics and density distribution of crystalline phases of hard spheres. *Mol. Phys.* 97:1273.

Verlet, L., and J.-J. Weis. 1972. Equilibrium theory of simple liquids. *Phys. Rev. A.* 5:939.

Weis, J.-J. 1974. Perturbation theory for solids. *Mol. Phys.* 28:187. See also Weis, J.-J. 1976. *Mol. Phys.* 32:296.

Wertheim, M. S. 1963. Exact solution of the Percus-Yevick integral equation for hard spheres. *Phys. Rev. Lett.* 10:321.

Woodcock, L. V. 1981. Glass transition of the hard-sphere model and Kauzmann's paradox. *Ann. NY Acad. Sci.* 371:274.

Zhou, S. 2007. Solid phase thermodynamic perturbation theory: Test and application to multiple solid phases. *J. Chem. Phys.* 127:084512.

6 Free Energy Perturbation Theories for Simple Fluids and Solids

This chapter deals with perturbation theories for simple fluids and solids based, in one way or another, on series expansions of the Helmholtz free energy. Some of these expansions lead, from a different viewpoint, to certain theories cited in Section 4.4 within the context of integral equation perturbation theories. This chapter includes the analysis of the performance of some of these theories for several potential models as compared with the simulation data.

6.1 SERIES EXPANSION OF THE FREE ENERGY

Let us consider a spherically symmetric pair potential $u(r)$ that can be split into two contributions, a reference potential $u_0(r)$ and a perturbation $u_1(r)$, like in Equation 4.40. The total potential energy of the system will be

$$\mathcal{U}_N = \sum_{i=1}^{N-1} \sum_{j=i+1}^{N} u\left(r_{ij}\right) = \mathcal{U}_0 + \mathcal{U}_1, \tag{6.1}$$

where $\mathcal{U}_0$ and $\mathcal{U}_1$ are the total potential energies due to the reference and perturbation potentials, respectively.

Let us introduce a parameter α, with $0 \leq \alpha \leq 1$, coupling the reference and perturbation potentials to form a family of potentials $u(r, \alpha)$ of the form

$$u\left(r, \alpha\right) = u_0\left(r\right) + u_1\left(r; \alpha\right) \tag{6.2}$$

and

$$\mathcal{U}_N\left(\alpha\right) = \mathcal{U}_0 + \mathcal{U}_1\left(\alpha\right), \tag{6.3}$$

with the condition that $u_1(r, 1) = u_1$ and $u_1(r, 0) = 0$, so that $u(r, \alpha)$ reduces to the reference potential u_0 for $\alpha = 0$ and for $\alpha = 1$ we recover the full potential $u(r)$.

Then, the Helmholtz free energy can be expanded in power series of α to give

$$F\left(\alpha\right) = F_0 + \left.\frac{\partial F}{\partial \alpha}\right|_{\alpha=0} \alpha + \frac{1}{2} \left.\frac{\partial^2 F}{\partial \alpha^2}\right|_{\alpha=0} \alpha^2 + \cdots, \tag{6.4}$$

where F_0 is the free energy of a system with the reference potential $u_0(r)$.

The corresponding expansion for the RDF $g(r; \alpha)$ will be

$$g(r; \alpha) = g_0(r) + \left. \frac{\partial g(r; \alpha)}{\partial \alpha} \right|_{\alpha=0} \alpha + \cdots \tag{6.5}$$

From Equations 2.16 and 2.18, the canonical partition function takes the form

$$Q(\alpha) = \frac{\Lambda^{-3N}}{N!} Q_c(\alpha) = \frac{\Lambda^{-3N}}{N!} \int_V \cdots \int_V e^{-\beta \mathcal{U}_N(\alpha)} dr_1 \ldots dr_N, \tag{6.6}$$

from which the derivatives of the free energy $F(\alpha) = -k_B T \ln Q(\alpha)$ involved in Equation 6.4 can easily be obtained with the result

$$\left. \frac{\partial(\beta F)}{\partial \alpha} \right|_{\alpha=0} = \beta \left\langle \mathcal{U}_1'(\alpha) \big|_{\alpha=0} \right\rangle_0, \tag{6.7}$$

$$\left. \frac{\partial^2(\beta F)}{\partial \alpha^2} \right|_{\alpha=0} = \beta \left\langle \mathcal{U}_1''(\alpha) \big|_{\alpha=0} \right\rangle_0 - \beta^2 \left[\left\langle \mathcal{U}_1'(\alpha) \big|_{\alpha=0}^2 \right\rangle_0 - \left\langle \mathcal{U}_1'(\alpha) \big|_{\alpha=0} \right\rangle_0^2 \right], \tag{6.8}$$

and so on. In these expressions, subscript 0 in the angular brackets means that the averages are performed in the reference system ensemble and primed quantities mean derivatives with respect to parameter α.

Let us now consider the particular case in which the reference and perturbation potentials are linearly coupled, that is to say

$$u(r, \alpha) = u_0(r) + \alpha u_1(r) \tag{6.9}$$

and

$$\mathcal{U}_N(\alpha) = \mathcal{U}_0 + \alpha \mathcal{U}_1, \tag{6.10}$$

so that $\mathcal{U}_1(\alpha) = \alpha \mathcal{U}_1$, $\mathcal{U}_1'(\alpha) = \mathcal{U}_1$, and $\mathcal{U}_1^{(n)} = 0$ for $n > 1$, and Equations 6.7 and 6.8 reduce to

$$\left. \frac{\partial(\beta F)}{\partial \alpha} \right|_{\alpha=0} = \beta \langle \mathcal{U}_1 \rangle_0, \tag{6.11}$$

$$\left. \frac{\partial^2(\beta F)}{\partial \alpha^2} \right|_{\alpha=0} = -\beta^2 \left[\langle \mathcal{U}_1^2 \rangle_0 - \langle \mathcal{U}_1 \rangle_0^2 \right], \tag{6.12}$$

and then Equation 6.4 leads to

$$\beta F = \beta F_0 + \beta \langle \mathcal{U}_1 \rangle_0 - \frac{1}{2} \beta^2 \left[\langle \mathcal{U}_1^2 \rangle_0 - \langle \mathcal{U}_1 \rangle_0^2 \right] + \cdots, \tag{6.13}$$

which is the *high temperature expansion* (HTE) derived by Zwanzig (1954) by expanding βF in power series of β, namely,

$$\beta F = \sum_{n=0}^{\infty} \frac{1}{n!} \left. \frac{\partial^n(\beta F)}{\partial \beta^n} \right|_{\beta=0} \beta^n. \tag{6.14}$$

From Equation 6.13, all the thermodynamic properties of the system can be expressed as power series in terms of the inverse of the reduced temperature. Thus, for example, the Helmholtz free energy F, the excess internal energy U^E, and the compressibility factor $Z = PV/Nk_BT$ are expressed in the form

$$\frac{F}{Nk_BT} = \sum_{n=0}^{\infty} \frac{F_n}{Nk_BT} \frac{1}{T^{*n}}, \tag{6.15}$$

$$\frac{U^E}{N\varepsilon} = \sum_{n=1}^{\infty} \frac{U_n}{N\varepsilon} \frac{1}{T^{*n-1}}, \tag{6.16}$$

$$Z = \sum_{n=0}^{\infty} Z_n \frac{1}{T^{*n}}, \tag{6.17}$$

respectively, where the zero-order terms correspond to the contributions due to the reference system, the HS fluid in many cases, and ε is the energy parameter of the potential. From standard thermodynamic relationships, it is easy to see that

$$\frac{U_n}{N\varepsilon} = n \frac{F_n}{Nk_BT}, \tag{6.18}$$

$$Z_n = \rho \left[\frac{\partial \left(F_n/Nk_BT \right)}{\partial \rho} \right]_T, \tag{6.19}$$

so that from the knowledge of either F or U we can readily obtain the other two quantities.

In a similar way, the radial distribution function $g(r)$ can be expressed as

$$g(r) = \sum_{n=0}^{\infty} g_n(r) \frac{1}{T^{*n}}. \tag{6.20}$$

The knowledge of the perturbation terms $g_n(r)$ provides an alternative route to Equation 6.19 to obtain the perturbation terms Z_n of the EOS through the pressure equation (2.36).

The form of Equation 6.13 is suitable for computer simulations but not for theoretical calculations. Expressions for the first and second perturbation terms, more appropriate for the latter purposes, were derived by Zwanzig (1954) and by Henderson and Barker (1971) as follows.

Rewriting Equation 3.77 for the system with potential (6.9) in the form

$$F(\alpha) = F_0 + 2\pi N\rho \int_0^\alpha d\alpha' \int_0^\infty u_1(r) \, g(r; \alpha') \, r^2 dr, \tag{6.21}$$

and taking for the radial distribution function $g(r, \alpha)$ of the fluid with pair potential (6.9) its series expansion (6.5), leads to

$$\frac{\partial F}{\partial \alpha}\bigg|_{\alpha=0} = 2\pi N\rho \int u_1\left(r\right) g_0\left(r\right) r^2 dr, \tag{6.22}$$

which involves only the RDF $g_0(r)$ of the reference fluid. Expression (6.22) was derived by Zwanzig (1954), who also reported an expression for the second derivative performed in the canonical ensemble. However, Henderson and Barker (1971) showed that for numerical calculation it is necessary to carry out the derivative in the grand canonical ensemble. The final result is

$$\begin{aligned}
\frac{\partial^2 F}{\partial \alpha^2}\bigg|_{\alpha=0} &= -\frac{1}{2}\beta N\rho \int g_0\left(\mathbf{r}_1, \mathbf{r}_2\right) \left[u_1\left(r_{12}\right)\right]^2 d\mathbf{r}_2 - \beta N\rho^2 \\
&\quad \times \int g_0\left(\mathbf{r}_1, \mathbf{r}_2, \mathbf{r}_3\right) u_1\left(r_{12}\right) u_1\left(r_{23}\right) d\mathbf{r}_2 d\mathbf{r}_3 \\
&\quad - \frac{1}{4}\beta N\rho^3 \int \left[g_0\left(\mathbf{r}_1, \mathbf{r}_2, \mathbf{r}_3, \mathbf{r}_4\right)\right. \\
&\quad \left. - g_0\left(\mathbf{r}_1, \mathbf{r}_2\right) g_0\left(\mathbf{r}_3, \mathbf{r}_4\right)\right] u_1\left(r_{12}\right) u_1\left(r_{34}\right) d\mathbf{r}_2 d\mathbf{r}_3 d\mathbf{r}_4 \\
&\quad + N\left(\frac{\partial \rho}{\partial P}\right)\bigg|_{\alpha=0} \left\{\frac{\partial}{\partial \rho}\left[\frac{1}{2}\rho^2 \int g_0\left(\mathbf{r}_1, \mathbf{r}_2\right) u_1\left(r_{12}\right) d\mathbf{r}_2\right]\right\}^2.
\end{aligned} \tag{6.23}$$

Therefore, whereas the first derivative involves only the pair correlation function of the reference system, which for the HS case is quite easily and accurately obtained as shown in Chapter 5, the second and higher-order derivatives involve higher-order correlation functions of the reference fluid, which cannot be easily obtained, and so some kind of approximation must be introduced.

Both HTE and α-expansion may diverge for steeply repulsive potentials at low temperatures, when $\beta u_1(r) \gg 1$. To overcome this drawback, it has been proposed to expand the free energy in terms of the Mayer function (Pelissetto and Hansen 2006, Sillrén and Hansen 2007). To this end, Sillrén and Hansen (2007) take for the α-dependence of $\mathcal{U}_1$ the expression

$$\beta \mathcal{U}_1\left(\alpha\right) = -\sum_{i=1}^{N-1}\sum_{j=i+1}^{N} \ln\left[1 + \alpha f\left(r_{ij}\right)\right], \tag{6.24}$$

where $f(r)$ is the Mayer function for the potential $u_1(r)$. Then,

$$\beta \mathcal{U}_1'\big|_{\alpha=0} = -\sum_{i=1}^{N-1}\sum_{j=i+1}^{N} f\left(r_{ij}\right), \tag{6.25}$$

$$\beta \mathcal{U}_1''\big|_{\alpha=0} = \sum_{i=1}^{N-1}\sum_{j=i+1}^{N} f^2\left(r_{ij}\right), \tag{6.26}$$

and, from Equations 6.7, 6.8, and 6.4 with $\alpha = 1$, we have

$$\beta F_1 = -2\pi N\rho \int_0^\infty g_0(r) f(r) r^2 dr, \tag{6.27}$$

$$\beta F_2 = \pi N\rho \int_0^\infty g_0(r) f^2(r) r^2 dr - \frac{1}{2}\beta^2 \left[\left\langle \left(\mathcal{U}_1'\big|_0\right)^2 \right\rangle_0 - \left\langle \mathcal{U}_1'\big|_0 \right\rangle_0^2 \right]. \tag{6.28}$$

Again, the last term in this equation involves higher-order correlation functions of the reference fluid and so it can be calculated only by introducing some approximation.

The result (6.27) was obtained before by Kincaid et al. (1976a,b) by expressing the RDF of the α-system in terms of the cavity function, $g(r, \alpha) = y(r, \alpha)\exp[-\beta u_1(\alpha)]$, then expanding the cavity function in power series of α, and inserting the result into Equation 6.21.

6.2 CALCULATION OF THE PERTURBATION TERMS BY COMPUTER SIMULATION

The term F_0/NkT in Equation 6.15 can be obtained from any suitable EOS Z_0 for the reference system by integration of Equation 6.19 for $n = 0$. Higher-order terms in Equation 6.15 can be obtained in the form (Barker and Henderson 1968, 1976)

$$\frac{F_1}{Nk_BT} = \frac{1}{N}\sum_i \langle N_i \rangle_0 \, u_1^*(r_i), \tag{6.29}$$

$$\frac{F_2}{Nk_BT} = -\frac{1}{2}\frac{1}{N}\sum_{ij} \left[\langle N_i N_j \rangle_0 - \langle N_i \rangle_0 \langle N_j \rangle_0 \right] u_1^*(r_i) \, u_1^*(r_j), \tag{6.30}$$

$$\frac{F_3}{Nk_BT} = \frac{1}{6}\frac{1}{N}\sum_{i,j,k} \left[\langle N_i N_j N_k \rangle_0 - 3\langle N_i N_j \rangle_0 \langle N_k \rangle_0 + 2\langle N_i \rangle_0 \langle N_j \rangle_0 \langle N_k \rangle_0 \right]$$

$$\times u_1^*(r_i) \, u_1^*(r_j) \, u_1^*(r_k), \tag{6.31}$$

$$\frac{F_4}{Nk_BT} = -\frac{1}{24}\frac{1}{N}\sum_{i,j,k,l} \left[\langle N_i N_j N_k N_l \rangle_0 - 4\langle N_i N_j N_k \rangle_0 \langle N_l \rangle_0 \right.$$

$$- 3\langle N_i N_j \rangle_0 \langle N_k N_l \rangle_0 + 12\langle N_i N_j \rangle_0 \langle N_k \rangle_0 \langle N_l \rangle_0$$

$$\left. -6\langle N_i \rangle_0 \langle N_j \rangle_0 \langle N_k \rangle_0 \langle N_k \rangle_0 \right] u_1^*(r_i) \, u_1^*(r_j) \, u_1^*(r_k) \, u_1^*(r_l), \tag{6.32}$$

and so on, where N_i is the number of intermolecular distances in the range (r_i, r_{i+1}), with $\Delta r = r_{i+1} - r_i \ll \sigma$, $i = 0, 1, \ldots$, angular brackets mean averages, subscript 0 indicates that the averages are performed in the reference system, and $u_1^*(r) = u_1(r)/\varepsilon$.

In a similar way can be obtained the perturbation terms in Equation 6.20 for the RDF (Smith et al. 1971, Barker and Henderson 1976). The zero- and first-order terms are

$$g_0 \left(\frac{r_i + \Delta r}{2} \right) = \frac{3 \langle N_i \rangle_0}{2\pi N \rho \left(r_{i+1}^3 - r_i^3 \right)}, \tag{6.33}$$

$$g_1 \left(\frac{r_i + \Delta r}{2} \right) = -\frac{3 \sum_j \left[\langle N_i N_j \rangle_0 - \langle N_i \rangle_0 \langle N_j \rangle_0 \right] u_1^* \left(r_j \right)}{2\pi N \rho \left(r_{i+1}^3 - r_i^3 \right)}. \tag{6.34}$$

For the particular case of the SW potential, $u_i^*(r) = -1$ within the well $0 < x \le \lambda$, where $x = r/\sigma$, and the expressions for the terms F_n simplify to (Alder et al. 1972)

$$\frac{F_1}{N k_B T} = -\frac{\langle M \rangle_0}{N}, \tag{6.35}$$

$$\frac{F_2}{N k_B T} = -\frac{1}{2} \frac{1}{N} \left\langle (M - \langle M \rangle_0)^2 \right\rangle_0, \tag{6.36}$$

$$\frac{F_3}{N k_B T} = -\frac{1}{6} \frac{1}{N} \left\langle (M - \langle M \rangle_0)^3 \right\rangle_0, \tag{6.37}$$

$$\frac{F_4}{N k_B T} = -\frac{1}{24} \frac{1}{N} \left[\left\langle (M - \langle M \rangle_0)^4 \right\rangle_0 - 3 \left\langle (M - \langle M \rangle_0)^2 \right\rangle_0^2 \right], \tag{6.38}$$

and so on, where M is the number of pairs separated a distance $x \le \lambda$ of each other. Expressions (6.29) through (6.38) can be evaluated from computer simulation in the reference HS system. Such calculations have been carried out by a number of authors mainly for the SW fluid (Barker and Henderson 1968, Smith et al. 1971, Alder et al. 1972, Largo and Solana 2003a,b, 2004, Espíndola-Heredia et al. 2009), but also for fluids and solids with other hard-core potential models (Díez et al. 2006, 2007, Betancourt-Cárdenas et al. 2007, Solana 2008, Zhou and Solana 2009).

For potential models with continuous tails, in practice, the simulations are limited to the first- and second-order terms in expansion (Equation 6.15), using Equations 6.29 and 6.30 respectively, as the calculation of the third-order term from computer simulation, Equation 6.31, is an extremely challenging problem and the fourth-order term, Equation 6.32, is virtually unfeasible. More affordable are these simulations for the SW fluid and so the simulation data for F_3 and F_4, using Equations 6.37 and 6.38 respectively, were reported by Alder et al. (1972) for $\lambda = 1.5$ and recently by Espíndola-Heredia et al. (2009) for $1.1 \le \lambda \le 3$. These simulations reveal that the second-order term is typically one order of magnitude smaller than the first-order term. The behavior of the third- and fourth-order terms, only known at present for the SW fluid, is more complex: they are the same order of magnitude than the second-order term at low densities, and even larger for large potential widths, but become negligible at high densities. For both continuous and discrete potential models, the calculations by computer simulations of the perturbation terms of the RDF are limited in practice to $n \le 1$, which, through the pressure equation (2.36), allows us to obtain the perturbation terms Z_0 and Z_1 of the EOS.

It is worth discussing here the extent to which an "exact" perturbation expansion, that is Equation 6.15 with the perturbation terms determined from computer simulation, truncated at $n = 2$ for the energy and at $n = 1$ for the EOS, gives reliable results for these properties by comparing with the simulation data. This sort of comparison has been carried out for SW fluids (Largo and Solana 2004) and for fluids (Díez et al. 2006) and solids (Díez et al. 2007) with the Sutherland potential. The results show that for these systems this approach is accurate enough for supercritical temperatures and medium to large potential ranges (say $\lambda = 1.5$ or larger for the SW potential and $n = 6$ or lower for the Sutherland potential), but becomes insufficient for shorter-ranged potentials at low temperatures and the accuracy worsens as the potential range and temperature decrease. It is to be noted that for low enough temperatures the convergence of the series (Equation 6.13) cannot be guaranteed. This behavior is illustrated in Figure 6.1 for two examples of SW fluids.

6.3 PERTURBATION THEORIES FOR HARD-CORE POTENTIALS

6.3.1 FIRST-ORDER PERTURBATION THEORY

Let us consider expansion Equation 6.13 or, equivalently Equation 6.15, truncated at $n = 1$, that is,

$$\frac{F}{Nk_BT} = \frac{F_0}{Nk_BT} + \frac{\langle U_1 \rangle_0}{Nk_BT} = \frac{F_0}{Nk_BT} + \frac{F_1}{Nk_BT}\frac{1}{T^*}$$

$$= \frac{F_0}{Nk_BT} + \frac{2\pi}{k_BT}\rho \int_0^\infty u_1(r)\, g_0(r)\, r^2 dr. \tag{6.39}$$

The last term is the so-called *high temperature approximation* (HTA), that is,

$$\frac{F_{HTA}}{Nk_BT} = \frac{2\pi}{k_BT}\rho \int_0^\infty u_1(r)\, g_0(r)\, r^2 dr. \tag{6.40}$$

Calculation of the free energy of a fluid within the first-order approximation requires only the EOS and the radial distribution function of the reference fluid. The free energy F_0 of the reference fluid is related to the corresponding EOS through the exact thermodynamic relationship

$$\frac{F_0}{Nk_BT} = \frac{F_{id}}{Nk_BT} + \int_0^\rho (Z_0 - 1)\frac{d\rho'}{\rho'}, \tag{6.41}$$

where

$$\frac{F_{id}}{Nk_BT} = 3\ln\Lambda + \ln\rho - 1 \tag{6.42}$$

is the free energy of the ideal gas.

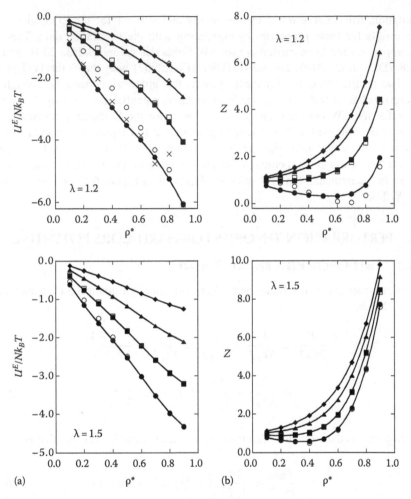

FIGURE 6.1 Excess energy (a) and EOS (b) for two examples of SW fluids from simulation-based perturbation theory. Filled circles, squares, triangles, and diamonds are, respectively, the simulation data from Largo and Solana (2003c) for $T^* = 0.7$, 1.0, 1.5, and 2.0 in the case of $\lambda = 1.2$, and for $T^* = 1.5$, 2.0, 3.0, and 5.0 in the case of $\lambda = 1.5$. Open symbols are the results from perturbation theory truncated at $n = 2$ in the excess energy, Equation 6.16, and at $n = 1$ in the EOS, Equation 6.17. Crosses are the results from perturbation theory truncated at $n = 4$. The perturbation terms for the excess energy were obtained from the simulation data of Espíndola-Heredia et al. (2009) for $F_1 - F_4$ using Equation 6.18. The perturbation terms for the EOS are from Largo and Solana (2004). The curves are guides for the eye.

For potentials with a spherical hard core, the natural reference system is the hard-sphere one. For the EOS of the HS fluid, we can use the Carnahan–Starling equation (5.21) and for the corresponding RDF either the RFA or the FMSA approximations analyzed in Sections 5.4 and 5.5, respectively. The values of F_1 predicted in this way are in excellent agreement with the simulation data (Largo and Solana 2003a,

Díez et al. 2006). The excess energy and the EOS in the first-order approximation can be obtained from F_1 using the relationships (Equations 6.18 and 6.19), together with Equations 6.16 and 6.17 truncated at $n = 1$.

The same approximation can be applied to crystalline solids. For the EOS and the averaged RDF, we can take those described in Section 5.6. In this case, however, the free energy F_0 of the HS crystalline solid cannot be obtained from Equation 6.41, as there is not a continuous path from the ideal gas to the crystalline solid. It is more advisable to choose as the reference state the crystalline solid at close packing density instead of the ideal gas. Taking for the solid an EOS of the form of Equation 5.95, the resulting free energy is (Alder et al. 1968)

$$\frac{F_0}{Nk_BT} = 3\ln\Lambda + \ln\rho_0 + 3\ln\left(\frac{3}{2\alpha}\right) - \left(S_0 + S_1\alpha + S_2\alpha^2 + \cdots\right), \qquad (6.43)$$

where $S_0 = -0.24$ is a correction constant determined from computer simulation and

$$S_1 = C_0 - 3, \quad S_2 = \frac{1}{2}(C_1 - C_0 + 3), \quad S_3 = \frac{1}{3}(C_2 - C_1 + C_0 - 3), \ldots \quad (6.44)$$

6.3.2 VAN DER WAALS AND RELATED APPROXIMATIONS

6.3.2.1 van der Waals Equation

Let us return to Equation 6.39. As a crude approximation we can take $g_0(r) = 1$, the ideal gas value, and denote

$$a = -2\pi \int_0^\infty u_1(r) r^2 dr. \qquad (6.45)$$

Then, from Equation 6.39, we have

$$\frac{F}{Nk_BT} = \frac{F_0}{Nk_BT} - \frac{a}{k_BT}\rho, \qquad (6.46)$$

which is the *van der Waals* (vdW) *mean field approximation* (MFA).

For the configurational partition function of the reference HS fluid, we can take a free-volume approach, Equation 5.91 divided by $N!$ to correct for the indistinguishability of the particles in the fluid. An expression of the free volume per particle suitable for low densities is

$$v_f = \frac{V}{N} - b, \qquad (6.47)$$

where $b = (2/3)\pi\sigma^3$ is the *covolume* per particle, that is, half the volume around any of the particles of the fluid, with diameter σ, from which is excluded the center of any other particle. Putting all together, we readily obtain

$$\frac{F}{Nk_BT} = 3\ln\Lambda - \ln\left(\frac{1}{\rho} - b\right) + \frac{1}{N}\ln N! - \frac{a}{k_BT}\rho, \qquad (6.48)$$

which leads to the well-known vdW EOS

$$P = \frac{Nk_BT}{V - Nb} - a\frac{N^2}{V^2}.$$ (6.49)

The vdW approximation is reasonable at very low densities. For higher densities, parameter a cannot be considered independent of density nor the excluded volume per particle is given by the simple form taken for b.

6.3.2.2 Haar–Shenker–Kohler Approximation

A better approximation than the vdW one for low densities consists in replacing a, as given by Equation 6.45, with

$$a = k_BT\left(B_2 - B_2^{(0)}\right),$$ (6.50)

where B_2 and $B_2^{(0)}$ are the second virial coefficients of the actual and reference systems, respectively. Then, the expression of the free energy reads (Haar and Shenker 1971, Kohler and Haar 1981)

$$\frac{F}{Nk_BT} = \frac{F_0}{Nk_BT} + \left(B_2 - B_2^{(0)}\right)\rho.$$ (6.51)

This guaranties the right low-density behavior, up to the level of the second virial coefficient, and yields a better performance than the vdW approximation at higher densities, provided that we take a better choice for F_0 than that used in the vdW approximation. Note that, in spite of the resemblance with Equations 6.39 and 6.48, the Haar–Shenker–Kohler expression is a different class of perturbation approximation because the last term in Equation 6.51 is not first order in $1/T$ in general, as B_2 may have a more or less complex temperature dependence.

The Haar–Shenker–Kohler approximation is frequently used for the calculation of the thermodynamic properties of gases at relatively low densities and, in particular, for the vapor properties in liquid–vapor equilibrium calculations at temperatures well below critical, eventually in combination with simulations for higher densities and liquid phase properties.

6.3.2.3 Generalized van der Waals Theory

For $\beta = 0$, or equivalently $T = \infty$, a fluid with a spherical hard-core potential behaves as a HS fluid. The relationship between the excess energy U^E and the configurational integral

$$U^E = k_BT^2\left(\frac{\partial \ln Q_c}{\partial T}\right)_{N,V} = -\left(\frac{\partial \ln Q_c}{\partial \beta}\right)_{N,V},$$ (6.52)

upon integration between $\beta = 0$ and β, yields

$$\ln Q_c = \ln Q_c^{(0)} - \int_0^\beta U^E d\beta' = \ln Q_c^{(0)} - 2\pi N\rho \int_0^\beta d\beta' \int_0^\infty u_1(r)\, g(r, \beta')\, r^2 dr,$$ (6.53)

where we have made use of the energy equation (2.46). Denoting

$$\Delta F = \frac{1}{\beta} \int_0^\beta U^E d\beta' = 2\pi N k_B T \rho \int_0^\beta d\beta' \int_0^\infty u_1(r) g(r, \beta') r^2 dr, \qquad (6.54)$$

we have

$$\frac{F}{N k_B T} = \frac{F_0}{N k_B T} + \frac{\Delta F}{N k_B T}, \qquad (6.55)$$

which is the expression of the free energy in the *generalized van der Waals* (GvdW) *theory* for simple fluids (Vera and Prausnitz 1972, Sandler 1985).

For the particular case of the SW fluid

$$U^E = 2\pi N \rho \int_0^\infty u_1(r) g(r, \beta) r^2 dr = -2\pi N \varepsilon \rho \int_\sigma^{\lambda\sigma} g(r, \beta) r^2 dr = -\frac{1}{2} N \varepsilon N_c(\rho, T), \qquad (6.56)$$

where $N_c(\rho, T)$ is the *coordination number*, that is, the average number of particles that interact with every particle in the fluid. In the low-density limit $g(r, \beta) \rightarrow e^{-\beta u(r)}$ and

$$N_c(\rho, T) = \frac{4}{3}\pi\rho^* (\lambda^3 - 1) e^{\varepsilon/k_B T}, \qquad \rho \rightarrow 0. \qquad (6.57)$$

With this expression for $N_c(\rho, T)$, the GvdW theory is exact up to the level of the second virial coefficient. For higher densities, several *coordination number models* for $N_c(\rho, T)$ have been derived (Lee et al. 1985, 1986, Heyes 1991, Vimalchand et al. 1992, Largo and Solana 2002), mainly based on lattice-gas theory, as well as a number of semi-empirical approximations. However, most of the existing coordination number models either are inaccurate, or predict the wrong low-density behavior, or introduce some kind of adjustable parameter. Therefore, there is room for further theoretical developments in this field.

The GvdW theory can be extended to fluids with spherical hard-core potentials other than the SW potential by defining an averaged perturbation potential $\langle u_1(r) \rangle$ (Sandler 1985) as

$$U^E = \frac{1}{2}N \langle u_1(r) \rangle N_c(\rho, T), \qquad (6.58)$$

where $N_c(\rho, T)$ is the coordination number for the SW fluid with $\lambda = R$, an effective range for the actual potential, and

$$\langle u_1(r) \rangle = \frac{\int_\sigma^{R\sigma} u_1(r) g(r, \beta) r^2 dr}{\int_\sigma^{R\sigma} g(r, \beta) r^2 dr}. \qquad (6.59)$$

6.3.3 BARKER–HENDERSON SECOND-ORDER PERTURBATION THEORY

6.3.3.1 Barker–Henderson Macroscopic and Local Approximations

To obtain an approximate expression for the second-order term, Equation 6.30, in the expansion of the free energy, Barker and Henderson (1967a) divide the space around any particle taken as a reference into concentric shells, with shell i containing the centers of N_i particles. For large macroscopic shells, the numbers N_i and N_j would be uncorrelated for $i \neq j$, and so the term between brackets in Equation 6.30 would be zero, whereas for $i = j$, from Equation 2.53, we would have

$$\langle N_i^2 \rangle_0 - \langle N_i \rangle_0^2 = \langle N_i \rangle_0 \, \rho k_B T \kappa_T^{(0)}, \tag{6.60}$$

where $\kappa_T^{(0)} = (1/\rho)(\partial \rho / \partial P)_0$ is the macroscopic isothermal compressibility of the reference system. The average involved in the preceding equation can be performed by means of the RDF of the reference fluid, with the result

$$\frac{F_2}{N k_B T} = -\pi k_B T \rho \left(\frac{\partial \rho}{\partial P} \right)_0 \int_0^\infty [u_1(r)]^2 \, g_0(r) \, r^2 dr, \tag{6.61}$$

which is the expression for the second-order perturbative term in the *macroscopic compressibility* (mc) *approximation*. Equation 6.61 gives the correct second-order contribution to the second virial coefficient. However, instead of the mc, it seems more appropriate to consider the local compressibility, that is, $(\partial[\rho g_0(r)]/\partial P)_0$, instead of $(\partial \rho / \partial P)_0$. This leads to the *local compressibility* (lc) *approximation*

$$\frac{F_2}{N k_B T} = -\pi k_B T \rho \int_0^\infty [u_1(r)]^2 \left(\frac{\partial [\rho g_0(r)]}{\partial P} \right)_0 r^2 dr$$

$$= -\pi k_B T \rho \left(\frac{\partial \rho}{\partial P} \right)_0 \int_0^\infty [u_1(r)]^2 \left(\frac{\partial [\rho g_0(r)]}{\partial \rho} \right)_0 r^2 dr. \tag{6.62}$$

The comparison of the results for $F_2/N k_B T$ obtained from the *Barker–Henderson* (BH) mc and lc approximations for SW fluids with $1.1 \leq \lambda \leq 2.0$ with the simulation data reveals that at low densities both approximations underestimate, agree, and overestimate that quantity for low, intermediate, and large values of λ (Largo and Solana 2003a). At high densities, both approximations give too low absolute values. This behavior is illustrated in Figure 6.2 for three different values of λ.

6.3.4 MORE ADVANCED APPROXIMATION FOR THE SECOND-ORDER TERM

A better approximation, though more involved, for the second-order term was derived by Smith et al. (1970). Starting from Equation 6.23 and using the Kirkwood superposition approximation (Kirkwood 1935),

$$g_0(\mathbf{r}_1, \mathbf{r}_2, \mathbf{r}_3) = g_0(\mathbf{r}_1, \mathbf{r}_2) \, g_0(\mathbf{r}_1, \mathbf{r}_3) \, g_0(\mathbf{r}_2, \mathbf{r}_3), \tag{6.63}$$

$$g_0(\mathbf{r}_1, \mathbf{r}_2, \mathbf{r}_3, \mathbf{r}_4) = g_0(\mathbf{r}_1, \mathbf{r}_2) \, g_0(\mathbf{r}_1, \mathbf{r}_3) \, g_0(\mathbf{r}_1, \mathbf{r}_4) \, g_0(\mathbf{r}_2, \mathbf{r}_3) \, g_0(\mathbf{r}_2, \mathbf{r}_4) \, g_0(\mathbf{r}_3, \mathbf{r}_4), \tag{6.64}$$

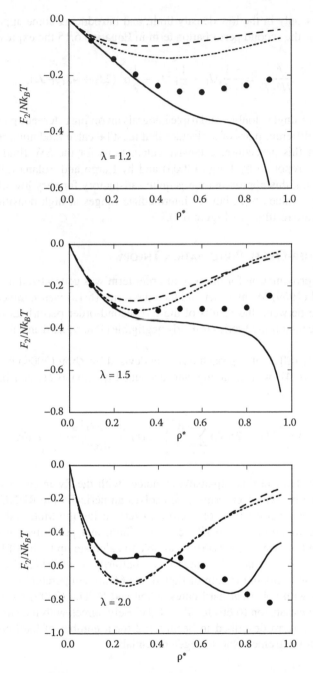

FIGURE 6.2 Second-order perturbation term in expansion (6.15) for SW fluids. Points: simulation data from Espíndola-Heredia et al. 2009. Dashed curve: mc approximation, Equation 6.61. Dotted curve: lc approximation, Equation 6.62. Continuous curve: approximation (6.65). For all theoretical calculations the RFA approximation (see Section 5.4) has been used for $g_0(r)$.

which is exact only in the low density limit, and introducing some approximations, they obtain for the second perturbation term in Equation 6.15 the expression

$$\frac{F_2}{Nk_BT} = -\frac{1}{4}\rho J_1 - \frac{1}{2}\rho^2 J_2 - \frac{1}{8}\rho^3 (2J_3 + 4J_4 + J_5), \qquad (6.65)$$

where the J_i are single, double, and triple integrals involving reference pair correlation functions for different pairs of molecules that must be calculated numerically.

Following this procedure, extensive calculations for the SW fluid with $1.1 \leq \lambda \leq 2.0$ were reported by Largo (2003) and by Largo and Solana (2003a). Their results show that this approximation is quite satisfactory for very low as well as for relatively large values of λ, but for intermediate ranges at high densities it fails to provide accurate results (see Figure 6.2).

6.3.5 HIGHER-ORDER PERTURBATION THEORY

The BH lc approximation for the second-order term was generalized by Praestgaard and Toxvaerd (1969) to any order and then the series (6.15) was summed. However, the difference between the results of the BH second-order perturbation theory and those from the re-summed series is nearly negligible (Praestgaard and Toxvaerd 1969, Largo 2003).

An entirely different approach has been devised by Zhou (2006a,b) on the basis of expansion (6.4), which, taking into account Equation 6.21, can be written in the form

$$\beta F(\alpha) = \beta F_0 + 2\pi N\rho \sum_{n=1}^{\infty} \frac{\beta}{n!} \int_0^{\infty} u_1(r) \left.\frac{\partial^{n-1} g(r; \alpha)}{\partial \alpha^{n-1}}\right|_{\alpha=0} r^2 dr. \qquad (6.66)$$

The first-order term in this expansion coincides with the Zwanzig result, Equation 6.22. To obtain higher-order terms, Zhou solves numerically the RHNC theory (see Section 4.5) for several values of α close to zero, using the Malijevský and Labík (1987) bridge function for the HS reference fluid, and performs numerically the derivatives. In this way, Zhou (2008) was able to calculate up to the fifth-order term in expansion (6.66). It is to be noted that, contrarily to the HTE, Equation 6.13, the perturbation terms in the sum of Equation 6.66 are temperature-dependent even for potentials with spherical hard cores. Zhou and Solana (2009) have shown that the term F_2 in expansion (6.66) for $T^* = 1.0$ closely agrees with the simulation data obtained in the form described in Section 6.2 for a number of hard-core potential models even for extremely short-ranged potentials.

6.4 PERTURBATION THEORIES FOR SOFT-CORE POTENTIALS

The theories discussed in Section 6.3 can be readily extended to systems with soft-core potentials provided that we split the potential into appropriate short-range repulsive and long-range, in most cases attractive, contributions and define a temperature-dependent, and eventually also density-dependent, effective hard-sphere diameter for the particles. This again allows us to use the HS system as the reference and proceed as for fluids with spherical hard-core potentials. Several possible criteria to determine the effective diameter were analyzed by Hess et al. (1998) and by Raineri et al. (2004). Other theories have been specifically developed for soft-core potentials. In this section, we will describe several approaches of both kinds.

6.4.1 BARKER–HENDERSON PERTURBATION THEORY FOR SOFT-CORE POTENTIALS

Barker and Henderson (1967b, 1976) split the soft-core potential in the form of Equation 4.40 with

$$
\begin{aligned}
u_0(r) &= \begin{cases} u(r), & r \le \sigma \\ 0, & r > \sigma \end{cases} \\
u_1(r) &= \begin{cases} 0, & r \le r_m \\ u(r), & r > \sigma \end{cases}
\end{aligned}
\tag{6.67}
$$

where σ is usually the distance such that $u(\sigma) = 0$. Then, the thermodynamic properties of the system can be determined from a second-order perturbation theory, using Equation 6.15 truncated at $n = 2$ together with Equation 6.22 for the first-order term and either Equation 6.61 or Equation 6.62 for the second-order term, provided that we know the radial distribution function and the EOS of the reference system. The problem is that these properties are not known, in general, for a system with the potential $u_0(r)$ of Equation 6.67. To overcome this drawback, Barker and Henderson approximate the RDF and the EOS by those of an HS fluid with a temperature-dependent effective diameter d given by

$$
d = \int_0^\sigma \left[1 - e^{-\beta u(r)}\right] dr.
\tag{6.68}
$$

For several potential models, this equation leads to analytic expressions for the effective diameter (see Heyes 1997 for a description of a number of these potentials).

The BH theory with d determined from Equation 6.68 gives poor results at high temperatures and very high densities. To improve the results for these states, Henderson and Barker (1970) introduced a temperature- and density-dependent effective diameter by replacing the upper limit of integration in Equation 6.68 with $\sigma_0(\rho, T)$. The latter was determined from the condition that the free energy is insensitive to small changes in σ_0, which does not necessarily mean to minimize the free energy. There may be several values of σ_0 for which F is an extreme for fixed density and temperature. The lowest value for which this occurs is the appropriate choice.

6.4.2 WEEKS–CHANDLER–ANDERSEN PERTURBATION THEORY

In the *Weeks–Chandler–Andersen* (WCA) theory (Chandler and Weeks 1970, Andersen et al. 1971, Weeks et al. 1971), the reference and perturbation parts of the potential are

$$u_0(r) = \begin{cases} u(r) - u(r_m), & r \leq r_m \\ 0, & r \geq r_m \end{cases}$$

$$u_1(r) = \begin{cases} u(r_m), & r \leq r_m \\ u(r), & r \geq r_m \end{cases}$$

(6.69)

where r_m is the position of the minimum of the potential. In this way, the reference includes all the contribution to the repulsive forces. Then, introducing the cavity distribution function $y_0(r)$ for the reference fluid

$$y_0(r) = g_0(r) e^{\beta u_0(r)},$$

(6.70)

for strongly repulsive reference potentials it is reasonable to approximate $y_0(r)$ by the cavity function of a fluid of HS with an appropriate diameter d, so that

$$g_0(r) \simeq y_{HS}(r,d) e^{-\beta u_0(r)}.$$

(6.71)

The condition used to determine d is

$$\int \left[y_{HS}(r,d) e^{-\beta u_0(r)} - 1 \right] d\mathbf{r} = \int \left[y_{HS}(r;d) e^{-\beta u_{HS}(r;d)} - 1 \right] d\mathbf{r},$$

(6.72)

or, equivalently,

$$\int y_{HS}(r;d) \left[e^{-\beta u_0(r)} - e^{-\beta u_{HS}(r;d)} \right] d\mathbf{r} = 0.$$

(6.73)

The integrand in the latter equation was called by Andersen et al. (1971) the *blip function*. According to the compressibility equation 2.54, Equation 6.72 states that the compressibility of the reference fluid and that of the equivalent HS fluid are equal or, from the viewpoint of Equation 2.63, for $k \to 0$ the static structure factor is equal for these two fluids. The effective diameter d can be obtained from Equation 6.72 by an iterative procedure. An essentially analytical expression for d, derived from Equation 6.72, is also available (Verlet and Weis 1972).

The free energy F_0 of the reference system can be expanded in power series of the Bolzmann factor $\exp[-\beta u_0(r)]$ around $\exp[-\beta u_{HS}(r;d)]$, the *blip function expansion*. To first order reads

$$\frac{F_0}{Nk_BT} = \frac{F_{HS}}{Nk_BT} - \frac{1}{2}\rho \int y_{HS}(r;d) \left[e^{-\beta u_0(r)} - e^{-\beta u_{HS}(r;d)} \right] d\mathbf{r} + \cdots.$$

(6.74)

The higher-order terms in the series involve an increasing number of products of blip functions with increasing the order of the term in the expansion. Andersen et al. (1971)

showed that for steeply repulsive reference potentials the terms beyond the first-order term, which is proportional to the integral in the left-hand side of Equation 6.73, are negligible. Therefore, with d determined from condition (6.73), the free energy of the reference system reduces to that of a system of HS of diameter d. Introducing the reference RDF, Equation 6.71, together with the corresponding free energy $F_0 = F_{HS}(d)$, into Equation 6.39, we obtain a first-order perturbation theory within the WCA approach.

In the WCA theory, the RDF of the actual fluid is approximated by that of the reference fluid. The reliability of this approach was confirmed for the LJ fluid at moderate to high densities by Rull et al. (1984) and by Cuadros et al. (1996) by means of computer simulation. These authors also showed that the thermodynamic properties of the LJ fluid are essentially linearly dependent on the parameter α coupling the WCA reference and perturbation parts of the potential. This means that, at least for the LJ potential, with the splitting of Equation 6.69, the contribution from the higher-order derivatives in expansion (6.4) must be negligible, which provides support to the first-order WCA perturbation theory.

At high densities, close to the freezing line, the diameters resulting from Equation 6.72 may be so large that the reduced density $\rho^* = \rho d^3$ is beyond the range of existence of the fluid phase of the HS system, and so there will not be a way of obtaining a reliable $y_{HS}(r)$. To overcome this drawback, Ree (1976) proposed to replace r_m in Equation 6.69 with a temperature- and density-dependent splitting distance $r_0(\rho, T)$, determined from the already mentioned Henderson and Barker (1970) criterion of insensitivity of the Helmholtz free energy with respect to small changes in r_0.

A different scheme with a similar purpose was devised by Kang et al. (1985). The reference and perturbation parts of the potential now are taken to be

$$u_0(r) = \begin{cases} u(r) - w(r), & r \le \sigma_0 \\ 0, & r > \sigma_0 \end{cases}$$

$$u_1(r) = \begin{cases} w(r), & r \le \sigma_0 \\ u(r), & r > \sigma_0 \end{cases}$$

$$\qquad (6.75)$$

with $\sigma_0 = \min(r_1, r_m)$, where r_1 is the nearest-neighbor distance for the lattice structure corresponding to the system considered. The choice for $w(r)$ is arbitrary; a simple form is

$$w(r) = u(\sigma_0) - u'(\sigma_0)(\sigma_0 - r), \qquad (6.76)$$

where $u'(\sigma_0) = du(r)/dr|_{r=\sigma_0}$.

The WCA theory, like other perturbation theories, suffers from inconsistency between the pressure obtained from the free energy and that resulting from the pressure equation (2.36). A way to determine d, other than Equation 6.72, in such a way to achieve consistency between these two routes has been developed by Lado (1984). The condition is now

$$\int \left[e^{-\beta u(r)} - e^{-\beta u_{HS}(r;d)} \right] \frac{\partial y_{HS}(r;d)}{\partial d} d\mathbf{r} = 0. \qquad (6.77)$$

The conditions for an arbitrary reference potential $u_0(r)$ depending on two parameters σ_0 and ε_0 are

$$\int \left[e^{-\beta u(r)} - e^{-\beta u_0(r)} \right] \frac{\partial y_0(r)}{\partial \sigma_0} d\mathbf{r} = 0 \qquad (6.78)$$

$$\int \left[e^{-\beta u(r)} - e^{-\beta u_0(r)} \right] \frac{\partial y_0(r)}{\partial \varepsilon_0} d\mathbf{r} = 0. \qquad (6.79)$$

A variant of the WCA theory, denoted HS-WCA, has been developed by Ben-Amotz and Stell (2004a) by replacing $g_0(r)$, the reference RDF, in the first-order term F_1 of the free energy with $g_{HS}(r)$, the HS RDF, and removing the WCA criterion (Equation 6.72) for determining d. Formulated in this way, the theory is less sensitive to the precise value of d, which allows for more flexibility in the choice of the criterion used to determine it, and so eventually may be only temperature-dependent and more easily determined than from Equation 6.72.

6.4.3 SONG–MASON APPROXIMATION

Song and Mason (1989) developed an analytical EOS based on the same splitting of the potential as in the WCA theory. Starting from the virial equation (2.36), it is easy to see that it can be rewritten in the form

$$Z = 1 + B_2 \rho + I\rho, \qquad (6.80)$$

in which B_2 is the second virial coefficient, and

$$I = \frac{2\pi}{3} \int_0^\infty f'(r) \left[y(r) - 1 \right] r^3 dr, \qquad (6.81)$$

where $y(r)$ is the cavity function, $f'(r) = df(r)/dr$, and $f(r)$ is the Mayer function.

For realistic soft-core potential models, the function $f'(r)$ exhibits a marked peak for some value of $r < r_m$ and its absolute value is quite small for $r > r_m$ and decays to zero for large distances. Moreover, $y(r)$ presents an oscillatory behavior around 1 for $r > r_m$. Therefore, the integrand in Equation 6.83 will be quite small for $r > r_m$ and so we can replace the upper limit of integration with r_m. In addition, in the region $0 \leq r \leq r_m$, the cavity functions $y(r)$ and $y_0(r)$ of the actual and reference systems, respectively, are quite similar so that the former can be replaced by the latter. Also, within the same region, according to Equation 6.69, we have

$$f'(r) = -\beta \frac{du(r)}{dr} e^{-\beta u(r)} = -\beta \frac{du_0(r)}{dr} e^{-\beta [u_0(r) + u(r_m)]} = f_0'(r) \left[1 - \beta u(r_m) + \cdots \right].$$
$$(6.82)$$

Putting all together within Equation 6.81, we obtain

$$I \approx \frac{2\pi}{3} \int_0^{r_m} f_0'(r) \left[1 - \beta u(r_m) \right] \left[y_0(r) - 1 \right] r^3 dr. \qquad (6.83)$$

Because of the strongly peaked shape of $f'(r)$ mentioned earlier, a region close to its maximum will give the main contribution to I, so that we are interested only in the values of $y_0(r)$ within this region. Denoting r_0 to the value of r for which the integrand in Equation 6.83 reaches its maximum value, we can expand $y_0(r)$ around this point, with the result

$$y_0(r) = y_0(r_0) + \frac{dy_0(r)}{dr}\bigg|_{r=r_0}(r - r_0) + \cdots, \qquad (6.84)$$

and so

$$I \approx \frac{2\pi}{3}[y_0(r_0) - 1]\left[\int_0^{r_m} f_0'(r)\, r^3 dr - \beta u(r_m)\int_0^{r_m} f_0'(r)\, r^3 dr\right.$$

$$\left. + \frac{dy_0(r)}{dr}\bigg|_{r=r_0}\int_0^{r_m} f_0'(r)(r - r_0)\, r^3 dr + \cdots\right]. \qquad (6.85)$$

The last two terms in this expression are negligible at moderate to high temperatures and largely cancel out with each other at low temperatures. Furthermore, within the spirit of the WCA theory, we can approximate $y_0(r_0)$ by $y_{HS}(d)$, where d is an effective HS diameter still undetermined, so that, as a good approximation

$$I \approx \frac{2\pi}{3}[y_{HS}(d) - 1]\int_0^{r_m} f_0'(r)\, r^3 dr = -2\pi[y_{HS}(d) - 1]\int_0^{r_m} f_0(r)\, r^2 dr, \qquad (6.86)$$

and the EOS (6.80) reduces to

$$Z = 1 + B_2\rho + \omega\rho\,[g_{HS}(d) - 1], \qquad (6.87)$$

where we have taken into account that $y_{HS}(d) = g_{HS}(d)$ and $\omega = -2\pi\int_0^{r_m} f_0(r)r^2 dr$.

The contact value $g_{HS}(d)$ of the RDF of a fluid of HS of diameter d can be obtained from any suitable EOS, such as the Carnahan–Starling equation (5.21), using the relationship (2.43). To this end, we must determine the effective diameter d or, equivalently, a closely related quantity: the covolume $b(T) = (2/3)\pi d^3$. For the latter, Song and Mason (1989) use the ansatz

$$b(T) = 2\pi\int_0^{r_m}\left\{1 - [1 + \beta u_0(r)]\, e^{-\beta u_0(r)}\right\}r^2 dr, \qquad (6.88)$$

which behaves properly at both low and high temperatures. A useful consequence of this choice is that

$$b = \omega + T\frac{d\omega}{dT}. \qquad (6.89)$$

Moreover, at high temperatures, where the repulsive forces are dominant, $\omega \to B_2$.

Although derived in a quite heuristic way, the Song–Mason approximation is appealing, because it only requires the knowledge of the interaction potential or, for real fluids, the second virial coefficient that can be experimentally measured. It is easy to see that for potentials with a spherical hard core the Song–Mason and the Haar–Shenker–Kohler approximations are equivalent.

6.4.4 VARIATIONAL PERTURBATION THEORY

The variational perturbation theory (Mansoori and Canfield 1969, 1970, Rasaiah and Stell 1970) is a first-order perturbation theory based on the inequality

$$\beta F \leq \beta F_0 + \beta \langle \mathcal{U}_1 \rangle_0 = \beta F_0 + 2\pi \beta N \rho \int_0^\infty u_1(r) g_0(r) r^2 dr, \qquad (6.90)$$

which can be derived from Equation 6.13 and is known as the *Gibbs–Bogoliubov inequality* (Isihara 1968). This suggests that we could bring the inequality in Equation 6.90 close to equality if we determine the potential parameters of the reference system by minimizing the right-hand side of this equation.

6.5 MODE EXPANSION

For a system with a potential of the form of Equation 4.40 taking into account Equation 6.1, it is easy to see that the configurational integral can be written in the form

$$Q_c = Q_c^{(0)} \langle e^{-\beta \mathcal{U}_1} \rangle_0, \qquad (6.91)$$

where $Q_c^{(0)}$ is the configurational integral of the reference system with pair potential $u_0(r)$ and total potential energy $\mathcal{U}_0$.

On the other hand, assuming that the perturbing pair potential has a Fourier transform, that the system consists of N particles in a cubic box with volume V, and considering periodic boundary conditions, Andersen and Chandler (1970) write

$$\mathcal{U}_1 = \frac{1}{2V} \sum_{\mathbf{k}} \hat{u}_1(k) \left[\hat{\rho}(\mathbf{k}) \hat{\rho}(-\mathbf{k}) - N \right]$$

$$= \hat{u}_1(0) \frac{N(N-1)}{2V} + \frac{1}{V} \sum_{\mathbf{k}}' \hat{u}_1(k) \left[\hat{\rho}(\mathbf{k}) \hat{\rho}(-\mathbf{k}) - N \right], \qquad (6.92)$$

where
  $\hat{u}_1(k)$ is the Fourier transform of $u_1(r)$
  $\hat{\rho}(\mathbf{k})$ is the Fourier transform of the single-particle density for a given configuration
  the primed sum denotes that the sum extends over half the **k**-space excluding
    $\mathbf{k} = 0$

Inserting this result into Equation 6.91 leads to

$$Q_c = Q_c^{(0)} \exp\left[-\frac{1}{2}\beta N\rho\left(1 - N^{-1}\right)\hat{u}_1\left(0\right)\right]\left\{\prod_{\mathbf{k}}{}' \exp\left[\beta\rho\hat{u}_1\left(k\right)\right]\right\}\left\langle\prod_{\mathbf{k}}{}' s\left(\mathbf{k}\right)\right\rangle_0,$$
(6.93)

where

$$s\left(\mathbf{k}\right) = \exp\left\{-\left[\frac{\beta\hat{u}_1\left(k\right)}{V}\right]\hat{\rho}\left(\mathbf{k}\right)\hat{\rho}\left(-\mathbf{k}\right)\right\}$$
(6.94)

is the $\mathbf{k}^{\text{th}}$ mode and $\rho = N/V$ the number density. The simplest approximation to the average in Equation 6.93 is the *RPA*

$$\left\langle\prod_{\mathbf{k}}{}' s\left(\mathbf{k}\right)\right\rangle_0 \approx \prod_{\mathbf{k}}{}' \left\langle s\left(\mathbf{k}\right)\right\rangle_0,$$
(6.95)

which would be exact if there were only one $\mathbf{k}$ vector in the half $\mathbf{k}$-space. Successive improvements can be obtained by replacing the average in Equation 6.93 with a product involving averages over products of an increasing number of modes. This is the essence of the mode expansion, which is formally exact. Then, the mode expansion of the Helmholtz free energy in the thermodynamic limit reads

$$\beta F = \beta F_0 + \frac{1}{2}\beta N\rho\hat{u}_1\left(0\right) + \sum_{n=1}^{\infty}\beta F_n,$$
(6.96)

where F_n is the contribution due to the correlations between n modes. In particular,

$$\beta F_1 = -\sum_{\mathbf{k}}{}'\left[\rho\beta\hat{u}_1\left(k\right) + \ln\left\langle s\left(\mathbf{k}\right)\right\rangle\right].$$
(6.97)

If the reference system is the ideal gas, Andersen and Chandler (1970) showed that

$$\frac{\beta F_1}{V} = -\frac{1}{2\left(2\pi\right)^3}\int\left\{\beta\rho\hat{u}_1\left(k\right) - \ln\left[1 + \beta\rho\hat{u}_1\left(k\right)\right]\right\}d\mathbf{k},$$
(6.98)

and Equation 6.96 takes the form

$$\frac{\beta F}{V} = \frac{\beta F_0}{V} + \frac{1}{2}\beta\rho^2\int u_1\left(r\right)d\mathbf{r}$$

$$- \frac{1}{2\left(2\pi\right)^3}\int\left\{\beta\rho\hat{u}_1\left(k\right) - \ln\left[1 + \beta\rho\hat{u}_1\left(k\right)\right]\right\}d\mathbf{k} + \sum_{n=2}^{\infty}\frac{\beta F_n}{V}.$$
(6.99)

The corresponding expression for an arbitrary reference system is (Andersen and Chandler 1971)

$$\frac{\beta F}{V} = \frac{\beta F_0}{V} + \frac{1}{2}\beta\rho^2 \int_0^\infty u_1(r)\, g_0(r)\, d\mathbf{r}$$

$$- \frac{1}{2(2\pi)^3} \int \left\{ \beta\rho\hat{u}_1(k)\, S_0(k) - \ln\left[1 + \beta\rho\hat{u}_1(k)\, S_0(k)\right]\right\} d\mathbf{k} + \sum_{n=2}^\infty \frac{\beta F_n}{V},$$

$$(6.100)$$

where $g_0(r)$ and $S_0(k)$ are the RDF and the structure factor of the reference system, respectively. Truncating expansion (6.100) at $n = 1$ gives the RPA, which can be written in the form

$$\frac{\beta F_{RPA}}{V} = \frac{\beta F_0}{V} + \frac{\beta F_{HTA}}{V} + \frac{\beta F_R}{V}, \tag{6.101}$$

where F_{HTA} is given by Equation 6.40 and

$$\frac{\beta F_R}{V} = -\frac{1}{2(2\pi)^3} \int \left\{ \beta\rho\hat{u}_1(k)\, S_0(k) - \ln\left[1 + \beta\rho\hat{u}_1(k)\, S_0(k)\right]\right\} d\mathbf{k}. \tag{6.102}$$

The functional derivative of the RPA approximation with respect to $\hat{u}_1(k)$ leads to the RPA expression (4.45) for $S(k)$, which, introducing the relationship (4.6) for the reference system, can be expressed in the form

$$S(k) = \frac{1}{1 - \rho\left[\hat{c}_0(k) - \beta\hat{u}_1(k)\right]}, \tag{6.103}$$

whence the result (4.44) for $c(r)$ in the RPA immediately follows.

On the other hand, for hard-core potential models, the rate of convergence of the series in Equation 6.100 depends on the choice of the perturbing potential within the core (Chandler and Andersen 1971). As the RDF must be zero for $r < \sigma$, which from the relationship

$$\frac{\delta F}{\delta u(r)} = \frac{1}{2}N\rho g(r), \tag{6.104}$$

is equivalent to $\delta F/\delta u_1(r) = 0$, the optimal choice for $u_1(r)$ for $r < \sigma$ will be the one minimizing the free energy, as given by the expansion (6.100) truncated at $n = 1$, with respect to $u_1(r)$, which is equivalent to the condition

$$\frac{\delta F_R}{\delta u_1(r)} = 0, \quad r < \sigma. \tag{6.105}$$

The RPA supplemented with this optimizing condition defines the ORPA cited in Section 4.4. The optimization can be performed by defining for $u_1(r)$ within the core a trial function depending on a number of parameters that are determined by minimizing the function (6.102) with respect to them (Andersen et al. 1976) or by calculating $u_1(r_i)$ in a number of points $r_i < \sigma$ by minimizing the function (6.102) with respect to all the $u_1(r_i)$ (Lang et al. 1999).

Also, applying the relationship (6.104) to Equation 6.101, with F_R given by Equation 6.102, and comparing the result with Equation 4.46, we easily obtain for the Fourier transform of $C(r)$ the expression

$$\hat{C}(k) = -\frac{[S_0(k)]^2 \beta \hat{u}_1(k)}{1 + \rho S_0(k) \beta \hat{u}_1(k)}, \tag{6.106}$$

which, upon inverse Fourier transforming, provides a way of obtaining $C(r)$ once the optimized potential $u_1(r)$ inside the core has been determined from condition (6.105).

The RPA results can also be derived from a different expansion, the *cluster expansion*, involving the renormalized potential $C(r)$ given by Equation 4.46 (Andersen and Chandler 1972). Introducing in addition the optimizing condition (4.50), or equivalently (6.105), the expansion yields, as particular cases, the ORPA, the EXP approximation, Equation 4.47, and the so-called ORPA $+ \mathscr{B}_2$. In the latter approximation, the free energy is expressed in the form (Andersen and Chandler 1972)

$$\frac{\beta}{V} F_{\text{ORPA}+\mathscr{B}_2} = \frac{\beta}{V} F_{\text{ORPA}} + \mathscr{B}_2, \tag{6.107}$$

where

$$\mathscr{B}_2 = -\frac{1}{2}\rho^2 \int h_0(r) \frac{1}{2}[C(r)]^2\, dr + \frac{1}{2}\rho^2 \int \left\{ g_0(r) \sum_{n=3}^{\infty} \frac{1}{n!}[C(r)]^n \right\} dr$$

$$= -\frac{1}{2}\rho^2 \int \left\{ g_0(r)\left[e^{C(r)} - C(r) - 1 \right] - \frac{1}{2}[C(r)]^2 \right\} dr. \tag{6.108}$$

This function is similar to a second virial coefficient (Sung and Chandler 1974), whence the name of this approximation. The ORPA $+ \mathscr{B}_2$ is a similar approximation for the free energy as the EXP approximation for the RDF (Andersen and Chandler 1972). The combination of the two is called the *optimized cluster theory* (OCT).

Therefore, we have two different routes to obtain the thermodynamic properties from ORPA and related theories: the free energy route, as described in this section, and the RDF route, as described in Section 4.4.

Closely related to the RPA and ORPA approximations is the perturbation theory devised by Khanpour (2010, 2011). Starting from Equation 6.66 and expressing explicitly the first-order term, we have

$$\beta F(\alpha) = \beta F_0 + 2\pi N \beta \rho \int_0^{\infty} u_1(r) g_0(r) r^2 dr$$

$$+ \frac{1}{2} N \beta \rho \sum_{n=2}^{\infty} \frac{1}{n!} \int_0^{\infty} u_1(r) \left. \frac{\partial^{n-1} h(r;\alpha)}{\partial \alpha^{n-1}} \right|_{\alpha=0} dr, \tag{6.109}$$

or, upon Fourier transforming the last term,

$$\beta F(\alpha) = \beta F_0 + 2\pi N \beta \rho \int_0^\infty u_1(r) g_0(r) r^2 dr$$

$$+ \frac{1}{2(2\pi)^3} N \beta \rho \sum_{n=2}^\infty \frac{1}{n!} \int_0^\infty \hat{u}_1(k) \left. \frac{\partial^{n-1} \hat{h}(k; \alpha)}{\partial \alpha^{n-1}} \right|_{\alpha=0} d\mathbf{k}, \qquad (6.110)$$

where $\hat{h}(k; \alpha)$ is the Fourier transform of $h(r; \alpha)$. From Equations 4.5 and 4.6, it is easy to see that the derivatives involved in Equation 6.110 can be obtained in terms of the derivatives of the Fourier transform $\hat{c}(k; \alpha)$ of the DCF $c(r; \alpha)$ and the static structure factor $S_0(k)$ of the reference fluid. Explicit expressions of these derivatives up to the fifth have been reported by Khanpour (2011). To proceed further, Khanpour assumes the following form for the DCF of fluids with spherical hard-core potentials

$$c(r, \alpha) = \begin{cases} c_{HS}(r), & r \le \sigma \\ c_{HS}(r) + e^{-\alpha \beta u_1(r)} - 1, & r > \sigma \end{cases}, \qquad (6.111)$$

where $c_{HS}(r)$ is the HS DCF. Then, these derivatives can be evaluated if we know the static structure factor of the HS reference fluid. In particular, taking the high-temperature limit in Equation 6.111 and proceeding as indicated, the last term in Equation 6.110 leads to the RPA result (Equation 6.102). In the general case, the RDF of the fluid obtained from the procedure just outlined will not satisfy $g(r) = 0$ for $r < \sigma$. To remedy this drawback, Khanpour adopts the same optimization procedure as in the ORPA.

6.6 HIERARCHICAL REFERENCE THEORY

The ORPA works well at moderate to high densities, but worsens at low densities and fails to provide the right second virial coefficient. Neither is the theory suitable to deal with critical phenomena (Andersen et al. 1972a), a shortcoming that is shared in common with most other theories.

Parola and Reatto (1984, 1985) developed a theory, the *hierarchical reference theory* (HRT), which provides the correct low-density behavior, approaches the ORPA at high densities, and also is able to give fairly good results for the critical properties. The theory considers a potential of the form of Equation 4.40 whose perturbation contribution is again assumed to have a Fourier transform. To account for the critical fluctuations of all length scales, these fluctuations are gradually switched on by considering a sequence of intermediate potentials $u^{(q)}(r)$ defined as

$$u^{(q)}(r) = u_0(r) + u_1^{(q)}(r), \qquad (6.112)$$

where the Fourier transform of the perturbation part $u_1^{(q)}(r)$ of this potential is

$$\hat{u}_1^{(q)}(k) = \begin{cases} \hat{u}_1(k), & k \ge q \\ 0, & k < q \end{cases}, \qquad (6.113)$$

with q varying between 0 and ∞, so that for $q = \infty$ the system reduces to the reference system and for $q = 0$ we recover the system with the full interacting potential (Equation 4.40). With the cut off of Equation 6.113, for every value of q the fluctuations corresponding to $k < q$, the largest ones, are suppressed. The remaining of the perturbation potential, $u_1(r) - u_1^{(q)}(r)$, is treated in the RPA in order to avoid a discontinuity in the pair correlation function which otherwise would occur at $k = q$. Every system with potential $u^{(q)}(r)$ is taken as the reference system for a system with potential $u^{(q-dq)}(r)$, where dq indicates an infinitesimal increase in q. The variation in the Helmholtz free energy $F^{(q)}$ obeys the exact equation (Parola and Reatto 1985, Meroni et al. 1990)

$$\frac{\partial \left[\beta \mathcal{F}^{(q)}/V \right]}{\partial q} = \frac{q^2}{4\pi^2} \ln \left[1 + \frac{\beta \hat{u}_1^{(q)}(q)}{\hat{C}^{(q)}(q)} \right], \tag{6.114}$$

with

$$\frac{\beta \mathcal{F}^{(q)}}{V} = \frac{\beta F^{(q)}}{V} + \frac{1}{2}\beta\rho \left[u_1(0) - u_1^{(q)}(0) \right] - \frac{1}{2}\beta\rho^2 \left[\hat{u}_1(0) - \hat{u}_1^{(q)}(0) \right], \tag{6.115}$$

$$\hat{C}^{(q)}(k) = \hat{c}^{(q)}(k) - \beta\hat{u}_1(k) + \beta\hat{u}_1^{(q)}(k), \tag{6.116}$$

where $\hat{c}^{(q)}(k)$ is the Fourier transform of the DCF of the system with potential $u^{(q)}(r)$, including the ideal gas contribution. For $q = 0$, the q-system coincides with the actual system, so that $\hat{u}_1 = \hat{u}_1^{(q)}$ and then $\mathcal{F} = F$ and $\hat{C}(k) = \hat{c}(k)$, the free energy and the Fourier transform of the DCF of the actual system, respectively. For $q = \infty$, we only have the reference system, so that $u_1^{(q)} = 0$, $F_q = F_0$, and $\hat{c}_q(k) = \hat{c}_0(k)$. Then, Equation 6.115 reduces to the MFA for the free energy in the RPA, and Equation 6.116 is the RPA approximation, Equation 4.44, for the DCF.

To complete Equation 6.114, we need some approximation for the Fourier transform $\hat{C}(k)$ of the modified DCF. Meroni et al. (1990), assuming for the reference system a fluid of HS with effective diameter d, proposed an expression of the form

$$\hat{C}^{(q)}(k) = \hat{c}_0(k) - a^{(q)}\beta\hat{u}_1(k)\hat{f}^{(q)}(k), \tag{6.117}$$

where $a^{(q)}$ is a parameter determined for every q from the condition of thermodynamic consistency between the compressibility and the free energy routes, which in terms of the modified quantities $\hat{C}^{(q)}(k)$ and $\mathcal{F}^{(q)}$ is expressed as

$$\hat{C}^{(q)}(0) = -\frac{\partial^2 \left[\beta \mathcal{F}^{(q)}/V \right]}{\partial\rho^2}, \tag{6.118}$$

and $\hat{f}^{(q)}(k)$ is a function determined from the condition that for every q the RDF $g^{(q)}(r)$ must be zero for $r < d$, whereas $f^{(q)}(r)$, the inverse Fourier transform of $\hat{f}^{(q)}(k)$, is zero for $r > d$. This is equivalent to the closure

$$c(r) = c_0(r) - a^{(0)}\beta u_1(r), \quad r > d \tag{6.119}$$

of the OZ equation for the fully interacting system, where $c_0(r)$ is the DCF of the reference HS fluid. Note the resemblance of Equation 6.119 with the SCOZA closure (4.26), but now the consistence is imposed through Equation 6.118 to every stage of the switching on of the perturbation.

6.7 USING NON-HARD-SPHERE REFERENCE SYSTEM

Although in most cases the HS system is taken as the reference system, other choices are possible and in some cases may be advantageous. Thus, Melnyk et al. (2007) have proposed to use a short-ranged HCY fluid as the reference system, and the difference between the actual and the reference potentials beyond the hard core as the perturbation. They showed that this choice is advantageous with respect to the HS reference for predicting the structure factor and the critical properties of longer-ranged HCY fluids within the RPA. To this end, the reference potential must be enough short-ranged as to prevent the existence of a stable critical point. A convenient feature of the HCY fluid is that there are approximate analytical solutions for the thermodynamic and structural properties based on the MSA, as mentioned in Chapter 4. Obviously there would be little advantage, if any, in using the HCY reference fluid to study a fluid with the same type of potential, but it might be useful to use such a reference fluid to study fluids with different, but similarly shaped, potential models (Melnyk et al. 2009, 2010).

On the other hand, Zhou (2009), within the context of his perturbation theory for hard-core potential models (see Equation 6.66 and the accompanying explanation), has proposed to include part of the potential for $r > \sigma$ within the reference potential and the remaining within the perturbation. This was motivated by the observation that, using the HS reference potential, for SW fluids with $\lambda = 1.25$, the higher-order terms $(n > 2)$ in expansion (6.66) become larger than the $n = 2$ term at low enough temperatures and, therefore, the series surely will not converge (Zhou 2008). The splitting of the potential in the new approach is arbitrary, provided that the reference includes the hard core, that the critical temperature of the reference fluid is sufficiently lower than the temperature of interest for the actual fluid, and that we are able to obtain the RDF and the EOS of the reference fluid with enough simplicity and accuracy. As usually the latter requires the use of some kind of integral equation theory, and in general their accuracy increases as temperature increases, the fact that the temperatures considered are strongly supercritical for the reference fluid makes easier to accurately obtain the properties of the latter. The lower the critical temperature of a fluid, the shorter the range of the attractive tail, and so it would be desirable to split the potential as closely to the hard core as possible. However, the larger the range of the perturbation potential, the slower will be the convergence of the perturbation expansion (6.66). Therefore, some kind of compromise must be adopted for the splitting distance of the potential. A detailed discussion on the performance and applications of this approach has been reported by Zhou (2011). In any case, as this perturbation theory requires numerical solution from IET, for every temperature and density of interest, for both the reference fluid and the fluid with a potential in which the perturbation is slightly switched on, the usefulness of the procedure, against the direct solution of the IET for the actual fluid, must be carefully weighed up in every situation.

With regard to soft-core potential models, as seen in Section 6.4, the hard-sphere system is usually taken as the reference. However, with this choice a region of the configuration space, accessible to the actual system, is inaccessible to the reference system, which introduces some error that increases with the softness of the core (Mon 2000, 2001). Mon (2001) developed a computer simulation procedure to correct the free energy for the neglected configurations, and showed that when the correction term is included within the context of the variational perturbation theory a considerable improvement is achieved for SS fluids. A theoretical procedure to approximately obtain the correction term has been developed by Ben-Amotz and Stell (2004b).

An alternative, and quite frequently used, strategy for soft-core potential models consists in the use of soft reference potentials. Among the latter, the soft-sphere potentials are the preferred, because their thermodynamic properties depend on a single parameter Γ (see Section 4.5) and so, once these properties are known for a single isotherm, they can be easily obtained for any other isotherm. The properties of the soft-sphere reference fluid can be obtained from computer simulation or from integral equation theory. The first few terms in expansion (6.15) for realistic potential models can also be obtained from computer simulation using the soft-sphere reference fluid in the form described in Section 6.2.

A different approach is used in the variational perturbation theory of Ross (1979), which also uses a soft-sphere reference system, whose free energy is determined from variational perturbation theory using the HS reference system with the packing fraction of the latter used as the variational parameter. A correction term ΔF_0 is added to the reference free energy thus obtained in order to bring the result into complete agreement with simulations, resulting in the expression

$$F_0 = F_{HS} + 2\pi N\rho \int_d^\infty u_0\,(r)\,g_{HS}\,(r)\,r^2 dr + \Delta F_0,   \tag{6.120}$$

where
   subscript 0 refers to the soft-sphere reference system
   d is the effective HS diameter determined from the variational procedure

Introducing this result into the expression (6.90) for the free energy of the actual fluid in the variational theory, and replacing in the latter expression the RDF of the reference soft-sphere fluid with the HS RDF, one finally obtains

$$F = F_0 + 2\pi N\rho \int_d^\infty u_1\,(r)\,g_{HS}\,(r)\,r^2 dr.   \tag{6.121}$$

Alternatively, F_0 can be obtained from integration of a suitable EOS for the soft-sphere reference fluid through the relationship (6.41).

6.8 RESULTS FOR SOME POTENTIAL MODELS

In the previous sections, we have described a number of perturbation theories mainly based on series expansions of the Helmholtz free energy, although some of them can also be considered from the viewpoint of the integral equation perturbation theories addressed in Section 4.4. In this section, we will analyze the results provided by some of these theories for a number of simple potential models. In general, as only the first-order term in these expansions can be obtained easily and accurately from theory, these theories are expected to give good results at high densities, where the repulsive forces are dominant, and poorer at moderately low densities, where the perturbation expansion often converges more slowly because the attractive forces play a more important role, and, moreover, some perturbation theories fail to provide the right second virial coefficient. Theories based on inverse temperature expansions will also fail at extremely low temperatures for short-ranged potentials, as the series is not guaranteed to converge in such circumstances, as described in Section 6.2.

6.8.1 SQUARE-WELL FLUIDS

The performance of the second-order BH perturbation theory for the SW fluid of variable width was extensively analyzed by Largo (2003). It was found that the theory provides very satisfactory results for the excess energy and the EOS of supercritical SW fluids with moderate to large well-widths, say $\lambda \gtrsim 1.5$, with little differences between the mc and lc approximations. This is illustrated in Figure 6.3 for

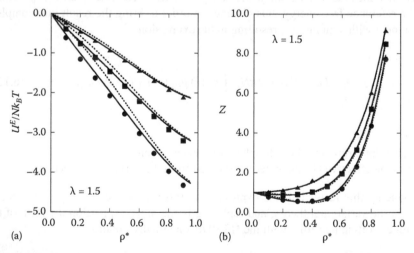

FIGURE 6.3 Excess energy (a) and EOS (b) for the SW fluid with $\lambda = 1.5$. Filled circles, squares, and triangles are the simulation data from Largo and Solana (2003c) for $T^* = 1.5$, 2.0, and 3.0, respectively. Dotted curves are the results from the first-order perturbation theory, Equation 6.39, and continuous curves are those from the second-order BH perturbation theory in the mc and lc approximations, Equations 6.61 and 6.62, respectively, indistinguishable at the scale of the figure. The Carnahan–Starling equation (5.21) was used for Z_0 and the RFA (see Section 5.4) for $g_0(r)$.

the typical value $\lambda = 1.5$. For the temperatures considered in Figure 6.3, the difference between the mc and lc approximations is nearly negligible and for the compressibility factor Z even the first-order approximation provides completely satisfactory results. For $1.2 \lesssim \lambda \lesssim 1.5$, the theory still works very satisfactorily for the EOS, whereas the results for the excess energy increasingly worsen as temperature and well-width decrease.

For extremely short-ranged SW potentials and very low temperatures, the BH theory fails to provide reliable results. In contrast, even in these extreme situations, the Zhou third-order perturbation theory continues to offer excellent performance (Zhou 2006b, Zhou and Solana 2008), as shown in Figure 6.4 for $\lambda = 1.05$. The accuracy of the Zhou theory extends to the liquid–vapor coexistence, as seen in Figure 6.5, although for temperatures close to the critical point the theory overestimates the simulation data. For comparison, the predicted temperatures from the BH theory are too high near the critical point and too low in the liquid region (Figure 6.5).

The performance of the ORPA, EXP, and LEXP theories for the SW fluid with variable width was analyzed by Kahl and Hafner (1982). Their results indicate that for $\lambda = 1.5$ all of these theories provide very good agreement with the simulation data for the energy as obtained from the energy equation, with slightly better performance of the ORPA at low temperatures. For lower values of λ, the ORPA considerably overestimates the energy, more markedly the lower is the temperature; better agreement is achieved with the EXP approximation and still better with the LEXP. Concerning the EOS derived from the virial equation, for $\lambda = 1.5$ the ORPA considerably overestimates the pressure at high densities, whereas the other two theories provide quite satisfactory results, except for extremely low temperatures. The situation is less clear

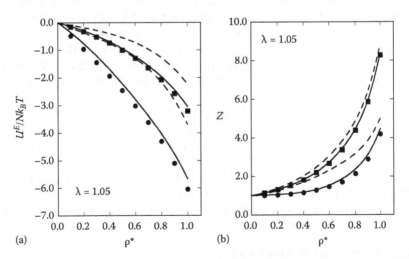

FIGURE 6.4 Excess energy (a) and EOS (b) for the SW fluid with $\lambda = 1.05$ after Zhou and Solana (2008). Filled circles and squares are the simulation data for $T^* = 0.5$ and 0.8, respectively. Dashed curves are the results from the second-order BH perturbation theory in the lc approximation. Continuous curves are the results from the third-order Zhou perturbation theory.

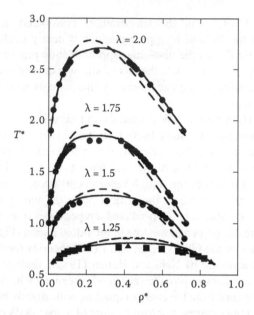

FIGURE 6.5 Liquid–vapor equilibrium for SW fluids with several values of the potential range λ. Points: simulation data from Singh et al. (2003) (circles), del Río et al. (2002) (squares), and Elliott and Hu (1999) (triangles). Dashed curves: BH second-order perturbation theory in the lc approximation. Continuous curves: Zhou's third-order perturbation theory, from Zhou (2006a, 2007; personal communication, 2010).

for smaller well-widths, but on the whole all the three theories provide comparable accuracy. The Khanpour (2011) theory, carried out to fifth order, also performs better than the ORPA for short-ranged SW potentials at low temperatures.

A key feature of these theories is that the structure factor and the RDF are quite easily obtained from those corresponding to the reference system (see Section 4.4). The optimization procedure of the ORPA gives rise to a reduction of the effect of the attractive forces with respect to the RPA predictions, which manifests in a remarkable depletion of the height of the first peak of the static structure factor of the SW fluid at high densities, thus approaching more closely the values $S_0(k)$ corresponding to the reference HS fluid (Kahl and Hafner 1982). As a consequence, the contact values $g(\sigma)$ of the RDF predicted by the ORPA for $\lambda = 1.5$ are too high as compared with the simulation data. Better results near contact are obtained from the EXP approximation, which however gives too high values near $r = \lambda\sigma$ and the result worsens as temperature and density decrease. The best overall results are obtained from the LEXP approximation.

6.8.2 HARD-CORE YUKAWA FLUIDS

The BH theory provides results comparable with those of the inverse temperature expansion of the MSA (see Figure 4.5) for the EOS and the excess energy of attractive HCY fluids with moderate values of the inverse-range parameter κ, with little difference between the mc and lc approximations for the excess energy, even at low temperatures, and negligible for the EOS. For shorter-ranged HCY fluids the theory deviates appreciably from the simulation data at low temperatures. Concerning the liquid–vapor coexistence, the results are quite good for $\kappa \lesssim 2.0$, whereas for larger values of κ the theory provides too high values for the coexistence curve in the $T^* - \rho^*$ plane in the neighborhood of the critical point, and the deviation from simulations increases with κ.

Better results for short-ranged attractive HCY fluids are obtained from the Zhou third-order perturbation theory (Zhou 2006b), but still the predicted liquid–vapor coexistence curve is somewhat high near the critical point. In this region, the results from SCOZA (see Figure 4.6) are slightly more accurate, although the situation might change by carrying out the Zhou theory to higher order (Zhou 2008). In any case, the predicted critical temperatures from the third-order approximation are in very good agreement with the simulation data.

Pini et al. (2002) analyzed the predictions of the ORPA and nonlinear ORPA for attractive HCY fluids with $\kappa = 4$ and 9. For the compressibility factor, they found that, for supercritical temperatures, the ORPA, from the compressibility (virial) route strongly overpredicts (underpredicts) the pressure, whereas the energy route provides excellent agreement with simulation. With the NL ORPA, the differences between the three routes reduce substantially, while the energy route continues to be the most accurate and compares favorably with the RHNC for very short ranges and low temperatures. Critical temperatures are accurately predicted by the nonlinear ORPA, with slightly better accuracy obtained from the energy route as compared with the compressibility route; somewhat worse are the values resulting from the ORPA with the energy route and much worse those from the compressibility route. For the radial distribution function, the nonlinear ORPA is much more accurate than the ORPA near contact for these values of κ, but less accurate than the RHNC.

The performance of the HRT for this kind of fluids was analyzed by Caccamo et al. (1999). Their results indicate that this theory is of similar accuracy as SCOZA for the energy and the compressibility factor, but even worse for the contact values of the RDF (see Section 4.5 for more details). The predicted critical temperatures are also in very good agreement with the simulation data, but not better than those from nonlinear ORPA-energy or from SCOZA. For large values of κ, the HRT slightly overpredicts the coexistence temperatures near the critical point.

6.8.3 CRYSTALLINE SOLIDS WITH THE SUTHERLAND POTENTIAL

It is worth analyzing the performance of perturbation theories for simple crystalline solids. For these systems, the possible choices in practice reduce to the simulation-based perturbation theory and to the BH perturbation theory. This kind of analysis was performed by Díez et al. (2007) and Díez (2009) for fcc solids with the Sutherland potential (1.13) with n ranging from 6 to 36. For the (angle-averaged) RDF, it was found that, for $n = 6$, the simulation-based first-order, in expansion (6.20), perturbation theory accurately reproduces the simulation data even for quite low temperatures and densities close to melting, as seen in Figure 6.6. A more careful inspection

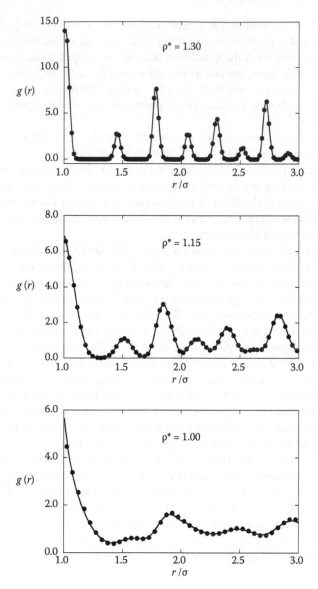

FIGURE 6.6 Angle-averaged radial distribution function of the Sutherland fcc solid with $n = 6$ as a function of the reduced distance r/σ for $T^* = 0.6$ and three different reduced densities. Solid lines are obtained from interpolation of the simulation data and points are the results from the simulation-based first-order perturbation theory, after Díez (2009).

reveals that at densities close to melting the theory underpredicts the values of $g(r)$ near contact and the agreement worsens and limits to higher densities as n increases. On the contrary, the agreement increases as density and temperature increase.

Similar accuracy is obtained for the excess energy and the EOS (Figure 6.7), except for the latter quantity at low temperatures near the solid–fluid transition.

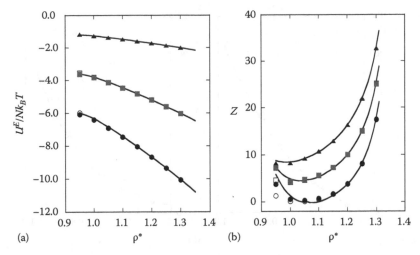

FIGURE 6.7 Excess energy (a) and EOS (b) for the Sutherland fcc solid with $n = 6$. Filled symbols are the simulation data and open symbols are the results from the simulation-based first-order perturbation theory (nearly indistinguishable from the formers at the scale of the figure), for $T^* = 0.6$ (circles), 1.0 (squares), and 3.0 (triangles) respectively, from Díez et al. (2007). The curves are the results from the first-order perturbation theory.

Excellent results are also obtained for the excess energy from the first-order perturbation theory with the RDF of the reference fcc HS solid determined from the procedure described in Section 5.6 using Equation 5.93 for the free volume, but the EOS is overpredicted.

Instead, the first-order perturbation theory provides very satisfactory results for these two quantities when using for the RDF and the EOS of the reference system the parameterizations quoted in Section 5.6, as shown in Figure 6.7. Even for shorter-ranged potentials (larger n), the perturbation theory provides quite satisfactory results, except at very low temperatures for extremely short-ranged potentials, with little difference between first- and second-order approximations (Díez et al. 2007).

6.8.4 SOFT-SPHERE FLUIDS

Hoover et al. (1970) analyzed the performance of the first-order perturbation theory, Equation 6.39, for the soft-sphere fluid with $n = 12$. Using for the break point σ_0 of the potential the criterion $u(\sigma_0) = k_B T$, they found that the theory considerably overpredicts the free energy and the EOS at high values of $\Gamma = \rho\sigma^3(\varepsilon/k_B T)^{3/n}$. Much better agreement is obtained with the criterion of insensitivity of the free energy with respect to the choice of the diameter (see the discussion following Equation 6.68), which for this potential implies minimizing the free energy, although still the predicted values for these quantities are slightly high (see Figure 6.8).

Andersen et al. (1971) showed that the WCA theory provides excellent agreement with simulations for the excess free energy of the SS fluid with $n = 12$, and also for the EOS obtained from the former quantity, whereas the results derived from

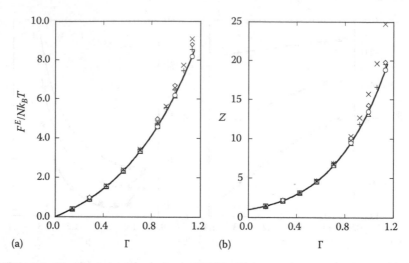

FIGURE 6.8 Excess free energy (a) and EOS (b) for the soft-sphere fluid with $n = 12$ as a function of $\Gamma = \rho\sigma^3(\varepsilon/k_B T)^{3/n}$. Curves: Young and Rogers (1984) parameterizations of the simulation data. Squares: WCA, from Andersen et al. (1971) (the values for Z displayed here are those obtained from the free energy). Triangles: self-consistent WCA, from Lado (1984). Circles: WCA with the Kang et al. (1985) splitting of the potential. Crosses and pluses: BH theory with the criterions $u(\sigma_0) = k_B T$ and of minimization of the free energy, respectively, from Hoover et al. (1970). Diamonds: VPT, from Ree (1976).

the virial route are somewhat high for high values of Γ. The Lado (1984) self-consistency criterion for choosing the effective diameter d solves the disagreement between these two routes to the EOS, while retaining the agreement between theory and simulation. Very good results are also obtained from the WCA with d obtained from the insensitivity criterion for F (Ree 1976) and still better with the Kang et al. (1985) splitting of the potential, Equations 6.75 and 6.76. The latter procedure also provides excellent accuracy for the excess energy of SS fluids with $n = 9$, 6, and 4. Figure 6.8 displays the results from these theories for $n = 12$.

Young and Rogers (1984) used an inverse 12th as the reference fluid for SS fluids with $n = 9$, 6, and 4, in combination with the variational perturbation theory, and compared the results for the energy with those obtained from the Ross (1979) VPT. To this end, the RDF of the reference fluid in the inverse 12th VPT was obtained from integral equation theory. They found that both theories agree quite well with simulations, although the inverse 12th VPT provides better results for low values of Γ, especially for the softer potentials.

Ben-Amotz and Stell (2004b) analyzed the predictions of the VPT, with the hard-sphere reference fluid, for the EOS and the free energy of SS fluids with $n = 12$, 9, 6, and 4, and compared the results with those obtained from the WCA and the Lado (1984) self-consistent WCA. They showed that the VPT slightly overestimates the free energy and the EOS for all these values of n. The inclusion of the Mon correction brings the theory in close agreement with simulations. In contrast, the Lado WCA theory, which provides quite satisfactory results for $n = 12$, increasingly deviates from simulations with increasing the softening of the potential, and the deviation is still more marked in the ordinary WCA.

The OCP was studied by Lee and Ree (1988) in both the fluid and solid phases. They used a splitting of the potential like in Equation 6.75 with a density-dependent σ_0 and $w(r)$ linearly varying with r, like in Equation 6.76, in combination with the WCA. For the OCP with this splitting, the first-order perturbation term of the free energy, Equation 6.40, takes the form

$$\frac{F_1}{Nk_BT} = \frac{2\pi}{k_BT}\rho \int_0^\infty \left[u_1(r)g_{HS}(r) - \frac{e^2}{r} \right] r^2 dr, \qquad (6.122)$$

where the term $-e^2/r$ arises from the uniform neutralizing background. The results obtained in this way for the thermal contribution to the excess energy of the fluid are too high for high values of $\Gamma = (q^2/k_BT)(4\pi\rho/3)^{1/3}$ as Figure 6.9 shows, and still worse for the solid (not shown in Figure 6.9).

The thermal contribution to the excess energy is $\Delta U^E = U^E - U_M$, where U_M is the Madelung energy for the solid, the electrostatic energy for the system with the particles placed at the lattice sites, which is $U_M/Nk_BT = -0.895873\Gamma$ for the OCP in the fcc lattice and 0.895929Γ in the bcc lattice. Taking into account that F_1, as given by Equation 6.122, is the Madelung energy for a system of particles with interacting potential $u_1(r)$ and distributed according to $g_{HS}(r)$, Lee and Ree replaced F_1 with U_M not only for the solid but also for the fluid, considering that the long range of the Coulomb potential makes F_1 nearly insensitive to the structure. The fact that $F_1 \simeq U_M$ even for large values of Γ was confirmed by numerical calculations.

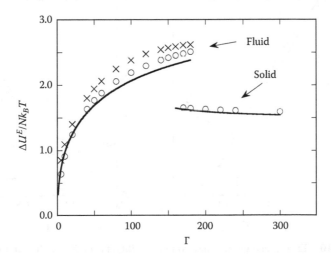

FIGURE 6.9 Thermal contribution to the excess energy for the OCP for the fluid and fcc solid phases. The curves are the simulation data as parameterized by Slattery et al. (1982) for the fluid and by Helfer et al. (1984) for the fcc solid. points: WCA with the splitting Equations 6.75 and 6.76 and F_1 as in Equation 6.122 (crosses), and replacing F_1 with U_M (circles), after Lee and Ree (1988). Compare this figure with Figure 4.7.

With this modification, the results for the solid come in close agreement with simulations and also considerably improve for the fluid, although in the latter some deviation from the simulation data remains for large values of Γ (see Figure 6.9).

6.8.5 LENNARD–JONES FLUID

Barker and Henderson (1976) showed that the second-order perturbation theory with F_1 and F_2 determined from computer simulation through Equations 6.29 and 6.30, respectively, provides very satisfactory results for the free energy and the EOS of LJ fluids at moderate temperatures.

The second-order BH perturbation theory, in either the mc or lc approximations, also predicts accurately the EOS at moderate temperatures, with the effective HS diameter determined from Equation 6.68, provided that we use accurate expressions for the EOS and the RDF of the reference HS fluid, such as the CS EOS and the RFA or the FMSA expressions for the RDF analyzed in Chapter 5, but underpredicts the excess energy at high densities. Neither the VPT with HS reference works satisfactorily. Other perturbation theories, including ORPA, soft-sphere reference and Ross versions of the VPT, and the WCA with the splitting of Equations 6.75 and 6.76, all of them yield quite good results, as shown in Figure 6.10, and remain accurate even at extremely high temperatures.

The WCA results for the RDF of the LJ fluid closely agree with simulations at high densities, but are too low near the first peak at lower densities. Better agreement is obtained from the ORPA, especially at moderate densities, and still better

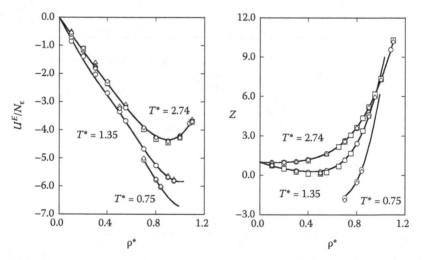

FIGURE 6.10 Excess energy U^E and compressibility factor Z for the LJ fluid at three reduced temperatures: $T^* = 0.75$ (subcritical), $T^* = 1.35$ (slightly supercritical), and $T^* = 2.74$ (supercritical). The curves are the results from the Johnson et al. (1993) fitting of the simulation data. Circles: ORPA, from Andersen et al. (1972a). Squares: Ross VPT, from Ross (1979). Triangles: soft-sphere reference VPT, from Young and Rogers (1984). Diamonds: WCA with the splitting of Equations 6.75 and 6.76, from Kang et al. (1985).

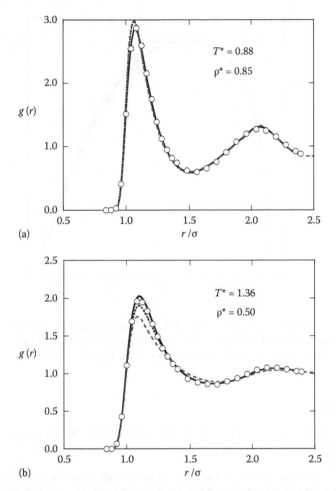

FIGURE 6.11   Two examples of the predictions for the radial distribution function of the LJ fluid from several perturbation theories. Points: simulation data from Verlet (1968). Dashed curves are the results from the WCA. Dotted and continuous curves are the results from the ORPA and EXP theories, after Andersen et al. (1972b). In the figure at the top, the WCA and ORPA results are hardly distinguishable from each other.

with the EXP approximation. These behaviors are illustrated in Figure 6.11 for two different states.

Sung and Chandler (1974) determined the liquid–vapor coexisting densities for the LJ fluid by means of the OCT. The results, as shown in Figure 6.12, are in close agreement with the simulation data except near the critical point. Tau et al. (1995) carried out the same kind of calculations from the HRT, obtaining still greater accuracy, (see Figure 6.12), particularly in the critical region, as one might have expected since this theory was designed to account for some special properties of the fluid near criticality, as mentioned in Section 6.6.

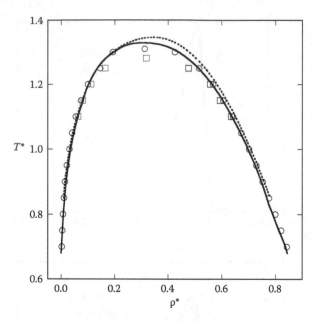

FIGURE 6.12   Liquid–vapor equilibrium for the LJ fluid. Points: simulation data from Lofti et al. (1992) (circles) and from Panagiotopoulos (1994) (squares). Dashed curve: OCT, from Sung and Chandler (1974). Continuous curve: HRT, from Tau et al. (1995).

REFERENCES

Alder, B. J., W. G. Hoover, and D. A. Young. 1968. Studies in molecular dynamics. V. High-density equation of state and entropy for hard disks and spheres. *J. Chem. Phys.* 49:3688.

Alder, B. J., D. A. Young, and M. A. Mark. 1972. Studies in molecular dynamics. X. Corrections to the augmented van der Waals theory for the square well fluid. *J. Chem. Phys.* 56:3013.

Andersen, H. C. and D. Chandler. 1970. Mode expansion in equilibrium statistical mechanics. I. General theory and application to the classical electron gas. *J. Chem. Phys.* 53:547.

Andersen, H. C., and D. Chandler. 1971. Mode expansion in equilibrium statistical mechanics. III. Optimized convergence and application to ionic solution theory. *J. Chem. Phys.* 55:1497.

Andersen, H. C. and D. Chandler. 1972. Optimized cluster expansions for classical fluids. I. General theory and variational formulation of the mean spherical model and hard-sphere Percus-Yevick equations. *J. Chem. Phys.* 57:1918.

Andersen, H. C., D. Chandler, and J. D. Weeks. 1972a. Roles of repulsive and attractive forces in liquids: The optimized random phase approximation. *J. Chem. Phys.* 56:3812.

Andersen, H. C., D. Chandler, and J. D. Weeks. 1972b. Optimized cluster expansions for classical fluids. III. Application to ionic solutions and simple liquids. *J. Chem. Phys.* 57:2626.

Andersen, H. C., D. Chandler, and J. D. Weeks. 1976. Roles of repulsive and attractive forces in liquids: The equilibrium theory of classical fluids. *Adv. Chem. Phys.* 34:105.

Andersen, H. C. J. D. Weeks, and D. Chandler. 1971. Relationship between the hard-sphere fluid and fluids with realistic repulsive forces. *Phys. Rev. A* 4:1597.

Barker, J. A. and D. Henderson. 1967a. Perturbation theory and equation of state of fluids: The square-well potential. *J. Chem. Phys.* 47:2856.

Barker, J. A. and D. Henderson. 1967b. Perturbation theory and equation of state of fluids. II. A successful theory of liquids. *J. Chem. Phys.* 47:4714.

Barker, J. A. and D. Henderson. 1968. Perturbation theory and the equation of state for fluids. III. Improved results for the square-well potential. *Proceedings of the Fourth Symposium on Thermophysical Properties*, ed. J. R. Moszynsky, pp. 30–36. New York: American Society of Mechanical Engineers.

Barker, J. A. and D. Henderson. 1976. What is "liquid"? Understanding the states of matter. *Rev. Mod. Phys.* 48:587.

Ben-Amotz, D. and G. Stell. 2004a. Reformulation of Weeks–Chandler–Andersen perturbation theory directly in terms of a hard-sphere reference system. *J. Phys. Chem. B* 108:6877.

Ben-Amotz, D. and G. Stell. 2004b. Hard sphere perturbation theory for fluids with soft-repulsive-core potentials. *J. Chem. Phys.* 120:4844.

Betancourt-Cárdenas, F. F., L. A. Galicia-Luna, and S. I. Sandler. 2007. Thermodynamic properties for the triangular-well fluid *Mol. Phys.* 105:2987.

Caccamo, C., G. Pellicane, D. Costa, D. Pini, and G. Stell. 1999. Thermodynamically self-consistent theories of fluids interacting through short-range forces. *Phys. Rev. E* 60:5533.

Chandler, D. and H. C. Andersen. 1971. Mode expansion in equilibrium statistical mechanics. II. A rapidly convergent theory for ionic solutions. *J. Chem. Phys.* 54:26.

Chandler, D. and J. D. Weeks. 1970. Equilibrium structure of simple liquids. *Phys. Rev. Lett.* 25:149.

Cuadros, F., W. Ahumada, and A. Mulero. 1996. The high temperature approximation and linearity of the thermodynamic properties on the WCA perturbation parameter. *Chem. Phys.* 204:41.

del Río, F., E. Ávalos, R. Espíndola, L. F. Rull, G. Jackson, and S. Lago. 2002. Vapour-liquid equilibrium of the square-well fluid of variable range via a hybrid simulation approach. *Mol. Phys.* 100:2531.

Díez, A. 2009. Teoría y simulación de las propiedades termodinámicas de equilibrio de sistemas con potencial de esferas duras y cola atractiva de potencia inversa. PhD dissertation, Universidad de Cantabria Santandev, Spain.

Díez, A., J. Largo, and J. R. Solana. 2006. Thermodynamic properties of van der Waals fluids from Monte Carlo simulations and perturbative Monte Carlo theory *J. Chem. Phys.* 125:074509.

Díez, A., J. Largo, and J. R. Solana. 2007. Thermodynamic properties of model solids with short-ranged potentials from Monte Carlo simulations and perturbation theory. *J. Phys. Chem. B* 111:10194.

Elliott, J. R. and L. Hu. 1999. Vapor-liquid equilibria of square-well spheres. *J. Chem. Phys.* 110:3043.

Espíndola-Heredia, R., F. del Río, and A. Malijevsky. 2009. Optimized equation of the state of the square-well fluid of variable range based on a fourth-order free-energy expansion. *J. Chem. Phys.* 130:024509.

Haar, L. and S. H. Shenker. 1971. Equation of state for dense gases. *J. Chem. Phys.* 55:4951.

Helfer, H. L., R. L. McCrory, and H. M. Van Horn. 1984. Further Monte Carlo calculations for the classical one-component plasma in the range $100 \le \Gamma \le 160$: The fcc lattice. *J. Stat. Phys.* 37:577.

Henderson, D. and J. A. Barker. 1970. Perturbation theory of fluids at high temperatures. *Phys. Rev. A* 1:1266.

Henderson, D. and J. A. Barker. 1971. Perturbation theories. In *Physical Chemistry. An Advanced Treatise*, Vol. VIIIA, ed. D. Henderson, pp. 377–412. New York: Academic Press.

Hess, S., M. Kröger, and H. Voigt. 1998. Thermomechanical properties of the WCA-Lennard-Jones model system in its fluid and solid states. *Physica A* 250:58.

Heyes, D. M. 1991. Coordination number and equation of state of square-well and square-shoulder fluids: Simulation and quasi-chemical model. *J. Chem. Soc. Faraday Trans.* 87:3373.

Heyes, D. M. 1997. Thermodynamics and elastic moduli of fluids with steeply repulsive potentials. *J. Chem. Phys.* 107:1963.

Hoover, W. G., M. Ross, K. W. Johnson, D. Henderson, J. A. Barker, and B. C. Brown. 1970. Soft-sphere equation of state. *J. Chem. Phys.* 52:4931.

Isihara, A. 1968. The Gibbs–Bogoliubov inequality. *J. Phys. A* 1:539.

Johnson, J. K., J. A. Zollweg, and K. E. Gubbins. 1993. The Lennard-Jones equation of state revisited. *Mol. Phys.* 78:591.

Kahl, G. and J. Hafner. 1982. Optimized cluster theory, optimized random phase approximation and mean spherical model for the square-well fluid with variable range. *Phys. Chem. Liq.* 12:109.

Kang, H. S., C. S. Lee, T. Ree, and F. H. Ree. 1985. A perturbation theory of classical equilibrium fluids. *J. Chem. Phys.* 82:414.

Khanpour, M. 2010. A simple derivation of random phase approximation expression in perturbation theory of fluids and its nonlinear generalization. *J. Mol. Liq.* 157:34.

Khanpour, M. 2011. Perturbation theory of liquids for short-ranged hard-core potentials: Structure and thermodynamics of short-ranged square-well fluids. *Phys. Rev. E* 83:021203.

Kincaid, J. M., G. Stell, and C. K. Hall. 1976a. Isostructural phase transitions due to core collapse. I. A one-dimensional model. *J. Chem. Phys.* 65:2161.

Kincaid, J. M., G. Stell, and E. Goldmark. 1976b. Isostructural phase transitions due to core collapse. II. A three-dimensional model with a solid–solid critical point. *J. Chem. Phys.* 65:2172.

Kirkwood, J. G. 1935. Statistical mechanics of fluid mixtures. *J. Chem. Phys.* 3:300.

Kohler, F. and L. Haar. 1981. A new representation for thermodynamic properties of a fluid. *J. Chem. Phys.* 75:388.

Lado, F. 1984. Choosing the reference system for liquid state perturbation theory. *Mol. Phys.* 52:871.

Lang, A., G. Kahl, C. N. Likos, H. Löwen, and M. Watzlawek. 1999. Structure and thermodynamics of square-well and square-shoulder fluids. *J. Phys.: Condens. Matter* 11:10143.

Largo, J. 2003. Teoría y simulación de las propiedades de equilibrio de fluidos de pozo cuadrado. PhD dissertation, Universidad de Cantabria Santandev, Spain.

Largo, J. and J. R. Solana. 2002. Theory and computer simulation of the coordination number of square-well fluids of variable width. *Fluid Phase Equilibr.* 193:277.

Largo, J. and J. R. Solana. 2003a. Theory and computer simulation of the first- and second-order perturbative contributions to the free energy of square-well fluids. *Mol. Simul.* 29:363.

Largo, J. and J. R. Solana. 2003b. Theory and computer simulation of the zero- and first-order perturbative contributions to the pair correlation function of square-well fluids. *Fluid Phase Equilibr.* 212:11.

Largo, J. and J. R. Solana. 2003c. Generalized van der Waals theory for the thermodynamic properties of square-well fluids. *Phys. Rev. E* 67:066112.

Largo, J. and J. R. Solana. 2004. First-order perturbative contribution to the compressibility factor of square-well fluids from Monte Carlo and integral equation theory. *J. Phys. Chem. B* 108:10062.

Lee, J. W. and F. H. Ree. 1988. Perturbation theory of a classical one-component plasma. *Phys. Rev. A* 38:5714.

Lee, K.-H., M. Lombardo, and S. I. Sandler. 1985. The generalized van der Waals partition function. II. Application to the square-well fluid. *Fluid Phase Equilibr.* 21:177.

Lee, K.-H., S. I. Sandler, and N. C. Patel. 1986. The generalized van der Waals partition function. III. Local composition models for a mixture of equal size square-well molecules. *Fluid Phase Equilibr.* 25:31

Lofti, A., J. Vrabec, and J. Fischer. 1992. Vapour liquid equilibria for the Lennard-Jones fluid from the *NpT* plus test particle method. *Mol. Phys.* 76:1319.

Malijevský, A. and S. Labík. 1987. The bridge function for hard spheres. *Mol. Phys.* 60:663.

Mansoori, G. A. and F. B. Canfield. 1969. Variational approach to the equilibrium thermodynamic properties of simple liquids. I. *J. Chem. Phys.* 51:4958.

Mansoori, G. A. and F. B. Canfield. 1970. Inequalities for the Helmholtz free energy. *J. Chem. Phys.* 53:1618.

Melnyk, R., F Mou£ka, I. Nezbeda, and A. Trokhymchuk. 2007. Novel perturbation approach for the structure factor of the attractive hard-core Yukawa fluid. *J. Chem. Phys.* 127:094510.

Melnyk, R., I. Nezbeda, D. Henderson, and A. Trokhymchuk. 2009. On the role of the reference system in perturbation theory: An augmented van der Waals theory of simple fluids. *Fluid Phase Equilibr.* 279:1.

Melnyk, R., P. Orea, I. Nezbeda, and A. Trokhymchuk. 2010. Liquid/vapor coexistence and surface tension of the Sutherland fluid with a variable range of interaction: Computer simulation and perturbation theory studies. *J. Chem. Phys.* 132:134504.

Meroni, A., A. Parola, and L. Reatto. 1990. Differential approach to the theory of fluids. *Phys. Rev. A* 42:6104.

Mon, K. K. 2000. Hard sphere perturbation theory of dense fluids with singular perturbation. *J. Chem. Phys.* 112:3245.

Mon, K. K. 2001. Hard sphere perturbation theory for thermodynamics of soft-sphere model liquid. *J. Chem. Phys.* 115:4766.

Panagiotopoulos, A. Z. 1994. Molecular simulation of phase coexistence: Finite-size effects and determination of critical parameters for two- and three-dimensional Lennard-Jones fluids. *Int. J. Thermophys.* 15:1057.

Parola, A. and L. Reatto. 1984. Liquid-state theory for critical phenomena. *Phys. Rev. Lett.* 53:2417.

Parola, A. and L. Reatto. 1985. Hierarchical reference theory of fluids and the critical point. *Phys. Rev. A* 31:3309.

Pelissetto, A. and J.-P. Hansen. 2006. An effective two-component description of colloid-polymer phase separation. *Macromolecules* 39:9571.

Pini, D., A. Parola, and L. Reatto. 2002. A simple approximation for fluids with narrow attractive potentials. *Mol. Phys.* 100:1507.

Praestgaard, E. and S. Toxvaerd. 1969. Perturbation theory for fluids. *J. Chem. Phys.* 51:1895.

Raineri, F. O., G. Stell, and D. Ben-Amotz. 2004. Progress in thermodynamic perturbation theory and self-consistent Ornstein-Zernike approach relevant to structural-arrest problems. *J. Phys.: Condens. Matter* 16:S4887.

Rasaiah, J. and G. Stell. 1970. Upper bounds on free energies in terms of hard-sphere results. *Mol. Phys.* 18:249.

Ree, F. H. 1976. Equilibrium properties of high-density fluids. *J. Chem. Phys.* 64:4601.

Ross, M. 1979. A high-density fluid-perturbation theory based on an inverse 12th power hard-sphere reference system. *J. Chem. Phys.* 71:1567.

Rull, L. F., F. Cuadros, and J. J. Morales. 1984. Deviation of the radial distribution function in the Weeks–Chandler–Andersen approximation. *Phys. Rev. A* 30:2781.

Sandler, S. I. 1985. The generalized van der Waals partition function. I. Basic theory. *Fluid Phase Equilibr.* 19:233.

Sillrén, P. and J.-P. Hansen. 2007. Perturbation theory for systems with strong short-ranged interactions. *Mol. Phys.* 105:1803.

Singh, J. K., D. A. Kofke, and J. R. Errington. 2003. Surface tension and vapor-liquid phase coexistence of the square-well fluid. *J. Chem. Phys.* 119:3405.

Slattery, W. L., G. D. Dolen, and H. E. DeWitt. 1982. N dependence on the classical one-component plasma Monte Carlo calculations. *Phys. Rev. A* 26:2255.

Smith, W. R., D. Henderson, and J. A. Barker. 1970. Approximate evaluation of the second-order term in the perturbation theory of fluids. *J. Chem. Phys.* 53:508.

Smith, W. R., D. Henderson, and J. A. Barker. 1971. Perturbation theory and the radial distribution function of the square-well fluid. *J. Chem. Phys.* 55:4027.

Solana, J. R. 2008. Thermodynamic properties of double square-well fluids: Computer simulations and theory. *J. Chem. Phys.* 129:244502.

Song, Y. and E. A. Mason. 1989. Statistical-mechanical theory of a new analytical equation of state. *J. Chem. Phys.* 91:7840.

Sung, S. H. and D. Chandler. 1974. Optimized cluster theory, the Lennard-Jones fluid, and the liquid-gas phase transition. *Phys. Rev. A* 9:1688.

Tau, M., A. Parola, D. Pini, and L. Reatto. 1995. Differential theory of fluids below the critical temperature: Study of the Lennard-Jones fluid and of a model of C_{60}. *Phys. Rev. E* 52:2644.

Vera, J. H. and J. M. Prausnitz. 1972. Generalized van der Waals theory for dense fluids. *Chem. Eng. J.* 3:1.

Verlet, L. 1968. Computer "experiments" on classical fluids. II. Equilibrium correlation functions. *Phys. Rev.* 165:201.

Verlet, L. and J.-J. Weis. 1972. Perturbation theory for the thermodynamic properties of simple liquids. *Mol. Phys.* 24:1013.

Vimalchand, P., A. Thomas, I. G. Economou, and M. D. Donoue. 1992. Effect of hard-sphere structure on pure-component equation of state calculations. *Fluid Phase Equilibr.* 73:39.

Weeks, J. D., D. Chandler, and H. C. Andersen. 1971. Role of repulsive forces in determining the equilibrium structure of simple liquids. *J. Chem. Phys.* 54:5237.

Young, D. A. and F. J. Rogers. 1984. Variational fluid theory with inverse 12th power reference potential. *J. Chem. Phys.* 81:2789.

Zhou, S. 2006a. Thermodynamic perturbation theory in fluid statistical mechanics. *Phys. Rev. E* 74:031119.

Zhou, S. 2006b. Improvement on macroscopic compressibility approximation and beyond. *J. Chem. Phys.* 155:144518.

Zhou, S. 2007. Performance evaluation of third-order thermodynamic perturbation theory and comparison with existing liquid state theories. *J. Phys. Chem. B* 111:10736.

Zhou, S. 2008. Fifth-order thermodynamic perturbation theory of uniform and nonuniform fluids. *Phys. Rev. E* 77:041110.

Zhou, S. 2009. How to make thermodynamic perturbation theory to be suitable for low temperature? *J. Chem. Phys.* 130:054103.

Zhou, S. 2011. Liquid theory with high accuracy and broad applicability: Coupling parameter series expansion and non hard sphere perturbation strategy. *AIP Adv.* 1:040703.

Zhou, S. and J. R. Solana. 2008. Third-order thermodynamic perturbation theory for effective potentials that model complex fluids. *Phys. Rev. E* 78:021503.

Zhou, S. and J. R. Solana. 2009. Comprehensive investigation about the second order term of thermodynamic perturbation expansion. *J. Chem. Phys.* 131:134106.

Zwanzig, R. W. 1954. High-temperature equation of state by a perturbation method. I. Nonpolar gases. *J. Chem. Phys.* 22:1420.

7 Perturbation Theories for Simple Fluid Mixtures

This chapter analyzes a number of perturbation theories for mixtures of simple fluids like those considered in Chapter 6. Apart from the quite straightforward extensions to mixtures of some of the integral equation and integral equation theories introduced in Chapter 4, which are cited in passing, the formulation for mixtures of several of the free energy perturbation theories addressed in Chapter 6 is described. In addition, this chapter describes a number of models, related to conformal solution theory, specifically designed for mixtures. Particular attention is paid to theories for HS mixtures, as they are often used as the reference system in perturbation theories for mixtures, in a similar way as the pure HS fluid is frequently used in perturbation theories for simple fluids. This chapter ends by discussing the performance of several of the considered theories.

7.1 REAL AND IDEAL MIXTURES

The thermodynamic properties of a mixture depend on its composition, which can be specified by means of the mole fractions x_i, with $i = 1, 2, \ldots, m$, of the m species forming the mixture. Thus, for example, the chemical potential can be expressed in the form

$$\mu_{mix}(T, P, x_1, \ldots, x_m) = \sum_i x_i \mu_i^0(T, P) + k_B T \sum_i x_i \ln x_i + \mu^E(T, P, x_i, \ldots, x_m),$$

$$(7.1)$$

where $\mu_i^0(T, P)$ is the chemical potential of the pure component i at the same pressure and temperature of the mixture. We can define an *ideal mixture* as that satisfying the condition

$$\mu^E(T, P, x_i, \ldots, x_m) = 0. \qquad (7.2)$$

Therefore, Equation 7.1 indicates that a real mixture consists of three contributions: first due to the chemical potentials μ_i^0 of the pure components, second due to the mixing process, and a third one, $\mu^E(T, P, x_i, \ldots, x_m)$, which is the excess chemical potential of the real mixture with respect to that corresponding to the ideal mixture.

Other thermodynamic properties can be expressed in a similar way. Thus, for example, the free energy F_{mix}, the energy U_{mix}, the enthalpy H_{mix}, the entropy S_{mix}, and the volume V_{mix} of the mixture are, respectively,

$$F_{mix}(T,P,x_1,\ldots,x_n) = \sum_i x_i F_i^0(T,P) + Nk_B T \sum_i x_i \ln x_i + F^E(T,P,x_i,\ldots,x_m),$$

$$(7.3)$$

$$U_{mix}(T,P,x_1,\ldots,x_m) = \sum_i x_i U_i^0(T,P) + U^E(T,P,x_i,\ldots,x_m), \qquad (7.4)$$

$$H_{mix}(T,P,x_1,\ldots,x_m) = \sum_i x_i H_i^0(T,P) + H^E(T,P,x_i,\ldots,x_m), \qquad (7.5)$$

$$S_{mix}(T,P,x_1,\ldots,x_m) = \sum_i x_i S_i^0(T,P) - Nk_B \sum_i x_i \ln x_i + S^E(T,P,x_i,\ldots,x_m),$$

$$(7.6)$$

$$V_{mix}(T,P,x_1,\ldots,x_m) = \sum_i x_i V_i^0(T,P) + V^E(T,P,x_i,\ldots,x_m). \qquad (7.7)$$

Denoting by $\Delta_{mix}X$ the variation of the property X in the mixing process, an ideal mixture satisfies

$$\Delta_{mix}U = 0, \quad \Delta_{mix}H = 0, \quad \Delta_{mix}V = 0, \qquad (7.8)$$

whereas

$$\Delta_{mix}S = -Nk_B \sum_i x_i \ln x_i, \quad \Delta_{mix}F = -Nk_B T \sum_i x_i \ln x_i. \qquad (7.9)$$

Instead of the thermodynamic functions or the excess properties of the mixture at constant pressure, we can use those corresponding to constant volume, which are defined in a similar way. The latter functions can be obtained from the former ones using standard thermodynamic relationships.

7.2 CONFORMAL MIXTURES

The theory of conformal mixtures or conformal solutions, initially developed by Longuet-Higgins (1951), can be considered as a perturbation theory based on the corresponding states principle. We will give here a brief account of this theory.

Let us consider an intermolecular potential of the form

$$u(r) = \varepsilon \varphi \left(\frac{r}{\sigma}\right) \qquad (7.10)$$

as is, for example, the LJ potential (1.7), and a fluid, which we will take as the reference fluid, consisting of N particles interacting with each other by means of a potential of the form (7.10) with parameters ε_0 and σ_0. Suppose now that we have another fluid with a potential of the same form (7.10), that is to say *conformal*, but with slightly different parameters $\varepsilon = \varepsilon_0 + \Delta\varepsilon$ and $\sigma = \sigma_0 + \Delta\sigma$ or, alternatively, suppose that the reference fluid experiences a small perturbation modifying its parameters in the indicated form, so that we will have

$$u(r) = u_0(r) + \Delta u(r) = \varepsilon \varphi \left(\frac{r}{\sigma} \right). \tag{7.11}$$

From the corresponding states principle, the thermodynamic properties of the reference and perturbed fluids, expressed in terms of suitable reduced variables, obey to the same expressions or, in other words, if the two fluids are in corresponding states their thermodynamic properties are equal. Therefore, if the two fluids are at the same reduced temperature $T^* = k_B T / \varepsilon$ and volume $V^* = V/\sigma^3$, their thermodynamic properties must be identical. Obviously, if the two fluids are at the same temperature T and volume V, their thermodynamic properties will be slightly different. In such circumstances, the thermodynamic properties of the perturbed fluid may be expressed as a series expansion in powers of the perturbed parameters around the values corresponding to the unperturbed fluid. Thus, for example, the configurational free energy will be

$$F_c(T, V) = F_c^{(0)}(T, V) + \left(\frac{\partial F_c(T, V)}{\partial \varepsilon} \right) \Bigg|_{\varepsilon = \varepsilon_0} \Delta \varepsilon + \left(\frac{\partial F_c(T, V)}{\partial \sigma} \right) \Bigg|_{\sigma = \sigma_0} \Delta \sigma + \cdots . \tag{7.12}$$

From the definition (2.18) of the configurational integral, one easily obtains

$$\left(\frac{\partial F_c(T, V)}{\partial \varepsilon} \right) \Bigg|_{\varepsilon = \varepsilon_0} = \left(\frac{\partial F_c(T, V)}{\partial \mathcal{U}(\mathbf{r}_1, \ldots, \mathbf{r}_N)} \right) \left(\frac{\partial \mathcal{U}(\mathbf{r}_1, \ldots, \mathbf{r}_N)}{\partial \varepsilon} \right) \Bigg|_{\varepsilon = \varepsilon_0}$$

$$= \left\langle \frac{\mathcal{U}(\mathbf{r}_1, \ldots, \mathbf{r}_N)}{\varepsilon} \right\rangle \Bigg|_{\varepsilon = \varepsilon_0} = \frac{\langle \mathcal{U}_0(\mathbf{r}_1, \ldots, \mathbf{r}_N) \rangle}{\varepsilon_0} = \frac{U_c^{(0)}}{\varepsilon_0}, \tag{7.13}$$

where $U_c^{(0)}$ is the configurational energy of the reference system. Analogously

$$\left(\frac{\partial F_c(T, V)}{\partial \sigma} \right) \Bigg|_{\sigma = \sigma_0} = \left(\frac{\partial F_c(T, V)}{\partial V} \right) \left(\frac{\partial V}{\partial \sigma} \right) \Bigg|_{\sigma = \sigma_0}$$

$$= - \left(P - \frac{N k_B T}{V} \right) \left(\frac{3V}{\sigma} \right) \Bigg|_{\sigma = \sigma_0} = \left(\frac{N k_B T}{V} - P_0 \right) \left(\frac{3V}{\sigma_0} \right), \tag{7.14}$$

where we have taken into account the definition of reduced volume mentioned earlier. Introducing the latter two results into Equation 7.12, one obtains

$$F_c(T, V) = F_c^{(0)}(T, V) + U_c^{(0)} \frac{\varepsilon - \varepsilon_0}{\varepsilon_0} + 3 \left(N k_B T - P_0 V \right) \frac{\sigma - \sigma_0}{\sigma_0} + \cdots , \tag{7.15}$$

where we have put $\Delta \varepsilon = \varepsilon - \varepsilon_0$ and $\Delta \sigma = \sigma - \sigma_0$.

Let us consider now an m-component mixture, in which the potential energy of interaction between two molecules belonging to any two species i and j is of the form

$$u_{ij}(r) = \varepsilon_{ij}\, \varphi\left(\frac{r}{\sigma_{ij}}\right). \tag{7.16}$$

Because of the common form of the interacting potential, this kind of mixtures are called *conformal mixtures* or *conformal solutions* (Longuet-Higgins 1951).*

Let us assume also that there exists a monocomponent fluid, which we will take as the reference fluid, such that its molecules interact with each other by means of a potential of the form (7.16) with the condition that $\varepsilon_{ij} \sim \varepsilon_0$ and $\sigma_{ij} \sim \sigma_0$, whatever be the species i and j. In this situation, we can apply a perturbative treatment analogous to that applied to the monocomponent fluid, from which we would obtain for the configurational free energy of the mixture[†]

$$F_c(T, V, x_1, \ldots, x_m)$$

$$= F_c^{(0)}(T, V) + \sum_{i,j=1}^{m} x_i x_j \left[U_c^{(0)} \frac{\varepsilon_{ij} - \varepsilon_0}{\varepsilon_0} + 3\left(N k_B T - P_0 V\right) \frac{\sigma_{ij} - \sigma_0}{\sigma_0} \right] + \cdots, \tag{7.17}$$

which allows us to obtain the thermodynamic properties of the mixture from those of the monodisperse reference fluid. If $\varepsilon_{ij} = \varepsilon_0$ and $\sigma_{ij} = \sigma_0$ for every i and j, the free energy of the mixture reduces to $F_c^{(0)}$, which can be considered as the free energy of an ideal mixture that includes the contribution from the ideal entropy of mixing (7.9).

Alternatively, instead of the length parameter σ, we could have taken the volume σ^3 as the perturbation parameter, which seems more logical if we are to take the reduced volume V^* or the reduced density ρ^* as one of the variables of the system. Then, instead of Equation 7.17, we would have obtained

$$F_c(T, V, x_1, \ldots, x_N)$$

$$= F_c^{(0)}(T, V) + \sum_{i,j=1}^{m} x_i x_j \left[U_c^{(0)} \frac{\varepsilon_{ij} - \varepsilon_0}{\varepsilon_0} + \left(N k_B T - P_0 V\right) \frac{\sigma_{ij}^3 - \sigma_0^3}{\sigma_0^3} \right] + \cdots \tag{7.18}$$

Obviously, Equations 7.17 and 7.18 become equivalent when σ and σ_0 are enough close to each other.

The theory of conformal mixtures is of a general character, in the sense that it is independent of the model chosen to describe the reference fluid and, moreover, it applies not only to fluid but also to solid mixtures. Truncating the expansion at first

---

* In its original formulation by Longuet-Higgins (1951), the term "*conformal solution*" was used in a more restrictive sense by imposing also the conditions that $\varepsilon_{ij}/\varepsilon_0$ must be close, to unity and that $\sigma_{ij} = \dfrac{1}{2}(\sigma_{ii} + \sigma_{jj})$ for every i and j.

[†] From now on we will suppress subscript or superscript *mix* when it be clear that we are referring to a mixture.

order has the inconvenience that the predicted excess functions are all positive or all negative, whereas experimentally it is found that for many mixtures some excess functions are positive and some others are negative. The theory of conformal mixtures has been extended to include second-order perturbation terms (Brown and Longuet-Higgins 1951, Brown 1957a–c), and in this case the predicted excess functions for a given mixture not necessarily all of them have the same sign. The problem is that the calculation of the second-order contribution involves n-body correlation functions of the reference fluid, with $n = 2$–4 (Brown 1957a), which prevents an exact calculation.

In a real mixture, the total potential energy $\mathcal{U}_N$ for a particular configuration depends not only on the positions of the N molecules that form the mixture but also on the species to which belongs the molecule assigned to each position. The *random mixing approximation* consists in replacing $\mathcal{U}_N$ in the configurational partition function of the mixture by its average over all possible assignments of the molecules to the N positions. In the randomized mixture, the probability that two molecules belonging to the species i and j are at position $\mathbf{r}_k$ and $\mathbf{r}_l$ depends only on the product x_i and x_j of the mole fractions for these components. Then, the total potential energy of the system for a given set of positions of the molecules in the random mixture approximation for a conformal m-component mixture will be

$$\langle \mathcal{U} \rangle_N \equiv \mathcal{U}(\mathbf{r}_1, \ldots, \mathbf{r}_N, x_1, \ldots, x_N) = \sum_{k=1}^{N-1} \sum_{l=k+1}^{N} \langle u(r_{kl}) \rangle = \sum_{i,j=1}^{m} \sum_{k=1}^{N-1} \sum_{l=k+1}^{N} x_i x_j u_{ij}(r_{kl})$$

$$= \sum_{i,j=1}^{m} \sum_{k=1}^{N-1} \sum_{l=k+1}^{N} x_i x_j \varepsilon_{ij} \varphi \left(\frac{r_{kl}}{\sigma_{ij}} \right). \tag{7.19}$$

The random mixing approximation becomes exact for an ideal mixture. As a matter of fact, the random mixture can be considered as an ideal mixture in which the pair potential for all the interactions is $\langle u(r) \rangle$ (Brown 1957a).

The configurational free energy of a mixture can be expressed as a series expansion in terms of the inverse of the reduced temperature, similar to Equation 6.13 for a monocomponent fluid but taking as the reference system the random mixture (Brown 1957b), namely

$$\beta F_c = \beta F_c^{(0)} - \sum_{n=2}^{\infty} \frac{K_n}{n!} \beta^n = \beta F_c^{(0)} - \frac{1}{2} \beta^2 \overline{\langle \mathcal{U}^2 \rangle_N - \langle \mathcal{U} \rangle_N^2} + \cdots, \tag{7.20}$$

where the angular brackets mean an average over assignments for a given set of N positions and the bar means an average over positions performed in the random mixture ensemble, and we have taken into account that the first-order term vanishes.

7.3 n-FLUID MODELS FOR CONFORMAL MIXTURES

The n-fluid models may be considered as zero-order conformal mixture theories or zero-order perturbation theories for mixtures since they consist in neglecting all terms but $F_c^{(0)}$ in expansions (7.17) or (7.18). From the viewpoint of the latter of these two expansions, this approximation seems more reasonable in advance if one chooses

$$\sigma_0^3 = \sum_i \sum_j x_i x_j \sigma_{ij}^3, \tag{7.21}$$

$$\varepsilon_0 = \sum_i \sum_j x_i x_j \varepsilon_{ij}, \tag{7.22}$$

because in this case the first-order term in Equation 7.18 vanishes. However, as we will see, other choices for the parameters of the reference system are also used.

On the other hand, restricting ourselves for simplicity to binary fluid mixtures, the reference fluid may be a monocomponent fluid, one-fluid model, or an ideal mixture of two or three fluids. Several of these models are described subsequently.

7.3.1 RANDOM MIXTURE MODEL

Considering that the reference fluid is a random mixture, this is equivalent to take

$$g_{ij}(r) = g_0(r) \tag{7.23}$$

for every i and j, where $g_0(r)$ is the RDF of the reference monocomponent fluid. Then, the energy equation (2.57) for the mixture leads to

$$U^E = 2\pi N\rho \int_0^\infty u_0(r) g_0(r) r^2 dr, \tag{7.24}$$

where

$$u_0(r) = \sum_{i,j=1}^m x_i x_j u_{ij}(r) \tag{7.25}$$

plays the role of $\langle u(r) \rangle$ in Equation 7.19.

Analogously, the pressure equation (2.55) takes the form

$$Z = 1 - \frac{2\pi}{3} \frac{\rho}{k_B T} \int_0^\infty g_0(r) \frac{du_0(r)}{dr} r^3 dr. \tag{7.26}$$

Other thermodynamic functions can be obtained from Equations 7.24 and 7.26. These expressions must be completed with the appropriate *mixing rules*, allowing us to obtain the parameters ε_0 and σ_0 of the reference system in terms of the parameters ε_i and σ_i of the pure fluids constituting the mixture. However, it is to be noted that approximation (7.23) is a quite poor one when the parameters σ_i are appreciably different from one species to another.

7.3.2   VAN DER WAALS ONE-FLUID MODEL

In this model, the thermodynamic properties of the mixture are approximated by those of a monocomponent reference fluid with parameters σ_0 and ε_0 determined from Equation 7.21 and from

$$\varepsilon_0 \sigma_0^3 = \sum_i \sum_j x_i x_j \varepsilon_{ij} \sigma_{ij}^3, \tag{7.27}$$

respectively (Leland et al. 1968). Then, for conformal mixtures with potentials of the form (7.10), the energy equation (2.57) takes the form

$$U^E = 2\pi N \rho \sum_{ij} x_i x_j \varepsilon_{ij} \sigma_{ij}^3 \int_0^\infty \varphi\left(\frac{r}{\sigma_{ij}}\right) g_{ij}\left(\frac{r}{\sigma_{ij}}\right) \left(\frac{r}{\sigma_{ij}}\right)^2 d\left(\frac{r}{\sigma_{ij}}\right)$$

$$= 2\pi N \rho \varepsilon_0 \sigma_0^3 \int_0^\infty \varphi\left(\frac{r}{\sigma_0}\right) g_0\left(\frac{r}{\sigma_0}\right) \left(\frac{r}{\sigma_0}\right)^2 d\left(\frac{r}{\sigma_0}\right). \tag{7.28}$$

This implies that the partial RDF for every pair ij is equal to that of the reference fluid, that is,

$$g_{ij}\left(\frac{r}{\sigma_{ij}}; \rho, T, \{x_i\}, \{\sigma_{ij}\}, \{\varepsilon_{ij}\}\right) = g_0\left(\frac{r}{\sigma_0}; \rho\sigma_0^3, \frac{k_B T}{\varepsilon_0}\right), \tag{7.29}$$

where $\{x_i\}$, $\{\sigma_{ij}\}$, and $\{\varepsilon_{ij}\}$ mean the whole sets of mole fractions of the components, and distance and energy parameters of the potential, respectively.

In a similar way, the pressure equation (2.55) now writes

$$Z = 1 - \frac{2}{3}\pi \frac{\rho}{k_B T} \sum_{ij} x_i x_j \varepsilon_{ij} \sigma_{ij}^3 \int_0^\infty \frac{d\varphi(r_{ij}/\sigma_{ij})}{d(r_{ij}/\sigma_{ij})} g_{ij}\left(\frac{r_{ij}}{\sigma_{ij}}\right) \left(\frac{r_{ij}}{\sigma_{ij}}\right)^3 d\left(\frac{r_{ij}}{\sigma_{ij}}\right) =$$

$$= 1 - \frac{2}{3}\pi \frac{\rho}{k_B T} \varepsilon_0 \sigma_0^3 \int_0^\infty \frac{d\varphi(r/\sigma_0)}{d(r/\sigma_0)} g_0\left(\frac{r}{\sigma_0}\right) \left(\frac{r}{\sigma_0}\right)^3 d\left(\frac{r}{\sigma_0}\right). \tag{7.30}$$

Equations 7.21 and 7.27, which together constitute the so-called *van der Waals mixing rules*, must still be complemented with some rules for the unlike mixed parameters σ_{ij} and ε_{ij} as, for example, the *Lorentz–Berthelot mixing rules*:

$$\sigma_{ij} = \frac{1}{2}\left(\sigma_i + \sigma_j\right), \tag{7.31}$$

$$\varepsilon_{ij} = \left(\varepsilon_i \varepsilon_j\right)^{1/2}. \tag{7.32}$$

In spite of its simplicity, the *van der Waals one-fluid model* (vdW1) provides quite satisfactory results for mixtures in which the potential parameters for the different components do not differ substantially from each other.

7.3.3 HARD-SPHERE EXPANSION

Proposed by Mansoori and Leland (1972), this version of the conformal theory uses a one-fluid model to obtain the excess properties of the mixture with respect to a reference HS mixture. The parameters ε_0 and d_0 of the monocomponent reference fluid are determined by solving the equations

$$\varepsilon_0 d_0^3 = \sum_i \sum_j x_i x_j \varepsilon_{ij} d_{ij}^3, \tag{7.33}$$

$$\varepsilon_0^2 d_0^3 = \sum_i \sum_j x_i x_j \varepsilon_{ij}^2 d_{ij}^3, \tag{7.34}$$

where d_{ij} are the effective HS diameters for the interaction between the components i and j. In the version of Shukla et al. (1986) of the *hard-sphere expansion* (HSE), the WCA approach (see Section 7.4) is used to determine the effective hard-sphere diameters as well as for the calculation of the attractive contribution to the thermodynamic properties.

7.3.4 VAN DER WAALS TWO-FLUID MODEL

Leland et al. (1969) proposed a two-fluid van der Waals model (vdW2) for conformal mixtures with the parameters σ_{0i} and ε_{0i} given by

$$\sigma_{0i}^3 = \sum_j x_j \sigma_{ij}^3, \tag{7.35}$$

$$\varepsilon_{0i} \sigma_{0i}^3 = \sum_j x_j \varepsilon_{ij} \sigma_{ij}^3. \tag{7.36}$$

The energy and pressure equations are now

$$U^E = 2\pi N\rho \sum_i x_i \varepsilon_{0i} \sigma_{0i}^3 \int_0^\infty \varphi\left(\frac{r}{\sigma_{0i}}\right) g_{ii}\left(\frac{r}{\sigma_{0i}}\right) \left(\frac{r}{\sigma_{0i}}\right)^2 d\left(\frac{r}{\sigma_{0i}}\right), \tag{7.37}$$

$$Z = 1 - \frac{2}{3}\pi \frac{\rho}{k_B T} \sum_i x_i \varepsilon_{0i} \sigma_{0i}^3 \int_0^\infty \frac{d\varphi(r/\sigma_{0i})}{d(r/\sigma_{0i})} g_{ii}\left(\frac{r}{\sigma_{0i}}\right) \left(\frac{r}{\sigma_{0i}}\right)^3 d\left(\frac{r}{\sigma_{0i}}\right). \tag{7.38}$$

Therefore, this model approximates a binary mixture by an ideal mixture of the two reference fluids (obviously, for a multicomponent mixture there will be as many reference fluids as components in the mixture). The preceding results can be derived from the energy (2.57) and pressure (2.55) equations for mixtures if one assumes that the partial RDFs satisfy the relationships

$$g_{ii}\left(\frac{r}{\sigma_{ii}}; \rho, T, \{x_i\}, \{\sigma_{ij}\}, \{\varepsilon_{ij}\}\right) = g_{0i}\left(\frac{r}{\sigma_{0i}}; \rho\sigma_{0i}^3, \frac{k_B T}{\varepsilon_{0i}}\right), \tag{7.39}$$

$$g_{ij}\left(\frac{r}{\sigma_{ij}}\right) = \frac{1}{2}\left[g_{ii}\left(\frac{r}{\sigma_{ii}}\right) + g_{jj}\left(\frac{r}{\sigma_{jj}}\right)\right]. \tag{7.40}$$

7.3.5   THREE-FLUID MODELS

A three-fluid model can be derived as a generalization of a two-fluid model by defining a reference fluid for each different type of interaction ij (three for a binary mixture). In one of such models, proposed by Scott (1956), a binary mixture is regarded as an ideal mixture of three components with parameters σ_{ij} and ε_{ij}, with $i,j = 1,2$.

Another three-fluid model is the *mean density approximation* (MDA) (Fisher and Leland 1970, Mansoori and Leland 1972) in which the three reference fluids have the same parameter σ_0, as given by Equation 7.21, but retain the three different ε_{ij}, so that the partial RDFs are taken to be

$$g_{ij}\left(\frac{r}{\sigma_{ij}}; \rho, T, \{x_i\}, \{\sigma_{ij}\}, \{\varepsilon_{ij}\}\right) = g_0\left(\frac{r}{\sigma_{ij}}; \rho\sigma_0^3, \frac{k_B T}{\varepsilon_{ij}}\right). \tag{7.41}$$

Instead, in the *scaling approximation* (SA) of Gazzillo et al. (1999), the three reference fluids retain the three different parameters σ_{ij} but have the same parameter ε_0, that is,

$$g_{ij}\left(\frac{r}{\sigma_{ij}}; \rho, T, \{x_i\}, \{\sigma_{ij}\}, \{\varepsilon_{ij}\}\right) = g_0\left(\frac{r}{\sigma_{ij}}; \rho\sigma_m^3, \frac{k_B T}{\varepsilon_0}\right), \tag{7.42}$$

with $\sigma_m^3 = \sum_i x_i \sigma_i^3$.

To some extent related to the n-fluid models may be considered the one proposed by Hamad (1996a, 1997). For the kind of mixtures we are considering here, although the model is not limited to conformal mixtures, Hamad proposes three different equations of state, depending on the information available on the second virial coefficient of the mixture, as shown next

$$Z = 1 + \rho \sum_i \sum_j x_i x_j B_2^{(ij)} \mathcal{F}_{ij}\left(\rho \sum_k x_k f_{ijk}, T; \{\varepsilon_{ij}\}\right), \tag{7.43}$$

$$Z = 1 + \rho \sum_i x_i \mathcal{F}_i\left(\rho \sum_k x_k f_{ik}, T; \{\varepsilon_i\}\right) \sum_j x_i B_2^{(ij)}, \tag{7.44}$$

$$Z = 1 + \rho B_2 \mathcal{F}\left(\rho \sum_k x_k f_k, T; \varepsilon\right). \tag{7.45}$$

Here, the function $\mathcal{F}$ is related to the reference fluid EOS Z_0 through the relationship $\mathcal{F} = (Z_0 - 1)/B_2\rho$, B_2 is the second virial coefficient of the mixture, $B_2^{(ij)}$ the contribution of the interaction between components i and j to B_2, and the parameters f_i, f_{ij}, and f_{ijk} are related to the second and third virial coefficients *of a hard-body fluid mixture* whose components have molecular shapes similar to those of the components of the actual mixture. For the spherically symmetric potential models considered here, the hard-body mixture will be the HS mixture, for which the second and third virial coefficients are known analytically (Kihara 1955, Kihara and Miyoshi 1975). In this case, $f_i = f_{ii} = f_{iii...} = \sigma_{ii}^3$, $f_{ij} = \sigma_{ij}^3$, and, for binary mixtures,

$$
\begin{aligned}
f_{iij} &= \frac{\left(\sigma_{ii}^3 - 12\sigma_{ii}\sigma_{ij}^2 + 16\sigma_{ij}^3\right)}{5}, & \sigma_{ij} &\geq \frac{\sigma_{ii}}{2}, \\
&= 0, & \sigma_{ij} &\leq \frac{\sigma_{ii}}{2}, \\
f_{iji} &= \frac{\sigma_{ii}^3\left(8\sigma_{ij} - 3\sigma_{ii}\right)}{5\sigma_{ij}}, & \sigma_{ij} &\geq \frac{\sigma_{ii}}{2}, \\
&= \frac{16\sigma_{ij}^3}{5}, & \sigma_{ij} &\leq \frac{\sigma_{ii}}{2},
\end{aligned}
\tag{7.46}
$$

where σ_{ij} is the distance of closest approach between the centers of two spheres of species i and j, $f_{iij} = f_{jij}$, and $f_{jii} = f_{iji}$. Equations 7.43 through 7.45 may be considered as corresponding to three-, two-, and one-fluid models, respectively.

7.4 EXTENSION TO MIXTURES OF PERTURBATION THEORIES FOR MONOCOMPONENT SYSTEMS

7.4.1 EXTENSION TO MIXTURES OF INTEGRAL EQUATION AND INTEGRAL EQUATION PERTURBATION THEORIES

The extension of the OZ equation (4.1) to mixtures reads

$$
c_{ij}(r) = h_{ij}(r) - \sum_k \rho_k \int c_{ik}(\mathbf{r}')h_{kj}(|\mathbf{r} - \mathbf{r}'|)d\mathbf{r}', \tag{7.47}
$$

where $\rho_k = \rho x_k$ is the number density of component k. In a similar way, the exact closure (4.10) now writes

$$
c_{ij}(r) = h_{ij}(r) - \ln y_{ij}(r) + B_{ij}(r). \tag{7.48}
$$

Some of the simple closures cited in Section 4.2 can be readily extended to mixtures, without further complications, apart from the fact that now there are more equations to solve. Those closures involving semi-empirical parameters, such as the VM closure (4.23), may require a re-definition of the parameters. Self-consistent integral equation theories will require additional constraints in order to reduce the number of parameters to be determined from consistency conditions, because otherwise the number of parameters needed in their formulation for simple fluids will be multiplied for mixtures by the number of different pairs ij. Several ways of extending the RHNC to simple mixtures were cited in Section 4.4. In case of taking a mixture as the reference system, the needed bridge function can be obtained from any suitable integral equation theory for the reference mixture.

7.4.2 EXTENSION TO MIXTURES OF FREE ENERGY PERTURBATION THEORIES

The extension to mixtures of some of the perturbation theories based on the series expansion of the free energy analyzed in Chapter 6 is straightforward. For the reference system, we may choose single- or multi-component systems. Thus, using a single-component HS reference system, Equation 6.39 for the free energy in the first-order perturbation theory for simple fluids, for mixtures transforms into (Leonard et al. 1970)

$$\frac{F}{Nk_BT} = \frac{F_0}{Nk_BT} + \sum_i x_i \ln x_i + 2\pi\beta\rho \sum_{i,j} x_ix_j \int_{\sigma_{ij}}^{\infty} u_{ij}(r)\, g_0(r)\, r^2 dr, \qquad (7.49)$$

where subscript 0 refers to a single-component fluid of HS with diameter

$$d = \sum_{i,j} x_ix_j\delta_{ij}, \qquad (7.50)$$

with

$$\delta_{ij} = \int_0^{\sigma_{ij}} \left[1 - e^{-\beta u_{ij}(r)}\right] dr. \qquad (7.51)$$

Leonard et al. (1970) also derived the expression corresponding to use a HS mixture as the reference system. For a binary mixture, it writes

$$\frac{F}{Nk_BT} = \frac{F_0}{Nk_BT} - 4\pi\rho x_1x_2 d_{12}^2 g_{12}^{(0)}(d_{12})(d_{12} - \delta_{12})$$

$$+ 2\pi\beta\rho \sum_{i,j} x_ix_j \int_{\sigma_{ij}}^{\infty} u_{ij}(r)\, g_{ij}^{(0)}(r)\, r^2 dr, \qquad (7.52)$$

where F_0 and $g_{ij}^{(0)}$ are the free energy and the partial RDFs, respectively, of an HS mixture with sphere diameters $d_{ii} = \delta_{ii}$ and, for additive hard-sphere (AHS) mixtures, $d_{ij} = (d_{ii} + d_{jj})/2$. The second term in the right-hand side arises from the fact that $\delta_{ij} \neq (\delta_{ii} + \delta_{jj})/2$, in general. Obviously, this term vanishes when all the components of the mixture have potentials with spherical hard cores with additive diameters.

The WCA theory is another perturbation theory which can be applied to mixtures as a direct extension of its pure component counterpart (Lee and Levesque 1973, Shukla et al. 1986). Denoting

$$b_{ij}(d_{ij}) = \int y_{ij}^{HS}(r, d_{ij}) \left[e^{-\beta u_{ij}^{(0)}(r)} - e^{-\beta u^{HS}(r,d_{ij})}\right] d\mathbf{r}, \qquad (7.53)$$

where $u_{ij}^{(0)}(r)$ is the reference potential for the interaction ij, defined in a similar way as in Equation 6.69, condition (6.73) is now replaced with the set of conditions

$$b_{ij}(d_{ij}) = 0. \qquad (7.54)$$

However, in determining the diameters d_{ij} in this way, again the additivity condition will not be satisfied, in general. The problem may be overcome by using a nonadditive hard-sphere (NAHS) mixture, for which $d_{ij} \neq (d_{ii} + d_{jj})/2$, to obtain $g_{ij}^{(0)}$ and F_0 (Kahl 1990). However, it is often preferred to use AHS fluid mixtures because their thermodynamic and structural properties are better known than those of NAHS mixtures. Lee and Levesque (1973) determined the diameters d_{ii} from condition (7.54) and the d_{ij}, with $i \neq j$, from the additivity condition and the same approximation was used in the Shukla et al. (1986) improved version of the Lee–Levesque theory. Fischer and Lago (1983) determine the diameters in the same way, but now include the contribution to the first-order blip expansion of the free energy of the reference system arising from the fact that the d_{ij} for unlike interactions obtained from the additivity condition does not satisfy condition (6.73), so that the Fischer–Lago expression for the free energy of the reference system for binary mixtures takes the form

$$\frac{F_0}{Nk_BT} = \frac{F_{mix}^{HS}}{Nk_BT} - \rho x_1 x_2 b_{12}(d_{12}).$$ (7.55)

Instead, Bohn et al. (1986), following an earlier suggestion by Perram (1984) (see also Perram et al. 1988), replace the conditions (7.54) for a binary mixture with

$$\begin{aligned} x_1 b_{11}(d_{11}) + x_2 b_{12}(d_{12}) = 0 \\ x_1 b_{12}(d_{12}) + x_2 b_{22}(d_{22}) = 0 \end{aligned},$$ (7.56)

which imply that $F_0 = F_{mix}^{HS}$, together with the additivity condition.

Another procedure was devised by Kahl and Hafner (1985). For binary mixtures, it is based on determining d_{11} and d_{22} from condition (7.54), with the reference potentials $u_{11}^{(0)}(r)$ and $u_{22}^{(0)}(r)$, while the splitting distance for the potential $u_{12}(r)$, instead of being the minimum of the potential as in Equation 6.69, is determined from the requirement that the additivity condition and Equation 7.54 must be satisfied simultaneously.

The Song–Mason approximation was extended to mixtures by Song (1990). The EOS of the mixture reads

$$Z = 1 + \rho \sum_{i,j} x_i x_j \left\{ B_{ij} + \omega_{ij} \left[g_{ij}^{HS}(d_{ij}) - 1 \right] \right\},$$ (7.57)

where $\omega_{ij} = -2\pi \int_0^{r_m(ij)} f_{ij}^{(0)}(r) r^2 dr$, with $f_{ij}^{(0)}(r)$ being the Mayer function for the reference potential $u_{ij}^{(0)}(r)$ and $r_m(i,j)$ the position of the minimum of the potential $u(r_{ij})$, and B_{ij} is the contribution to the second virial coefficient from the pairs of components ij. d_{ij} is determined from the covolume for the interaction ij, defined by

$$b_{ij}(T) = 2\pi \int_0^{r_m(i,j)} \left\{ 1 - \left[1 + \beta u_{ij}^{(0)}(r) \right] e^{-\beta u_{ij}^{(0)}(r)} \right\} r^2 dr.$$ (7.58)

The extension of the variational theory to mixtures is also straightforward and was carried out by Mansoori and Leland (1970). The inequality (6.90) now becomes

$$\beta F \leq \beta F_0 + 2\pi \beta N \rho \sum_{i,j} x_i x_j \int_0^\infty u_{ij}^{(1)}(r) g_{ij}^{(0)}(r) r^2 dr, \qquad (7.59)$$

where $u_{ij}^{(1)}(r)$ is the perturbative part of the pair potential for the interaction between particles i and j in a splitting similar to that in Equation 6.69.

We will end this section summarizing the extension of the RPA to mixtures. Further details may be found in Andersen and Chandler (1971, 1972). The free energy is given as in Equation 6.101, where now F_0 is the free energy of the reference HS mixture,

$$\frac{\beta F_{HTA}}{V} = 2\pi \beta \rho^2 \sum_{i,j} x_i x_j \int u_{ij}^{(1)}(r) g_{ij}^{(0)}(r) d\mathbf{r}, \qquad (7.60)$$

and

$$\frac{\beta F_R}{V} = -\frac{1}{2(2\pi)^3} \int \left\{ \text{Tr} \mathbf{R}(k) \mathbf{S}^{(0)}(k) - \ln \det \left[\mathbf{1} + \mathbf{R}(k) \mathbf{S}^{(0)}(k)\right]\right\} d\mathbf{k}. \qquad (7.61)$$

In the last expression, $\mathbf{R}(k)$ and $\mathbf{S}^{(0)}(k)$ are matrices whose elements are

$$R_{ij}(k) = \beta \rho x_i \hat{u}_{ij}(k) \qquad (7.62)$$

and

$$S_{ij}^{(0)}(k) = \delta_{ij} + x_i \rho \hat{h}_{ij}^{(0)}(k), \qquad (7.63)$$

respectively, where $\hat{h}_{ij}^{(0)}(k)$ is the Fourier transform of the total correlation function for the pair of components ij, "Tr" means the trace of the matrix, "det" means a determinant, δ_{ij} is the Kronecker delta, and **1** is the unit matrix.

The optimizing condition now writes

$$\frac{\delta F_R}{u_{ij}^{(1)}(r)} = 0, \quad r < \sigma_{ij}, \qquad (7.64)$$

where σ_{ij} is the distance of closest approach between the centers of spheres with diameters σ_i and σ_j.

7.5 MIXTURES OF ADDITIVE HARD SPHERES

In the preceding section, we have seen that most of the perturbation theories require the knowledge of the EOS and structural properties of a reference HS mixture. In this section, we will describe some theories for AHS mixtures. Denoting σ_i and σ_j the diameters of the spheres of species i and j, for AHS, the distance σ_{ij} of closest approach between their centers satisfies condition (7.31).

7.5.1 SCALED PARTICLE THEORY FOR ADDITIVE HARD-SPHERE FLUID MIXTURES

The derivation of the SPT for AHS mixtures (Lebowitz et al. 1965) is quite similar
to that for the pure HS fluid (see Section 5.2). The work $W(r, \{x_i\}, \{r_i\})$ needed to
insert a sphere of radius r in a HS fluid mixture consisting of m species with different
radius $r_i = \sigma_i/2$ and mole fractions x_i is related to the probability of having a cavity
of radius r from which it is excluded any point of any particle of the fluid by means
of an expression similar to Equation 5.1, namely,

$$p_0(r) = e^{-W(r, \{x_i\}, \{r_i\})/k_B T}. \tag{7.65}$$

Note that the minimum distance from the center of a sphere of radius r_i to the center of
the cavity is $r + r_i$, which depends on component i. Denoting by $p_1(r)$ the probability
that inside the cavity be located the center of a particle, for $r \le 0$ at most the center
of one particle can be within the cavity, so that

$$p_1(r) = \frac{4}{3}\pi\rho \sum_{i=1}^{m} x_i(r + r_i)^3 = 1 - p_0(r), \quad r \le 0, \tag{7.66}$$

and therefore

$$W(r) = -k_B T \ln\left[1 - \frac{4}{3}\pi\rho \sum_{i=1}^{m} x_i(r + r_i)^3\right], \quad r \le 0. \tag{7.67}$$

The equivalent of Equation 5.3 for mixtures is

$$\frac{\partial W(r)}{\partial r} = 4\pi\rho k_B T \sum_{i=1}^{m} x_i(r + r_i)^2 G_i(r + r_i), \tag{7.68}$$

and, taking into account Equation 7.67, one obtains

$$G_i(r + r_i) = \frac{1}{1 - (4/3)\pi\rho \sum_{j=1}^{m} x_j(r + r_j)^3}, \quad r \le 0. \tag{7.69}$$

In a similar way as for the monocomponent HS fluid, it holds true

$$\sum_{i=1}^{m} x_i G_i(\infty) = \frac{\beta}{\rho} \sum_{i=1}^{m} P_i = Z, \tag{7.70}$$

where
   P_i is the partial pressure of the component i
   Z is the compressibility factor of the mixture

As in the pure HS fluid, for $r \geq 0$ a polynomial form is assumed for $G(r + r_i)$:

$$G_i(r + r_i) = \beta \frac{P_i}{x_i \rho} + \frac{a_i}{r + r_i} + \frac{b_i}{(r + r_i)^2}, \tag{7.71}$$

where the coefficients a_i and b_i are determined from the condition that for $r = 0$ Equations 7.69 and 7.71 must be equal, as well as must be their first derivatives. This gives

$$a_i = -\frac{2\beta P_i r_i}{x_i \rho} + \frac{2r_i}{1 - \xi_3} + \frac{6r_i^2 \xi_2}{(1 - \xi_3)^2}, \tag{7.72}$$

$$b_i = \frac{\beta P_i r_i^2}{x_i \rho} - \frac{r_i^2}{1 - \xi_3} - \frac{6r_i^3 \xi_2}{(1 - \xi_3)^2}, \tag{7.73}$$

where

$$\xi_l = \frac{\pi}{6} \rho \sum_{i=1}^{m} x_i \sigma_i^l. \tag{7.74}$$

Introducing the results (Equations 7.72 and 7.73) into Equation 7.71, taking into account that $G_i(r_i + r_j) = G_j(r_j + r_i) = g_{ij}(\sigma_{ij})$, after some algebra one obtains

$$g_{ij}(\sigma_{ij}) = \frac{1}{1 - \xi_3} + \frac{3}{2} \frac{\xi_2}{(1 - \xi_3)^2} \left(\frac{\sigma_{ii} \sigma_{jj}}{\sigma_{ij}} \right) + \frac{3}{4} \frac{\xi_2^2}{(1 - \xi_3)^3} \left(\frac{\sigma_{ii} \sigma_{jj}}{\sigma_{ij}} \right)^2, \tag{7.75}$$

which, from the virial theorem (2.56) for HS mixtures, yields

$$Z = \frac{6}{\pi \rho} \left[\frac{\xi_0}{1 - \xi_3} + 3 \frac{\xi_1 \xi_2}{(1 - \xi_3)^2} + 3 \frac{\xi_2^2}{(1 - \xi_3)^3} \right], \tag{7.76}$$

where $\sigma_{ii} = \sigma_i$ and σ_{ij} is given by Equation 7.31.

Introducing the expression (7.71) for $G_i(r + r_i)$, with coefficients a_i and b_i given by Equations 7.72 and 7.73, respectively, into Equation 7.68 and integrating with respect to r, we would obtain the expression of $W(r)$. Alternatively, we might approximate $W(r)$ with a third-degree polynomial, like in Equation 5.11 for the monocomponent HS fluid, and determine the coefficients from the condition of continuity of $W(r)$ and its first two derivatives at $r = 0$. Both procedures lead to the same result (Lebowitz et al. 1965).

7.5.2   PERCUS–YEVICK THEORY FOR ADDITIVE HARD-SPHERE FLUID MIXTURES

The solution of the PY theory for the RDF and the EOS of a pure HS fluid was generalized by Lebowitz (1964) to an m-component mixture of AHS. The PY closure (4.15) for mixtures transforms into

$$c_{ij}(r) = f_{ij}(r) y_{ij}(r), \tag{7.77}$$

where $f_{ij} = e^{-\beta u_{ij}(r)} - 1$ is the Mayer function for the pairs ij. For an AHS mixture,

$$f_{ij}(r) = \begin{cases} 0, & r < \sigma_{ij} \\ 1, & r > \sigma_{ij} \end{cases}. \tag{7.78}$$

Taking into account this relationship, together with the definition of the cavity function $y_{ij}(r) = g_{ij}(r)e^{\beta u_{ij}(r)}$ for pairs ij, from Equation 7.77 the following conditions, similar to those in Equation 5.13, arise

$$\left. \begin{array}{ll} g_{ij}(r) = 0, & r < \sigma_{ij} \\ g_{ij}(r) = y_{ij}(r), & r > \sigma_{ij} \\ c_{ij}(r) = -y_{ij}(r), & r < \sigma_{ij} \\ c_{ij}(r) = 0, & r > \sigma_{ij} \end{array} \right\}. \tag{7.79}$$

Defining the functions

$$\begin{array}{ll} s_{ij}(r) = 12(\eta_i\eta_j)^{1/2} r g_{ij}(r), & r > \sigma_{ij} \\ s_{ij}(r) = -12(\eta_i\eta_j)^{1/2} r c_{ij}(r), & r < \sigma_{ij} \end{array}, \tag{7.80}$$

where $\eta_i = (\pi/6)\rho\sigma_i^3$ is the packing fractions for the species i, the OZ equation (7.47) for mixtures can be expressed in terms of the functions $s_{ij}(r)$. Lebowitz (1964) solved the resulting equation and obtained the expressions for the contact values $g_{ij}(\sigma_{ij})$ of the partial RDFs, which may be written in the unified form:

$$g_{ij}(\sigma_{ij}) = \frac{1}{1 - \xi_3} + \frac{3\xi_2\sigma_i\sigma_j / (\sigma_i + \sigma_j)}{(1 - \xi_3)^2}, \tag{7.81}$$

where the ξ_l have the same meaning as in Equation 7.74 and, in particular, $\xi_3 \equiv \eta_{mix}$, where $\eta_{mix} = \sum_i \eta_i$ is the packing fraction of the mixture. Introduction of expression (7.81) into the virial equation (2.56) for HS mixtures leads to

$$Z_{PY}^v = \frac{6}{\pi\rho} \left[\frac{\xi_0}{1 - \xi_3} + 3\frac{\xi_1\xi_2}{(1 - \xi_3)^2} + 3\frac{\xi_2^3(1 - \xi_3)}{(1 - \xi_3)^3} \right]. \tag{7.82}$$

On the other hand, from the compressibility equation, which for mixtures may be written in the form

$$\frac{1}{k_B T} \left(\frac{\partial P}{\partial \rho_j} \right)_T = 1 - 4\pi\rho \sum_i x_i \int_0^\infty r^2 c_{ij}(r)\, dr, \tag{7.83}$$

Lebowitz (1964) obtained

$$Z_{PY}^c = \frac{6}{\pi\rho} \left[\frac{\xi_0}{1 - \xi_3} + 3\frac{\xi_1 \xi_2}{(1 - \xi_3)^2} + 3\frac{\xi_2^3}{(1 - \xi_3)^3} \right], \tag{7.84}$$

which, as in the case of the monocomponent fluid, coincides with the SPT result (7.76).

The PY compressibility equation (7.84) was also derived by Baxter (1970) using the procedure based on Equations 4.8 and 4.9 and outlined in Section 5.3 for the pure HS fluid, Equations 5.19 and 5.20. For HS mixtures, Equations 4.8 and 4.9 transforms into Equations 7.85 and 7.86

$$rc_{ij}(|r|) = -q'_{ij}(r) + 2\pi\rho \sum_k x_k \int_{\Delta\sigma_{ik}}^{\sigma_{ik}} q_{ik}(t) q'_{jk}(r+t) \, dt, \quad \Delta\sigma_{ij} < r < \sigma_{ij}, \tag{7.85}$$

$$rh_{ij}(|r|) = -q'_{ij}(r) + 2\pi\rho \sum_k x_k \int_{\Delta\sigma_{ik}}^{\sigma_{ik}} q_{ik}(t)(r-t) h_{jk}(|r-t|) \, dt, \quad r > \Delta\sigma_{ij}, \tag{7.86}$$

where $\Delta\sigma_{ij} = (\sigma_i - \sigma_j)/2$ and $q_{ij}(z) = 0$ for $z > \sigma_{ij}$ by definition. Then, in a similar way as for the pure HS fluid

$$q_{ij}(r) = \frac{1}{2} a_i \left(r^2 - \sigma_{ij}^2 \right) + b_i \left(r - \sigma_{ij} \right), \quad \Delta\sigma_{ij} < r < \sigma_{ij}, \tag{7.87}$$

with

$$a_i = 1 - 2\pi\rho \sum_k x_k \int_{\Delta\sigma_{ik}}^{\sigma_{ik}} q_{ik}(t) \, dt, \quad b_i = 2\pi\rho \sum_k x_k \int_{\Delta\sigma_{ik}}^{\sigma_{ik}} q_{ik}(t) t \, dt. \tag{7.88}$$

Substituting Equation 7.87 into Equations 7.88 and solving the system yields (Baxter 1970)

$$a_i = \frac{1 - \xi_3 + 3\sigma_i \xi_2}{(1 - \xi_3)^2}, \quad b_i = -\frac{3}{2} \frac{\sigma_i^2 \xi_2}{(1 - \xi_3)^2}. \tag{7.89}$$

Once $q_{ij}(r)$ is completely specified by Equations 7.87 and 7.89, Equation 7.85 can be integrated to obtain $c_{ij}(r)$ and, from the compressibility equation (7.83) for mixtures, the PY compressibility EOS (7.84) results.

Lebowitz (1964) also reported analytical expressions for the Laplace transforms of the partial RDFs $g_{ij}(r)$. Inversion of these expressions was performed by Throop and Bearman (1965) by means of the same procedure previously used by Wertheim (1963) for the monocomponent HS fluid, although they did not report analytical expressions or any numerical data of the partial RDFs. Subsequently,

Leonard et al. (1971) derived essentially analytical expressions, though quite complex, for the RDFs.

7.5.3 Some Improved Approximations

In a similar way as for the monocomponent HS fluid, neither the SPT (7.76) nor the PY-v (7.82) equations provide satisfactory results at high packing fractions, as illustrated in Figure 7.1 for a binary mixture with diameter ratio $R \equiv \sigma_1/\sigma_2 = 3/2$ and two different mole fractions x_1, where subscript 1 refers to the bigger spheres. Figure 7.1 shows that the SPT, or equivalently the PY-c, and the PY-v equations overestimate and underestimate, respectively, the EOS, like in the monocomponent fluid. Similar results are obtained for higher diameter ratios, up to at least $R = 3$, and mole fractions, with little relative differences with diameter ratio and/or mole fraction. Obviously, neither the contact values $g_{ij}(\sigma_{ij})$ of the partial RDFs obtained from these theories at high densities are satisfactory.

For these reasons, there have been developed some heuristic improvements similar to those that for the monocomponent HS fluid allows to obtain the Carnahan–Starling equation (5.21) from either the SPT or the combination of the two PY equations (see Section 5.3). Thus, multiplying the SPT expression (7.75) by a factor 2/3, Boublík (1970) obtains for the contact RDFs the expression

$$g_{ij}(\sigma_{ij}) = \frac{1}{1 - \xi_3} + \frac{3}{2}\frac{\xi_2}{(1 - \xi_3)^2}\left(\frac{\sigma_{ii}\sigma_{jj}}{\sigma_{ij}}\right) + \frac{1}{2}\frac{\xi_2^2}{(1 - \xi_3)^3}\left(\frac{\sigma_{ii}\sigma_{jj}}{\sigma_{ij}}\right)^2, \qquad (7.90)$$

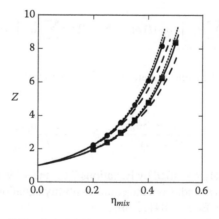

FIGURE 7.1 Compressibility factor Z for binary HS fluid mixtures with diameter ratio $R = 3$ as a function of the packing fraction η_{mix} of the mixture. Points: simulation data from Barrio and Solana (2005, 2006) for $x_1 = 0.75$ (circles) and $x_1 = 0.0625$ (squares). Dotted curves: SPT equation (7.76), or equivalently PY-c. Dashed curves: PY-v equation (7.82). Continuous curves: BMCSL equation (7.91).

and, introducing this result into the virial equation (2.56), the corresponding EOS reads

$$Z = \frac{6}{\pi \rho} \left[\frac{\xi_0}{1 - \xi_3} + \frac{3\xi_1 \xi_2}{(1 - \xi_3)^2} + \frac{(3 - \xi_3)\xi_2^3}{(1 - \xi_3)^3} \right]. \tag{7.91}$$

The latter expression was also obtained independently by Mansoori et al. (1971) as the sum of the PY virial (7.82) and compressibility (7.84) equations weighted by 1/3 and 2/3, respectively, and for this reason both the expression (7.90) for the contact RDFs and the EOS (7.91) are usually referred to as the *Boublík–Mansoori–Carnahan–Starling–Leland* (BMCSL) results, although strictly speaking explicit expressions of the $g_{ij}(\sigma_{ij})$ were not reported by Mansoori et al. (1971). The performance of the BMCSL EOS is excellent, as shown in the example displayed in Figure 7.1, and extends over wide ranges of diameter ratios and compositions.

A similar conclusion applies to the contact RDFs, as is clear from Figure 7.2, although some deviation of $g_{11}(\sigma_{11})$ can be observed for large values of the diameter ratio R and extremely small values of the mole fraction x_1. Several improvements for the BMCSL RDFs at contact, mainly for $g_{11}(\sigma_{11})$, have been proposed in the literature for HS mixtures with large values of R and/or small values of x_1, often based on some known exact results for $T \rightarrow \infty$ and/or $x_1 \rightarrow 0$ (the colloidal limit) together with heuristic considerations. A number of these expressions have been collected and analyzed by Barrio and Solana (2008).

Concerning the partial RDFs, the results from the PY theory for HS mixtures deviate from the simulation data in a similar way as does the PY theory for the mono-component fluid. Therefore, different approaches have been devised to improve the accuracy. Thus, Grundke and Henderson (1972) and Lee and Levesque (1973) intro-duced empirical corrections to the partial RSFs in a similar way as done previously by Verlet and Weis (1972) for the monocomponent fluid, as mentioned in Section 5.3, using the BMCSL expressions (7.90) for the contact RDFs as constraints. The FMSA theory (see Section 5.5) and the RFA method (see Section 5.4) were also extended to AHS mixtures by Tang and Lu (1995) and Bravo Yuste et al. (1998), respectively. Both theories provide analytical expressions for the Laplace transforms of the partial RDFs using the BMCSL results as an input and provide nearly the same accuracy. The RFA theory provides excellent agreement with simulations for those mixtures for which the BMCSL equation (7.90) is accurate, as illustrated in Figure 7.3 for a binary mixture of HS quite asymmetric in size. Similar accuracy is achieved with the FMSA theory.

On the other hand, several of the perturbation theories cited in the preceding sections might be applied to AHS mixtures. However, concerning the EOS and the contact values of the partial RDFs, there is no point to resort to these theories as the BMCSL equations (7.90) and (7.91) are more simple and accurate. With regard to the partial RDFs beyond contact, one might think of using a three-fluid model, for a binary mixture, in combination with the RDF of the monocomponent HS fluid obtained from a suitable theory, which will be simpler than the corresponding counterpart for the mixture; however, this procedure seems not to yield very satisfactory results (Chen et al. 1987, Barrio and Solana 2005).

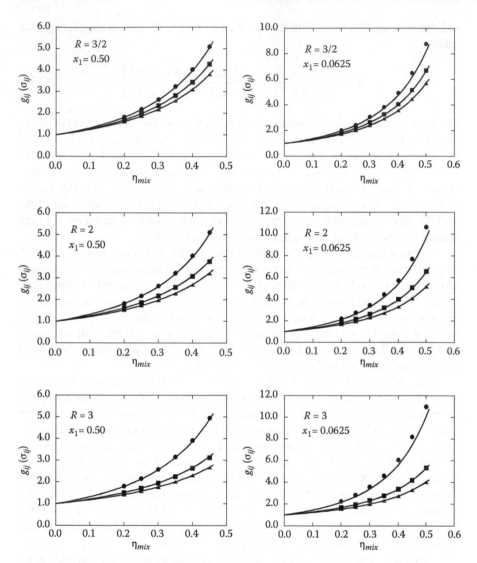

FIGURE 7.2 Partial radial distribution functions at contact distance $g_{ij}(\sigma_{ij})$ for binary mixtures of HS as a function of the packing fraction η_{mix} of the mixture for several diameter ratios R and mole fractions x_1 of the bigger spheres. Points: simulation data from Barrio and Solana (2005, 2006) for $g_{11}(\sigma_{11})$ (circles), $g_{12}(\sigma_{12})$ (squares), and $g_{22}(\sigma_{22})$ (triangles). Curves: BMCSL expression (7.90).

7.6   MIXTURES OF NONADDITIVE HARD SPHERES

In some mixtures, instead of an AHS mixture as the reference system in perturbation theory, it is better to consider a mixture of NAHS. In contrast to AHS, for NAHS mixtures, the distance σ_{ij} of closest approach between the centers of

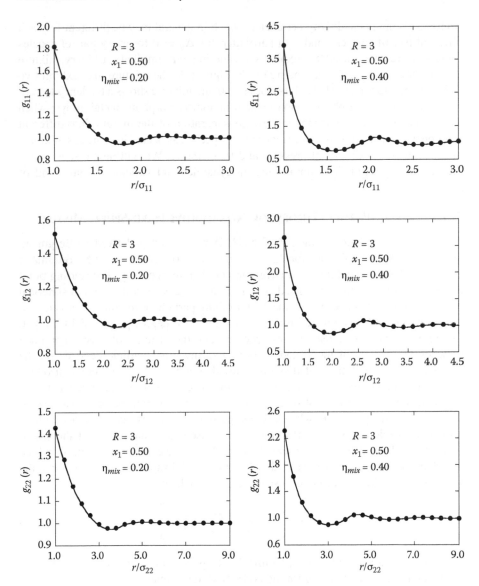

FIGURE 7.3 Partial radial distribution functions for an equimolar binary mixture of HS as a function of the reduced radial distance r/σ_{ij} for diameter ratio $R = 3$ and two different packing fractions. Points: simulation data from Barrio and Solana (2002). Curves: RFA theory.

spheres belonging to different species i and j is not the arithmetic mean of their diameters, but

$$\sigma_{ij} = \frac{1}{2}(\sigma_i + \sigma_j)(1 + \Delta_{ij}), \qquad (7.92)$$

where Δ_{ij} is the nonadditivity parameter, which in general may be different for each different pair of species i and j. In particular, for $\Delta_{ij} = 0$ for every pair of species i and j we recover an AHS mixture. Considering for simplicity a binary mixture we will have a single Δ. Depending on the sign of Δ, the mixture may exhibit two different thermodynamic behaviors: for $\Delta < 0$, the mixture shows a tendency toward heterocoordination, a phenomenon present in some amorphous metal alloys; when $\Delta > 0$ one observes a tendency toward the separation of the mixture into different fluid phases, enriched in one or another of the species, as occurs in some gas mixtures under extreme conditions of pressure and temperature. We will next discuss some theoretical approximations for the thermodynamics and structure of this kind of mixtures.

7.6.1 Scaled-Particle Theory for Nonadditive Hard-Sphere Mixtures

The general formulation of the SPT for NAHS mixtures was derived by Bergmann (1976) and explicit solutions for $\Delta_{ij} < 0$ and $\Delta_{ij} > 0$ were reported by Bergmann and Tenne (1978) and Tenne and Bergmann (1978), respectively. The starting point is the expression of the work needed to introduce a cavity in the fluid, a spherical region from which it is excluded the center of any particle of the fluid, that is to say, the equivalent of Equation 7.67 for $r \leq 0$ and the third-degree polynomial for $r \geq 0$ for AHS, but now we must take into account that the distance of closest approach between the centers of the scaled particle and any other particle depends on the species to which both the scaled and the unscaled particles belong. Let us introduce a parameter s determining the scaling of the distance σ_{ij} of closest approach between the centers of two particles of species i and j so that $\sigma_{ij}(s)$ will be such distance for the value s of the parameter when a scaled particle of species i is in contact with an unscaled particle of species j. Parameter s is chosen such that for $s = 1$ the scaled i-particle will be identical to unscaled i-particle, whereas for some value s_i, which will depend on the species i, the scaled i-particle will disappear. Then, for $s < s_i$ the work $W_i(s)$ required to introduce a scaled i-cavity will be

$$W_i(s) = -k_B T \ln \left[1 - \frac{4}{3}\pi\rho \sum_{j=1}^{m} x_j \sigma_{ij}^3(s) \right], \quad s \leq s_i. \tag{7.93}$$

For $s > s_i$ it is assumed for $W_i(s)$ a third-degree polynomial, similar to those for the monocomponent HS fluid and for the AHS mixtures, namely

$$W_i(s) = W_i(s_i) + W_i'(s - s_i) + \frac{1}{2}W_i''(s - s_i)^2 + W_C(s), \quad s \geq s_i, \tag{7.94}$$

where $W_C(s)$ is the work needed to create a macroscopic i-cavity. The coefficients in the expansion are determined, as usual, from the condition of continuity of $W_i(s)$ and its first two derivatives at $s = s_i$. To do so we need an explicit expression for the scaling function $\sigma_{ij}(s)$ as well as for $W_C(s)$. There is some room for the choice of the scaling function; denoting $a_{ij} = (\sigma_i + \sigma_j)\Delta_{ij}/2$, Tenne and Bergmann (1978) chose

$$\sigma_{ij}(s) = \frac{\sigma_{ii} + a_{ij}}{2}s + \frac{\sigma_{jj} + a_{ij}}{2}, \quad s \geq s_i, \tag{7.95}$$

and for s_i they took the minimum of the values satisfying the condition $\sigma_{ij}(s_i) = 0$, with $j = 1, \ldots, m$. With this choice the size of a macroscopic i-cavity depends on the interaction ij. Therefore, to calculate the work needed to create a macroscopic i-cavity, we must consider different spherical shells surrounding the cavity, with each shell being permeable for some species but not for others and the innermost shell being impenetrable. As a consequence, an osmotic pressure will develop across each of the concentric shells. The work $W_C(s)$ needed to create the macroscopic i-cavity will consist of a contribution due to the expansion work against the pressure in the innermost shell plus additional contributions due to expansion work of each of the spherical shells against the osmotic pressures. The EOS may then be obtained from the exact relationship (Bergamann 1976)

$$Z = 1 + \frac{1}{k_B T} \sum_{i=1}^{m} \left. \frac{\partial W_i(s)}{\partial s} \right|_{s=1}. \tag{7.96}$$

Alternatively, the pressure can be obtained by integrating the Gibbs–Duhem equation

$$\frac{\partial P}{\partial \rho} = \sum_{i=1}^{m} x_i \rho \frac{\partial \mu_i}{\partial \rho}, \tag{7.97}$$

where μ_i is the chemical potential for component i, which is related to $W_i(s)$ through the relationship

$$\mu_i = k_B T \ln\left[x_i \rho \Lambda_i^3\right] + W_i(1), \tag{7.98}$$

in which the first term of the right-hand side is the ideal gas contribution. Then, Equation 7.97 leads to

$$\frac{\partial P}{\partial \rho} = 1 + \sum_{i=1}^{m} x_i \rho \frac{\partial W_i(1)}{\partial \rho}. \tag{7.99}$$

Any of these two procedures leads to a set of equations that may be solved numerically.

Instead of the relationship (7.95), Mazo and Bearman (1990) chose the scaling function to be

$$\sigma_{ij}(s) = \frac{\sigma_{ii}}{2} s + a_{ij} + \frac{\sigma_{jj}}{2}, \quad s \geq s_i, \tag{7.100}$$

which has the advantage with respect to Equation 7.95 that now the size of a large i-cavity depends only on σ_{ii}. With this approach, the term $W_C(s)$ in Equation 7.94 takes the form

$$W_C(s) = P \frac{\pi}{6} \sigma_{ii}^3 (s - s_i)^3. \tag{7.101}$$

The value of s_i is obtained from the same condition as before. Note that in the last expression only the total pressure P of the fluid is needed, so that this procedure avoids the need of considering the osmotic pressures across the different shells and the shells themselves, in contrast to the Tenne and Bergmann approach. Using the Gibbs–Duhem equation (Equation 7.97), Mazo and Bearman (1990) derived an analytical expression for the EOS of NAHS.

Although simpler to use than the Tenne and Bergmann approach, the Mazo and Bearman scaling (7.100) has the drawback that it is not applicable to the limiting case of the Widom–Rowlinson binary mixture (Widom and Rowlinson 1970), for which $\sigma_{11} = \sigma_{22} = 0$ and $\sigma_{12} = a > 0$, because in the scaling law (7.100) a_{12} is not scaled. On the other hand, none of the two approaches reduces, as a particular case, to the monocomponent SPT EOS (Mazo and Bearman 1970).

The Bergmann and Tenne approach for binary symmetric $(\sigma_1 = \sigma_2 = \sigma)$ equimolar $(x_1 = x_2 = 0.5)$ mixtures with $\Delta = 1$ was compared with simulations by Ehrenberg et al. (1990). They found that the SPT results obtained from the virial route are in quite good agreement with simulations, whereas those obtained from the Gibbs–Duhem equation are too high. As the value of Δ is reduced, approaching to zero but still positive, the Gibbs–Duhem results tend to be closer to the simulations than the virial ones, which become too low for Δ close to zero, and even so in this case the agreement is limited to low densities. For negative nonadditivities, Bergmann and Tenne (1978) showed that their SPT theory works quite well for moderate to low densities, whereas for high densities it provides too high results, especially from the Gibbs–Duhem route, and the deviation increases as the value of Δ becomes more negative.

7.6.2 INTEGRAL EQUATION THEORIES FOR NONADDITIVE HARD-SPHERE MIXTURES

In contrast to the monocomponent HS fluid and to the AHS mixtures, to the best of our knowledge, at present there are no available analytical solutions for the EOS and the RDFs for NAHS from integral equation theories. For the particular case of equimolar symmetric mixtures of NAHS with positive nonadditivities, an approximate analytical solution was derived by Gazzillo (1987) from the PY IET; the results obtained for the compressibility factor and the contact values of the RDFs for $\Delta \le 0.5$ were in very close agreement with those provided by the exact PY theory. Nixon and Silbert (1984) compared the results for the EOS obtained from the PY and SPT-virial theories with the simulation data for the same kind of mixtures; they found that the PY theory yields quite accurate results for Δ close to zero, but as the nonadditivity parameter becomes more negative the predicted values at high densities are increasingly low as compared with the simulations. In contrast, the SPT values are systematically high for all values of Δ. A significant improvement over the PY results for the EOS and the RDFs near contact was achieved by Ballone et al. (1986) with the MS and BPGG closures (see Section 4.2) for both positive and negative nonadditivities. The excellent performance of the latter closure was confirmed by Jung et al. (1994a) by comparing with their extensive simulation data. Some evidence also exists (Lomba et al. 1996) on the good performance for this kind of fluids of the VM closure (see Section 4.2) as well as that of a doubly self-consistent VM closure introduced by these authors. The latter imposes thermodynamic consistency between the virial and compressibility routes and also between the virial EOS and the chemical potential, for example through the Gibbs–Duhem equation 7.97.

The demixing transition of NAHS mixtures of equal-sized diameters with positive nonadditivities was analyzed by Gazzillo (1991) from the PY, MS, and BPGG closures. For equimolar mixtures with $0 < \Delta \le 1$, it was shown that the PY theory considerably overestimates the critical density obtained from simulation. Better agreement was achieved with the MS and BPGG closures, but still the predicted values were somewhat high. For the BPGG closure, in particular, this may be due to the fact that the theory considerably underestimates the excess chemical potential at high densities, at least for low values of Δ. It is to be noted that the critical density increases with decreasing Δ and that low values of Δ are more appropriate, in general, to modelize real fluids. A modification to the BPGG closure introduced by Gazzillo (1991), by imposing good agreement at $\Delta = 0$ with the results for Z and μ^E obtained from the Carnahan–Starling equation (5.21), removes the inaccuracy mentioned earlier. Similar considerations apply to non-equimolar mixtures. The VM and doubly self-consistent VM closures also provide very good predictions of the chemical potential, particularly the second of these closures as one might expect; the coexistence curve is also very accurately predicted by these theories, the second of them being more accurate for the region not too close to the critical point (CP) and vice versa near the CP. The later situation is due to the inadequacy of the consistency condition between the virial and compressibility routes near the CP.

7.6.3 PERTURBATION THEORIES FOR NONADDITIVE HARD-SPHERE MIXTURES

Contrarily to the case of AHS mixtures, for which we have available simple and accurate expressions for the contact RDFs and the EOS derived from the SPT or PY theories, namely the BMCSL expressions (7.90) and (7.91) respectively, the situation is very different in the case of NAHS mixtures, as we have seen in this section. Therefore, in this case, we should not discard in advance the field of free energy perturbation theories, and so the performance of some of these theories is discussed next.

Henderson and Barker (1968) developed, within the general framework of the conformal mixture theory, a perturbation theory for HS mixtures, either additive or nonadditive, with single-component HS reference fluid. To first order, the expression of the free energy reduces to the first two terms of the right-hand side of Equation 7.49, with δ_{ij} in Equation 7.50 replaced with σ_{ij} and $d \equiv \sigma_0$, the diameter of the spheres in the reference system. If instead of Equation 7.50 one takes Equation 7.21 for σ_0, one obtains the vdW1 result for this kind of mixtures. The second-order term was derived independently by Henderson and Leonard (1971) and Smith (1971).

Using for the reference fluid an AHS mixture, the Leonard et al. (1970) expression for the free energy in the first-order perturbation theory for a NAHS binary mixture reduces to the first two terms in the right-hand side of Equation 7.52 with $\delta_{ij} = \sigma_{ij}$ again and $d_{ij} \equiv \sigma_{ij}^0 = (\sigma_{ii} + \sigma_{jj})/2$. Then, to first order, the free energy of the NAHS mixture may be rewritten as

$$\frac{F}{Nk_BT} = \frac{F_0}{Nk_BT} + 2\pi\rho \sum_{i,j} x_i x_j \left(\sigma_{ij}^0\right)^3 g_{ij}^{(0)}\left(\sigma_{ij}^0\right) \Delta_{ij}. \tag{7.102}$$

For the particular case of symmetric binary mixtures, $\sigma_1 = \sigma_2 = \sigma$, the reference AHS mixture is a single-component HS fluid and, using for the latter the expression of the free energy and the contact RDF obtained from the CS equation (5.21), expression (7.102) reduces to

$$\frac{F}{Nk_BT} = 3\ln\Lambda + \ln\rho - 1 + \sum_i x_i \ln x_i + \frac{(4-3\eta)\eta}{(1-\eta)^2} + 12 x_1 x_2 \frac{(2-\eta)\eta}{(1-\eta)^3}\Delta, \tag{7.103}$$

which is denoted as *MIX1 approximation* by Melnyk and Sawford (1975). Henderson et al. (1972) developed the second-order perturbation theory for mixtures with multicomponent reference fluid and Melnyk and Sawford (1975) derived a particular solution, which they termed *MIX2 approximation*, for symmetric NAHS mixtures with the AHS mixture as the reference system.

The MIX1 approximation was generalized by Schaink and Hoheisel (1992) to asymmetric mixtures by assuming for the contact RDFs the expression

$$g_{ij}\left(\sigma_{ij}\right) = \left(\frac{\sigma_{ij}^0}{\sigma_{ij}}\right)^3 \left\{ g_{ij}^{(0)}\left(\sigma_{ij}^0\right) + 3\Delta_{ij} \frac{1}{(1-\xi_3)^2} \right.$$

$$\left. \times \left[1 + 3\frac{\sigma_{ii}\sigma_{jj}}{\sigma_{ij}^0} \frac{\xi_2}{1-\xi_3} + \frac{3}{2}\left(\frac{\sigma_{ii}\sigma_{jj}}{\sigma_{ij}^0}\right)^2 \frac{\xi_2^2}{(1-\xi_3)^2}\right] \right\}, \tag{7.104}$$

where

$g_{ij}^{(0)}(\sigma_{ij}^0)$ are the contact RDFs for the reference AHS mixture, as given by the BMCSL expression (7.90)

$g_{ij}(\sigma_{ij})$ are those for NAHS

σ_{ij}^0 and σ_{ij} are the contact distance between the centers of two particles of species i and j in the reference AHS mixture and in the NAHS mixture, respectively

The preceding equation, when introduced into the virial equation (2.56) for HS mixtures, leads to the same EOS as that derived from Equation 7.102 for the free energy. For the symmetric case, Equation 7.104 leads to the MIX1 EOS and for AHS mixtures reduces to the BMCSL expression (7.90).

Adams and McDonald (1975) analyzed the predictions of several of these first- and second-order perturbation theories, with either one-component or multicomponent HS fluid as the reference fluid, for the compressibility factor of symmetric equimolar NAHS mixtures with $\Delta < 0$. They found that none of them yields satisfactory accuracy for the whole range of densities and nonadditivities. A similar situation occurs for positive nonadditivities. Several attempts have been conducted to improve the MIX1 approximation (Paricaud 2008, Santos et al. 2010), but the results are still far from being satisfactory for a wide variety of NAHS mixtures.

Neither is the Hamad equation (Equation 7.43) accurate, in general, except for low to moderate densities.

The Mayer function perturbation theory of Sillrén and Hansen (2007) (see Section 6.1) was extended by the same authors to NAHS mixtures. To this end, the deviation from nonadditivity is taken as the perturbation so that, for a binary mixture of N_A particles of species A and N_B particles of species B, we will have

$$\mathcal{U}_1 = \sum_{i=1}^{N_A} \sum_{j=1}^{N_B} u_1^{AB} \left(r_{ij} \right). \tag{7.105}$$

Then, Equations 6.24, 6.27, and 6.28 transform into

$$\beta \mathcal{U}_1^{AB} \left(\alpha \right) = - \sum_{i=1}^{N_A} \sum_{j=1}^{N_B} \ln \left[1 + \alpha f_{AB} \left(r_{ij} \right) \right], \tag{7.106}$$

$$\beta F_1 = -4\pi N x_A x_B \rho \int_0^\infty g_{AB}^{(0)} (r) f_{AB} (r) r^2 dr, \tag{7.107}$$

$$\beta F_2 = 2\pi N x_A x_B \rho \int_0^\infty g_{AB}^{(0)} (r) f_{AB}^2 (r) r^2 dr - \frac{1}{2} \beta^2 \left[\left\langle (\mathcal{U}_1'|_0)^2 \right\rangle_0 - \langle \mathcal{U}_1'|_0 \rangle_0^2 \right], \tag{7.108}$$

where $f_{AB}(r_{ij}) = e^{-\beta u_1^{AB}(r_{ij})} - 1$. As in the case of monocomponent fluids, the last term in Equation 7.108 cannot be calculated exactly in general and, moreover, for mixtures there is not available an approximate expression such as the BH lc approximation Equation 6.62, so that in practice we are limited to use a first-order perturbation theory.

Figure 7.4 compares the predictions from the first-order Mayer function perturbation theory and the MIX1 theory with simulations for the critical density ρ_c for the demixing transition of NAHS mixtures as a function of the nonadditivity parameter Δ. It is seen that the former of these theories provides quite good accuracy for $\Delta \leq 1$. Also, the MIX1 theory is remarkably accurate for $\Delta \lesssim 0.4$. The situation is quite different for the coexistence curve, as Figure 7.5 shows. Now, the MIX1 theory is more accurate for $\Delta = 0.2$ and in very close agreement with simulations, although this seems to be somewhat fortuitous, as none of the two is accurate for $\Delta = 0.1$ nor are expected to be accurate at coexistence for $\Delta > 0.2$ away from the CP. Concerning the EOS, both theories are quite accurate for small positive nonadditivities ($\Delta \lesssim 0.2$), as illustrated in Figure 7.6, but fail completely for negative nonadditivities, although the situation for the Mayer function perturbation theory might change if the second-order term could be accurately obtained.

Another approximation for the contact RDFs and the EOS of NAHS mixtures was developed by Hamad (1996b, 1997) on the basis of the SPT for AHS mixtures and

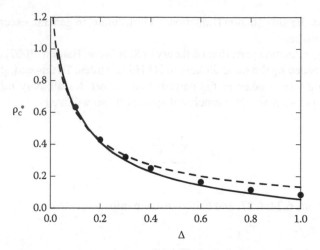

FIGURE 7.4   Reduced critical density $\rho_c^* = \rho\sigma^3$ for the demixing transition of symmetric NAHS fluid mixtures as a function of the nonadditivity parameter Δ. Points: simulation data from Góźdź (2003). Continuous curve: first-order Mayer function perturbation theory. Dashed curve: MIX1 theory.

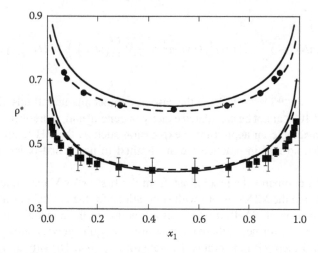

FIGURE 7.5   Reduced density $\rho^* = \rho\sigma^3$ at coexistence for symmetric NAHS mixtures as a function of the mole fraction. Points: simulation data from Rovere and Pastore (1994) for $\Delta = 0.1$ (circles) and from Amar (1989) for $\Delta = 0.2$ (squares). Continuous curves: first-order Mayer function perturbation theory. Dashed curves: MIX1 theory.

the analytically known (the second and third) virial coefficients for NAHS mixtures. The contact RDFs are assumed to be

$$g_{ij}\left(\sigma_{ij}\right) = \frac{1}{1-\zeta_3} + \frac{3}{2}\frac{\zeta_2}{(1-\zeta_3)^2} + A\frac{\zeta_2^2}{(1-\zeta_3)^3}, \qquad (7.109)$$

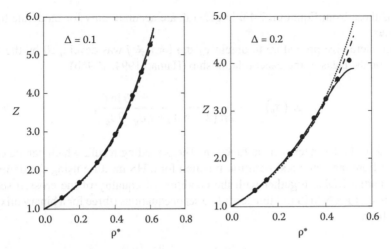

FIGURE 7.6 Compressibility factor Z of two binary equimolar symmetric mixtures of NAHS as a function of the reduced density $\rho^* = \rho\sigma^3$. Points: simulation data from Saija et al. (1998) for $\Delta = 0.1$ and from Jung et al. (1995) for $\Delta = 0.2$. Continuous curves: Equation 7.109 with $A = 1/2$ and $g_{12}(\sigma_{12})$ calculated from Equation 7.116. Dashed curves: first-order Mayer function perturbation theory. Dotted curves: MIX1 theory, which for $\Delta = 0.1$ is indistinguishable from the continuous curve at the scale of the figure.

with

$$\zeta_2 = \frac{\pi}{6}\rho \sum_k x_k a_{ijk}, \quad \zeta_3 = \frac{\pi}{6}\rho \sum_k x_k b_{ijk}. \tag{7.110}$$

Coefficients a_{ijk} and b_{ijk} can be obtained from the second and third virial coefficients for NAHS which, as said before, are analytical. The constraint that Equation 7.109 must reduce to the SPT expression (7.75) for AHS as a particular case yields $A = 3/4$. Then, coefficients a_{ijk} and b_{ijk} can be explicitly determined. For binary mixtures, they are

$$a_{111} = \sigma_{11}^3, \quad a_{112} = \sigma_{11}(2\sigma_{12} - \sigma_{11})^2, \quad a_{121} = \sigma_{11}^3(2\sigma_{12} - \sigma_{11})/\sigma_{12},$$
$$a_{122} = \sigma_{22}^3(2\sigma_{12} - \sigma_{22})/\sigma_{12}, \quad a_{221} = \sigma_{22}(2\sigma_{12} - \sigma_{22})^2, \quad a_{22} = \sigma_{22}^3,$$
$$b_{111} = \sigma_{11}^3, \quad b_{112} = (2\sigma_{12} - \sigma_{11})^3, \quad b_{121} = \sigma_{11}^3, \tag{7.111}$$
$$b_{122} = \sigma_{22}^3, \quad b_{221} = (2\sigma_{12} - \sigma_{22})^3, \quad b_{222} = \sigma_{22}^3.$$

However, as in the BMCSL expression (7.90) for AHS mixtures, it is better to replace $A = 3/4$ with $A = 1/2$ in Equation 7.109. In this way, the latter equation reduces to the BMCSL expression when $\Delta = 0$. The EOS corresponding to Equation 7.109 can be obtained using the pressure equation (2.56) for HS mixtures. The approximation just outlined might be considered as belonging to the class of the three-fluid models. However, the predicted values of the EOS from this approximation for NAHS, as well as those from Equation 7.43, in general, are accurate only for moderate to low densities.

An alternative procedure to obtain $g_{ij}(\sigma_{ij})$ for $i \neq j$ was developed by the same author on the basis of the exact relationship (Hamad 1994, 1996b)

$$g_{ij}\left(\sigma_{ij}\right) = -\frac{1}{2\pi\left(2 - \delta_{ij}\right)\rho x_i x_j \sigma_{ij}^2}\frac{\partial \ln Q}{\partial \sigma_{ij}}, \qquad (7.112)$$

where δ_{ij} is the Kronecker delta function. The preceding result, which can be easily derived from the canonical partition function for a HS mixture using the definition of the partial RDFs, together with the condition of equality of the crossed second derivatives, for a NAHS mixture leads to a set of equations (three for a binary mixture) of the form

$$x_i\sigma_{ii}^2\frac{\partial g_{ii}\left(\sigma_{ii}\right)}{\partial \sigma_{ij}} = 2x_j\sigma_{ij}^2\frac{\partial g_{ij}\left(\sigma_{ij}\right)}{\partial \sigma_{ii}}, \qquad (7.113)$$

$$x_j\sigma_{jj}^2\frac{\partial g_{jj}\left(\sigma_{jj}\right)}{\partial \sigma_{ij}} = 2x_i\sigma_{ij}^2\frac{\partial g_{ij}\left(\sigma_{ij}\right)}{\partial \sigma_{jj}}, \qquad (7.114)$$

$$x_i^2\sigma_{ii}^2\frac{\partial g_{ii}\left(\sigma_{ii}\right)}{\partial \sigma_{jj}} = x_j^2\sigma_{jj}^2\frac{\partial g_{jj}\left(\sigma_{jj}\right)}{\partial \sigma_{ii}}. \qquad (7.115)$$

with $i \neq j$. The integration of the two former of these equations for the general case of a multicomponent mixture leads to

$$g_{ij}\left(\sigma_{ij}\right) = \frac{x_i}{2x_j\sigma_{ij}^2}\int_{\sigma_{ij}}^{\sigma_{ii}}\frac{\partial g_{ii}\left(\sigma_{ii}\right)}{\partial \sigma_{ij}}\sigma_{ii}^2 d\sigma_{ii} + \frac{x_j}{2x_i\sigma_{ij}^2}\int_{\sigma_{ij}}^{\sigma_{jj}}\frac{\partial g_{jj}\left(\sigma_{jj}\right)}{\partial \sigma_{ij}}\sigma_{jj}^2 d\sigma_{jj} + g_0\left(\sigma_{ij}\right), \quad (7.116)$$

where $g_0(\sigma_{ij})$ is the contact RDF of a pure fluid of HS with diameter σ_{ij} and the same number density as the mixture. Equation 7.116, which must be numerically solved, allows us to obtain the partial RDF $g_{ij}(\sigma_{ij})$ at contact for the unlike interaction ij from the knowledge of the RDFs $g_{ii}(\sigma_{ii})$ and $g_{jj}(\sigma_{jj})$ at contact for the like interactions ii and jj, respectively. Using for the latter two contact RDFs the expression 7.109 with $A = 1/2$, quite good accuracy is achieved for the compressibility factor Z for low positive nonadditivities as well as for negative nonadditivities, as illustrated in the examples of Figures 7.6 and 7.7, respectively. The accuracy in the predicted $g_{ij}(\sigma_{ij})$ is, in general, considerably lower.

7.7 OTHER SIMPLE MIXTURES

Integral equation theories and integral equation perturbation theories are used sometimes for simple mixtures, although they are not the preferred theories for mixtures because of their increased complexity as compared with pure fluids, as said before.

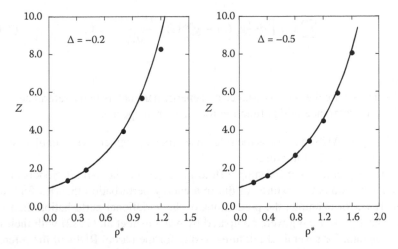

FIGURE 7.7 Compressibility factor Z of two binary equimolar symmetric mixtures of NAHS as a function of the reduced density $\rho^* = \rho\sigma^3$. Points: simulation data from Jung et al. (1994b). Continuous curves: Equation 7.109 with $A = 1/2$ and $g_{12}(\sigma_{12})$ calculated from Equation 7.116.

As a general rule, we can say that theories that work well for pure fluids also do so for simple fluid mixtures. Thus, for example, Schöl-Paschinger et al. (2005) obtained very good accuracy for the predicted phase diagrams of HCY binary mixtures using SCOZA. They considered symmetric binary mixtures with $\varepsilon_{11} = \varepsilon_{22}$, and several values of the ratio $\varepsilon_{12}/\varepsilon_{11}$ and of the parameter κ, and found from simulations as well as from theory that, depending on the values of κ and of the ratio mentioned earlier, the mixtures may exhibit a CP for the liquid–vapor (L–V) transition, a critical end point (CEP) for the demixing transition, where the L–L and the L–V lines intersect to each other, a tricritical point, where they merge the CEP and the L–V CP, and a triple point. The temperature and density obtained from SCOZA for each of these points were in excellent agreement with the simulations. Quite accurate results were also obtained by Caccamo et al. (1998) for the L–V coexistence of the same kind of mixtures from the RHNC theory.

Kambayashi and Hiwatari (1990) successfully applied their version of the modified HNC (see Section 4.5 for more details) to supercooled binary mixtures of soft spheres with $n = 12$. The extremely soft case of strongly coupled plasma mixtures was studied by Kang and Ree (1998) by means of their PHNC theory (see Section 4.4 for more details); the results for the excess energy and the partial RDFs were in excellent agreement with the simulation data.

Enciso et al. (1987) extended to mixtures the RHNC with the equivalent for mixtures of the optimizing conditions (4.55) and (4.56), namely,

$$\sum_{i,j} x_i x_j \int d\mathbf{r} \left[g_{ij}(r) - g_{ij}^{(0)}(r) \right] \frac{\partial B_{ij}^{(0)}(r)}{\partial d_k} = 0, \qquad (7.117)$$

$$\sum_{i,j} x_i x_j \int d\mathbf{r} \left[g_{ij}(r) - g_{ij}^{(0)}(r) \right] \frac{\partial B_{ij}^{(0)}(r)}{\partial \varepsilon_l} = 0, \qquad (7.118)$$

where

$\{d_k\}$ and $\{\varepsilon_l\}$ are the sets of distance and energy parameters in the reference system

$B_{ij}^{(0)}$ is the reference bridge function for the pair of species ij

Considering an AHS mixture as the reference, good agreement with simulations was obtained for several LJ mixtures.

More frequently used for simple fluid mixtures are the n-fluid models and the different extensions to mixtures of the free energy perturbation theories for simple fluids. Thus, for example, the predictions of the variational perturbation theory for the free energy of mixing were compared by Nakanishi et al. (1982) with their own simulation data for several LJ mixtures. Using for the partial RDFs of the reference HS fluid mixture the PY solution, the theoretical results were in excellent agreement with the simulations, as shown in Figure 7.8.

The vapor pressures for LJ mixtures with different size and energy ratios and compositions were investigated by Vrabec et al. (1995), who found very satisfactory results from the WCA theory with condition (7.56). Shukla et al. (1986) compared with simulations the predictions of the vdW1, HSE, MDA, and WCA theories, as well as the first-order perturbation theory of Leonard et al. (1970), Equation 7.52, for

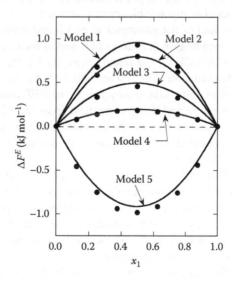

FIGURE 7.8 Excess free energy of mixing at $T = 120$ K and $\rho^* = 0.75$ for LJ mixtures with $\sigma_{11} = \sigma_{22} = \sigma_{12} = 3.405$ Å, atomic mass 39.948 amu, values corresponding to the pure LJ fluid, and values of the ratios $\varepsilon_{22}/\varepsilon_{11}$ and $\varepsilon_{12}/\varepsilon_{11}$ equal to 2 and 1 for Model 1, to 2 and $\sqrt{2}$ for Model 2, to 3 and $\sqrt{3}$ for Model 3, to 4 and 2 for Model 4, and to 2 and 2 for Model 5, respectively. Points are simulation data and curves are the predictions from the variational perturbation theory, after Nakanishi et al. (1982).

several LJ mixtures mimicking mixtures of simple real fluids. They found that the WCA theory was more accurate than the first-order perturbation theory of Leonard et al. for the total volume and enthalpy, whereas for the excess properties, V^E, H^E, and G^E, both theories yielded comparable accuracy, which was attributed to the cancellation of errors in the second of these theories between the thermodynamic functions of the mixture and those of the pure component fluids. For these excess properties, the WCA theory was also found to be more accurate than the vdW1 and HSE theories. In any case, the accuracy of the WCA theory extended only to ratios of the σ parameters smaller than 1.3. The same authors analyzed the predictions of the vdW1, MDA, and WCA theories for the excess free energy of mixtures of equal-sized LJ molecules with different compositions, ratios of the ε parameters, and mixing rules. For Lorentz–Berthelot mixtures, good agreement with simulations was obtained from the WCA and MDA theories, even for quite large ε ratios, whereas for non-Lorentz–Berthelot mixtures the accuracy of the second of these two theories was found to be considerably superior. The predictions from the vdW1, MDA, and WCA theories for the energy and the EOS of LJ mixtures were analyzed by Miyano (1991, 1994), who showed that all these theories yield fairly accurate values of the energy even for quite extreme values of the energy and size ratios of the interactions; the WCA also yields very satisfactory results for the EOS, whereas the agreement of the other two theories with the simulation data at high densities is quite poor. However, the performance of the vdW1 model for the EOS much improves when the size parameter of the pure reference fluid is determined in such a way that the EOS of an HS mixture is reproduced by that of a pure HS fluid calculated in the vdW1 approximation, as proposed by Takamiya and Nakanishi (1988). With this choice, the accuracy of the vdW1 approximation is in general superior to that of the other cited theories and extends also to the predicted pressures for the vapor–liquid equilibria. Also, the performance of the HSE theory can be improved by properly determining the effective HS diameter (Chen et al. 1987). Blas and Fujihara (2002) compared the excess volume and enthalpy of several binary LJ mixtures determined from the vdW1 theory with a pure LJ reference fluid with their own simulation data. For the Helmholtz free energy of the reference fluid, they used the equation of Johnson et al. (1993) obtained from the fitting of the simulation data. They found excellent agreement between theory and simulation.

The picture that emerges from the preceding discussion is that for non-hard-sphere mixtures the preferred choice for obtaining the thermodynamic properties should be one of the *n*-fluid models, especially the vdW1 or the HSE, because of their simplicity and accuracy, provided that the reference fluid is properly determined.

REFERENCES

Adams, D. J. and I. R. McDonald. 1975. Fluids of hard spheres with nonadditive diameters. *J. Chem. Phys.* 63:1900.

Amar, J. G. 1989. Application of the Gibbs ensemble to the study of fluid-fluid phase equilibrium in a binary mixture of symmetric non-additive hard spheres. *Mol. Phys.* 67:739.

Andersen, H. C. and D. Chandler. 1971. Mode expansion in equilibrium statistical mechanics. III. Optimized convergence and application in ionic solution theory. *J. Chem. Phys.* 55:1497.

Andersen, H. C. and D. Chandler. 1972. Optimized cluster expansions for classical fluids. I. General theory and variational formulation of the mean spherical model and hard-sphere Percus-Yevick equations. *J. Chem. Phys.* 57:1918.

Ballone, P., G. Pastore, G. Galli, and D. Gazzillo. 1986. Additive and non-additive hard sphere mixtures. Monte Carlo simulation and integral equation results. *Mol. Phys.* 59:275.

Barrio, C. and J. R. Solana. 2002. Test of a scaling approximation for the radial distribution functions of additive hard sphere fluid mixtures. *Proceedings of the 16th European Conference on Thermophysical Properties (ECTP-2002).* Imperial College, London, U.K.

Barrio, C. and J. R. Solana. 2005. Mapping a hard-sphere fluid mixture onto a single component hard-sphere fluid. *Physica A* 351:387.

Barrio, C. and J. R. Solana. 2006. Scaled hard-sphere equation of state and contact pair correlation functions for additive hard-sphere fluid mixtures. *Proceedings of the VII Iberoamerican Conference on Phase Equilibria and Fluid Properties for Process Design,* Univ. Michoacana de San Nicolás de Hidalgo, Morelia, México, pp. 606–619.

Barrio, C. and J. R. Solana. 2008. Binary mixtures of additive hard spheres. Simulations and theories. In *Theory and Simulation of Hard-sphere Fluids and Related Systems,* ed. A. Mulero, *Lect. Notes Phys.* 753:133–182. Berlin, Germany: Springer-Verlag.

Baxter, R. J. 1970. Ornstein-Zernike relation and Percus-Yevick approximation of fluid mixtures. *J. Chem. Phys.* 52:4559.

Bergmann, E. 1976. Scaled particle theory for non-additive hard spheres. General theory and solution of the Widom-Rowlinson model. *Mol. Phys.* 32:237.

Bergmann, E. and R. Tenne. 1978. Scaled particle theory of mixtures of hard spheres with negatively non-additive diameters. *Chem. Phys. Lett.* 56:310.

Blas, F. J. and I. Fujihara. 2002. Excess properties of Lennard-Jones binary mixtures from computer simulation and theory. *Mol. Phys.* 100:2823.

Bohn, M., J. Fischer, and F. Kohler. 1986. Prediction of excess properties for liquid mixtures: Results from perturbation theory for mixtures with linear molecules. *Fluid Phase Equilibr.* 31:233.

Boublík, T. 1970. Hard-sphere equation of state. *J. Chem. Phys.* 53:471.

Bravo Yuste, S., A. Santos, and M. López de Haro. 1998. Structure of multi-component hard-sphere mixtures. *J. Chem. Phys.* 108:3683. A Mathematica® program to calculate the partial RDFs of additive HS mixtures from the RFA is available at the URL: http://www1.unex.es/eweb/fisteor/santos/filesRFA.html (accessed February 28, 2011).

Brown, W. B. 1957a. The statistical thermodynamics of mixtures of Lennard-Jones molecules. I. Random mixtures. *Phil. Trans. Roy. Soc. Lond. A* 250:175.

Brown, W. B. 1957b. The statistical thermodynamics of mixtures of Lennard-Jones molecules. II. Deviation from random mixing. *Phil. Trans. Roy. Soc. Lond. A* 250:221.

Brown, W. B. 1957c. The second-order theory of conformal solutions. *Proc. Roy. Soc. A* 240:561.

Brown, W. B. and H. C. Longuet-Higgins. 1951. The statistical thermodynamics of multicomponent systems. *Proc. Roy. Soc. A* 209:416.

Caccamo, C., D. Costa, and G. Pellicane. 1998. A comprehensive study of the phase diagram of symmetrical hard-core Yukawa mixtures. *J. Chem. Phys.* 109:4498.

Chen, L.-J., J. F. Ely, and G. A. Mansoori. 1987. Mean density approximation and hard sphere expansion theory: A review. *Fluid Phase Equilibr.* 37:1.

Ehrenberg, V., H. M. Schaink, and C. Hoheisel. 1990. Pressure and coexistence curve of two- and three-dimensional nonadditive hard core mixtures. Exact computer calculation results compared with scaled particle theory predictions. *Physica A* 169:365.

Enciso, E., F. Lado, M. Lombardero, J. L. F. Abascal, and S. Lago. 1987. Extension of the optimized RHNC equation to multicomponent liquids. *J. Chem. Phys.* 87:2249.

Fischer, J. and S. Lago. 1983. Thermodynamic perturbation theory for molecular liquid mixtures. *J. Chem. Phys.* 78:5750.

Fisher, G. D. and T. W. Leland. 1970. Corresponding states principle using shape factors. *Ind. Eng. Chem. Fundam.* 9:537.

Gazzillo, D. 1987. Symmetric mixtures of hard spheres with positively nonadditive diameters: An approximate analytic solution of the Percus-Yevick integral equation. *J. Chem. Phys.* 87:1757.

Gazzillo, D. 1991. Fluid-fluid phase separation of nonadditive hard-sphere mixtures as predicted by integral-equation theories. *J. Chem. Phys.* 95:4565.

Gazzillo, D., A. Giacometti, R. G. Della Valle, E. Venuti, and F. Carsughi. 1999. A scaling approximation for structure factors in the integral equation theory of polydisperse nonionic colloidal fluids. *J. Chem. Phys.* 111:7636.

Góźdź, W. R. 2003. Critical-point and coexistence curve properties of a symmetric mixture of nonadditive hard spheres: A finite size scaling study. *J. Chem. Phys.* 119:3309.

Grundke, E. W. and D. Henderson. 1972. Distribution functions of multi-component fluid mixtures of hard spheres. *Mol. Phys.* 24:269.

Hamad, E. 1994. Consistency test for mixture pair correlation function integrals. *J. Chem. Phys.* 101:10195.

Hamad, E. Z. 1996a. A general mixture theory. I. Mixtures of spherical molecules. *J. Chem. Phys.* 105:3229.

Hamad, E. Z. 1996b. Contact pair correlation functions and equation of state for nonadditive hard-sphere mixtures. *J. Chem. Phys.* 105:3222.

Hamad, E. Z. 1997. Simulation and model testing of size asymmetric non-additive hard spheres. *Mol. Phys.* 91:371.

Henderson, D. and J. A. Barker. 1968. Perturbation theory and the equation of state of mixtures of hard spheres. *J. Chem. Phys.* 49:3377.

Henderson, D. J., A. Barker, and W. R. Smith. 1972. Perturbation theory in classical statistical mechanics. *Utilitas Math.* 1:211.

Henderson, D. and P. J. Leonard. 1971. Conformal solution theory: Hard-sphere mixtures. *Proc. Natl. Acad. Sci. USA* 68:2354.

Johnson, J. K., J. A. Zollweg, and K. E. Gubbins. 1993. The Lennard-Jones equation of state revisited. *Mol. Phys.* 78:591.

Jung, J., M. S. Jhon, and F. H. Ree. 1994a. An analytic equation of state and structural properties of nonadditive hard sphere mixtures. *J. Chem. Phys.* 100:9064.

Jung, J., M. S. Jhon, and F. H. Ree. 1994b. Homo- and heterocoordination in nonadditive hard-sphere mixtures and a test of the van der Waals one-fluid model. *J. Chem. Phys.* 100:528.

Jung, J., M. S. Jhon, and F. H. Ree. 1995. Fluid-fluid phase separations in nonadditive hard sphere mixtures. *J. Chem. Phys.* 102:1349.

Kahl, G. 1990. Nonadditive hard-sphere reference system for a perturbative liquid state theory of binary systems. *J. Chem. Phys.* 93:5105.

Kahl, G. and J. Hafner. 1985. A blip-function calculation of the structure of liquid binary alloys. *J. Phys. F: Met. Phys.* 15:1627.

Kambayashi, S. and Y. Hiwatari. 1990. Theory of supercooled liquids and glasses for binary soft-sphere mixtures via a modified hypernetted-chain integral equation. *Phys. Rev. A* 42:2176.

Kang, H. S. and F. H. Ree. 1998. Thermodynamic and structural properties of strongly coupled plasma mixtures from the perturbative HNC-equation. In *Strongly Coupled Coulomb Systems, Proceedings of the Strongly Coupled Coulomb Systems Conference*, ed. G. J. Kalman, J. M. Rommel, and K. Blagoev, pp. 153–158. New York: Plenum Press.

Kihara, T. 1955. Virial coefficients and models of molecules in gases. B. *Rev. Mod. Phys.* 27:412.

Kihara, T. and K. Miyoshi. 1975. Geometry of three convex bodies applicable to three-molecule clusters in polyatomic gases. *J. Stat. Phys.* 13:337.

Lebowitz, J. L. 1964. Exact solution of generalized Percus-Yevick equation for a mixture of hard spheres. *Phys. Rev.* 133:A895.

Lebowitz, J. L., E. Helfand, and E. Praestgaard. 1965. Scaled particle theory for fluid mixtures. *J. Chem. Phys.* 43:774.

Lee, L. L. and D. Levesque. 1973. Perturbation theory for mixtures of simple liquids. *Mol. Phys.* 26:1351.

Leland, T. W., J. S. Rowlinson, and G. A. Sather. 1968. Statistical thermodynamics of mixtures of molecules of different sizes. *Trans. Far. Soc.* 64:1447.

Leland, T. W., J. S. Rowlinson, G. A. Sather, and I. D. Watson. 1969. Statistical thermodynamics of two-fluid models of mixtures. *Trans. Far. Soc.* 65:2034.

Leonard, P. J., D. Henderson, and J. A. Barker. 1970. Perturbation theory and liquid mixtures. *Trans. Faraday Soc.* 66:2439.

Leonard, P. J., D. Henderson, and J. A. Barker. 1971. Calculation of the radial distribution function of hard-sphere mixtures in the Percus-Yevick approximation. *Mol. Phys.* 21:107.

Lomba, E., M. Alvarez, L. L. Lee, and N. G. Almarza. 1996. Phase stability of binary non-additive hard-sphere mixtures: A self-consistent integral equation study. *J. Chem. Phys.* 104:4180.

Longuet-Higgins, H. C. 1951. The statistical thermodynamics of multi-component systems. *Proc. Roy. Soc. A* 205:247.

Mansoori, G. A., N. F. Carnahan, K. E. Starling, and T. W. Leland. 1971. Equilibrium thermodynamic properties of the mixture of hard spheres. *J. Chem. Phys.* 54:1523.

Mansoori, G. A. and T. W. Leland. 1970. Variational approach to the equilibrium thermodynamic properties of simple fluid mixtures. III. *J. Chem. Phys.* 53:1931.

Mansoori, G. A. and T. W. Leland. 1972. Statistical thermodynamics of mixtures. A new version for the theory of conformal solutions. *J. Chem. Soc. Faraday Trans. II* 68:320.

Mazo, R. M. and R. J. Bearman. 1990. Scaled particle theory for mixtures of nonadditive hard disks or hard spheres: An alternative scaling. *J. Chem. Phys.* 93:6694.

Melnyk, T. W. and B. L. Sawford. 1975. Equation of state of a mixture of hard spheres with non-additive diameters. *Mol. Phys.* 29:891.

Nakanishi, K., S. Okazaki, K. Ikari, T. Higuchi, and H. Tanaka. 1982. Free energy of mixing, phase stability, and local composition in Lennard-Jones mixtures. *J. Chem. Phys.* 76:629.

Miyano, Y. 1991. Equation of state for Lennard-Jones fluid mixtures. *Fluid Phase Equilibr.* 66:125.

Miyano, Y. 1994. Equation of state calculations and Monte Carlo simulations of internal energies, compressibility factors and vapor-liquid equilibria for Lennard-Jones fluid mixtures. *Fluid Phase Equilibr.* 95:1.

Nixon, J. H. and M. Silbert. 1984. Percus-Yevick results for a binary mixture of hard spheres with non-additive diameters. I. Negative non-additive parameter. *Mol. Phys.* 52:207.

Paricaud, P. 2008. Phase equilibria in polydisperse nonadditive hard-sphere systems. *Phys. Rev. E* 78:021202.

Perram, J. W. 1984. Perturbation theory of molecular fluids. Lecture at Herbstschule: *Struktur und Eigenschaften der Flüssigkeiten*, St. Georgen, Austria.

Perram, J. W., J. Rasmussen, and E. Praestgaard. 1988. On the determination of effective hard convex body parameters for liquid mixtures. *Mol. Phys.* 64:617.

Rovere, M. and G. Pastore. 1994. Fluid-fluid phase separation in binary mixtures of asymmetric non-additive hard spheres. *J. Phys.: Condens. Matter* 6:A163.

Saija, F., G. Pastore, and P. V. Giaquinta. 1998. Entropy and fluid-fluid separation in nonadditive hard-sphere mixtures. *J. Phys. Chem. B* 102:10368.

Santos, A., M. López de Haro, and S. B. Yuste. 2010. Virial coefficients, thermodynamic properties, and fluid-fluid transition of nonadditive hard-sphere mixtures. *J. Chem. Phys.* 132:204506.

Schaink, H. M. and C. Hoheisel. 1992. The phase-behavior of Lennard-Jones mixtures with nonadditive hard cores: Comparison between molecular dynamic calculations and perturbation theory. *J. Chem. Phys.* 97:8561.

Schöl-Paschinger, E., D. Levesque, J.-J. Weis, and G. Kahl. 2005. Phase diagram of a binary symmetric hard-core Yukawa mixture. *J. Chem. Phys.* 122:024507.

Scott, R. L. 1956. Corresponding states treatment of nonelectrolyte solutions. *J. Chem. Phys.* 25:193.

Shukla, K. P., M. Luckas, H. Marquardt, and K. Lucas. 1986. Conformal solutions: Which model for which application? *Fluid Phase Equilibr.* 26:129.

Sillrén, P. and J.-P. Hansen. 2007. Perturbation theory for systems with strong short-ranged interactions. *Mol. Phys.* 105:1803.

Smith, W. R. 1971. Perturbation theory and conformal solutions. I. Hard-sphere mixtures. *Mol. Phys.* 21:105.

Song, Y. 1990. Statistical-mechanical theory for mixtures. *J. Chem. Phys.* 92:2683.

Takamiya, M. and K. Nakanishi. 1988. A modified van der Waals one-fluid approximation based on a simple perturbed hard-sphere equation of state. *Fluid Phase Equilibr.* 41:215.

Tang, Y. and B. C.-Y. Lu. 1995. Improved expressions for the radial distribution function of hard spheres. *J. Chem. Phys.* 103:7463.

Tenne, R. and E. Bergmann. 1978. Scaled particle theory for nonadditive hard spheres: Solutions for general positive nonadditivity. *Phys. Rev. A* 17:2036.

Throop, G. J. and R. J. Bearman. 1965. Radial distribution functions for mixtures of hard spheres. *J. Chem. Phys.* 42:2838.

Verlet, L. and J.-J. Weis. 1972. Equilibrium theory of simple liquids. *Phys. Rev. A* 5:939.

Vrabec, J., A. Lofti, and J. Fischer. 1995. Vapour liquid equilibria of Lennard-Jones model mixtures from the *NpT* plus test particle method. *Fluid Phase Equilibr.* 112:173.

Wertheim, M. S. 1963. Exact solution of the Percus-Yevick integral equation for hard spheres. *Phys. Rev. Lett.* 10:321.

Widom, B. and J. S. Rowlinson. 1970. New model for the study of liquid-vapor phase transitions. *J. Chem. Phys.* 52:1670.

8 Perturbation Theories for Molecular Fluids

The extension of integral equation and integral equation perturbation theories to molecular fluids was discussed in Chapter 4. Here, after outlining the general framework of the free energy perturbation theories for fluids with anisotropic interactions, several approaches to obtain the properties of different kinds of reference hard-body molecular fluids are described. The performance of these theories is discussed for a number of illustrative examples. This chapter continues with the description of a number of theoretically based approaches for the EOS of real molecular fluids and ends with a section devoted to the theoretical description of the nonisotropic phases and the related phase transitions.

8.1 EXTENSION OF THE FREE ENERGY PERTURBATION THEORY TO FLUIDS WITH ANISOTROPIC INTERACTIONS

Let us consider an anisotropic and axially symmetric pair potential $u(r_{12}, \Omega_1, \Omega_2)$, where r_{12} is the distance between the centers of any two molecules 1 and 2 and the angles Ω_1 and Ω_2 define the orientations of these molecules. In an analogous way as done in Section 6.1, we will assume that the potential can be split into reference $u_0(r_{12}, \Omega_1, \Omega_2)$ and perturbation $u_1(r_{12}, \Omega_1, \Omega_2)$ parts and introduce a parameter α, with $0 \leq \alpha \leq 1$, to define a family of potentials $u(r_{12}, \Omega_1, \Omega_2; \alpha)$ in such a way that

$$u(r_{12}, \Omega_1, \Omega_2; \alpha) = u_0(r_{12}, \Omega_1, \Omega_2) + u_1(r_{12}, \Omega_1, \Omega_2; \alpha), \qquad (8.1)$$

with $u(r_{12}, \Omega_1, \Omega_2; 0) = u_0(r_{12}, \Omega_1, \Omega_2)$ and $u(r_{12}, \Omega_1, \Omega_2; 1) = u(r_{12}, \Omega_1, \Omega_2)$. Then, the thermodynamic and structural properties can be expanded in power series of α like in Equations 6.4 and 6.5. In particular, the first- and second-order terms in the expansion of the free energy are given by Equations 6.7 and 6.8, respectively, or by Equations 6.11 and 6.12 if the coupling is similar to that in Equation 6.9, that is to say, if

$$u(r_{12}, \Omega_1, \Omega_2; \alpha) = u_0(r_{12}, \Omega_1, \Omega_2) + \alpha u_1(r_{12}, \Omega_1, \Omega_2). \qquad (8.2)$$

In the latter case, the first-order term, similarly to Equation 6.22, may be written as

$$\left. \frac{\partial F}{\partial \alpha} \right|_{\alpha=0} = \frac{1}{2(4\pi)^2} N\rho \int d\mathbf{r} \int \int u_1(r, \Omega_1, \Omega_2) g_0(r, \Omega_1, \Omega_2) d\Omega_1 d\Omega_2. \qquad (8.3)$$

The second-order term is given by a straightforward generalization of Equation 6.23 to include integration over angles, and therefore in general involves triplet and quadruplet correlation functions and can be calculated only in an approximate way.

For an asymmetric potential with a spherically symmetric core taken as the reference potential $u_0(r)$, expression (8.3) simplifies to a form similar to Equation 6.22, namely,

$$\left.\frac{\partial F}{\partial \alpha}\right|_{\alpha=0} = 2\pi N\rho \int \langle u_1(r, \Omega_1, \Omega_2)\rangle g_0(r) r^2 dr, \qquad (8.4)$$

where the angular brackets mean an unweighted average over orientations. For some intermolecular potentials, $\langle u_1(r, \Omega_i, \Omega_j)\rangle = 0$ and so the first-order perturbation term vanishes. In this case, the last two terms in the right-hand side of the (generalized) Equation 6.23 also vanish and the second-order perturbation term reduces to

$$\left.\frac{\partial^2 F}{\partial \alpha^2}\right|_{\alpha=0} = -\frac{1}{2(4\pi)^2}\beta N\rho \int g_0(r_1, r_2)\, d\mathbf{r}_2 \int [u_1(r_{12}, \Omega_1, \Omega_2)]^2\, d\Omega_1 d\Omega_2$$

$$- \frac{1}{(4\pi)^3}\beta N\rho^2 \int g_0(r_1, r_2, r_3)\, d\mathbf{r}_2 d\mathbf{r}_3$$

$$\times \int u_1(r_{12}, \Omega_1, \Omega_2)\, u_1(r_{23}, \Omega_2, \Omega_3)\, d\Omega_1 d\Omega_2 d\Omega_3. \qquad (8.5)$$

For a potential with either symmetric or asymmetric core, a particular choice for the reference and perturbation contributions in Equation 8.2 is

$$u_0(r_{12}) = \langle u(r_{12}, \Omega_1, \Omega_2)\rangle, \qquad (8.6)$$

$$u_1(r_{12}, \Omega_1, \Omega_2) = u(r_{12}, \Omega_1, \Omega_2) - u_0(r_{12}), \qquad (8.7)$$

so that condition $\langle u_1(r, \Omega_i, \Omega_j)\rangle = 0$ is obviously fulfilled.

Gubbins and Gray (1972) and Perram and White (1974) reported the expressions of the zero- and first-order perturbation terms of the pair correlation function $g(r_{12}, \Omega_1, \Omega_2)$ for the coupling (8.2) and for the more general coupling (8.1), respectively. The latter authors also proposed a definition of the reference and perturbation contributions to $u(r_{ij}, \Omega_i, \Omega_j; \alpha)$ that included some of the angular dependence of the potential into the reference part.

In the most general case, the reference fluid will be one consisting of molecules with a similar shape as the actual fluid, most frequently hard-body molecules. Therefore, we will need to know accurately the EOS and the pair correlation function of the reference hard-body fluid. Sections 8.2 through 8.6 are devoted to describe a number of theoretical approaches developed to this end.

8.2 SCALED-PARTICLE-LIKE APPROACHES FOR HARD-BODY MOLECULAR FLUIDS

8.2.1 Scaled Particle Theory for Hard Convex Body Fluids

The SPT was extended by Gibbons (1969) and Boublík (1974) to *hard convex body* (HCB) fluids. To obtain the contact pair correlation function and the EOS of HCB fluids in the SPT approach, we will follow a similar procedure as that used in Section 5.2 for the HS fluid. Let us introduce a parameter s scaling the size of a HCB particle in such a way that for $s = 1$ we have an ordinary particle and for $s = 0$ the scaled particle reduces to a point. Let $v^e(s)$ be the volume, averaged over orientations, from which is excluded the center of an s-particle because of the presence of an ordinary particle of the fluid, so that $v^e(0) = v$, the volume of an ordinary particle of the fluid, and $v^e(-1) = 0$. In the most general case, the averaged volume from which the center of a particle j is excluded because of the presence of a particle i is (Kihara 1953)

$$v_{ij}^e = v_i + v_j + R_i S_j + R_j S_i, \tag{8.8}$$

where v_i, S_i, and R_i are the volume, surface area, and $1/4\pi$ times the mean curvature integrated over the surface of the body, respectively, of particle i. Expressions of these quantities for different HCBs can be found in the literature (Kihara 1953, Boublík and Nezbeda 1986). Considering that i is an ordinary particle of the fluid with parameters v, S, and R, and j is the s-particle with parameters $s^3 v$, $s^2 S$, and sR, we will have

$$v^e(s) = v + s^3 v + Rs^2 S + sRS. \tag{8.9}$$

Then, Equation 5.2 can be rewritten in the form

$$W(s) = W(0) + \rho k_B T \int_0^s G(s') \frac{\partial v^e(s')}{\partial s'} ds', \tag{8.10}$$

where now $G(s)$ is averaged over orientations, and from Equation 8.9

$$\frac{\partial v^e(s)}{\partial s} = 3s^2 v + 2sRS + RS. \tag{8.11}$$

In this way, Equations 5.3 through 5.7 become

$$G(s) = \frac{1}{(3s^2 v + 2sRS + RS)\rho k_B T} \frac{\partial W(s)}{\partial s}, \tag{8.12}$$

$$W(s) = -k_B T \ln[1 - \rho v^e(s)], \quad s \leq 0, \tag{8.13}$$

$$G(s) = \frac{1}{1 - \rho v^e(s)}; \quad s \leq 0, \tag{8.14}$$

$$Z = G(\infty) = 1 + \frac{1}{2}\rho v_{11}^e G(1). \tag{8.15}$$

The last equation, in which v_{11}^e is given by Equation 8.8 for $i = j = 1$, is the expression of the virial theorem for HCB fluids and reduces as a particular case to the virial equation (2.43) for the HS fluid.

For $s \geq 0$, Boublík (1974) assumes for $G(s)$ the expression

$$G(s) = a_0 + a_1 \frac{sRS + RS}{3s^2v + 2sRS + RS} + a_2 \frac{RS}{3s^2v + 2sRS + RS}, \quad s \geq 0. \quad (8.16)$$

In a similar way as for the HS fluid (see Section 5.2), parameters a_i are determined from Equation 8.15 together with the conditions that for $s = 0$ both $G(s)$ and $\partial G(s)/\partial s$ must give the same result either calculated from Equation 8.14 or from Equation 8.16. This yields

$$G(s) \equiv g(1) = \frac{1}{1-\eta} + \frac{3\alpha(1+\alpha)}{1+3\alpha} \frac{\eta}{(1-\eta)^2} + \frac{3\alpha^2}{1+3\alpha} \frac{\eta^2}{(1-\eta)^3}, \quad (8.17)$$

$$Z = \frac{1}{1-\eta} + 3\alpha \frac{\eta}{(1-\eta)^2} + 3\alpha^2 \frac{\eta^2}{(1-\eta)^3}, \quad (8.18)$$

where $\eta = \rho v$ is the packing fraction and $\alpha = RS/3v$ is the *nonsphericity parameter*, indicating the departure of the HCB particle from the spherical shape. For HS $\alpha = 1$ and Equations 8.17 and 8.18 reduce to Equations 5.9 and 5.10, respectively. The preceding derivation uses a single scaling parameter s, but more scaling parameters might be considered, as done by Cotter and Martire (1970a–c) who derived a SPT for hard spherocylinders with two scaling parameters, one for the length and another for the breadth.

The procedure described for pure HCB fluids was also extended by Gibbons (1969) and Boublík (1974) to HCB fluid mixtures in a similar way as the SPT for the pure HS fluid was extended, as indicated in Section 7.5, to HS fluid mixtures. The result for the EOS is as follows:

$$Z = \frac{1}{1-\eta} + \rho \frac{\langle R \rangle \langle S \rangle}{(1-\eta)^2} + \frac{1}{3}\rho^2 \frac{\langle R^2 \rangle \langle S \rangle^2}{(1-\eta)^3}, \quad (8.19)$$

where

$$\langle R \rangle = \sum_i x_i R_i, \quad \langle R^2 \rangle = \sum_i x_i R_i^2, \quad \langle S \rangle = \sum_i x_i S_i,$$

$$\eta = \rho \langle v \rangle, \quad \langle v \rangle = \sum_i x_i v_i. \quad (8.20)$$

Boublík (1974) also reported expressions for the contact values of the averaged partial correlation functions $g_{ij}(1)$ leading to Equation 8.19 obtained from a similar procedure as that used to derive the expressions $g_{ij}(\sigma_{ij})$ for additive HS (AHS) fluid mixtures (see Section 7.5). In a subsequent paper, Boublík (1975) improved the SPT expressions for the $g_{ij}(1)$ in a similar way as done previously for AHS fluid mixtures. These improved expressions, combined with the pressure equation for HCB fluids,

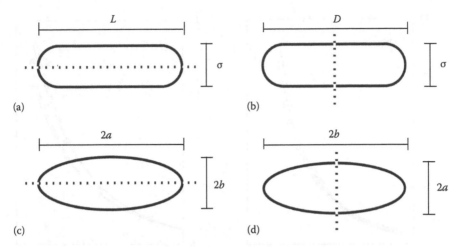

FIGURE 8.1 Planar representation of several typical hard convex molecules showing their defining geometrical parameters. The dashed lines represent the axis of rotation that generates the three-dimensional bodies. (a) *Prolate spherocylinder* (PSC), $\gamma = L/\sigma$; (b) *Oblate spherocylinder* (OSC), $\gamma = D/\sigma$; (c) *Prolate ellipsoid of revolution* (PER), $\kappa = a/b$; and (d) *Oblate ellipsoid of revolution* (OER), $\kappa = a/b$.

led to the *improved SPT* (or ISPT) equations of state

$$Z = \frac{1}{1-\eta} + 3\alpha\frac{\eta}{(1-\eta)^2} + \alpha^2\frac{\eta^2(3-\eta)}{(1-\eta)^3} \tag{8.21}$$

and

$$Z = \frac{1}{1-\eta} + \rho\frac{\langle R \rangle \langle S \rangle}{(1-\eta)^2} + \frac{1}{9}\rho^2\frac{\langle R^2 \rangle \langle S \rangle^2 (3-\eta)}{(1-\eta)^3}, \tag{8.22}$$

for monocomponent and multicomponent HCB fluids, respectively. As particular cases, Equation 8.21 reduces to the CS equation (5.21) for the pure HS fluid and Equation 8.22 reduces to the BMCSL equation (7.91) for AHS fluid mixtures.

Figure 8.1 displays several models of HCB molecules with their geometrical characteristic parameters, and Figure 8.2 compares the predictions of the SPT and ISPT equations of state with simulations for some examples of these model fluids. In Figure 8.2, it is seen that the ISPT results are more accurate than those from SPT; however, for more anisotropic HCB molecules, both theories strongly overestimate the EOS at high densities.

Because of the modest performance of the SPT and ISPT equations of state for HCB fluids, there have been proposed a number of semi-empirical approximations, more or less reminiscent with the SPT, based on the known virial coefficients (see, e.g., Nezbeda 1976, Boublík 1981, 1986, 1994, 2004). The drawback of these approximations is their dependence on the knowledge of the virial coefficients that,

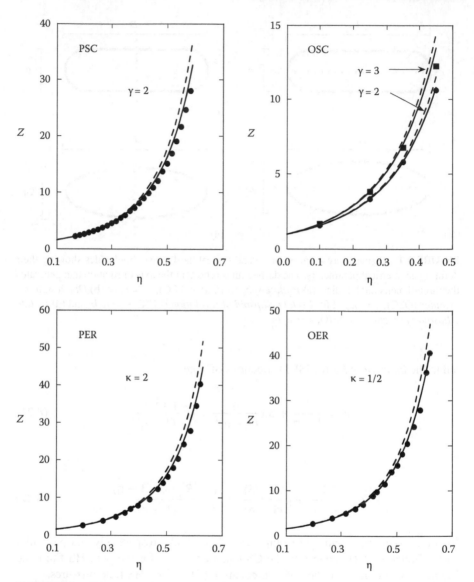

FIGURE 8.2 Compressibility factor Z for several HCB fluids (see Figure 8.1 for descriptions) as a function of the packing fraction η. Points: simulation data from Veerman and Frenkel (1990), Kadlec et al. (2000), and Frenkel and Mulder (1985), for PSC, OSC, and PER-OER, respectively. Dashed curves: SPT equation (8.18). Continuous curves: ISPT equation (8.21).

apart from the second one for HCB fluids, which is given exactly by

$$B_2^* = 1 + 3\alpha \tag{8.23}$$

in units of the molecular volume v, must be calculated numerically or estimated from more or less heuristic considerations. This is particularly disappointing for HCB

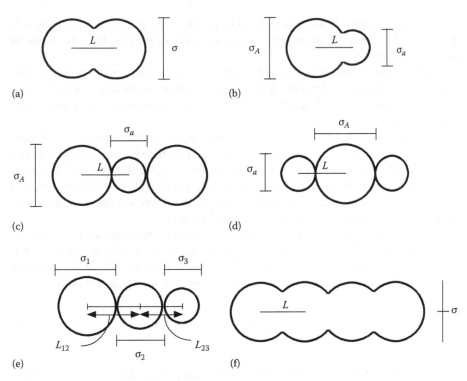

FIGURE 8.3   Planar representation of several typical fused HS molecules showing their defining geometrical parameters. (a) *Homonuclear diatomics* (HMD), (b) *heteronuclear diatomics* (HTD), (c) *linear symmetric triatomics* (LST), (d) *linear symmetric triatomics* (LST), (e) *linear nonsymmetric triatomics* (LNST), and (f) *linear homonuclear polyatomics* (LHP).

fluid mixtures, for which the availability of virial coefficients is less than for pure HCB fluids.

For hard nonconvex molecules R is not well defined. To apply Equation 8.22 to these fluids too, Boublík and Nezbeda (1977) proposed to approximate R with that corresponding to the convex envelopes of the nonconvex molecules, while still retaining for S and v their actual values. Expressions to obtain α from this convention for the *fused hard-sphere* (FHS) molecular models displayed in Figure 8.3 have been reported by several authors (Boublík and Nezbeda 1977, Lago et al. 1983, Amos and Jackson 1991, Archer and Jackson 1991). Lago et al. (1983) also devised an alternative way of obtaining the nonsphericity parameter α from the expression

$$\alpha = \frac{1}{3\pi} \frac{v'v''}{v}, \tag{8.24}$$

where v is again the molecular volume,

$$v' = \sum_{i=1}^{n} \frac{\partial v}{\partial \sigma_i}, \qquad v'' = \sum_{i,j=1}^{n} \frac{\partial^2 v}{\partial \sigma_i \partial \sigma_j}, \tag{8.25}$$

and n is the number of beads in the molecule. For the molecular models in Figure 8.3, but for model c, this definition yields the same values of α as the Boublík–Nezbeda approximation.

An alternative procedure to obtain an effective nonsphericity parameter for nonconvex molecules, provided that the second virial coefficient is known, consists in obtaining α from Equation 8.23, like for convex molecules, as proposed by Rigby (1976).

The ISPT equation (8.21) with the Boublík–Nezbeda approximation for R provides excellent agreement with simulations for hard diatomic and linear triatomic molecules, as shown in Figure 8.4. For larger linear molecules, the ISPT equation increasingly overestimates the pressure with increasing length, or with increasing the number n of beds for a given center-to-center distance. For nonlinear molecules in general, R must be determined from numerical methods.

On the other hand, Rosenfeld (1988) developed an approach unifying the PY theory and the SPT for hard-body fluid mixtures. Starting from the *fundamental measures* R_i, S_i, and v_i of a HCB, Rosenfeld defined a set of fundamental measures $(R_i^{(0)}, R_i^{(1)}, R_i^{(2)}, R_i^{(3)}) \equiv (1, R_i, S_i, v_i)$ and a set of composition-averaged measures

$$\xi^{(l)} = \sum_i \rho_i R_i^{(l)}, \quad l = 0, 1, 2, 3, \tag{8.26}$$

which for HSB mixtures are proportional to those defined by Equation 7.74. In terms of the latter fundamental measures, the PY-c, PY-v, SPT, and other related equations of state for HS pure fluids and mixtures all share the common form

$$\beta P = \frac{\xi^{(0)}}{1 - \xi^{(3)}} + \frac{a_2}{(1 - \xi^{(3)})^2} + \frac{a_3}{(1 - \xi^{(3)})^3}. \tag{8.27}$$

With the choice $a_2 = \xi^{(1)}\xi^{(2)}$, as is the case of the SPT and several related equations of state, Equation 8.27 reproduces exactly the second virial coefficient of pure HCB fluids. The a_3 coefficient changes from an EOF to another and might be determined from some consistency condition. The before-mentioned SPT EOS derived by Cotter and Martire (1970a–c) using two scaling parameters was not expressed in terms of fundamental measures, but Rosenfeld (1988) noted that when expressed in this way, and for those cases for which it reproduces exactly the second virial coefficient, which is not always the case, the resulting EOS is of the form of the PY-c equation.

8.3   PERCUS–YEVICK THEORY FOR HARD-SPHERE CHAIN FLUIDS

A theory for the EOS and structure of fluids with linear chain molecules made of freely jointed HS was derived by Chiew (1990a,b, 1991) on the basis of the PY theory for the site–site interactions. Before tackling this theory, we will outline the PY solution for a closely related system: the adhesive or SHS fluid mixture, which was derived by Perram and Smith (1975) and Barboy (1975). Therefore, let us consider an

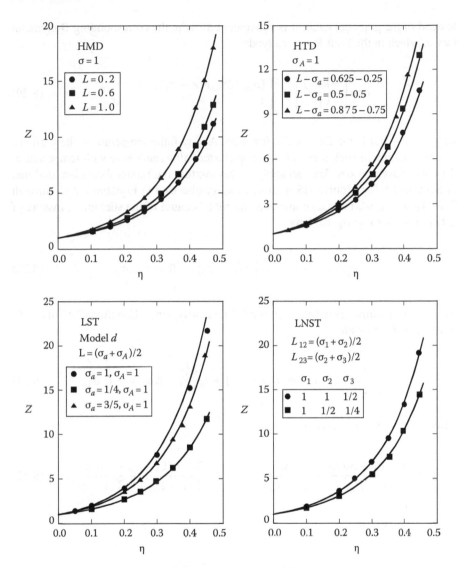

FIGURE 8.4 Compressibility factor for several FHS fluids (see Figure 8.3 for descriptions) as a function of the packing fraction η. Points: simulation data from Tildesley and Streett (1980) for HMD, from Archer and Jackson (1991) for HTD, and from Amos and Jackson (1991) for all other models. The geometrical parameters are given in units of the largest sphere diameter. Curves: ISPT equation (8.21).

m-component mixture of particles interacting with each other by means of a potential of the form (1.11), that is to say

$$\frac{u_{ij}(r)}{k_B T} = \begin{cases} \infty, & r < \sigma_{ij} \\ \ln\left[12\tau_{ij}\left(R_{ij} - \sigma_{ij}\right)/R_{ij}\right], & \sigma_{ij} < r < R_{ij} \,. \\ 0, & r > R_{ij} \end{cases} \qquad (8.28)$$

Instead of the potential itself, it is convenient to write the corresponding Boltzmann factor, which in the limit $R_{ij} \to \sigma_{ij}$ reads

$$\exp\left[\frac{-u_{ij}(r)}{k_B T}\right] = \begin{cases} (\sigma_{ij}/12\tau_{ij})\delta(r - \sigma_{ij}), & r \leq \sigma_{ij} \\ 1, & r > \sigma_{ij} \end{cases}, \tag{8.29}$$

where $\delta(r - \sigma_{ij})$ is the Dirac delta function. At any finite temperature, the particles in this system will stick together forming clusters of variable size with temperature-dependent average size. We can apply to this mixture the Baxter formalism outlined in Section 7.5 for additive HS mixtures and synthesized in Equations 7.85 through 7.89. However, we must take into account that, because of the stickiness, instead of $h_{ij}(r) = -1$ for $r < \sigma_{ij}$, we have

$$h_{ij}(r) = -1 + l_{ij}\frac{\sigma_{ij}}{12}\delta\left(r - \sigma_{ij}\right), \quad 0 < r < \sigma_{ij}, \tag{8.30}$$

where l_{ij} are parameters to be determined. As a consequence, Equations 7.87 and 7.89 must be replaced with

$$q_{ij}(r) = \frac{1}{2}a_i\left(r^2 - \sigma_{ij}^2\right) + b_i\left(r - \sigma_{ij}\right) + l_{ij}\frac{\sigma_{ij}^2}{12}, \quad \Delta\sigma_{ij} < r < \sigma_{ij}, \tag{8.31}$$

where $\Delta\sigma_{ij} = (\sigma_i - \sigma_j)/2$, and

$$a_i = \frac{1 - \xi_3 + 3\sigma_i\xi_2}{(1 - \xi_3)^2} - \frac{X_i}{(1 - \xi_3)}, \quad b_i = -\frac{3}{2}\frac{\sigma_i^2\xi_2}{(1 - \xi_3)^2} + \frac{1}{2}\frac{\sigma_i X_i}{(1 - \xi_3)}, \tag{8.32}$$

respectively, where

$$\xi_l = \frac{\pi}{6}\rho\sum_{j=1}^{m} x_j\sigma_j^l, \quad X_i = \frac{\pi}{6}\rho\sum_{j=1}^{m} x_j l_{ij}\sigma_{ij}^2\sigma_j. \tag{8.33}$$

On the other hand, by combining Equations 7.85, 7.86, and 8.30 with the PY closure, Perram and Smith (1975) showed that the l_{ij} are given by the solutions of the equation

$$l_{ij}\tau_{ij} = a_i + \frac{b_i}{\sigma_{ij}} + \frac{\pi}{6}\rho\sum_{k=1}^{m} x_k\frac{l_{jk}\sigma_{jk}^2}{\sigma_{ij}}q_{ik}(\Delta\sigma_{ik}). \tag{8.34}$$

The resulting compressibility EOS is (Barboy 1975, Perram and Smith 1975)

$$Z = \frac{6}{\pi\rho}\left[\frac{\xi_0}{1-\xi_3} + 3\frac{\xi_1\xi_2}{(1-\xi_3)^2} + 3\frac{\xi_2^3}{(1-\xi_3)^3}\right]$$

$$-\frac{\pi}{12}\frac{\rho}{1-\xi_3}\sum_{i,j} x_i x_j l_{ij}\sigma_{ij}^2\left(\sigma_i + \sigma_j + 3\sigma_i\sigma_j\frac{\xi_2}{1-\xi_3}\right)$$

$$+\left(\frac{\pi\rho}{36}\right)^2\sum_{i,j,k} x_i x_j x_k l_{ij}l_{ik}l_{jk}\sigma_{ij}^2\sigma_{ik}^2\sigma_{jk}^2, \tag{8.35}$$

which reduces to the PY compressibility equation (7.84) for HS mixtures when $l_{ij} = 0$ for every i and j. On the other hand, for a pure fluid of SHS, Equation 8.35 reduces to

$$Z = \frac{1+\eta+\eta^2}{(1-\eta)^3} - l\eta\frac{1(2+\eta)}{2(1-\eta)^2} + \frac{l^3\eta^2}{36}. \tag{8.36}$$

where

$$l = \frac{1}{1-\eta}\left\{6 - \tau + \tau\eta^{-1} - \left[(6 - \tau + \tau\eta^{-1})^2 - 6(1+2\eta^{-1})\right]^{1/2}\right\}. \tag{8.37}$$

Equation 8.36, which reduces to the PY compressibility equation (5.17) for a pure HS fluid when $l = 0$, was first derived by Baxter (1968), who also derived the EOS that arises from the virial route. However, the latter showed unphysical behavior in the region of phase coexistence.

We are now in a situation of dealing with chains made of freely jointed HS. The main differences with the SHS model are the following: (1) now all the molecular chains have a fixed number n of spheres and so the thermodynamic and structural properties of the fluid are temperature-independent and (2) as we are interested in linear chains, spheres in the inner positions of the chain must form two and only two bonds and particles located in the extremes must form one and only one bond; bonding is not allowed between spheres belonging to different chains. Following the Chiew (1990a,b, 1991) approach, the system consisting of N chains, each of them with n HS is viewed as a mixture of nN HS with diameters σ_i and partial densities ρ_i, $i = 1,\ldots,nN$. The spheres (monomers) stick together to form the n-mer molecules in such a way that, if the spheres are labeled from 1 to nN, the first n-mer is formed by the spheres 1 to n, the second one by those from $n + 1$ to $2n$, and so on. Each sphere i can only bond the spheres $i + 1$ and $i - 1$, but for the spheres in the extremes of the chain; the first sphere in the chain only forms a bond with the second one and the last sphere only forms a bond with the last but one. The center-to-center distance between the neighboring particles i and j is $\sigma_{ij} = (\sigma_i + \sigma_j)/2$. If the volume of the system is V, the number density of spheres of the species i is obviously $\rho_i = 1/V$. For this mixture, Equation 8.30 applies, provided that $j = i + 1$, where now the l_{ij} must satisfy the obvious condition

$$4\pi\rho_j \int_0^{\sigma_{ij}} r^2 g_{ij}(r)\, dr = 2l_{ij}\eta_j\left(\frac{\sigma_{ij}}{\sigma_j}\right)^3 = 1, \tag{8.38}$$

where $\eta_j = \pi\rho_j/6$ is the packing fraction for species j. The EOS of this system from the compressibility route is then given by Equation 8.35 with the l_{ij} determined from condition (8.38). For the particular case of homonuclear chains, for which $\sigma_i = \sigma$ for all i, this yields (Chiew 1990a)

$$\frac{P}{\rho k_B T} = Z_{HS} - \frac{n-1}{n} g_{HS}(\sigma), \tag{8.39}$$

where

ρ is the monomer number density

Z_{HS} is the EOS of the HS fluid at density ρ or packing fraction $\eta = (\pi/6)\rho\sigma^3$

the last term, in which $g_{HS}(\sigma)$ is the contact value of the radial distribution function (RDF) of the HS fluid, accounts for the effect of bonding

As for Z_{HS}, the expression obtained from this procedure would be the PY-c equation (5.17), but Chiew (1990a) proposed to replace it with the more accurate Carnahan–Starling equation (5.21). In a similar way, $g_{HS}(\sigma)$ would be the PY result

$$g_{PY}(\sigma) = \frac{1 + \frac{1}{2}\eta}{(1-\eta)^2}, \tag{8.40}$$

but it may be replaced with

$$g_{CS}(\sigma) = \frac{1 - \eta/2}{(1-\eta)^3}, \tag{8.41}$$

which corresponds to the CS EOS.

In a similar way, starting from Equation 8.35, Chiew (1990a) derived the EOS for linear heteronuclear chains, made of n freely jointed HS of diameters σ_i, with $i = 1, \ldots, n$, in the form

$$\frac{P}{\rho k_B T} = Z_{mix}^{HS} - \frac{1}{2n} \frac{1}{1-\xi_3} \sum_{i=1}^{n-1} \frac{1}{\sigma_{i,i+1}} \left(\sigma_i + \sigma_{i+1} + \sigma_i \sigma_{i+1} \frac{\xi_2}{1-\xi_3} \right), \tag{8.42}$$

where Z_{mix}^{HS} is the EOS of a HS mixture with number density $\rho = n\rho_n$, where ρ_n is the number density of chains. Similarly as before, although the expression for Z_{mix}^{HS} which results from Equation 8.35 would be the PY-c EOS (7.84) for HS mixtures, it is advisable to replace it with the more accurate BMCSL equation (7.91), as proposed by Chew.

Within the same framework, Chiew (1990b, 1991) numerically solved the OZ equation to obtain the site–site pair correlation functions $g_{ij}(r)$ for homonuclear chains as well as the average pair correlation function of the chains

$$g(r) = \frac{1}{n^2} \sum_{i=1}^{n} \sum_{j=1}^{n} g_{ij}(r), \tag{8.43}$$

and the site-averaged correlation functions

$$g_i(r) = \frac{1}{n} \sum_{j=1}^{n} g_{ij}(r). \tag{8.44}$$

The latter functions represent the average correlation between site i of a molecule and all the sites of the other molecules. The expressions for the contact values of these functions are (Chiew 1991)

$$g(\sigma) = \frac{2 + (3n - 2)\eta}{2n(1 - \eta)^2}, \tag{8.45}$$

for the average RDF,

$$g_i(\sigma) = \frac{n + 2 + (5n - 2)\eta}{4n(1 - \eta)^2}, \tag{8.46}$$

for the sites at the ends of the chains, and

$$g_i(\sigma) = \frac{1 + (3n - 1)\eta}{2n(1 - \eta)^2}, \tag{8.47}$$

for the inner sites of the chains. Later, Tang and Lu (1996) developed an analytical expression for the averaged RDF of homonuclear hard chains within the Chew approach using a similar procedure as that used by them (Tang and Lu 1997a) within the context of the FMSA (see Section 4.4) to the same end.

The Chiew EOS is readily extended to mixtures of homonuclear hard-sphere chains. Considering an m-component mixture of chains in which component i, with mole fraction x_i, consists of n_i HS with diameter σ_i, the EOS for the mixture may be written in the form (Song et al. 1994a)

$$Z = 1 + \rho_c \sum_{i,j=1}^{m} x_i x_j n_i n_j B_{ij} g_{ij}(\sigma_{ij}) - \sum_{i=1}^{m} x_i (n_i - 1) [g_{ii}(\sigma_{ii}) - 1], \tag{8.48}$$

where
  ρ_c is the number density of chains
  $B_{ij} = (2/3)\pi\sigma_{ij}^3$ is the contribution to the second virial coefficient from the pair ij of HS
  $g_{ij}(\sigma_{ij})$ are the contact values of the HS partial RDFs, for which we can use the BMCSL expression (7.90), with

$$\xi_l = \frac{\pi}{6} \rho_c \sum_{i=1}^{m} n_i x_i \sigma_i^l. \tag{8.49}$$

Chiew (1990b) also reported analytical expressions for the contact values of the site–site pair correlation functions in a mixture of monomers and n-mers with equal monomer diameter. In particular, the contact value $g_{MD}(\sigma)$ of the site–site pair correlation function between monomers and dimers may be written as

$$g_{MD}(\sigma) = g_{HS}(\sigma) - \frac{1}{4(1-\eta)}. \tag{8.50}$$

The expression for $g_{HS}(\sigma)$ that arises from the theory is the PY result (8.40), but it may be replaced with the CS expression (8.41).

8.4 GENERALIZED FLORY THEORIES FOR HARD-SPHERE CHAIN FLUIDS

The *generalized Flory* (GF) theory was developed by Dickman and Hall (1986) on the basis of previous work on lattice models from Flory (1941, 1942) and Huggins (1941) for polymer solutions. Therefore, we will start by summarizing the Flory and Flory–Huggins (FH) models for lattice systems and then we will describe several extensions to off-lattice systems.

8.4.1 FLORY AND FLORY–HUGGINS LATTICE MODELS FOR CHAIN MOLECULAR FLUIDS

Let us consider a system consisting of N chains formed by n monomers in a lattice with N_s sites, each of which may be either empty or occupied by a segment, so that the fraction of occupied sites is $\phi = nN/N_s$. The probability of insertion of an additional n-mer is

$$p(N) = \langle e^{-\beta \Delta \mathcal{U}} \rangle, \tag{8.51}$$

where

$\Delta \mathcal{U}$ is the change in the potential energy of the system upon the introduction of the test particle for a given configuration of the $(N+1)$-particle system

angular brackets mean an average over all the configurations of the N-particle system and of the test particle

In terms of the partition functions, the probability may be written as

$$p(N) = \frac{Q_c(N+1)}{Q_c(N)Q_c(1)} = (N+1)\frac{Q(N+1)}{Q(N)Q(1)}, \tag{8.52}$$

where $Q(1)$, $Q(N)$, and $Q(N+1)$ are the canonical partition functions of the test-particle, N-particle, and $(N+1)$-particle systems, respectively, and $Q_c(1)$, $Q_c(N)$, and $Q_c(N+1)$ the corresponding configurational integrals. Equation 8.52 can be applied to a system of $N-1$ particles, from which we have

$$Q(N) = \frac{1}{N}Q(N-1)Q(1)p(N-1). \tag{8.53}$$

If we apply the same relation to $Q(N - 1)$, $Q(N - 2),\ldots,$ $Q(2)$, we will finally arrive at

$$Q(N) = \frac{1}{N!} [Q(1)]^N \prod_{i=1}^{N-1} p(i). \tag{8.54}$$

On the other hand, from the thermodynamic relationships $F = -k_B T \ln Q$ and $PV = \mu N - F$, with V replaced with N_s, which plays the role of the volume in this system, we can obtain the pressure. The chemical potential may be obtained from Equation 8.52 taking into account that

$$\mu = -k_B T \ln \left[\frac{Q(N+1)}{Q(N)} \right], \tag{8.55}$$

and F may be obtained from Equation 8.54. This yields

$$\frac{P}{k_B T} = \frac{1}{N_s} \left[\sum_{i=1}^{N-1} \ln p(i) - N \ln p(N) + N \ln(N+1) - \ln N! \right], \tag{8.56}$$

which, in the thermodynamic limit $N, N_s \to \infty$ with ϕ fixed, transforms into the so-called *osmotic EOS*

$$Z \equiv \frac{P}{\rho_n k_B T} = 1 - \ln p(\phi) + \frac{1}{\phi} \int_0^\phi \ln p(\phi') \, d\phi', \tag{8.57}$$

where $\rho_n = \phi/n$ is the number density of chains.

Flory considers that the monomers of the chains are randomly distributed in the lattice, as if they were free monomers, so that each site is occupied with probability ϕ and the segments of the test chain are inserted with probability $1 - \phi$. In this case,

$$p(\phi) = (1 - \phi)^n, \tag{8.58}$$

and from Equation 8.57

$$Z = 1 - \frac{n}{\phi} [\phi + \ln(1 - \phi)], \tag{8.59}$$

which is the Flory EOS.

In the FH theory, which takes into account in an approximate way the fact that the monomers are linked into chains, the first monomer of the test chain is inserted with probability $1 - \phi$ and the remaining with probabilities $1 - \phi_0$, with

$$\phi_0 = \phi \left[1 - \frac{2}{z} \left(1 - \frac{1}{n} \right) \right] \left[1 - \frac{2\phi}{z} \left(1 - \frac{1}{n} \right) \right]^{-1}, \tag{8.60}$$

where z is the coordination number of the lattice, so that the FH insertion probability is (Dickman and Hall 1986)

$$p(\phi) = (1 - \phi)(1 - \phi_0)^{n-1} = (1 - \phi)^n \left[1 - \frac{2\phi}{z}\left(1 - \frac{1}{n}\right)\right]^{1-n}. \tag{8.61}$$

Introducing this probability into Equation 8.57 yields

$$Z = \frac{n}{\phi}\left\{\frac{z}{2}\ln\left[1 - \frac{2\phi}{z}\left(1 - \frac{1}{n}\right)\right] - \ln(1 - \phi)\right\}. \tag{8.62}$$

8.4.2 Extension of the Flory and Flory–Huggins Theories to Continuous Space

The Flory and FH theories were extended by Dickman and Hall (1986) to continuous-space fluids. In this case, the osmotic EOS (Equation 8.57) takes the form

$$Z = 1 - \ln p(\eta) + \frac{1}{\eta}\int_0^\eta \ln p(\eta')\, d\eta', \tag{8.63}$$

where

$\eta = \rho_n v_n$ is the packing fraction of the chains with number density $\rho_n = N/V$ and molecular volume v_n

$p(\eta)$ is the insertion probability for the chains

The latter is not directly related to the insertion probability $p(\phi)$ in the lattice case. For homonuclear chains, the monomer insertion probability $p_1(\eta)$ in continuous space can be obtained from any suitable EOS for the HS fluid by integrating the osmotic EOS (Equation 8.63) to give

$$\ln p_1(\eta) = \int_0^\eta \left[1 - v_1\frac{dP_{HS}^*}{d\eta'}\right]\frac{d\eta'}{\eta'}, \tag{8.64}$$

where

$P_{HS}^* \equiv P_{HS}/k_B T$, P_{HS} is the pressure of a HS fluid with the same packing fraction η as the chain molecular fluid

v_1 is the volume of a sphere

The probability $p_1(\eta)$ plays a similar role in continuous space as $1 - \phi$ in the lattice model. Now, instead of the exponent n, or equivalently the volume ratio v_n/v_1, in the Flory insertion probability (8.58), it is more advisable to consider the ratio $n' = v_n^e/v_1^e$ of the corresponding excluded volumes, to take into account that the volume that a monomer excludes to the center of any other monomer is greater when it is a bonded monomer than when it is a free monomer. Then, the continuous space equivalent of the Flory insertion probability is

$$p_n(\eta) = [p_1(\eta)]^{n'}, \tag{8.65}$$

where, in general, $v_n^e \neq nv_1^e$ and must be obtained as an average over all possible configurations of the n-mer. For freely jointed HS, Dickman and Hall (1986) derived the approximate expression

$$v_n^e \approx v_3^e + (n-3)\left(v_3^e - v_2^e\right), \tag{8.66}$$

in which $v_1^e = (4/3)\pi\sigma^3$, $v_2^e = (27/16)v_1^e$, and $v_3^e \simeq 9.82605\sigma^3$ are the excluded volumes for monomers, dimers, and trimers, respectively, the latter averaged over all configurations (Honnell and Hall 1989).

Inserting the probability (8.65) into Equation 8.63 yields the simple expression

$$Z = 1 + n'\left(Z_{HS} - 1\right), \tag{8.67}$$

for the GF EOS.

In a similar way, the continuous space analogous of the FH probability (8.61) is

$$p_n(\eta) = p_1(\eta) \left[p_1(\eta_0)\right]^{n'-1}, \tag{8.68}$$

where η_0, which plays the role of ϕ_0 in Equation 8.60, is given by

$$\eta_0 = \eta \left[1 - \frac{2}{z_{RCP}}\left(1 - \frac{1}{n}\right)\right]\left[1 - \frac{2\eta}{z_{RCP}}\left(1 - \frac{1}{n}\right)\right]^{-1} = \frac{(1-a)\eta}{1 - a\eta}, \tag{8.69}$$

with $z_{RCP} = 9$, which is approximately the coordination number appropriate for HS at RCP (Bernal and Mason 1960, Scott 1962), and $a = (2/z_{RCP})(1 - 1/n)$. The *generalized Flory–Huggins* (GFH) EOS can be obtained from Equations 8.63, 8.64, and 8.68, for which we need an EOS for the pure HS fluid. Using for the latter the CS equation (5.21), Dickman and Hall (1986) reported the GFH EOS

$$Z = Z_{CS} + \frac{1}{\eta}\left(n' - 1\right)(1 - a)$$

$$\times \left\{2b\ln(1-\eta) + \frac{\eta}{(1-\eta)^3}\left[2b + (4 - 5b)\eta - \left(2 + a - 11a^2\right)\eta^2\right]\right\}, \tag{8.70}$$

with $b = a(1 + 3a)$.

8.4.3 GENERALIZED FLORY-DIMER THEORY

The GH and GFH theories are based on the monomer insertion probability, as if they were free monomers instead of being linked into a chain. For linear chains, it is required that each new inserted monomer, but the first one, must be next to one of the ends of the chain being inserted. A way to approximately correct for this drawback was devised by Honnell and Hall (1989) by using an EOS for dimers, in addition to that for monomers, to determine the insertion probability that appears in the osmotic EOS (8.63). They termed *generalized Flory-dimer* (GFD) to this theory. In a similar way as for the insertion probability of a monomer in the GF and GFH theories, the probability of insertion of a dimer into a chain fluid was approximated by the probability of insertion of a dimer into a fluid of dimers with the same packing fraction. If the probability of insertion of the first monomer is $p_1(\eta)$, the conditional probability of inserting the second will be $p_2(\eta)/p_1(\eta)$. The conditional probability of insertion of a third monomer is approximated by that of the second raised to a power $(v_3^e - v_2^e)/(v_2^e - v_1^e)$, to correct for the fact that the excluded volume for the third monomer may be different to that for the second one. The conditional probability of insertion of the ith monomer, with $i > 2$, will be that of the second raised to a power $(v_i^e - v_2^e)/(v_2^e - v_1^e)$, and the insertion probability for the whole chain

$$p_n(\eta) = p_1(\eta) \prod_{i=2}^{n} \left[\frac{p_2(\eta)}{p_1(\eta)}\right]^{(v_i^e - v_{i-1}^e)/(v_2^e - v_1^e)} = p_1(\eta) \left[\frac{p_2(\eta)}{p_1(\eta)}\right]^{(v_n^e - v_2^e)/(v_2^e - v_1^e)}. \quad (8.71)$$

Inserting this result into the continuous version (Equation 8.63) of the osmotic equation and eliminating the probabilities $p_1(\eta)$ and $p_2(\eta)$ in favor of the corresponding pressures, one readily obtains the EOS for the chain fluid in the form (Honnell and Hall 1989, Denlinger and Hall 1990)

$$Z = Z_{HD} + \frac{v_n^e - v_2^e}{v_2^e - v_1^e}(Z_{HD} - Z_{HS}), \quad (8.72)$$

or, equivalently,

$$Z = Z_{HS} + \frac{v_n^e - v_1^e}{v_2^e - v_1^e}(Z_{HD} - Z_{HS}), \quad (8.73)$$

where Z_{HS} and Z_{HD} are the equations of state of the HS and hard-dimer fluids. For the former, one can use CS equation (5.21) and for the latter, the EOS derived by Tildesley and Streett (1980) by fitting their simulation data for hard diatomics with the result

$$Z_{HD} = \frac{1 + \left(1 + a_1 L^* + a_2 L^{*3}\right)\eta + \left(1 + a_3 L^* + a_4 L^{*3}\right)\eta^2 - \left(1 + a_5 L^* + a_6 L^{*3}\right)\eta^3}{(1 - \eta)^3}, \quad (8.74)$$

where $L^* = L/\sigma$ is the reduced center-to-center distance ($L^* = 1$ for tangent spheres) and $a_1 = 0.37836$, $a_2 = 1.07860$, $a_3 = 1.30376$, $a_4 = 1.80010$, $a_5 = 2.39803$, $a_6 = 0.35700$.

Figure 8.5 compares the predictions from the GF, GFH, and GFD theories with the simulation data for linear flexible chains made of freely jointed HS with different number of monomers. It is seen that the GH theory overestimates the EOS at low densities and slightly underestimates it at high densities. The GFH theory is more accurate at low densities, but strongly underestimates

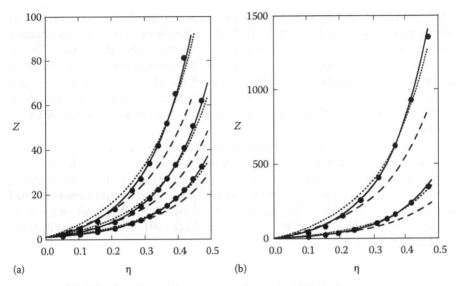

FIGURE 8.5 Compressibility factor $Z = P/\rho_n k_B T$ for linear flexible chains made of freely jointed HS as a function of the packing fraction $\eta = \rho_n v_n$. Points: (a) simulation data from Chang and Sandler (1994a) in and (b) from Gao and Weiner (1989) in figure. Dotted curves: GF theory, Equation 8.67. Dashed curves: GFH theory, Equation 8.70. Continuous curves: GFD theory, Equation 8.72. The curves and data correspond to $n = 4, 8,$ and 16 from down upwards in the left figure, and to $n = 51$ and 201 from down upwards in the right figure.

the EOS at high densities. In contrast, the GFD theory provides satisfactory accuracy at all densities even for extremely long chains.

A version of the GF theory incorporating an EOS for trimers, thus giving rise to the *generalized Flory-trimer* (GFT) theory, was developed by Yethiraj and Hall (1993), but little difference was found with respect to the GFD version. Honnell and Hall (1989) also generalized the procedure leading to the GFD equation by relating the EOS of the hard-chain molecular fluid to the equations of state of two reference fluids with $i - 1$ and $i < n$ monomers per chain, instead of the monomer and dimer fluids. Denoting by Z_i and Z_j the compressibility factors of the reference fluids, with $j = i - 1$, and by Z_n that for the n-mer fluid, the generalized GF EOS reads

$$Z_n = Z_i + \frac{v_n^e - v_i^e}{v_i^e - v_j^e}(Z_i - Z_j)$$

$$= Z_j + \frac{v_n^e - v_j^e}{v_i^e - v_j^e}(Z_i - Z_j). \tag{8.75}$$

The GFD theory was extended by Yethiraj and Hall (1991a) to star and branched molecules made of freely jointed HS by means of two different procedures to obtain approximately the excluded volume. In one of them, Equation 8.66 was used as for linear molecules. In the other, first the excluded volume v_c^e of the central part of the molecule, the central bead plus the first two beads of each arm, was obtained numerically by a Monte Carlo procedure, and then the excluded volume v_n^e for the whole molecule was obtained by adding to v_c^e the contributions of the remaining of the arms in a similar way as in Equation 8.66. Good agreement was found with simulations for 3- and 4-arm stars with 3 to 5 beads per arm, with little difference in the predictions using either of the two procedures to determine the excluded volume.

The GF, GFH, and GFD theories can be applied to rigid molecules made of tangent HS, provided that we obtain the appropriate value for v_3^e, which enters in Equation 8.66 and will depend on the bond angle. Gulati et al. (1996) extended the GF and GFD theories to heteronuclear hard chains and several authors derived expressions for the EOS of different kinds of mixtures of hard chains on the basis of these theories (Hall et al. 1989, Honnell and Hall 1991, Wichert and Hall 1994, Wichert et al. 1996).

8.5 WERTHEIM'S PERTURBATION THEORY FOR HARD-SPHERE CHAIN FLUIDS

One of the most successful theories for polyatomic molecular fluids consisting of bonded monomers with spherically symmetric potentials is the *thermodynamic perturbation theory* (TPT), derived by Wertheim (1984a,b, 1986a,b, 1987). In this theory, the interactions between the monomers forming the molecules are considered to consist of repulsive and attractive contributions. For nonbranched molecules, the attractive interactions are associated to sites A and B, the bonding sites that define a bonding angle ω with the center of the hard core of the monomer. The attractive interaction potential satisfies $\phi_{AA} = 0$, $\phi_{BB} = 0$, $\phi_{AB}(12) \neq 0$, and is so short-ranged that each site can bond to only another site. Due to the attractive interactions, the monomers tend to associate to form chains of variable length so that, at equilibrium, the fluid will consist of a mixture of molecules with different number of beads. A fluid of free monomers with only the repulsive interactions is taken as the reference fluid, usually the HS system. The thermodynamic properties of the molecular fluid are then expressed in terms of those corresponding to the reference fluid. The detailed derivation of the theory is too extensive and complex to be reproduced here. We will just give the main results of the theory, including the equations of state obtained at first and second order, as well as those from other refined versions and extensions.

8.5.1 WERTHEIM'S FIRST-ORDER PERTURBATION THEORY

In the Wertheim first-order TPT (TPT1), only two-body interactions between the monomers are considered. Wertheim (1984b) showed that the free energy of the fluid may be expressed as

$$\frac{F}{Nk_BT} = \frac{F_0}{Nk_BT} + \frac{F_{bond}}{Nk_BT}, \tag{8.76}$$

where

N is the total number of monomers, either bonded or unbonded

F_0 is the free energy of the reference fluid

F_{bond} is the bonding contribution to the free energy

The simplest case is a system of HS with one bonding site A, so that only dimers can be formed. For this system, Wertheim (1984b) showed that at equilibrium

$$\frac{F_{bond}}{Nk_BT} = \ln X - \frac{1}{2}X + \frac{1}{2}, \tag{8.77}$$

where $X = \rho_0/\rho$, the ratio of the number density of unbonded monomers ρ_0 to the total number density ρ of monomers either bonded or unbonded. The two densities are related by the equation

$$\rho = \rho_0 + 4\pi\rho_0^2 \int g_0(r) \langle f_A(r) \rangle r^2 dr, \tag{8.78}$$

in which $f_A(r)$ is the Mayer function and the angular brackets mean an average over orientations. Equation 8.78 may be obtained by minimizing the free energy, as given by a first-order Mayer function perturbation theory, with respect to ρ_0. If the monomers are HS in the sticky limit of the bonding sites the last expression reduces to

$$\rho = \rho_0 + \rho_0^2 K g_0(\sigma). \tag{8.79}$$

where K is a constant.

From Equation 8.77 the contribution to the pressure due to bonding is

$$P_{bond} = \beta(P - P_0) = \left(\rho - \frac{1}{2}\rho_0\right)\left[\frac{\rho}{\rho_0}\frac{\partial\rho_0}{\partial\rho} - 1\right]. \tag{8.80}$$

Substituting Equation 8.79 into Equation 8.80, we obtain

$$\beta P = \beta P_0 - \frac{1}{2}(\rho - \rho_0)\left(1 + \frac{\rho}{g_0(\sigma)}\frac{\partial g_0(\sigma)}{\partial\rho}\right), \tag{8.81}$$

where ρ_0 is given by the positive root of Equation 8.79 and, in the limit of complete dimerization $\rho_0 = 0$ and Equation 8.81 allows us to obtain the EOS of a fluid of diatomic molecules made of tangent HS. This result was generalized by Wertheim (1987) to a fluid of linear molecules consisting of tangent hard spheres (LTHS), formed by polymerization of monomers with two bonding sites A and B, assuming that no rings are formed. The corresponding compressibility factor $Z = P/\rho_n k_B T$ in the first-order approximation is

$$Z = nZ_0 - (n-1)\left[1 + \frac{\rho}{g_0(\sigma)}\frac{\partial g_0(\sigma)}{\partial\rho}\right], \tag{8.82}$$

where

$g_0(\sigma)$ is the contact value of the RDF of the HS reference fluid

Z_0 is the corresponding compressibility factor, both $g_0(\sigma)$ and Z_0 at number density ρ

n is the average number of beads per molecule

The same equation can be applied to a fluid with all its molecules having the same number n of beads. The result (8.82) was generalized by Ghonasgi and Chapman (1994a) to associating r-mer chains with two bonding sites at the ends of the chains.

If we take for the HS fluid the Carnahan–Starling equation (5.21) and the corresponding contact value of the RDF Equation 8.41, then the Werteim TPT1 EOS for LTHS molecules takes the form

$$Z = n\frac{1 + \eta + \eta^2 - \eta^3}{(1 - \eta)^3} - (n - 1)\frac{1 + \eta - \eta^2/2}{(1 - \eta)(1 - \eta/2)}. \tag{8.83}$$

Equation 8.82 applies to linear chains, thus excluding the possibility of intramolecular bonding, that is, the bonding of the ends of the chain with each other to form rings. Ghonasgi et al. (1994) modified Equation 8.82 for ring molecules by writing

$$Z = nZ_0 - (n - 1)\left[1 + \rho\frac{\partial \ln g_0(\sigma)}{\partial \rho}\right] - \rho\frac{\partial \ln g_{intra}(\sigma)}{\partial \rho}, \tag{8.84}$$

where $g_{intra}(\sigma)$ is the contact value of the site–site intramolecular pair correlation function for the sites at the ends of the chain. If one assumes that $g_{intra}(\sigma) = g_0(\sigma)$, a reasonable approximation for long chains, one obtains

$$Z = nZ_0 - (n - 1) - n\frac{\rho}{g_0(\sigma)}\frac{\partial g_0(\sigma)}{\partial \rho}. \tag{8.85}$$

The Wertheim first-order perturbation theory was extended by Chapman et al. (1986) and Joslin et al. (1987) to multicomponent associating fluid mixtures in which each species i may be present either as unbonded monomers or as bonded monomers forming dimers.

Thus far, we have considered monomers having only one or two bonding sites. The theory can be readily extended to the case of multiple bonding sites. Considering monomers with M bonding sites in a one-component fluid, according to Chapman (1988) and Jackson et al. (1988), the bonding contribution to the free energy may be written in the form

$$\frac{F_{bond}}{Nk_BT} = \sum_A \left(\ln X_A - \frac{X_A}{2}\right) + \frac{1}{2}M, \tag{8.86}$$

where N is the number of monomers, the sum extends over the whole set M of bonding sites, and

$$X_A = \frac{1}{1 + \rho \sum_B X_B \Delta_{AB}} \tag{8.87}$$

is the fraction of spheres *not bonded* at site A. In this expression, ρ is the total number density, the sum again extends over the whole set M of sites, and

$$\Delta_{AB} = 4\pi \int g_0(r) \langle f_{AB}(r) \rangle r^2 dr. \tag{8.88}$$

Solving Equation 8.87 we can obtain X_A. For the particular case of spheres with two bonding sites, Jackson et al. (1988) showed that the average number of beads per chain is $\langle n \rangle = 1/X_A$ and that the fraction of molecules with n beads is $nX_A^2(1 - X_A)^{n-1}$. Once X_A is known, from the density derivative of Equation 8.86 one obtains the bonding contribution to the pressure, which added to the contribution from the hard-sphere reference fluid will finally yield the pressure of the system. The theory was tested by Jackson et al. (1988) against simulation for HS with one or two SW bonding sites. The theoretical predictions for the EOS, the configurational free energy, and the mole fraction of monomers were in excellent agreement with the simulations.

The theory was further generalized by Chapman et al. (1988) to associating mixtures of monomers with multiple bonding sites. Now the change in the free energy due to bonding is

$$\frac{F_{mix}^{bond}}{Nk_BT} = \sum_i x_i \left[\sum_A \left(\ln X_A(i) - \frac{X_A(i)}{2} \right) + \frac{1}{2} M(i) \right], \tag{8.89}$$

where the first sum extends over the different species in the mixture, the second sum extends over the different sites of species i, x_i and $M(i)$ are the mole fraction of monomers and the total number of bonding sites, respectively, of species i, and

$$X_A(i) = \left[1 + \sum_j \sum_B \rho x_j X_B(j) \Delta_{AB}(ij) \right]^{-1}, \tag{8.90}$$

is the fraction of monomers of species i unbonded at site A, with

$$\Delta_{AB}(ij) = 4\pi \int g_{ij}^{(0)}(r) \langle f_{ij}^{AB}(r) \rangle r^2 dr, \tag{8.91}$$

where
$f_{ij}^{AB}(r)$ is the Mayer function for the interaction between the site A of species i and the site B of species j
$g_{ij}^{(0)}(r)$ is the RDF of the reference fluid

The EOS for the mixture can be obtained from the density derivative of Equation 8.89 by adding an EOS suitable for the reference mixture. In particular, when the reference mixture is an additive HS mixture, we can use the BMCSL equation (7.91).

For the particular case of an m-component mixture of linear molecules made of homonuclear tangent HS in which the molecules of species i have $n(i)$ equal-sized spheres with diameter σ_{ii}, proceeding in a similar way as that used to derive Equation 8.82, Chapman et al. (1988) obtain

$$\frac{P}{\rho k_B T} = Z_{mix}^{HS} - \sum_{i=1}^{m} \frac{\rho_{n(i)}}{\rho} [n(i) - 1] \left[1 + \frac{\rho}{g_{ii}^{HS}(\sigma_{ii})} \frac{\partial g_{ii}^{HS}(\sigma_{ii})}{\partial \rho} \right], \qquad (8.92)$$

where

$\rho_{n(i)}$ is the number density of the chains of species i

ρ is the total number density of the monomers

$g_{ii}^{HS}(\sigma_{ii})$ and Z_{mix}^{HS} are the contact RDF and the EOS for HS mixtures as given, for example, by the BMCSL expressions (7.90) and (7.91), respectively

In case the monomers are not HS, Z_{mix}^{HS} must be replaced with an appropriate EOS for the reference monomer mixture and $g_{ii}^{HS}(\sigma_{ii})$ must be replaced with $\Delta(ii)$.

Another extension of the Wertheim first-order TPT was developed by Archer and Jackson (1991) and Amos and Jackson (1991) by considering pure fluids with molecules made of tangent HS of different sizes. For linear molecules made of n tangent HS with different diameters σ_i, the EOS may be written in the form

$$\frac{P}{\rho k_B T} = Z_{mix}^{HS} + \sum_{i=1}^{n-1} Z_{i,i+1}^{bond}, \qquad (8.93)$$

where, in a similar way as before,

$$Z_{i,i+1}^{bond} = -\frac{1}{n} \left[1 + \frac{\rho}{g_{i,i+1}^{HS}(\sigma_{i,i+1})} \frac{\partial g_{i,i+1}^{HS}(\sigma_{i,i+1})}{\partial \rho} \right], \qquad (8.94)$$

with $\sigma_{i,i+1} = (\sigma_i + \sigma_{i+1})/2$. This approach is known as the *bonded hard-sphere* (BHS) theory. For the diatomic and linear triatomic molecules considered in Figure 8.4, the BHS theory yields results nearly indistinguishable from those of the ISPT at the scale of the figure. The BHS theory has the advantage that, for molecules made of tangent HS do not need the knowledge of the nonsphericity parameter that, except for relatively simple molecular geometries, may require complex numerical calculation, as said before. The theory may be extended to chains made of fused HS, as we will see in Section 8.6, but in this case it may be necessary to have knowledge of the nonsphericity parameter and/or other geometrical parameters characterizing the molecules.

8.5.2 Wertheim's Second-Order Perturbation Theory

Since there is not dependence on the bond angle ω in the TPT1, this approximation is inappropriate for low values of ω. If we consider a system of equal-sized HS with two bonding sites A and B and bond angle ω, polymerization can only take place for $\omega > \pi/3$, whereas for $\omega < \pi/3$ only dimers can be formed, because bonding of one site prevents the bonding of the other. TPT1 is unable of accounting for this fact. The situation is changed in the second-order TPT (TPT2), which includes in an approximate way three-body interactions between the monomers. For LTHS fluids, the TPT2 EOS is given by (Wertheim 1987)

$$Z_{TPT2} = Z_{TPT1} - (n - n^*) \frac{\rho}{y} \frac{\partial y}{\partial \rho}, \tag{8.95}$$

where Z_{TPT1} is given by Equation 8.82,

$$n^* = \frac{n}{2} + \frac{2y}{1 + 4y} + \frac{\left[n^2 (1 + 4y) - 4y\right]^{1/2}}{2 (1 + 4y)}, \tag{8.96}$$

and for fixed bond angle $\omega > \pi/3$, that is, for rigid molecules

$$y = \frac{g_0 \left(\sigma, \sigma, 2\sigma \sin (\omega/2)\right)}{\left[g_0 (\sigma)\right]^2} - 1, \tag{8.97}$$

where $g_0(\sigma, \sigma, 2\sigma \sin(\omega/2))$ is the value of the triplet correlation function for HS at the center-to-center distances corresponding to the three adjacent monomers in the molecule, whereas for flexible molecules

$$y = \int_{\pi/3}^{\pi} \left[\frac{g_0 \left(\sigma, \sigma, 2\sigma \sin (\omega/2)\right)}{\left[g_0 (\sigma)\right]^2} - 1\right] \zeta (\omega) \sin \omega \, d\omega, \tag{8.98}$$

where $\zeta(\omega)$ is a distribution of bond angles which must satisfy the condition

$$\int_0^{\pi} \zeta (\omega) \sin \omega \, d\omega = 1, \tag{8.99}$$

and, for the totally flexible case,

$$\zeta (\omega) = \begin{cases} 0, & 0 < \omega < \pi/3 \\ 2/3, & \pi/3 < \omega < \pi \end{cases}, \tag{8.100}$$

from which

$$y = \frac{2}{3} \frac{1}{\sigma^2} \int_{\sigma}^{2\sigma} \left[\frac{g_0 (\sigma, \sigma, x)}{\left[g_0 (\sigma)\right]^2} - 1\right] x \, dx. \tag{8.101}$$

To determine y from this expression, we cannot use the Kirkwood superposition approximation, Equation 6.63, because it is not enough accurate. From the simulation data for $g(\sigma, \sigma, x)$ as well as the low-density behavior, an expression for y was obtained by Wertheim (1987) for the totally flexible case

$$y = 0.233633\eta \, (1 + 1.482\eta) \,. \tag{8.102}$$

A slightly different expression, namely,

$$y = 0.233633\eta \, (1 + 0.472\eta) \,, \tag{8.103}$$

was derived by Müller and Gubbins (1993), and Phan et al. (1993) reported a very similar parametrization. To arrive at the result of Equation 8.103, the triplet correlation function for the three BHS was written as

$$g_0 \left[\sigma, \sigma, 2\sigma \sin \left(\frac{\omega}{2} \right) \right] = g_0 \, (\sigma) \, g_0 \, (\sigma) \, g_0^* \left[\sigma, \sigma, 2\sigma \sin \left(\frac{\omega}{2} \right) \right], \tag{8.104}$$

where

$$g_0^* \left[\sigma, \sigma, 2\sigma \sin \left(\frac{\omega}{2} \right) \right] = \frac{1 + c_1\eta + c_2\eta^2}{(1 - \eta)^3}. \tag{8.105}$$

This expression was proposed to fit the values of g_0^*, for packing fractions $\eta \leq 0.47$, obtained by Attard and Stell (1992) from the PY approximation, which was found to be in close agreement with the simulation data. Parameters c_1 and c_2 were determined from fitting the theoretical results and were reported by Müller and Gubbins (1993) for different values of the bond angle ω. Expression (8.105), together with (8.104), can be used also to determine y for rigid molecules from (8.97).

Once the expression for the density dependence of y is known, the EOS can be obtained from Equation 8.95. Figure 8.6 compares with simulations the results achieved with the TPT2 for the EOS of hard homonuclear nonlinear triatomic fluids with two different values of the bond angle ω. Similar accuracy is obtained for other bond angles.

8.5.3 RESUMMED THERMODYNAMIC PERTURBATION THEORY

When $\omega < \pi/3$ and $\zeta(\omega) \neq 0$, the TPT must be modified giving rise to the so-called resummed TPT or RTPT. Wertheim (1987) treated this problem in detail and reported an explicit expression for the resummed TPT1 or RTPT1 in the form

$$Z_{RTPT1} = nZ_0 - (n - 1) \left[1 + \frac{\rho}{g_0 \, (d)} \frac{\partial g_0 \, (d)}{\partial \rho} \right], \tag{8.106}$$

where d is some distance that depends on $\zeta(\omega)$ for $\omega < \pi/3$. The two extreme cases correspond to (1) double bonding is never prevented, in which case $d = \sigma$ and the EOS (8.82) of the TPT1 is recovered; and (2) double bonding is always prevented,

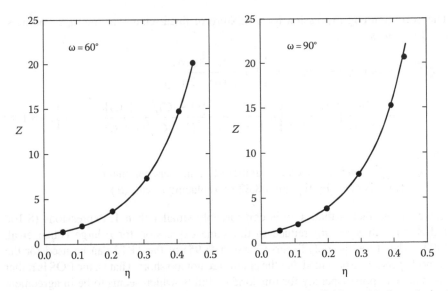

FIGURE 8.6 Compressibility factor Z for nonlinear triatomic fluids made of equal-sized tangent HS with two different values of the bond angle ω. Points: simulation data from Müller and Gubbins (1993). Continuous curves: Wetheim's second-order perturbation theory, Equation 8.95 with Equations 8.96, 8.97, 8.104, and 8.105.

so that only dimers can be formed (see Wertheim 1987 for details). Intermediate situations are of interest in polymerization, when double bonding of a monomer may be prevented in some cases by a first bonding. In considering polymerization, however, we must take into account that molecules with different number of beads are usually present. In these cases, n must be considered as the average number of monomers per molecule.

8.5.4 Correction to Account for Monodispersity

Strictly speaking, Wertheim's TPT was derived for fluids of associating HS with two bonding sites, which results in a polydisperse mixture of chains or rings with different number of monomers. The use of the Wertheim TPT EOS for a monodisperse fluid by identifying the average number of monomers per chain with the number n of monomers in any chain of the monodisperse fluid is appropriate in the TPT1, but not in the TPT2. Phan et al. (1993) have developed a version of the TPT2 for a monodisperse fluid of chains, each of them consisting of n, BHS. The resulting EOS is

$$Z_{TPT2} = Z_{TPT1} - \frac{2\eta}{1+4y}\left(\frac{\partial y}{\partial \eta}\right)\left\{n\frac{C}{1+C}\frac{1+[(1-C)/(1+C)]^{n-1}}{1-[(1-C)/(1+C)]^{n}} - 1\right\},$$

$$(8.107)$$

where $C = (1+4y)^{1/2}$ and Z_{TPT1} is given by Equation 8.82.

The corresponding expression for a mixture of linear chains having n_i monomers, $i = 1, \ldots, m$ is

$$Z_{mix}^{TPT2} = Z_{TPT1}\left(\langle n_i \rangle\right) - \left(\langle n_i \rangle - 1\right)\left[1 + \frac{\eta}{g(\sigma)}\frac{\partial g(\sigma)}{\partial \eta}\right]$$

$$- \frac{2\eta}{1+4y}\left(\frac{\partial y}{\partial \eta}\right)\sum_{i=1}^{m}\left\{n_i \frac{C}{1+C}\frac{1 + \left[(1-C)/(1+C)\right]^{n_i-1}}{1 - \left[(1-C)/(1+C)\right]^{n_i}} - 1\right\}, \quad (8.108)$$

where

$\langle n_i \rangle = \sum_{i=1}^{m} x_i n_i$ is the average number of monomers per chain

$Z_{TPT1}(\langle n_i \rangle)$ is given by Equation 8.82 by replacing n with $\langle n_i \rangle$

However, the fact is that, as y is considerably smaller than 1, expressions (8.108 and 8.95) with $n = \langle n_i \rangle$ reduce to the same expression for n large (Phan et al. 1993). Therefore, the polydisperse version (8.95) of the TPT2 can be used for the monodisperse case provided that the chains are not too short. That is, the EOS is rather insensitive to polydispersity for this kind of fluids, which seems to be in agreement with the simulation results.

8.5.5 DIMER CORRECTION

The Wertheim's formulation of the TPT2 requires the use of the triplet correlation function. This poses a problem even for LTHS fluids, because the triplet correlation function for the HS reference fluid is available only in numerical form and for a limited number of states. The situation would be worse if we were dealing with more realistic linear polyatomic fluids consisting of monomers interacting by means of site–site potentials more realistic than the HS one. To overcome these problems, Chang and Sandler (1994a) derived a formulation that uses information about the dimer fluid, in addition to the monomer fluid, instead of using the triplet correlation function. Their starting point is the free energy F_{bond}^{mono} needed to form a dimer of HS in a fluid of HS monomers, which is the work against the mean force acting on the two dimerizing monomers initially placed at infinite distance from each other, that is

$$F_{bond}^{mono} = -\int_{\infty}^{\sigma} \mathbf{f}(\mathbf{r}_1) \cdot d\mathbf{r}, \quad (8.109)$$

where $\mathbf{f}(\mathbf{r}_1)$ is the mean force defined in Equation 2.32. Taking into account this equation together with Equation 2.31, one easily arrives at

$$F_{bond}^{mono} = -kT \ln g_0(\sigma). \quad (8.110)$$

If this equation were valid for further bondings, resulting in molecules with increasing number of monomers, finally giving rise to a LTHS fluid in which each molecule would consist of n monomers, we would obtain the TPT1 result for the free energy

$$\frac{F}{N_n kT} = \frac{F_0}{N_n kT} - (n-1)\ln g_0(\sigma), \quad (8.111)$$

which leads to the TPT1 EOS (Equation 8.82). The fact that the TPT1 works well for LTHS fluids consisting of freely jointed HS for low values of n indicates that this approximation is quite right for these fluids. However, the agreement between TPT1 and simulation data worsens as n increases, especially in the low-density region. This means that, as the number of bonds increases, further bonding is affected by previous existing bonds. This may be attributed to excluded volume effects.

To account in an approximate way for the effect on bonding of previously formed bonds, Chang and Sandler (1994a) include information on the structure of the dimer fluid in the TPT, giving rise to the TPT-dimer or TPT-D theory. Similarly to Equation 8.110, the excess free energy required to form a tetramer molecule starting from two dimer molecules is

$$F_{bond}^{dimer} = -kT \ln g_d(\sigma),$$ (8.112)

where $g_d(\sigma)$ is the angle-averaged site–site contact pair correlation function of the dimer. If the starting HS fluid consists of N spheres, once the dimerizing process is completed, the resulting fluid will consist of $N/2$ dimers. This is the new reference system which polymerizes to form the final fluid consisting of N_n LTHS molecules, each of them formed by n monomers. If it is assumed that further bondings beyond dimers are not affected by previously formed bonds, taking into account that the total number of bonds between dimers needed to form an n-mer is $n/2 - 1$, or $N_n(n/2 - 1) = (n - 2)N_n/2$ for the n-mer fluid consisting of N_n LTHS molecules, the resulting excess free energy for the n-mer fluid is

$$\frac{F}{N_n kT} = \frac{F_0}{N_n kT} - \frac{n}{2} \ln g_0(\sigma) - \frac{n-2}{2} \ln g_d(\sigma),$$ (8.113)

and the EOS

$$Z = nZ_0 - \frac{n}{2}\eta \frac{\partial \ln g_0(\sigma)}{\partial \eta} - \frac{n-2}{2}\eta \frac{\partial \ln g_d(\sigma)}{\partial \eta}.$$ (8.114)

Note that if n were an odd number, the factor $(n - 2)/2$ would be noninteger. On the other hand, Equation 8.113 reduces to the TPT1 result (8.111) if we replace $g_d(\sigma)$ with $g_0(\sigma)$.

Chang and Sandler (1994a) use for the averaged contact pair correlation function of the dimers $g_d(\sigma)$ either the Chiew (1991) analytical solution of the PY theory, which, from Equation 8.45 for $n = 2$, is

$$g_d(\sigma) = \frac{1/2 + \eta}{(1 - \eta)^2},$$ (8.115)

or the expression derived by Yetiraj and Hall (1990, 1992)

$$g_d(\sigma) = \frac{(1 - 0.5\eta)(0.534 + 0.414\eta)}{(1 - \eta)^3},$$ (8.116)

by numerically evaluating the zero- and first-order terms in the density expansion of the contact pair correlation function of the dimer in combination with some heuristic considerations on $g_d(\sigma)$ based on simulations.

Introducing either expression (8.115) or expression (8.116) for $g_d(\sigma)$, together with expression (8.41) for $g_0(\sigma)$, into Equation 8.113, we can obtain two different expressions for the excess free energy, giving rise to two different equations of state

$$Z_{TPT-D1} = 1 + n\frac{2\eta\,(2-\eta)}{(1-\eta)^3} - \frac{n}{2}\frac{\eta\left(\frac{5}{2}-\eta\right)}{(1-\eta)\left(1-\frac{1}{2}\eta\right)} - \frac{n-2}{2}\frac{\eta\,(2+\eta)}{(1-\eta)\left(\frac{1}{2}+\eta\right)},$$

(8.117)

and

$$Z_{TPT-D2} = 1 + n\frac{2\eta\,(2-\eta)}{(1-\eta)^3} - \frac{n}{2}\frac{\eta\left(\frac{5}{2}-\eta\right)}{(1-\eta)\left(1-\frac{1}{2}\eta\right)}$$

$$- \frac{n-2}{2}\frac{\eta\,(1.749 - 0.120\eta - 0.207\eta^2)}{(1-\eta)\left(1-\frac{1}{2}\eta\right)(0.534 + 0.414\eta)},$$

(8.118)

respectively. Both expressions apply only for $n \geq 2$. Equation 8.117 was also derived independently by Ghonasgi and Chapman (1994a), who also proposed an empirical correction to the Chiew expression (8.115) to render it in closer agreement with the simulations at high density.*

Zhou et al. (1995b) generalized the TPT-D result by expressing the compressibility factor Z_n of the n-mer fluid in terms of those corresponding to shorter i-mer and j-mer fluids. Their result may be written in the form

$$Z_n = Z_i + \frac{n-i}{i-j}\left(Z_i - Z_j\right)$$

$$= Z_j + \frac{n-j}{i-j}\left(Z_i - Z_j\right),$$

(8.119)

which becomes equivalent to Equation 8.75, obtained from a generalization of the GFD theory, if one assumes a linear dependence of the excluded volume of the n-mer with the number n of beads in the chain. This is indeed the case for large n, as revealed by the computer simulations of Denlinger and Hall (1990).

8.5.6 SEQUENTIAL POLYMERIZATION

Let us consider the process of forming an n-mer starting from independent monomers. First, two monomers bond together to form a dimer and the change in the free energy involved will be of the form of Equation 8.112, which we rewrite as $F_{bond}(2) = -k_B T \ln g(1,1)$. Then, one new monomer is bonded to the dimer to

---

* Ghonasgi and Chapman termed SAFT-dimer or SAFT-D to the TPT-D theory. See Section 8.7 for details on SAFT and related theories.

form a trimer with a change in the free energy $F_{bond}(3) = -k_B T \ln g(1,2)$. Next, an additional monomer is bonded at one of the ends of the trimer to form a tetramer with $F_{bond}(4) = -k_B T \ln g(1,3)$, and so on, so that the formation of an n-mer involves the change in free energy

$$F_{bond}(n) = -k_B T \sum_{i=1}^{n-1} \ln [g(1,i)]. \qquad (8.120)$$

Let us now assume $g(1,i) \simeq g(1,2)$ for $i \geq 2$, an approximation that is supported by simulations of Kumar et al. (1996). Then, for the whole system with N_n chains with n-monomers each, we will have

$$\frac{F_{bond}}{N_n k_B T} = -\ln g(1,1) - (n-2)\ln g(1,2). \qquad (8.121)$$

Here, $g(1,1)$ is the contact RDF $g_{HS}(\sigma)$ of the reference HS fluid, and $g(1,2)$ is the contact site–site pair correlation function for monomer–dimer interaction, for which we can use the result (8.50) derived by Chiew, so that we will have

$$\frac{F_{bond}}{N_n k_B T} = -\ln g_{HS}(\sigma) - (n-2)\ln g_{MD}(\sigma). \qquad (8.122)$$

This result was derived by Yeom et al. (2002) within the framework of the TPT considering the process of sequential polymerization as described here, and so we will denote by TPT-SP this theory. The expression for the total free energy of the system is obtained by adding the monomer and bonding contributions. The corresponding EOS will consist of three contributions: one from the monomers, for which one can use the CS equation (5.21), another for the formation of dimers in the first step of the sequential polymerization, which will have the form of the last term in Equation 8.83 with $n = 2$, and a term accounting for the rest of the process of bonding, that is

$$Z_{TPT-SP} = n\frac{1+\eta+\eta^2-\eta^3}{(1-\eta)^3} - \frac{1+\eta-\eta^2/2}{(1-\eta)(1-\eta/2)} + (n-2)Z_{MD}. \qquad (8.123)$$

As for Z_{MD}, two different results may be obtained depending on the choice for $g_{HS}(\sigma)$ in Equation 8.50. With the PY Equation 8.40 choice

$$Z_{MD}^{PY} = -\frac{\eta(3+\eta)}{(1-\eta)(\eta+1)}, \qquad (8.124)$$

and with the CS Equation (8.41) choice

$$Z_{MD}^{CS} = -\frac{\eta(\eta^2+2\eta-9)}{(1-\eta)(\eta^2-3)}. \qquad (8.125)$$

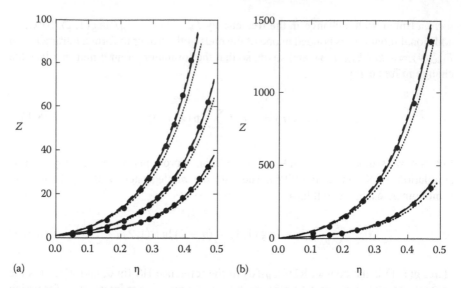

FIGURE 8.7 Compressibility factor $Z = P/\rho_n k_B T$ for linear flexible chains made of freely jointed HS as a function of the packing fraction $\eta = \rho_n v_n$. Points: (a) simulation data from Chang and Sandler (1994a) and (b) from Gao and Weiner (1989) in Dotted curves: Chiew's theory, Equation 8.39. Dashed curves: Wertheim's TPT1, Equation 8.83. Continuous curves: Wertheim's TPT2, Equation 8.95 with Equation 8.103, indistinguishable at the scale of the figure of those from Equation 8.107, and from those of the TPT-D2, Equation 8.118. The curves and data correspond to $n = 4$, 8, and 16 from down upwards in the left figure, and to $n = 51$ and 201 from down upwards in the right figure.

Figure 8.7 compares the results provided by the Chiew theory and by different versions of the Wertheim TPT with the simulations for the EOS of linear flexible chains made of freely jointed HS. The chains considered are the same as in Figure 8.5. The figure shows that the Chiew theory yields quite low values of the EOS, whereas, at first sight, the Wertheim theory provides satisfactory accuracy even at first order. There is little difference between the TPT2, the TPT2 corrected for monodispersity, and the TPT-D2 results. Comparison of Figures 8.5 and 8.7 reveals that also the GFD theory yields similar accuracy as the TPT2.

More insight into the performance of these theories can be achieved by numerically comparing their predictions with the simulation data. This is done in Tables 8.1 through 8.5 for different values of the number n of beads. One can see that for low values of n the GFD, TPT1, TPT2, TPT-D1, and TPT-D2 theories yield quite accurate results, although all of them, and especially the first two, slightly overestimate the EOS at moderate densities. The deviations increase as n increases, with the best overall agreement with simulations achieved with the two TPT-D versions. The difference between the TPT2, Equation 8.95 with Equation 8.103, and the TPT2 with the correction for monodispersion, Equation 8.107, is negligible for the cases considered in the tables.

An extension of the TPT2 theory, as well as several extensions of the TPT-D theory to mixtures of heteronuclear hard chains, was developed by Shukla and

TABLE 8.1

Compressibility Factor $Z = P/\rho_n k_B T$ for the Flexible Homonuclear LTHS Fluid with $n = 4$ as a Function of the Reduced Density of Monomers $\rho^* = \rho\sigma^3$

ρ^*	sim[a]	GFD Equation 8.72	TPT1 Equation 8.83	TPT2 Equation 8.95	TPT-D1 Equation 8.117	TPT-D2 Equation 8.118
0.10	1.49	1.53	1.54	1.51	1.48	1.50
0.20	2.22	2.30	2.33	2.27	2.21	2.25
0.30	3.28	3.40	3.45	3.36	3.31	3.34
0.40	4.84	4.97	5.04	4.92	4.89	4.90
0.50	7.09	7.21	7.31	7.16	7.17	7.14
0.55	8.56	8.67	8.79	8.63	8.66	8.61
0.60	10.26	10.42	10.57	10.39	10.46	10.37
0.65	12.49	12.52	12.71	12.52	12.62	12.51
0.70	15.00	15.06	15.31	15.10	15.24	15.09
0.75	18.26	18.14	18.46	18.23	18.42	18.22
0.80	22.10	21.90	22.31	22.07	22.31	22.06
0.85	27.02	26.50	27.04	26.78	27.08	26.78
0.90	32.49	32.19	32.89	32.62	32.99	32.62
ARD (%)		1.70	2.57	1.04	1.04	0.77

The average relative deviation is defined as ARD (%) $= 100 \times |Z_{calc} - Z_{sim}|/Z_{sim}$.

[a] Chang and Sandler (1994a).

TABLE 8.2

As in Table 8.1 for $n = 8$

ρ^*	sim[a]	GFD Equation 8.72	TPT1 Equation 8.83	TPT2 Equation 8.95	TPT-D1 Equation 8.117	TPT-D2 Equation 8.118
0.10	1.76	1.91	1.95	1.87	1.75	1.83
0.20	2.99	3.28	3.36	3.20	3.02	3.13
0.30	4.91	5.29	5.42	5.18	5.00	5.09
0.40	7.75	8.20	8.40	8.09	7.96	7.98
0.50	11.95	12.40	12.71	12.31	12.29	12.20
0.55	14.83	15.15	15.54	15.10	15.15	14.99
0.60	18.26	18.47	18.95	18.48	18.61	18.36
0.65	22.32	22.48	23.08	22.57	22.80	22.46
0.70	27.14	27.34	28.10	27.54	27.90	27.44
0.75	33.34	33.26	34.22	33.62	34.11	33.52
0.80	40.85	40.49	41.72	41.08	41.72	40.98
0.85	50.64	49.37	50.95	50.27	51.09	50.18
0.90	62.03	60.37	62.41	61.69	62.71	61.61
ARD (%)		3.57	5.33	2.65	1.90	1.78

[a] Chang and Sandler (1994a).

TABLE 8.3

As in Table 8.1 for $n = 16$

ρ^*	sim[a]	GFD Equation 8.72	TPT1 Equation 8.83	TPT2 Equation 8.95	TPT-D1 Equation 8.117	TPT-D2 Equation 8.118
0.10	2.25	2.67	2.75	2.58	2.29	2.48
0.20	4.47	5.24	5.42	5.06	4.64	4.89
0.30	8.09	9.06	9.36	8.83	8.39	8.60
0.40	13.59	14.65	15.12	14.41	14.09	14.14
0.50	21.96	22.77	23.50	22.60	22.52	22.32
0.55	27.13	28.12	29.03	28.04	28.12	27.75
0.60	34.05	34.59	35.72	34.63	34.92	34.35
0.65	41.90	42.41	43.83	42.65	43.17	42.37
0.70	51.80	51.91	53.69	52.43	53.21	52.14
0.75	65.15	63.49	65.74	64.39	65.48	64.11
0.80	81.28	77.66	80.53	79.08	80.53	78.82
ARD (%)		6.62	9.04	5.25	2.61	3.74

[a] Chang and Sandler (1994a).

TABLE 8.4

As in Table 8.1 for $n = 51$

η	sim[a]	GFD Equation 8.72	TPT1 Equation 8.83	TPT2 Equation 8.95	TPT-D1 Equation 8.117	TPT-D2 Equation 8.118
0.105	11.46	13.85	14.48	13.25	11.75	12.64
0.157	23.04	25.53	26.59	24.74	23.21	23.93
0.195	34.87	37.41	38.89	36.58	35.30	35.67
0.243	56.56	57.96	60.15	57.26	56.62	56.27
0.309	101.04	100.79	104.60	100.91	101.72	99.86
0.340	130.17	129.25	134.21	130.15	131.90	129.10
0.367	160.56	160.00	166.28	161.89	164.61	160.86
0.419	238.12	240.53	250.60	245.59	250.60	244.59
0.471	346.51	362.56	379.05	373.42	381.44	372.50
ARD (%)		5.38	9.39	4.56	2.71	3.26

Gao and Weiner (1989).
[a] Here $\eta = (\pi/6)\rho^*$.

Chapman (2000). These theories were successfully applied to hard models of block copolymers, with the best overall agreement achieved with the TPT-D theory with the improved $g_d(\sigma)$ derived by Ghonasgi and Chapman (1994a). Blas and Vega (2001a) extended the Wertheim TPT to branched chain fluids using the TPT1 approximation for the arms and the TPT2 for the articulation monomers.

TABLE 8.5

As in Table 8.1 for $n = 51$

η	sim[a]	GFD Equation 8.72	TPT1 Equation 8.83	TPT2 Equation 8.95	TPT-D1 Equation 8.117	TPT-D2 Equation 8.118
0.105	36.80	50.72	53.26	48.28	42.15	45.77
0.157	79.44	96.16	100.46	92.95	86.74	89.66
0.209	152.11	163.10	169.85	159.80	155.20	155.98
0.262	256.20	263.10	273.56	260.91	259.67	256.76
0.314	407.16	406.69	422.85	407.66	411.48	403.37
0.367	621.54	622.41	647.96	630.18	641.16	625.92
0.419	927.79	938.25	979.13	958.83	979.15	954.73
0.471	1354.76	1417.13	1484.03	1461.23	1493.70	1457.41
ARD (%)		9.35	14.10	8.48	5.89	6.52

Gao and Weiner (1989).

[a] η has the same meaning as in Table 8.5.

8.6 EXTENSIONS TO LINEAR FUSED HARD-SPHERE CHAINS

The extension of the GF and GFD theories developed for LTHS fluids to fluids consisting of *linear fused hard-sphere* (LFHS) molecules is based on defining suitable reference monomer and (eventually) dimer fluids, characterized by their respective effective number of monomers n_m and dimers n_d per n-mer, monomer σ_m and dimer σ_d diameters, and (eventually) dimer bond length L. Several ways of doing so will be reviewed here.

To apply the GFD theory to LFHS fluids, Honnell and Hall (1989) proposed to take $\sigma_m = \sigma_d = \sigma$, and determine n_m and n_d from the condition that the three— monomer, dimer, and n-mer—fluids must have the same packing fraction. This procedure was called *approximation A* by Yethiraj et al. (1993).

Another approximation, proposed by Yethiraj et al. (1993) and denoted by them as *approximation B*, consists in cutting the LFHS molecules into n monomers, each of them consisting in a sphere with one of two caps missing, and then replacing each of these monomers with a sphere with diameter σ_m, determined from the condition that the $n_m = n$ monomers have the same total volume as one n-mer. This defines the monomer reference fluid. In a similar way, to define the dimer reference fluid, the chain is cut up into $n/2$ segments, each of them consisting in two fused spheres with one of two caps missing, and then each segment is replaced with a complete dimer with the same bond length L as in the n-mer and an effective diameter σ_d, which in general will be different from that of the monomers in the reference monomer fluid, determined from the condition that the resulting $n_d = n/2$ dimers have the same volume as the parent n-mer.

Costa et al. (1995) introduced several new approximations. The *approximation C* considers $n_m = n_d = n$ and the reduced bond length L^* for the dimers is taken to be equal to that of the n-mers, whereas the diameters of monomers and dimers are determined from the condition that their packing fractions be equal to that of the

n-mer fluid. In *approximation AB*, the bond length L is equal for dimers and the n-mers, and n_m, σ_m, n_d, and σ_d are determined from the conditions that the surface area and packing fraction must be equal for monomers, dimers, and n-mers. The *approximation AC* considers the reduced bond length L^* for the dimers equal to that of the n-mers and determines σ_m, σ_d, n_m, and n_d in the same way just described for the AB approximation. The A, AB, and AC approximations, contrarily to the B and C approximations, fulfil the condition that when $L \to 0$ the n-mer fluid reduces to a HS fluid, as it should be. A drawback common to these approximations is that the use of Equation 8.66 requires the knowledge of v_3^e, which cannot be easily calculated for flexible LFHS molecules, and so v_n^e must be determined from simulation.

The criterions described earlier to determine the reference monomer fluid can be applied to those theories, like the GF and TPT1 ones, that only require a HS reference fluid. In practice, to apply the TPT1 theory to LFHS fluids, the actual LFHS molecule is replaced with an equivalent LTHS molecule with effective number of tangent spheres n_{eff} and effective diameter σ_{eff} determined by means of one of the criterions cited for the monomer reference fluid in the context of the GFD.

Costa et al. (1995) analyzed the performance of the GF-A, GF-B, GF-AB, GFD-A, GFD-B, GFD-C, GFD-AB, and GFD-AC approximations by comparing with their simulation data for the EOS of LFHS fluids of flexible n-mer chains with $L^* \simeq 0.4$ and $n = 4$, 8, and 16. They found the GF-AB approximation to be in very good agreement with the simulations, whereas the GF-A and GF-B approximations strongly overestimated the compressibility factor. Among the GFD approximations, the most satisfactory results were those from the GFD-AC approximation, which however yielded somewhat high values of Z. Still somewhat higher values were obtained from the GFD-AB approximation and too low values from the remaining ones. Zhou et al. (1995a) used the AB criterion to determine the reference monomer fluid within the context of the TPT1 and found that the resulting TPT1-AB theory performs comparably with the GFD-AC, but slightly worse than the GF-AB theory. The performance of the GF-AB, GFD-AC, and TPT1-AB theories for the EOS is illustrated in Figure 8.8 for several n-mer fluids.

Boublík (1989), and independently Walsh and Gubbins (1990), compare the second virial coefficient for hard-body fluids with that arising from the TPT. For HCB fluids the former is given exactly by Equation 8.23, as said before, and the same expression holds approximately also for nonconvex molecules, provided that R, which is ill defined in this case, is determined as that corresponding to the convex envelope of the molecule, according to the Boublík–Nezbeda prescription, as explained in Section 8.2. In the same section we mentioned that an alternative, but equivalent, procedure to determine the nonsphericity parameter of LFHS molecules is Equation 8.24. Still another way to determine an approximate value of α for a nonconvex hard-body fluid, provided that we know the corresponding second virial coefficient, is the Rigby (1976) procedure that, as explained in Section 8.2, consists in solving Equation 8.23 for α, as if it were a HCB fluid, taking for B_2^* the actual value.

On the other hand, the value of the second virial coefficient arising from the TPT equations of state (Equations 8.82 and 8.95) is

$$B_2^* = \tfrac{1}{2}\,(3n + 5)\,. \tag{8.126}$$

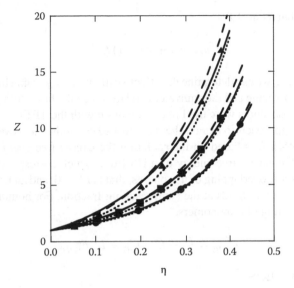

FIGURE 8.8 Compressibility factor Z for LFHS fluids with flexible molecules. Points: simulation data from Costa et al. (1995) for $n = 4$ and $L^* = 0.4058$ (circles), $n = 8$ and $L^* = 0.3948$ (squares), and $n = 16$ and $L^* = 0.3948$ (triangles). Curves: GF-AB (continuous), GFD-AC (dashed), and TPT1-AB (dotted).

Comparison of Equations 8.23 and 8.126 gives

$$n = 2\alpha - 1. \tag{8.127}$$

Substituting this result into Equation 8.82, the modified Wertheim or modified TPT1 (MTPT1) EOS results in the form

$$Z_{MTPT1} = (2\alpha - 1)\frac{1 + \eta + \eta^2 - \eta^3}{(1 - \eta)^3} - (2\alpha - 2)\frac{1 + \eta - \eta^2/2}{(1 - \eta)\left(1 - \eta/2\right)}. \tag{8.128}$$

This approach, first derived for pure fluids of rigid molecules, was later extended to mixtures and flexible chains by Boublík et al. (1990).

Two other ways of determining the effective number of spheres n_{eff} and the effective diameter σ_{eff}, in the TPT1 applied to LFHS fluids, are based on the two limiting cases of LFHS molecules with reduced center-to-center distance $L^* = L/\sigma$, namely, $L^* = 0$ and $L^* = 1$. In the first case, the LFHS molecule reduces to a single sphere, that is, $n_{eff} = 1$ for $L^* = 0$, whereas in the second case, the LFHS molecule is in fact an LTHS molecule, that is, $n_{eff} = n$ for $L^* = 1$. A simple way of interpolating between these two limits is (Jackson and Gubbins 1989)

$$n_{eff} = 1 + (n - 1)L^*, \tag{8.129}$$

and another (Phan et al. 1994)

$$n_{eff} = 1 + (n-1) L^{*3}. \tag{8.130}$$

Once n_{eff} is known, we can determine the effective diameter σ_{eff} so that both molecules, the real LFHS molecule and the equivalent LTHS molecule, have the same molecular volume. Both procedures were used in combination with the TPT1.

A different approach to extend the TPT to LFHS fluids was developed also by Phan et al. (1994). It is based on interpolating the excess free energy of an LFHS fluid between the excess free energy of a HS fluid, which corresponds to an LFHS fluid with complete overlapping of its spheres, that is $L^* = 0$, and an LTHS fluid, that is $L^* = 1$, all the three fluids at the same packing fraction. For homonuclear LFHS molecules consisting of n monomers,

$$F_{LFHS}^E = \left(1 - L^{*3}\right) F_{HS}^E + L^{*3} F_{LTHS}^E, \tag{8.131}$$

or, in terms of the EOS,

$$Z_{LFHS} = \left(1 - L^{*3}\right) Z_{HS} + L^{*3} Z_{LTHS}. \tag{8.132}$$

If we take for Z_{LFHS} the TPT1 expression (8.82), after rearranging, we easily arrive at

$$Z_{LFHS} = 1 + \left[1 + (n-1) L^{*3}\right](Z_0 - 1) - (n-1) L^{*3} \frac{\eta}{g_0(\sigma)} \frac{\partial g_0(\sigma)}{\partial \sigma}. \tag{8.133}$$

In particular, if we use the CS equation (5.21) for the reference HS fluid, we will have

$$Z_{LFHS} = 1 + \left[1 + (n-1) L^{*3}\right] \left[\frac{1 + \eta + \eta^2 - \eta^3}{(1-\eta)^3} - 1\right]$$

$$- (n-1) L^{*3} \frac{\eta (5/2 - \eta)}{(1-\eta)(1-\eta/2)}. \tag{8.134}$$

Expression (8.133) can be easily generalized to heteronuclear LFHS fluids.

Also, the BHS theory has been extended to FHS fluids by Amos and Jackson (1992). To this end, the actual molecule is replaced with an equivalent one formed by tangent HS. This equivalent molecule is determined from the conditions that it must have both molecular volume v_n and nonsphericity parameter α equal to those corresponding to the real molecule. The nonsphericity parameter is obtained from the Boublík–Nezbeda criterion, described in Section 8.2. Using this procedure, the BHS theory was applied by Amos and Jackson (1992) to FHS fluids consisting of hard diatomic, triatomic, and tetrahedral pentatomic molecules. For more complicated FHS molecules, either linear or not, they suggested to determine the equivalent molecule consisting of tangent spheres by means of applying the earlier-mentioned criterion to each diatomic segment in the molecule.

Alternatively, for homonuclear LFHS fluids, instead of dividing the molecule into diatomic segments, Largo et al. (2003) proposed to replace the actual molecule with an equivalent LTHS molecule with the same number of spheres and their diameters σ_i determined from the condition that the molecular volume v_n and nonsphericity parameter α must be equal for the two molecules. To obtain the nonsphericity parameter, expression (8.24) was used. As we have only two conditions to determine the diameters of the spheres in the equivalent LTHS molecule, only two diameters σ_1 and σ_2 can be determined, and therefore the monomers in the equivalent molecule will have either diameter σ_1 or diameter σ_2. For $n = 2$, this procedure leads to the same results as the one proposed by Amos and Jackson (1992). For $n > 2$ there will be, in general, more than one solution corresponding to different number and positions of the spheres of each size in the equivalent molecule. Largo et al. (2003) applied this procedure to rigid LFHS molecules with $L^* = 0.5$ and 0.6 and up to $n = 15$. Considering for convenience that $\sigma_1 > \sigma_2$, they found that a particular solution providing good results in all the considered cases consists in a sphere with diameter σ_1 placed in one of the ends of the equivalent chain and the remaining spheres with diameter σ_2. As shown in Figure 8.9, the agreement between this version of the BHS theory and simulations for the EOS is excellent.

The GFD theory also yields very good results for the same kind of fluids (Mehta and Honnell 1996a, Largo et al. 2003). For rigid LFHS molecules, Mehta and Honnell (1996a) showed that

$$v_n^e = v_1^e + (n - 1)\left(v_2^e - v_1^e\right), \tag{8.135}$$

and so from Equation 8.73,

$$Z = Z_{HS} + (n - 1)\left(Z_{HD} - Z_{HS}\right), \tag{8.136}$$

which coincides with the result of the generalized TPT equation (8.119) for $i = 2$ and $j = 1$. The comparison in Figure 8.9 with simulations reveals that both the GFD and BHS theories are of similar accuracy.

Yethiraj (1995) applied the GFD-A and GFD-B approximations to athermal n-alkanes, modeled as fused HS freely rotating chains with fixed bond lengths and bond angles. To this end, the excluded volumes that enter in Equation 8.72 were determined from simulations. Good agreement with simulations for the compressibility factor with n ranging from 4 to 32 was found from both approximations, with the greater accuracy for large n achieved with the B approximation. Vega et al. (1994) considered a similar model of fluids but with discretized torsional angles according to the *rotational isomeric state* (RIS) model, in which the allowed torsional angles are $0°$, $+120°$, and $-120°$, for the *trans* (*t*), *gauche*$^+$ (*g*$^+$), and *gauche*$^-$ (*g*$^-$) conformations, respectively, and intramolecular potential energies depending on the rotational configurations of the successive bonds. They obtained numerically the virial coefficients B_2 to B_5 for the different conformers of n-alkanes from n-butane to n-octane and found that the MTPT1, Equation 8.128, yields a quite accurate prediction for these coefficients. The problem with this approach is the difficulty of numerically calculating the second virial coefficient of n-alkanes with longer chains. A quite simple procedure to obtain an

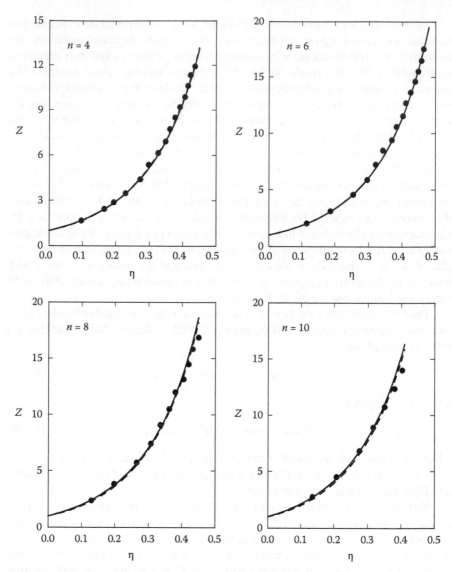

FIGURE 8.9 Compressibility factor Z for LFHS fluids with rigid molecules and reduced center-to-center distance $L^* = 0.5$ in the isotropic phase. Points: simulation data from Largo et al. (2003). Curves: GFD (dashed) and BHS (continuous), nearly indistinguishable to each other at the scale of the figure.

approximate second virial coefficient for hard models of n-alkanes was devised by Vega et al. (1996), later extended to branched chains by MacDowell and Vega (1998). With this approximation, the MTPT1 was found to accurately agree with the simulation data for the EOS for models of n-alkanes with interactions of the repulsive WCA form for the LJ potential and carbon number up to $C = 30$. Mehta and Honnell (1996b) extended the GFD theory to the same models of

n-alkanes with the volume and excluded volume of each conformer determined from an analytical algorithm developed by Dodd and Theodorou (1991). The results obtained for the virial coefficients B_2 to B_5 and the EOS were also in quite good agreement with the simulation data, although the predictions for the latter quantity for the longer chains were slightly less accurate than those achieved with the MTPT1 approximation. MacDowell et al. (2001) applied the MTPT1 Equation 8.128, with α determined either from Equation 8.23, with B_2^* numerically calculated, or from Equation 8.127, by replacing n with n_{eff} as determined from the AB criterion, to hard models of branched alkanes consisting in isomers of n-hexane, n-heptane, and n-octane, and found excellent agreement with simulations, with little difference in most cases between the two procedures used to determine α.

8.7 PERTURBATION THEORIES FOR MOLECULAR FLUIDS WITH DISPERSIVE FORCES

Several of the theories for athermal fluids analyzed in the preceding sections have been extended to molecular fluids with dispersive forces by adding to the free energy, or to the EOS, a suitable term to account for the attractive contributions to these thermodynamic quantities. Some of these extensions are reviewed in this section. As we will see, in their application to real fluids, a number of parameters need to be determined, usually by fitting some kind of experimental data for each particular fluid, which complicates the use of these theories and limits their practical usefulness. A way of simplifying the problem is to resort to *group contribution* (GC) *methods*. The GC methods for equations of state are based on the fact that, whereas the number of existing organic compounds is enormous, all of them may be built up with a relatively small number of functional groups of atoms, and it is assumed that each different group contributes in the same way to the parameters in a given EOS for different compounds belonging to the same family. Thus, once the contributions from the different groups are known and a suitable correlation between the group contributions and the molecular parameters is established, one can apply the EOS to many compounds without the need of determining the parameters for each compound separately. Therefore, the development of GC extensions of molecular-based theories is of particular interest in this field and some of the literature published on this subject will be specially cited in the discussion that follows.

8.7.1 PERTURBED HARD-CHAIN THEORY

The EOS in the perturbed hard-chain theory (PHCT), proposed by Beret and Prausnitz (1975) and Donohue and Prausnitz (1978), consists of a hard-chain contribution Z_{HC} plus a perturbation term Z_{pert} of the vdW *mean field* (MF) form, that is

$$Z = Z_{HC} + Z_{pert}, \tag{8.137}$$

where Z_{HC} is related to the CS equation (5.21) in the form

$$Z_{HC} = 1 + c\,(Z_{CS} - 1) = 1 + c\frac{4\eta - 2\eta^2}{(1 - \eta)^3}, \qquad (8.138)$$

in which $3c$ is the effective number of degrees of freedom, so that for a free monomer $c = 1$ and for a chain $c > 1$. This quantity is different from the number n of monomers in the chain and reflects the constraints imposed to the movement of the monomers because of bonding. Therefore, the chain contribution to the compressibility factor may be considered as a combination of the bonding contribution and the excess compressibility factor of the reference HS fluid. The perturbation contribution was derived from a parametrization reported by Alder et al. (1972), on the basis of a fourth-order perturbation theory with the perturbation terms fitted to simulation data for the SW fluid with $\lambda = 1.5$, and reads

$$Z_{pert} = c \sum_i \sum_j j A_{ij} \left(\frac{V_0}{V}\right)^j \left(\frac{q\varepsilon}{ck_B T}\right)^i, \qquad (8.139)$$

where V_0 is the close packing volume, which, in the particular case of a fluid of LTHS chains made of n monomers with diameter σ and N_n chains, is $V_0 = nN_n\sigma^3/\sqrt{2}$, q is a parameter that depends on the surface area of the molecule, and parameters A_{ij} were reported by Alder et al. (1972). To improve the performance of the PHCT, Donohue and Prausnitz (1978) fitted an expression of the form of Equation 8.139 to experimental data for methane. For a fluid of independent monomers $c = 1$, $q = 1$, and Equation 8.139 reduces to the expression derived by Alder et al. (1972). For molecular fluids $q > 1$ and $q/c > 1$, so that the effect of this quantity is to rescale the usual reduced temperature $T^* = k_B T/\varepsilon$, and the product $q\varepsilon$ acts as an effective potential depth. In a similar way as $k_B T/\varepsilon$ is a measure of the ratio of the kinetic to potential energy in a pure fluid of SW monomers, $ck_B T/q\varepsilon$ is a measure of the same ratio for a molecular fluid with site–site SW interactions. The fact that $q/c > 1$ implies a relative increase of the interaction energy, with respect to kinetic energy, of the molecular fluid with respect to a fluid of monomers with the same relative volume V/V_0.

The PHCT EOS depends on three parameters: the effective diameter d of a segment, or equivalently the effective close packing volume per segment $v_0 = V_0/nN = d^3/\sqrt{2}$, the effective potential depth $q\varepsilon$, and the effective number c of degrees of freedom. For real fluids, the PHCT EOS is used to correlate experimental data by fitting these parameters. The theory was extended to mixtures by Donohue and Prausnitz (1978) by introducing suitable mixing rules. Vimalchand and Donohue (1985) and Morris et al. (1987) replaced the attractive contribution Equation 8.139 with one based on the BH second-order perturbation theory for the LJ potential. Other modifications to the PHCT have been conducted to deal with polar and associating molecules and to combine GC methods with the PHCT. A summary of these and other developments were reported by Donohue and Vimalchand (1988).

8.7.2 Perturbed Hard-Sphere-Chain Theory

In the *perturbed hard-sphere-chain* (PHSC) theory (Song et al. 1994b), the Chiew equation (8.39), with the CS expressions for the reference Z_{HS} and $g_{HS}(\sigma)$, is used for Z_{HC} in Equation 8.137, and again a van der Waals-like expression is assumed for the perturbation contribution, which now for n-mer chains is written as

$$Z_{MF} = -n^2 \frac{a}{k_B T} \rho_n. \tag{8.140}$$

The PHSC EOS thus depends on two parameters, the effective diameter d and the vdW parameter a. The effective diameter is obtained from the effective covolume $b(T)$ as given by Equation 6.88, and parameter a is determined by equating the general expression (2.38) for the second virial coefficient with

$$B_2 = b - \frac{a}{k_B T}, \tag{8.141}$$

as yields the van der Waals equation. Song et al. (1994b) showed that for simple fluids these parameters are universal functions of the reduced temperature T^*. For molecular fluids, however, the reduced temperature in these universal function must be rescaled by means of a parameter $s(n)$ to account for the effect of bonding on the temperature scaling. Therefore, in this theory, the covolume b, or equivalently the molecular volume, and the potential energy parameter a are expressed as universal functions of the rescaled temperature. This leads to a corresponding states principle, and the PHSC EOS, expressed in terms of the appropriate reduced quantities, provides a universal value for the rescaled critical temperature $k_B T/s\varepsilon = 1.1020$, which can be used to determine s if we know the critical temperature T_c of the fluid. Therefore, there is no need of obtaining the parameters in this theory by fitting the experimental data, provided that we know the potential parameters and the critical temperature of the fluid.

Another version of the PHSC theory was developed by Hino and Prausnitz (1997). Again the Chiew EOS (Equation 8.39) for pure fluids, or the corresponding equation (8.48) for mixtures, was taken for the reference hard-chain EOS Z_{HC}, but now for the perturbation Z_{pert} of pure fluids an analytical expression derived by Chang and Sandler (1994b) was used on the basis of the second-order BH perturbation theory for SW fluids of variable width. The perturbation contribution for mixtures was obtained from that of pure fluids using a one-fluid approximation. The parameters in the EOS were obtained for a number of pure fluids, mixtures, and polymer solutions by fitting experimental data.

8.7.3 Boublík–Alder–Chen–Kreglewski Equation of State

Chen and Kreglewski (1977) combined the ISPT equation (8.21) with a perturbation term of the form of Equation 8.139, namely,

$$Z_{pert} = \sum_i \sum_j j D_{ij} \left(\frac{V_0}{V} \right)^j \left(\frac{\varepsilon}{k_B T} \right)^i, \tag{8.142}$$

where ε is the maximum potential depth. Constants D_{ij} were fitted to the experimental data for argon. The resulting equation is known as the *Boublík–Alder–Chen–Kreglewski* (BACK) EOS.

To apply the BACK EOS to real molecular fluids, it is considered that both ε and V_0 are temperature-dependent in the form

$$\varepsilon = \varepsilon_0 \left(1 + \frac{C_1}{k_B T}\right), \tag{8.143}$$

and

$$V_0 = V_{00} \left[1 - C_2 e^{-3u_0/k_B T}\right]^3. \tag{8.144}$$

The expression for ε accounts for the temperature dependence of the orientation-dependent interactions between nonspherical molecules (Kreglewski and Wilhoit 1975), and the expression for V_0 was determined on the basis of the BH formula (6.68). Parameter $C_2 \approx 0.12$ except for strongly associating fluids. Therefore, the BACK EOS depends on four parameters—α, ε_0, V_{00}, and C_2—that, in general, must be determined from the experimental data. The application of the BACK equation to mixtures may be carried out within the framework of a one-fluid approximation with the corresponding mixing rules (Aim and Boublík 1986). Further proposed refinements include the use of a Z_{pert} of the form of Equation 8.142 with parameters fitted to the experimental data for ethane (Aim and Boublík 1986, Saager et al. 1992), the use of a temperature-dependent effective molecular volume based on the WCA perturbation theory (Saager et al. 1992), and the extension to polar fluids (Saager and Fischer 1992).

8.7.4 CHAIN-OF-ROTATORS EQUATION OF STATE

To derive the so-called *chain-of-rotators* (COR) EOS, Chien et al. (1983) express the configurational integral as a product of repulsive and attractive (perturbative) contributions, namely, $Q_c = Q_{rep}Q_{pert}$. The repulsive contribution in turn is written as the product of translational and rotational contributions, that is $Q_{rep} = Q_{tr}Q_{rot}$. The translational contribution is obtained from integration of the CS EOS (5.21) using standard thermodynamic relationships. The rotational contribution is derived from that of a hard diatomic fluid, as results from integration of the ISPT Equation 8.21 with the Boublík–Nezbeda criterion to obtain α, after separating the translational contribution, taking into account that a diatomic molecule has three translational and two rotational degrees of freedom, and considering that the chain consists of $c/2$ dimers, where c may be considered as an effective number of monomers per chain. The EOS corresponding to the repulsive contribution is obtained in the usual way, and adding the attractive contribution one obtains the whole EOS as

$$Z = Z_{CS} + Z_{rot} + Z_{pert}, \tag{8.145}$$

with

$$Z_{rot} = \frac{c}{2} (\alpha - 1) \frac{3\eta + 3\alpha\eta^2 - (\alpha + 1)\eta^3}{(1 - \eta)^3}.$$ (8.146)

For Z_{pert} an expression similar to Equation 8.142 was assumed by Chien et al., with a slightly different expression for the temperature dependence of ε and refitted coefficients. Group contribution methods (Pults et al. 1989a,b) and extensions to polar and associating pure fluids and mixtures (Novenario et al. 1998) have also been developed for the COR EOS.

8.7.5 STATISTICAL ASSOCIATING FLUID THEORY

The SAFT was developed by Jackson et al. (1988) and Chapman et al. (1988, 1989) for chain molecular pure fluids and mixtures, either with or without dispersive forces, on the basis of an extension of Wertheim's TPT to mixtures of associating molecules (see Section 8.5). In its simplest version (Jackson et al. 1988), considering associating molecules with spherical hard cores and dispersive interactions, the SAFT expression for the free energy of pure fluids is obtained by adding a MF term F_{MF} to the TPT1 expression (8.76), that is

$$\frac{F}{Nk_BT} = \frac{F_0}{Nk_BT} + \frac{F_{bond}}{Nk_BT} + \frac{F_{MF}}{Nk_BT},$$ (8.147)

where
F_{bond} is given by either Equation 8.77 or Equation 8.86 for monomers with single or multiple bonding sites, respectively
F_{MF} is of the vdW form

$$\frac{F_{MF}}{Nk_BT} = -\frac{a}{k_BT}\rho.$$ (8.148)

The SAFT expression for the EOS in the case of monomers with one or two bonding sites is readily obtained by adding the MF contribution to Equation 8.82. Instead, Huang and Radosz (1990) replaced the MF contribution 8.148 with the Chen and Kreglewski expression for Z_{per}, as given by Equations 8.142 through 8.144, and reported the involved parameters regressed from experimental data of the coexistence curve for a number of non-associating fluids.

Alternatively, the reference and dispersion contributions to the EOS may be grouped together into a monomer contribution and use for this term an EOS suitable for the monomers. This procedure, with parametrizations determined either from experimental data for argon or from the simulation data for the pure LJ fluid, and using for the bonding contribution the contact RDF of HS mixtures, was adopted by Chapman et al. (1989, 1990), who obtained in this way good correlations for the experimental data of the liquid–vapor coexistence properties of several nonassociating (n-alkanes) as well as associating (methanol and acetic acid) fluids, considering ε, the effective number of beads per chain n_{eff}, and the effective diameter d of the monomers, as fitting parameters.

The theory is readily extended to associating mixtures of spherical monomers with dispersion interactions and multiple bonding sites. In this case, the free energy takes the form (Chapman et al. 1988)

$$\frac{F_{mix}}{Nk_BT} = \frac{F_{mix}^{(0)}}{Nk_BT} + \frac{F_{mix}^{bond}}{Nk_BT} + \frac{F_{mix}^{MF}}{Nk_BT},$$ (8.149)

where F_{mix}^{bond} is given by Equation 8.89 and the van der Waals MF contribution by

$$\frac{F_{mix}^{MF}}{Nk_BT} = -\frac{\rho}{k_BT} \sum_{i=1}^{m} \sum_{j=1}^{m} x_i x_j a_{ij},$$ (8.150)

where m is the number of components in the mixture. Now, the SAFT expression for the EOS in the case of monomers with a HS core and one or two bonding sites is obtained by adding to Equation 8.92 the MF contribution, in a similar way as for pure fluids.

The SAFT theory can be applied to associating mixtures of homonuclear chains, by taking for $F_{mix}^{(0)}$ and F_{mix}^{bond} in Equation 8.149, and for the corresponding contributions to the EOS, the expressions appropriate for chains obtained as indicated earlier and in Section 8.5. The EOS for mixtures of nonassociating chains is recovered by setting to zero the bonding contribution. Proceeding in this way, but replacing the MF term with the Chen and Kreglewski expression for Z_{per}, Huang and Radosz (1991) extended the theory to mixtures of nonassociating and associating chains by introducing appropriate mixing rules for the involved parameters.

For molecular fluids with known interactions between monomers belonging to different chains, it may be possible to apply the SAFT theory without using adjustable parameters. Thus, for example, Johnson and Gubbins (1992) and Walsh and Gubbins (1993) combined the SAFT with the WCA theory for associating fluids modeled as LJ monomers with one or two SW bonding sites. They found good agreement between theory and simulation for the liquid–vapor coexistence densities, the structure, the average number of bonded monomers, and the EOS for pure fluids and mixtures. A similar approach was successfully applied by Ghonasgi and Chapman (1994b) to models of polymer solutions and blends made of up to 25 LJ monomers.

Several other versions of the SAFT have been termed in specific ways. A number of them are described next.

8.7.5.1 SAFT-D

The SAFT-dimer or SAFT-D theory for nonassociating hard-chain fluids developed by Ghonasgi and Chapman (1994a) derives from the SAFT theory for the same kind of fluids by introducing a dimer correction. For LTHS molecules, the resulting EOS is equivalent to the TPT-D approximation (see Section 8.5 and explanations concerning Equations 8.113 and 8.117). Blas and Vega (2001b) used the SAFT-D approximation for the chain contribution to the free energy of LJ chains with pure LJ reference fluid and obtained a much better agreement with simulations, as compared with the monomer version, in the predicted liquid–vapor coexistence densities and vapor pressures.

8.7.5.2   SAFT-VR

In this version of the SAFT, developed by Gil-Villegas et al. (1997) for pure fluid
and mixtures, the perturbation contribution, frequently approximated by the Alder
et al. (1972) parametrization for the SW fluid with $\lambda = 1.5$, is replaced with an
expression derived on the basis of the BH second-order perturbation theory in the
lc approximation for potentials with variable range (VR). To this end, they use the
mean value theorem that, considering the HS fluid as the reference system, allows us
to write the first-order perturbation contribution to the free energy, the last term in
Equation 6.39, in the form

$$F_1 = 2\pi N \rho g_{HS}\left(\sigma; \rho_{eff}\right) \int_0^\infty u_1\left(r\right) r^2 dr = g_{HS}\left(\sigma; \rho_{eff}\right) F_{MF}\left(u_1\left(r\right)\right) = -Na\rho, \quad (8.151)$$

where
   ρ_{eff} is some effective density
   $F_{MF}(u_1(r))$ is the MF approximation for the contribution of the perturbation
      potential, which corresponds to set $g_0(r) = 1$ in the expression of F_1
   a is the vdW constant (6.45)

Explicit expressions of η_{eff} were reported by Gil-Villegas et al. for SW, Sutherland,
Yukawa, and Mie potential models. On the other hand, the same authors noted that
the second-order perturbation contribution to the free energy in the lc approximation
(6.62) may be written as

$$F_2 = -\frac{1}{2}\left(k_B T\right)^2 \rho^2 \kappa_T^{HS} \left[\frac{\partial F_1\left(\left[u_1\left(r\right)\right]^2\right)}{\partial \rho}\right]_T, \quad (8.152)$$

which allows to easily obtain F_2 from the expression (8.151) by replacing $u_1(r)$ with
$[u_1(r)]^2$.

   The SW version of the SAFT-VR was applied by the same authors to predict
the liquid–vapour coexistence of n-alkanes, with carbon number C from 1 to 8, and
n-perfluoroalkanes with $C = 1 - 3$, with the parameters σ, λ, ε, and n_{eff} obtained from
the fitting of the predicted liquid–vapor coexistence and saturated liquid densities to
experimental data. The results showed a considerable improvement over the SAFT
with MF perturbation term. The SAFT-VR was extended to mixtures by Galindo et al.
(1998) by introducing mixing rules for conformal and nonconformal mixtures, and
to electrolyte solutions (SAFT-VRE) by Galindo et al. (1999) and Gil-Villegas et al.
(2001). The theory was modified to improve its performance in the critical region by
Jiang and Prausnitz (1999, 2000) and by McCabe and Kiselev (2004a,b), who termed
SAFT-VRX to their approach. Another modification (SAFT-VR Mie), proposed by
Lafitte et al. (2006), was found to provide better performance than other SAFT-VR
approximations for predicting the saturation densities and pressures and compressed
liquid bulk properties, including derivative properties, of n-alkanes, and later was
extended to associating fluids by Lafitte et al. (2007).

8.7.5.3 PC-SAFT

Gross and Sadowski (2000) combined Equation 8.92 with the second-order BH perturbation theory to obtain the thermodynamic properties of chain fluids with dispersive forces. For the reference hard-chain fluid, instead of the site–site pair correlation functions, they used an analytical expression for the averaged pair correlation function, defined by Equation 8.43, derived by Tang and Lu (1996) from the Chiew theory. Using this approach, denoted perturbed-chain (PC) SAFT, Gross and Sadowski obtained good agreement with simulations for the EOS of SW chain fluids and fluid mixtures, using the vdW one-fluid mixing rules for the perturbative contributions in the mixtures. The procedure was later applied to real pure fluids and mixtures by Gross and Sadowski (2001), considering a LJ perturbation potential and fitting the needed parameters to experimental data. Correlation results for a number of pure compounds were reported by these authors and the PC-SAFT showed a considerable improvement over ordinary SAFT, which uses the Chen–Kreglewski expression for the attractive contribution, for the correlation and prediction of thermodynamic properties and phase equilibria of pure fluid and mixtures. Further applications and extensions to polymers (Gross and Sadowski 2002a), associating fluids (Gross and Sadowski 2002b), and polar systems (Tumakaka and Sadowski 2004, Gross 2005, Dominik et al. 2005, Karakatsani et al. 2005, Karakatsani and Economou 2006) have been developed. Dominik et al. (2007) used the SAFT-D approximation for the reference hard chain and the perturbed chain dispersion therm of the PC-SAFT to improve the description of the phase behavior of polymer solutions. Corrections to improve the behavior of the PC-SAFT near the critical point have been introduced by Bymaster et al. (2008) for pure fluids and by Tang and Gross (2010) for mixtures.

8.7.5.4 Soft-SAFT

Johnson et al. (1994) applied the SAFT to nonassociating LJ chains using a parametrization for the EOS Z_0 of the reference fluid of independent LJ monomers. For the bonding contribution, they used also a parametrization for the reference LJ fluid $g_0(\sigma)$ based on simulation data instead of the RDF for HS mixtures as in the above-mentioned approximation of Chapman et al. (1990). They obtained good agreement with their simulation data for the EOS and the internal energy of freely jointed LJ chains with up to 100 monomers. The approach was later successfully extended to associating chains by Blas and Vega, who termed soft-SAFT to this approximation, and used to predict the thermodynamic properties and phase behavior of nonassociating mixtures of LJ chains (Blas and Vega 1997) and of pure hydrocarbons as well as their binary and ternary mixtures (Blas and Vega 1998). A further correction was introduced by Llovell et al. (2004) to improve the performance of the theory in the critical region.

Within the same general framework, Müller and Gubbins (1995) developed an EOS for a simplified model of water with dipolar LJ (Stockmayer) interactions. The reference potential was the LJ potential with dipole–dipole interactions and the corresponding contribution to the free energy and EOS was obtained from that of the ordinary LJ fluid, as given by an analytical expression derived by Kolafa and Nezbeda (1994) on the basis of perturbation theory, plus a polar contribution. For the association term, they considered the interactions to be LJ too, and for the LJ RDF needed for the calculation of this term they used a parametrization of the simulation data. The parameters involved in the theory were fitted lo liquid–vapor coexistence data for water. The resulting theory showed reasonable agreement with simulations for the pressure, fraction of unbonded molecules, phase diagram, vapor pressure, and energy of vaporization. This approach was later extended to chain molecular nonassociating, associating, and polar pure fluids and mixtures by Kraska and Gubbins (1996a,b).

8.7.5.5 GC-SAFT

Group contribution methods have been developed for different SAFT versions. Thus, a number of group contributions to the parameters of the above-mentioned SAFT version of Chapman et al. (1990) and of the SAFT-VR, as well as the corresponding correlations with the molecular parameters, were reported by Tamouza et al. (2004, 2005) for pure fluids and mixtures. Lymperiadis et al. (2007) developed a GC method, which they called SAFT-γ, based on an extension of the SAFT-VR to heteronuclear molecules formed by different types of fused segments, with each segment modeled as a SW sphere. Emami et al. (2008) presented a number of group contributions to the parameters of the SAFT version of Huang and Radosz (1990) (see earlier) and PC-SAFT. NguyenHuynh et al. (2008) extended the Tamouza et al. (2004) GC method to mixtures of polar fluids in combination with the Chapman et al. (1990) version of the SAFT, as well as the SAFT-VR and PC-SAFT. Another GC-SAFT-VR version for pure fluids and fluid mixtures was developed by Peng et al. (2009) and later extended to polymers by Peng et al. (2010) and to associating functional groups by dos Ramos et al. (2011).

Among the theories reviewed in this section, the most widely used today are the different versions of the SAFT, perhaps because this theory has a sounded theoretical basis and so it is particularly suitable to introduce further refinements and extensions. We cannot dwell more on this matter, but the interested reader may consult the specialized reviews by Müller and Gubbins (2001), Economou (2002), Chapman et al. (2004), and Tan et al. (2008).

On the other hand, the application of these theories to predict the thermodynamic properties of real fluids relies on the use of some kind of experimental data to fit the involved parameters. One may wonder about the reliability of these, and other, theoretical approaches without using adjustable parameters. Here, we will discuss some of the results achieved for two models of chain fluids, with square-well and LJ monomer–monomer interactions respectively, for which the parameters can be determined, for most of the considered theories, from the molecular geometry and interaction potential.

Yethiraj and Hall. (1991b) used the first-order perturbation theory, Equation 6.39, to obtain the EOS of SW n-mer fluids with $n = 4$, 8, and 16. For the compressibility factor and the average site–site pair correlation function of the reference hard-chain fluid, they used the GFD equation (8.72) and PRISM theory (see Section 4.2), respectively. Good agreement with simulations was found for 4-mers, but for 8-mers and 16-mers the theoretical results were found to be too high at moderate densities. Much better agreement was achieved by these authors (Yethiraj and Hall 1991c) for the same fluids with the GFD theory with the SW monomer and dimer reference equations of state obtained from IET with the MSA closure. Banaszak et al. (1993) applied the TPT1 to SW chains, by using for the reference EOS Z_0 and contact RDF $g_0(\sigma)$ in Equation 8.82 semi-empirical expressions obtained for the reference fluid of independent SW monomers. The analytical expression for the RDF of the SW fluid (and other potential models) derived by Tang and Lu (1997a) on the basis of the FMSA (see Section 4.4) may be useful in this context. Bokis et al. (1994) compared the predictions of the PHCT, GF, GFD, and SAFT theories for the EOS of SW chain fluids and found that the best overall agreement with simulation was achieved with the GFD theory using MF approximations for the monomer and dimer-attractive contributions. Similar accuracy was obtained by Tavares et al. (1995) from the TPT1 and TPT-D theories, especially from the latter, with the SW monomer and dimer properties obtained from simulation in order to avoid errors arising from theoretical approximations. Later, the same authors (Tavares et al. 1997) showed that the TPT1 theory, Equation 8.82, with the SW reference EOS and contact RDF obtained, respectively, from the second-order BH perturbation theory and from an approximate expression, also derived from perturbation theory (Barker and Henderson 1976), provides quite good agreement with simulations for the EOS and the excess energy of the above-mentioned SW chain fluids; this approach was extended by Paredes et al. (2001) with similar success to SW chain mixtures using a one-fluid approximation. Gulati and Hall (1998) compared the predictions from the GFD, TPT-D, and PHSC theories for the compressibility factor of SW homonuclear chains with $n = 16$ and 32; both the GFD and TPT-D theories yielded excellent agreement with simulations, with slightly better results from the first of them for $n = 32$, whereas the PHSC considerably underestimated the compressibility factor at low to moderate densities. In addition, the same authors tested the accuracy of the GFD theory for the EOS of copolymers modeled as chains made of two different kinds of SW monomers; again the theoretical predictions were in excellent agreement with the simulations for different copolymer models with up to $n = 16$.

With regard to LJ chain fluids, Curro et al. (1993) applied the first-order BH and WCA perturbation theories to RIS models of n-alkane and polyethylene fluids considered as overlapping spheres with LJ site–site interactions. The reference RDF in the perturbation term was obtained from the PRISM theory (see Section 4.2) and for the reference hard-chain EOS either PRISM theory or GFD theory was used. The best overall agreement was achieved with the GFD-B approximation combined with the BH perturbation theory. Quite satisfactory results were also obtained by O'Lenick and Chiew (1995) using variational perturbation theory with the hard-chain reference system solved in the PY approximation; this procedure was later extended to mixtures of LJ chains by von Solms et al. (1999). On this basis, Chiew et al. (1999) developed a *perturbed Lennard–Jones chain* (PLJC) EOS for simple chainlike fluids and polymers, which in turn was extended to mixtures by Lee et al. (2000). Banaszak et al. (1994) applied the TPT1 theory to LJ chains using a fluid of unbounded LJ monomers as the reference system, with the reference EOS and density derivative of the RDF in Equation 8.82 obtained from simulations in order to avoid errors due to theoretical approximations. Good agreement was achieved in this way with the simulation data for the EOS of chains with up to 32 monomers at different temperatures. Using the WCA reference system instead of simulation data to obtain the density derivative of the RDF of the reference LJ fluid resulted in poorer accuracy. Johnson (1996) compared the predictions of the TPT1 and TPT1-D theories for the EOS and the configurational internal energy of linear flexible LJ chains with $N = 200$ and flexible LJ ring chains with $n = 3, 8$, and 20. For the reference Z_0, $g_0(\sigma)$, and $g_d(\sigma)$ in Equations 8.82 and 8.114, where σ is the LJ parameter, accurate parametrizations of the corresponding simulation data were used. Very good agreement with simulations was obtained with the TPT1-D theory, although it predicted a spurious two-phase region for ring polymers at supercritical temperatures. The TPT1 version yielded, in general, somewhat high values of the pressure and also overestimated the internal energy at low densities. Fairly good results were reported by Davies et al. (1998) for the EOS of chains with up to $n = 8$ monomers using the SAFT-VR approach, although the predicted liquid–vapor coexistence curves were too high, as compared with the simulation data, and the deviations increased with increasing n. MacDowell et al. (2000) compared the results obtained from the TPT1 theory for chains with 10 LJ monomers using for the reference LJ monomer fluid either the RHNC theory or the FMSA. Both choices for the calculation of the reference system properties yielded quite satisfactory results for the excess chemical potential and the pressure, with slightly better accuracy achieved with the TPT1-RHNC choice, whereas both theories overestimated the critical temperature. The FMSA has the before-mentioned advantage that analytical or nearly analytical solutions are available for a number of potential models, including the LJ potential (Tang and Lu 1997a). Also based on the FMSA is the approach devised by Tang and Lu (2000), whose starting point is to map the LJ potential into a sum of two Yukawa functions, closely mimicking the LJ potential, for which the FMSA provides an analytical solution for the RDF (Tang and Lu 1997b). Then, the RDF of the monomer LJ fluid is obtained in a similar way as in the EXP approximation Equation 4.47, namely, $g_{LJ}(r) = g_{HS}(r) \exp[g_1(r)]$ (called by Tang and Lu simplified EXP or *SEXP approximation*), where $g_1(r)$ is the first-order contribution to the RDF of the two-Yukawa potential according to expansion (4.59). This leads to analytical expressions for the reference system properties, which was combined by Tang and Lu (2000) with the TPT1 leading to very accurate predictions of the pressure of non-associating LJ chain fluids with up to 100 monomers as well as for the pressure of associating LJ chains and the fraction of nonassociated molecules.

8.8   NON-ISOTROPIC PHASES

Thus far we have considered only isotropic fluids, but the fact is that systems consisting of elongated or oblate molecules may present different nonisotropic phases with intermediate degrees of ordering between the fully disordered isotropic fluid and the fully ordered crystalline phases. Restricting ourselves for simplicity to hard-particle

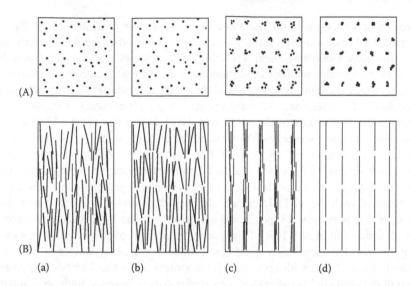

FIGURE 8.10   Illustration of several nonisotropic phases in a system of hard spherocylinders represented as thin rods. If the z axis is taken along the direction of the average orientation of the molecules, (A) drawing for each phase corresponds to a xy section and (B) corresponds to xz or yz sections. (a) Nematic, (b) Smectic, (c) columnar, and (d) crystalline.

systems, isotropic, nematic, smectic, columnar, and crystalline phases have been found in computer simulations of hard spherocylinders (Strobrants et al. 1986, 1987, Veerman and Frenkel 1990). Figure 8.10 illustrates the disposition of the molecules in these phases. In the nematic phase, the molecules are preferently aligned along a given direction, here taken as the z axis, so that there is a considerable degree of orientational order, but no positional order is revealed by the pair correlation function. In the smectic phase, in addition to orientational order, the correlation function $g_{\parallel}(z)$ in the direction of the z axis presents periodic oscillations, indicating a considerable positional order in that direction, whereas the correlation function $g_{\perp}(r)$ in the transverse direction displays a behavior similar to that of an isotropic fluid. The columnar phase, in addition to orientational order, possesses two-dimensional long-range positional order along the xy plane that manifests in the crystalline-like shape of $g_{\perp}(r)$ whereas $g_{\parallel}(z)$ remains fluid-like. Finally, the crystalline phase exhibits orientational order and three-dimensional positional order, so that both $g_{\perp}(r)$ and $g_{\parallel}(z)$ have crystalline-like structure.

Isotropic, nematic, ordered solid, and plastic solid have been observed in computer simulations of both prolate and oblate ellipsoids of revolution (Frenkel and Mulder 1985, Frenkel et al. 1985). The structure of a plastic solid, only appearing in molecules with axis ratio relatively close to one, is characterized by long-range positional order but no orientational order. A transition from fluid to a stable plastic crystal phase has been found also in simulations of hard HMD with reduced center-to-center distance $L^* \equiv L/\sigma \leq 0.4$ (Singer and Mumaugh 1990), and isotropic, nematic, smectic, and crystalline phases have been reported for linear tangent HS molecules with $n \geq 5$ (Vega et al. 2001). Obviously all these phases may arise in real fluids with molecular shapes similar to those of these hard-body fluids.

Perturbation theories may be used to obtain the thermodynamic properties of nonisotropic phases of realistic molecular systems. This requires, as usual, the knowledge of the thermodynamic properties of the corresponding reference system, generally one made of hard-body molecules. The SPT, extended to nonisotropic hard spherocylinder fluids by Cotter and Martire (1970a–c) and Cotter (1974, 1977), may be used to this end. However, most frequently approximations for nematic fluids are based, in one way or another, on the virial expansion of some reference fluid. We will discuss here a number of the latter approaches.

Let us assume a fluid with some degree of orientational order along a given direction, so that the molecular orientations are distributed according to the probability density $f(\Omega)$ with the normalization condition

$$\int f(\Omega)\, d\Omega = 1. \tag{8.153}$$

The degree of orientational order is measured by means of the *uniaxial order parameter*

$$S_2 = \left\langle \frac{3}{2}\cos^2\theta - \frac{1}{2} \right\rangle, \tag{8.154}$$

where θ is the angle between the molecular axis and the direction of the average orientation of the molecules, and the angular brackets mean an average weighted by the probability density $f(\Omega)$. This parameter is zero for a perfect isotropic fluid and one for a system with perfectly aligned molecules.

The free energy of the fluid can be expressed in the form

$$\frac{F}{Nk_BT} = \ln\left(\rho/q_k\right) - 1 + \int f(\Omega)\ln\left[4\pi f(\Omega)\right] d\Omega + \frac{F^E}{Nk_BT}, \tag{8.155}$$

where the q_k represents the kinetic contribution, and eventually that of internal degrees of freedom, per molecule to the partition function, so that the first two terms in the right-hand side account for the contribution of an isotropic fluid of noninteracting molecules, the ideal contribution. The third term in the right-hand side is the entropy contribution due to the orientational order and may be considered as the entropy of mixing if the fluid is viewed as a mixture with one species for each different orientation; for the perfectly isotropic fluid $f(\Omega) = 1/4\pi$, so that this term vanishes. The last term is the contribution to the excess free energy arising from the molecular interactions and can be obtained from the integration of the excess compressibility factor. If the latter is expressed as a virial expansion, the excess free energy contribution will be

$$\frac{F^E}{Nk_BT} = \int_0^\rho (Z-1)\,\frac{d\rho'}{\rho'} = \sum_{n=2}^{\infty} \frac{B_n\rho^{n-1}}{n-1}, \tag{8.156}$$

where B_n is the nth virial coefficient of the nonisotropic fluid. Truncating the sum at $n = 2$ and introducing the result into Equation 8.155 yields the Onsager (1949) result, which becomes exact for infinitely long molecules, because the isotropic-nematic transition is increasingly displaced toward lower densities as the molecular length increases. For moderately elongated molecules, the contribution of higher-order virial coefficients is not negligible.

In the *decoupling approximation*, proposed by Parsons (1979), it is assumed that, for intermolecular potentials of the form $u(r, \Omega) = \varphi(r/\sigma)$, where $\sigma = \sigma(\mathbf{n}_{12}, \Omega_1, \Omega_2)$ is an angle-dependent range parameter with $\mathbf{n}_{12} = \mathbf{r}_{12}/r_{12}$, the pair correlation function can be approximated by $g(\mathbf{r}, \Omega_1, \Omega_2) \approx g(r/\sigma)$, which becomes exact in the low-density limit where $g(r/\sigma) = e^{-\varphi/k_B T}$, and in this case the orientational contribution to the excess free energy can be decoupled from the translational one, which will then be the same as that for a fluid of spherically shaped particles. This is easily seen from the virial equation (2.58) for molecular fluids, where now the average is weighted with the probability density $f(\Omega)$, that is

$$Z = 1 - \frac{1}{6} \frac{\rho}{k_B T} \int \int \int g(r_{12}, \Omega_1, \Omega_2) \frac{\partial u(r_{12}, \Omega_1, \Omega_2)}{\partial r_{12}} r_{12} d\mathbf{r}_{12} f(\Omega_1) f(\Omega_2) \, d\Omega_1 d\Omega_2. \tag{8.157}$$

Performing the variable change $x = r/\sigma$ yields

$$Z = 1 + \left[-\frac{1}{2} \frac{\rho}{k_B T} \int_0^\infty g(x) \frac{\partial \varphi(x)}{\partial x} x^3 dx \right]$$
$$\times \left[\int \int f(\Omega_1) f(\Omega_2) \, d\Omega_1 d\Omega_2 \frac{1}{3} \int \sigma^3(\mathbf{n}_{12}, \Omega_1, \Omega_2) \, d\mathbf{n}_{12} \right]. \tag{8.158}$$

The term in the first bracket is, but for a factor $(4/3)\pi\sigma^3$, the excess compressibility factor of a reference fluid with a spherically symmetric potential $\varphi(x)$ and, in the case of a nonisotropic hard-body fluid, the reference will be the HS fluid, for which the compressibility factor is conveniently given by the Carnahan–Starling equation (5.21). The last integral, in the case of a fluid of HS with constant diameter σ, will give $4\pi\sigma^3$, which is three times the excluded volume per particle. In a similar way, for a hard-body fluid, the integral will be

$$\int \sigma^3(\mathbf{n}_{12}, \Omega_1, \Omega_2) \, d\mathbf{n}_{12} = 3v^e(\Omega_1, \Omega_2), \tag{8.159}$$

where $v^e(\Omega_1, \Omega_2)$ is the excluded volume for fixed orientations Ω_1 and Ω_2, so that the second bracket in Equation 8.158 is

$$\int \int f(\Omega_1) f(\Omega_2) \, d\Omega_1 d\Omega_2 \frac{1}{3} \int \sigma^3(\mathbf{n}_{12}, \Omega_1, \Omega_2) \, d\mathbf{n}_{12} = \langle v^e(\Omega_1, \Omega_2) \rangle, \tag{8.160}$$

Integration of the excess compressibility factor thus obtained, as in Equation 8.156, will give the excess free energy $F^E/Nk_B T$ as a product of two terms: one of them, the translational contribution, will be the excess free energy of a HS fluid; the other, the orientational contribution, will be the ratio of the excluded volumes v^e for the molecules, averaged over orientations, to that for the spheres or, equivalently, the ratio of the corresponding second virial coefficients, namely,

$$\frac{F^E}{Nk_BT} = \frac{B_2}{B_2^{HS}} \frac{(4 - 3\eta)\eta}{(1 - \eta)^2},$$
(8.161)

which together with Equation 8.155 yields the free energy of the nematic phase. The same result was arrived at by Lee (1987) in a somewhat more heuristic way. This requires the knowledge of the second virial coefficient of the nonisotropic fluid that depends on the probability density $f(\Omega)$ through the expression

$$B_2 = \frac{1}{2}\langle v^e(\Omega_1, \Omega_2)\rangle = \frac{1}{2}\int\int v^e(\Omega_1, \Omega_2) f(\Omega_1) f(\Omega_2)\, d\Omega_1 d\Omega_2.$$
(8.162)

To perform this integration, the orientation-dependent excluded volume $v^e(\Omega_1, \Omega_2)$ is needed. Analytical expressions for this quantity are available, among other hard molecules, for hard prolate spherocylinders (Onsager 1949, Esposito and Evans 1994), hard prolate and oblate ellipsoids of revolution (Isihara 1951), hard oblate spherocylinders (Mulder 2005), and linear chains of tangent HS (Williamson and Jackson 1995). On the other hand, the probability density $f(\Omega)$ is determined for each density by minimizing the free energy functional $F[f(\Omega)]$ with respect to $f(\Omega)$ with the restriction (8.153).

McGrother et al. (1996) analyzed the performance of the decoupling approximation for the isotropic-nematic (I-N) transition of a fluid of hard prolate spherocylinders with γ ranging from 4.2 to 6, and found that the theory accurately predicts the EOS of both the isotropic and nematic branches, as well as the densities and pressures of the coexisting phases, but fails to accurately predict the order parameter near the transition. Camp et al. (1996) compared the predictions of the Onsager and Parsons–Lee approximations for the I-N transitions of hard prolate ellipsoids of revolution with aspect ratios κ ranging from 5 to 20, and showed that both theories fairly accurately predict the variation of the order parameter at coexistence with elongation, and the second of these theories also provides quite satisfactory results for the coexistence densities, whereas the Onsager theory considerably overestimates them except for very high elongations; instead, the I-N coexistence pressure is overpredicted by the two theories, with increasing deviation from simulation with lowering κ and higher deviations from the Onsager theory than from the decoupling approximation.

In contrast with the case of HCB fluids, for rigid linear HS chains, the Parsons–Lee theory gives quite poor results, as showed by Williamson and Jackson (1998). However, a considerable improvement was achieved by Varga and Szalai (2000), with a modification consisting in replacing the actual volume of the molecules with an effective molecular volume to account for the nonconvex shape of the molecules. This generally resulted in an increased accuracy in the predicted EOS of both the isotropic and nematic phases, as well as in the order parameter in the latter phase, a slight improvement in the coexistence densities, and a considerable worsening in the coexistence pressure.

The Parsons–Lee decoupling approximation uses the exact second virial coefficient for the nematic fluid and approximates the higher-order virial coefficients with those of the HS fluid scaled by B_2/B_2^{HS}. Tjipto-Margo and Evans (1990) determined the exact third virial coefficient for hard ellipsoids of revolution and used it, in addition to the second one, to obtain the free energy from Equations 8.155 and 8.156, with the latter truncated at $n = 3$. They concluded that this approximation is accurate for predicting the density dependence of the order parameter for aspect ratio $\kappa \geq 5$, whereas for lower values of κ the influence of higher-order virial coefficients is not negligible. To incorporate their effect in an approximate way, these authors also used for the excess free energy a closed form obtained on the basis of the exact B_2 and B_3. Vega and Lago (1994) considered for the excess free energy an approximation similar to Equation 8.161 but replacing B_2^{HS} and the HS excess free energy term with those for the isotropic hard molecular fluid; for hard PSC, PER, and OER they found excellent agreement with the simulation data for the coexistence densities, pressure, and order parameter at the isotropic–nematic transition.

Once a suitable theory for the free energy of the reference hard-body fluid is available, the application of perturbation methods to nonisotropic fluids with other kinds of interactions can be carried out using similar approaches as those for isotropic fluids. Thus, for example, McGrother et al. (1997) determined from computer simulation in the reference hard spherocylinder fluid the first four perturbation terms of the free energy (see Section 6.2) of dipolar spherocylinders with $\gamma = 6$ to study, by means of the simulation-based perturbation theory, the effect of introducing dipolar interactions on the stability of the isotropic phase of the reference hard spherocylinder fluid. Williamson and del Río (1998) successfully used an approach similar to that of Vega and Lago (1994) to study the isotropic-nematic transition in a SW spherocylinder fluid, dividing the excess free energy of the isotropic phase into repulsive and attractive contributions, and scaling them with the ratios $B_2^{rep}(\text{nem})/B_2^{rep}(\text{iso})$ and $B_2^{att}(\text{nem})/B_2^{att}(\text{iso})$, respectively, where $B_2^{rep}(\text{nem})$ and $B_2^{att}(\text{nem})$ are the repulsive and attractive contributions to the second virial coefficient of the nematic fluid and $B_2^{rep}(\text{iso})$ and $B_2^{att}(\text{iso})$ those for the second virial coefficient of the isotropic fluid. The same authors analyzed the reliability of a direct extension of the decoupling approximation for the same fluid, using as the reference excess free energy that for a spherical SW fluid, instead of a HS fluid, but this approach gave poor results. Cuetos et al. (2005) applied the decoupling approximation to a soft spherocylinder fluid by obtaining an effective hard core, but the theory was found to overestimate the pressure in the nematic phase as well as in the isotropic phase near the I-N transition. Therefore, it seems that, in contrast with the situation for HCB fluids, for which the Parsons–Lee theory provides quite satisfactory results, this theory is inaccurate for more realistic potential models with soft repulsive or attractive interactions.

Concerning hard-body crystalline solids, they are frequently treated within the context of an extension of the cell or free volume theories (see Section 5.6) to include the rotational degrees of freedom. The configurational integral q_c for a single particle in the solid can be expressed as

$$q_c = \frac{1}{4\pi} \int\int e^{-\beta\phi(\mathbf{r},\Omega)} d\mathbf{r} d\Omega, \qquad (8.163)$$

where

$\phi(\mathbf{r}, \Omega)$ is the potential energy of interaction between a molecule and the remaining molecules

$\mathbf{r}$ is the position vector of the molecule

Ω its orientation

Contrarily to the case of solids with spherically shaped particles, for which there are theoretical approximations, the integral in Equation 8.163 must be calculated from simulation. This was done by Paras et al. (1992) for several crystalline phases of hard diatomic solids with values of the reduced center-to-center distance $L^* = L/\sigma$ ranging from 0.15 to 1.0. For relatively low values of L^*, stable plastic crystal phases are possible for this system, as said before, and they were dealt with by introducing a sphericalized reference pair potential defined in the form

$$\beta u_0(r_{12}) = -\ln\left[\frac{1}{4\pi}\int\int e^{-\beta u(r_{12},\Omega_1,\Omega_2)} d\Omega_1 d\Omega_2\right]. \qquad (8.164)$$

The EOS obtained from this procedure was found to be in very good agreement with the simulations for the orientationally ordered crystalline phases, whereas the predicted pressures for the plastic crystal phases were too low. In addition, the fluid–solid coexistence was determined using the cell theory for the solid and the Tildesley–Streett equation (8.74) for the fluid. The calculated coexistence densities, pressures, and chemical potentials were also quite accurate, except for plastic crystals with intermediate values of L^*.

The Wertheim TPT1 was extended by Sear and Jackson (1995) to a tangent hard diatomic solid by using for the free energy and the EOS of the HS reference solid either the result from the cell theory or an accurate fitting of the simulation data and obtaining the contact RDF from the virial equation (2.43) for HS. Vega and MacDowell (2001) applied a similar approach to the solid phase of flexible LTHS and obtained good agreement with the simulations for the EOS of LTHS with $n \lesssim 6$ beads. Vega and McBride (2002) analyzed by computer simulation the influence of the molecular flexibility on the EOS of LTHS systems. No difference was found between rigid and flexible molecules in the isotropic phase, in contrast to the nonisotropic phases. They also found that the EOS scales with the number n of beads of the molecule for both the isotropic and nonisotropic phases, but in a different way for the isotropic phase than for the nonisotropic phases. These scalings were used to derive the EOS for the solid phase of rigid LTHS on the basis of the Wertheim TPT1 solution for the solid phase of flexible LTHS, obtaining excellent agreement with the simulations. The reliability of this approach was later confirmed by Blas et al. (2003) by showing that the theory was able to accurately predict the solid–fluid coexistence densities and pressures for several rigid and flexible hard-chain molecular systems.

REFERENCES

Aim, K. and T. Boublík. 1986. Vapor-liquid equilibrium calculations with the BACK equation of state. *Fluid Phase Equilibr.* 29:583.

Alder, B. J., D. A. Young, and M. A. Mark. 1972. Studies in molecular dynamics. X. Corrections to the augmented van der Waals theory for the square well fluid. *J. Chem. Phys.* 56:3013.

Amos, M. D. and G. Jackson. 1991. BHS theory and computer simulations of linear heteronuclear triatomic hard-sphere molecules. *Mol. Phys.* 74:191.

Amos, M. D. and G. Jackson. 1992. Bonded hard-sphere (BHS) theory for the equation of state of fused hard-sphere polyatomic molecules and their mixtures. *J. Chem. Phys.* 96:4604.

Archer, A. L. and G. Jackson. 1991. Theory and computer simulations of heteronuclear diatomic hard-sphere molecules (hard dumbbells). *Mol. Phys.* 72:881.

Attard, P. and G. Stell. 1992. Three-particle correlations in a hard-sphere fluid. *Chem. Phys. Lett.* 189:128.

Banaszak, M., Y. C. Chiew, R. O'Lenick, and M. Radosz. 1994. Thermodynamic perturbation theory: Lennard-Jones chains. *J. Chem. Phys.* 100:3803.

Banaszak, M., Y. C. Chiew, and M. Radosz. 1993. Thermodynamic perturbation theory: Sticky chains and square-well chains. *Phys. Rev. E* 48:3760.

Barboy, B. 1975. Solution of the compressibility equation of the adhesive hard-sphere model for mixtures. *Chem. Phys.* 11:357.

Barker, J. A. and D. Henderson. 1976. What is "liquid"? Understanding the states of matter. *Rev. Mod. Phys.* 48:587.

Baxter, R. J. 1968. Percus-Yevick equation for hard spheres with surface adhesion. *J. Chem. Phys.* 49:2770.

Beret, S. and J. M. Prausnitz. 1975. Perturbed hard-chain theory: An equation of state for fluids containing small or large molecules. *AIChE J.* 21:1123.

Bernal, J. D. and J. Mason. 1960. Co-ordination of randomly packed spheres. *Nature* 188:910.

Blas, F. J., E. Sanz, C. Vega, and A. Galindo. 2003. Fluid-solid equilibria of flexible and linear rigid tangent chains from Wertheim's thermodynamic perturbation theory. *J. Chem. Phys.* 119:10958.

Blas, F. J. and L. F. Vega. 1997. Thermodynamic behaviour of homonuclear and heteronuclear Lennard-Jones chains with association sites from simulation and theory. *Mol. Phys.* 92:135.

Blas, F. J. and L. F. Vega. 1998. Prediction of binary and ternary diagrams using the statistical associating fluid theory (SAFT) equation of state. *Ind. Eng. Chem. Res.* 37:660.

Blas, F. J. and L. F. Vega. 2001a. Thermodynamic properties and phase equilibria of branched chain fluids using first- and second-order Wertheim's thermodynamic perturbation theory. *J. Chem. Phys.* 115:3906.

Blas, F. J. and L. F. Vega. 2001b. Improved vapor-liquid equilibria predictions for Lennard-Jones chains from the statistical associating fluid dimer theory: Comparison with Monte Carlo simulations. *J. Chem. Phys.* 115:4355.

Bokis, C. P., M. D. Donohue, and C. K. Hall. 1994. Application of a modified generalized Flory dimer theory to normal alkanes. *Ind. Eng. Chem. Res.* 33:1290.

Boublík, T. 1974. Statistical thermodynamics of convex molecule fluids. *Mol. Phys.* 27:1415.

Boublík, T. 1975. Hard convex body equation of state. *J. Chem. Phys.* 63:4084.

Boublík, T. 1981. Equation of state of hard convex body fluids. *Mol. Phys.* 42:209.

Boublík, T. 1986. Equations of state of hard body fluids. *Mol. Phys.* 59:371.

Boublík, T. 1989. Equation of state of linear fused hard-sphere models. *Mol. Phys.* 68:191.

Boublík, T. 1994. Third virial coefficient and the hard convex body equation of state. *Mol. Phys.* 83:1285.

Boublík, T. 2004. Third and fourth virial coefficients and the equation of state of hard prolate spherocylinders. *J. Phys. Chem. B* 108:7424.

Boublík, T. and I. Nezbeda. 1977. Equation of state for hard dumbbells. *Chem. Phys. Lett.* 46:315.

Boublík, T. and I. Nezbeda. 1986. *P-V-T* behaviour of hard body fluids. Theory and experiment. *Coll. Czech. Chem. Commun.* 51:2301.

Boublík, T., C. Vega, and M. Díaz-Peña. 1990. Equation of state of chain molecules. *J. Chem. Phys.* 93:730.

Bymaster, A., C. Emborsky, A. Dominik, and W. G. Chapman. 2008. Renormalization-group corrections to a perturbed-chain statistical associating fluid theory for pure fluids near to and far from the critical region. *Ind. Eng. Chem. Res.* 47:6264.

Camp, P. J., C. P. Mason, M. P. Allen, A. A. Khare, and D. A. Kofke. 1996. The isotropic-nematic phase transition in uniaxial hard ellipsoid fluids: Coexistence data and the approach to the Onsager limit. *J. Chem. Phys.* 105:2837.

Chang, J. and S. I. Sandler. 1994a. An equation of state for the hard-sphere chain fluid: Theory and Monte Carlo simulation. *Chem. Eng. Sci.* 49:2777.

Chang, J. and S. I. Sandler. 1994b. A completely analytic perturbation theory for the square-well fluid of variable well width. *Mol. Phys.* 81:745.

Chapman, W. G. 1988. Theory and simulation of associating fluid mixtures. PhD Dissertation, Cornell University, Ithaca, NY.

Chapman, W. G., K. E. Gubbins, C. G. Joslin, and C. G. Gray. 1986. Theory and simulations of associating liquid mixtures. *Fluid Phase Equilibr.* 29:337.

Chapman, W. G., G. Jackson, and K. E. Gubbins. 1988. Phase equilibria of associating fluids. Chain molecules with multiple bonding sites. *Mol. Phys.* 65:1057.

Chapman, W. G., K. E. Gubbins, G. Jackson, and M. Radosz. 1989. SAFT: Equation-of-state solution model for associating fluids. *Fluid Phase Equilibr.* 52:31.

Chapman, W. G., K. E. Gubbins, G. Jackson, and M. Radosz. 1990. New reference equation of state for associating liquids. *Ind. Eng. Chem. Res.* 29:1709.

Chapman, W. G., S. G. Sauer, D. Ting, and A. Ghosh. 2004. Phase behavior applications of SAFT based equations of state—From associating fluids to polydisperse, polar copolymers. *Fluid Phase Equilibr.* 217:137.

Chen, S. S. and A. Kreglewski. 1977. Applications of the augmented van der Waals theory of fluids. I. Pure fluids. *Ber. Bunsenges. Phys. Chem.* 81:1048.

Chien, C. H., R. A. Greenkorn, and K.-C. Chao. 1983. Chain-of-rotators equation of state. *AIChE J.* 29:560.

Chiew, Y. C. 1990a. Percus-Yevick integral equation theory for athermal hard-sphere chains. I. Equations of state. *Mol. Phys.* 70:129.

Chiew, Y. C. 1990b. Intermolecular site-site correlation functions of athermal hard-sphere chains: Analytic integral equation theory. *J. Chem. Phys.* 93:5067.

Chiew, Y. C. 1991. Percus-Yevick integral equation theory for athermal hard-sphere chains. II. Average intermolecular correlation functions. *Mol. Phys.* 73:359.

Chiew, Y. C., D. Chang, J. Lai, and G. H. Wu. 1999. A molecular-based equation of state for simple and chainlike fluids. *Ind. Eng. Chem. Res.* 38:4951.

Costa, L. A., Y. Zhou, C. K. Hall, and S. Carrà. 1995. Fused hard-sphere chain molecules: Comparison between Monte Carlo simulation for the bulk pressure and generalized Flory theories. *J. Chem. Phys.* 102:6212.

Cotter, M. A. 1974. Hard-rod fluid: Scaled particle theory revisited. *Phys. Rev. A* 10:625.

Cotter, M. A. 1977. Hard spherocylinders in an anisotropic mean field: A simple model for a nematic liquid crystal. *J. Chem. Phys.* 66:1098.

Cotter, M. A. and D. E. Martire. 1970a. Statistical mechanics of rodlike particles. I. A scaled particle treatment of a fluid of perfectly aligned rigid cylinders. *J. Chem. Phys.* 52:1902.

Cotter, M. A. and D. E. Martire. 1970b. Statistical mechanics of rodlike particles. II. A scaled particle investigation of the aligned → isotropic transition in a fluid of rigid spherocylinders. *J. Chem. Phys.* 52:1909.

Cotter, M. A. and D. E. Martire. 1970c. Statistical mechanics of rodlike particles. III. A fluid of rigid spherocylinders with restricted orientational freedom. *J. Chem. Phys.* 53:4500.

Cuetos, A. B. Martínez-Haya, S. Lago, and L. F. Rull. 2005. Parsons–Lee Monte Carlo study of soft repulsive nematogens. *J. Phys. Chem. B* 109:13729.

Curro, J. G., A. Yethiraj, K. S. Schweizer, J. D. McCoy, and K. G. Honnell. 1993. Microscopic equations-of-state for hydrocarbon fluids: Effect of attractions and comparison with polyethylene experiments. *Macromolecules* 26:2655.

Davies, L. A., A. Gil-Villegas, and G. Jackson. 1998. Describing the properties of chains of segments interacting via soft-core potentials of variable range with the SAFT-VR approach. *Int. J. Thermophys.* 19:675.

Denlinger, M. A. and C. K. Hall. 1990. Molecular-dynamics simulation results for the pressure of hard-chain fluids. *Mol. Phys.* 71:541.

Dickman, R. and C. K. Hall. 1986. Equation of state for chain molecules: Continuous-space analog of Flory theory. *J. Chem. Phys.* 85:4108.

Dodd, L. R. and D. N. Theodorou. 1991. Analytical treatment of the volume and surface area of molecules formed by an arbitrary collection of unequal spheres intersected by planes. *Mol. Phys.* 72:1313.

Dominik, A., W. G. Chapman, M. Kleiner, and G. Sadowski. 2005. Modeling of polar systems with the perturbed-chain SAFT equation of state. Investigation of the performance of two polar terms. *Ind. Eng. Chem. Res.* 44:6928.

Dominik, A., S. Jain, and W. G. Chapman. 2007. New equation of state for polymer solutions based on the statistical associating fluid theory (SAFT)-dimer equation for hard-chain molecules. *Ind. Eng. Chem. Res.* 46:5766.

Donohue, M. D. and J. M. Prausnitz. 1978. Perturbed hard chain theory for fluid mixtures: Thermodynamic properties for mixtures in natural gas and petroleum technology. *AIChE J.* 24:849.

Donohue, M. D. and P. Vimalchand. 1988. The perturbed-hard-chain theory. Extensions and applications. *Fluid Phase Equilibr.* 40:185.

dos Ramos, M. C., J. D. Haley, J. R. Westwood, and C. McCabe. 2011. Extending the GC-SAFT-VR approach to associating functional groups: Alcohols, aldehydes, amines and carboxylic acids. *Fluid Phase Equilibr.* 306:97.

Economou, I. G. 2002. Statistical associating fluid theory: A successful model for the calculation of thermodynamic and phase equilibrium properties of complex fluid mixtures. *Ind. Eng. Chem. Res.* 41:953.

Emami, F. S., A. Vahid, J. R. Elliott, and F. Feyzi. 2008. Group contribution prediction of vapor pressure with statistical associating fluid theory, perturbed-chain statistical associating fluid theory, and Elliott-Suresh-Donohue equations of state. *Ind. Eng. Chem. Res.* 47:8401.

Esposito, M. and G. T. Evans. 1994. Isotropic, nematic and smectic A phases in fluids of hard spherocylinders. *Mol. Phys.* 83:835.

Flory, P. J. 1942. Thermodynamics of high polymer solutions. *J. Chem. Phys.* 9:660.

Flory, P. J. 1941. Thermodynamics of high polymer solutions. *J. Chem. Phys.* 10:51.

Frenkel, D. and B. M. Mulder. 1985. The hard ellipsoid-of-revolution fluid I. Monte Carlo simulations. *Mol. Phys.* 55:1171.

Frenkel, D., B. M. Mulder, and J. P. MacTague. 1985. Phase diagram of hard ellipsoids of revolution. *Mol. Cryst. Liq. Cryst.* 123:119.

Galindo, A., L. A. Davies, A. Gil-Villegas, and G. Jackson. 1998. The thermodynamics of mixtures and the corresponding mixing rules in the SAFT-VR approach for potentials of variable range. *Mol. Phys.* 93:241.

Galindo, A., A. Gil-Villegas, G. Jackson, and A. N. Burgess. 1999. SAFT-VRE: Phase behavior of electrolyte solutions with the statistical associating fluid theory for potentials of variable range. *J. Phys. Chem. B* 103:10272.

Gao, J. and J. H. Weiner. 1989. Contribution of covalent bond force to pressure in polymer melts. *J. Chem. Phys.* 91:3168.

Ghonasgi, D. and W. G. Chapman. 1994a. A new equation of state for hard chain molecules. *J. Chem. Phys.* 100:6633.

Ghonasgi, D. and W. G. Chapman. 1994b. Prediction of the properties of model polymer solutions and blends. *AIChE J.* 40:878.

Ghonasgi, D., V. Perez, and W. G. Chapman. 1994. Intramolecular association in flexible hard chain molecules. *J. Chem. Phys.* 101:6880.

Gibbons, R. M. 1969. The scaled particle theory for particles of arbitrary shape. *Mol. Phys.* 17:81.

Gil-Villegas, A., A. Galindo, and G. Jackson. 2001. A statistical associating fluid theory for electrolyte solutions (SAFT-VRE). *Mol. Phys.* 99:531.

Gil-Villegas, A., A. Galindo, P. J. Whitehead, S. J. Mills, G. Jackson, and A. N. Burgess. 1997. Statistical associating fluid theory for chain molecules with attractive potentials of variable range. *J. Chem. Phys.* 106:4168.

Gross, J. 2005. An equation-of-state contribution for polar components: Quadrupolar molecules. *AIChE J.* 51:2556.

Gross, J. and G. Sadowski. 2000. Application of perturbation theory to a hard-chain reference fluid: An equation of state for square-well chains. *Fluid Phase Equilibr.* 168:183.

Gross, J. and G. Sadowski. 2001. Perturbed-chain SAFT: An equation of state based on a perturbation theory for chain molecules. *Ind. Eng. Chem. Res.* 40:1244.

Gross, J. and G. Sadowski. 2002a. Modeling polymer systems using perturbed-chain statistical associating fluid theory equation of state. *Ind. Eng. Chem. Res.* 41:1084.

Gross, J. and G. Sadowski. 2002b. Application of the perturbed-chain SAFT equation of state to associating systems. *Ind. Eng. Chem. Res.* 41:5510.

Gubbins, K. E. and C. G. Gray. 1972. Perturbation theory for the angular pair correlation function in molecular fluids. *Mol. Phys.* 23:187.

Gulati, H. S. and C. K. Hall. 1998. Generalized Flory equations of state for copolymers modeled as square-well chain fluids. *J. Chem. Phys.* 108:7478.

Gulati, H. S., J. M. Wichert, and C. K. Hall. 1996. Generalized Flory equations of state for hard heteronuclear chain molecules. *J. Chem. Phys.* 104:5220.

Hall, C. K., M. A. Denlinger, and K. G. Honnell. 1989. Generalized Flory theories for predicting properties of fluids containing long chain molecules. *Fluid Phase Equilibr.* 53:151.

Hino, T. and J. M. Prausnitz. 1997. A perturbed hard-sphere-chain equation of state for normal fluids and polymers using the square-well potential of variable width. *Fluid Phase Equilibr.* 138:105.

Honnell, K. G. and C. K. Hall. 1989. A new equation of state for athermal chains *J. Chem. Phys.* 90:1841.

Honnell, K. G. and C. K. Hall. 1991. Theory and simulation of hard-chain mixtures: Equations of state, mixing properties, and density profiles near hard walls. *J. Chem. Phys.* 95:4481.

Huang, S. H. and M. Radosz. 1990. Equation of state for small, large, polydisperse, and associating molecules. *Ind. Eng. Chem. Res.* 29:2284.

Huang, S. H. and M. Radosz. 1991. Equation of state for small, large, polydisperse, and associating molecules: Extension to fluid mixtures. *Ind. Eng. Chem. Res.* 30:1994.

Huggins, M. L. 1941. Solutions of long chain compounds. *J. Chem. Phys.* 9:440.

Isihara, A. 1951. Theory of anisotropic colloidal solutions. *J. Chem. Phys.* 19:1142.

Jackson, G., W. G. Chapman, and K. E. Gubbins. 1988. Phase equilibria of associating fluids. Spherical molecules with multiple bonding sites. *Mol. Phys.* 65:1.

Jackson, G. and K. E. Gubbins. 1989. Mixtures of associating spherical and chain molecules. *Pure Appl. Chem.* 61:1021.

Jiang, J. and J. M. Prausnitz. 1999. Equation of state for thermodynamic properties of chain fluids near-to and far-from the vapor-liquid critical region. *J. Chem. Phys.* 111: 5964.

Jiang, J. and J. M. Prausnitz. 2000. Phase equilibria for chain-fluid mixtures near to and far from the critical region. *AIChE J.* 46:2525.

Johnson, J. K. 1996. Perturbation theory and computer simulations for linear and ring model polymers. *J. Chem. Phys.* 104:1729.

Johnson, J. K., E. A. Müller, and K. E. Gubbins. 1994. Equation of state for Lennard-Jones chains. *J. Phys. Chem.* 98:6413.

Johnson, J. K. and K. E. Gubbins. 1992. Phase equilibria for associating Lennard-Jones fluids from theory and simulation. *Mol. Phys.* 77:1033.

Joslin, C. G., C. G. Gray, W. G. Chapman, and K. E. Gubbins. 1987. Theory and simulation of associating liquid mixtures. II. *Mol. Phys.* 62:843.

Kadlec, P., J. Janeček, and T. Boublík. 2000. Systems of oblate molecules. Monte Carlo study. *Mol. Phys.* 98:473.

Karakatsani, E. K. and I. G. Economou. 2006. Perturbed chain-statistical associating fluid theory extended to dipolar and quadrupolar molecular fluids. *J. Phys. Chem. B* 110:9252.

Karakatsani, E. K., T. Spyriouni, and I. G. Economou. 2005. Extended statistical associating fluid theory (SAFT) equations of state for dipolar fluids. *AIChE J.* 51:2328.

Kihara, T. 1953. Virial coefficients and models of molecules in gases. *Rev. Mod. Phys.* 25:831.

Kolafa, J. and I. Nezbeda. 1994. The Lennard-Jones fluid: An accurate analytic and theoretically-based equation of state. *Fluid Phase Equilibr.* 100:1.

Kraska, T. and K. E. Gubbins. 1996a. Phase equilibria calculations with a modified SAFT equation of state. 1. Pure alkanes, alkanols, and water. *Ind. Eng. Chem. Res.* 35:4727.

Kraska, T. and K. E. Gubbins. 1996b. Phase equilibria calculations with a modified SAFT equation of state. 2. Binary mixtures of *n*-alkanes, 1-alkanols, and water. *Ind. Eng. Chem. Res.* 35:4738.

Kreglewski, A. and R. C. Wilhoit. 1975. Thermodynamic properties of systems with specific interactions calculated from the hard-sphere equation of state. I. Binary systems with one inert component. *J. Phys. Chem.* 78:1961.

Kumar, S. K., I. Szleifer, C. K. Hall, and J. M. Wichert. 1996. Computer simulation study of the approximations associated with the generalized Flory theories. *J. Chem. Phys.* 104:9100.

Laffite, T., D. Bessières, M. M. Piñeiro, and J.-L. Daridon. 2006. Simultaneous estimation of phase behavior and second-derivative properties using the statistical associating fluid theory with variable range approach. *J. Chem. Phys.* 124:024509.

Laffite, T., M. M. Piñeiro, J.-L. Daridon, and D. Bessières. 2007. A comprehensive description of chemical association effects on second derivative properties of alcohols through a SAFT-VR approach. *J. Phys. Chem. B* 111:3447.

Lago, S., J. L. F. Abascal, and A. Ramos. 1983. Generalized Boublík equation: An accurate expression for the equation of state of hard fused spheres. *Phys. Chem. Liq.* 12:183.

Largo, J., M. J. Maeso, J. R. Solana, C. Vega, and L. G. MacDowell. 2003. Bonded hard-sphere theory and computer simulation of the equation of state of linear fused-hard-sphere fluids *J. Chem. Phys.* 119:9633.

Lee, S.-D. 1987. A numerical investigation of nematic ordering based on a simple hard-rod model. *J. Chem. Phys.* 87:4972.

Lee, Y. P., Y. C. Chiew, and G. P. Rangaiah. 2000. A molecular-based model for normal fluid mixtures: Perturbed Lennard-Jones chain equation of state. *Ind. Eng. Chem. Res.* 39:1497.

Llovell, F., J. C. Pàmies, and L. F. Vega. 2004. Thermodynamic properties of Lennard-Jones chain molecules: Renormalization-group corrections to a modified statistical associating fluid theory. *J. Chem. Phys.* 121:10715.

Lymperiadis, A., C. S. Adjiman, A. Galindo, and G. Jackson. 2007. A group contribution method for associating chain molecules based on the statistical associating fluid theory (SAFT-γ). *J. Chem. Phys.* 127:234903.

MacDowell, L. G., M. Müller, C. Vega, and K. Binder. 2000. Equation of state and critical behavior of polymer models: A quantitative comparison between Wertheim's thermodynamic perturbation theory and computer simulations. *J. Chem. Phys.* 113:419

MacDowell, L. G. and C. Vega. 1998. The second virial coefficient of hard alkane models. *J. Chem. Phys.* 109:5670.

MacDowell, L. G., C. Vega, and E. Sanz. 2001. Equation of state of model branched alkanes: Theoretical predictions and configurational bias Monte Carlo simulations. *J. Chem. Phys.* 115:6220.

McCabe, C. and S. B. Kiselev. 2004a. A crossover SAFT-VR equation of state for pure fluids: Preliminary results for light hydrocarbons. *Fluid Phase Equilibr.* 219:3.

McCabe, C. and S. B. Kiselev. 2004b. Application of crossover theory to the SAFT-VR equation of state: SAFT-VRX for pure fluids. *Ind. Eng. Chem. Res.* 43:2839.

McGrother, S. C., G. Jackson, and D. J. Photinos. 1997. The isotropic-nematic transition of dipolar spherocylinders: Combining thermodynamic perturbation with Monte Carlo simulation. *Mol. Phys.* 91:751.

McGrother, S. C., D. C. Williamson, and G. Jackson. 1996. A re-examination of the phase diagram of hard spherocylinders. *J. Chem. Phys.* 104:6755.

Mehta, S. D. and K. G. Honnell. 1996a. Equations of state and virial coefficients for rigid linear chains. *J. Phys. Chem.* 100:10408.

Mehta, S. D. and K. G. Honnell. 1996b. Generalized Flory theory for hard alkane fluids. *Mol. Phys.* 87:1285.

Morris. W. O., P. Vimalchand, and M. D. Donohue. 1987. The perturbed-soft-chain theory: An equation of state based on the Lennard-Jones potential. *Fluid Phase Equilibr.* 32:103.

Mulder, B. M. 2005. The excluded volume of hard sphero-zonotopes. *Mol. Phys.* 103:1411.

Müller, E. A. and K. E. Gubbins. 1993. Simulation of hard triatomic and tetratomic molecules. A test of associating fluid theories. *Mol. Phys.* 80:957.

Müller, E. A. and K. E. Gubbins. 1995. An equation of state for water from a simplified intermolecular potential. *Ind. Eng. Chem. Res.* 34:3662.

Müller, E. A. and K. E. Gubbins. 2001. Molecular-based equations of state for associating fluids: A review of SAFT and related approaches. *Ind. Eng. Chem. Res.* 40:2193.

Nezbeda, I. 1976. Virial expansion and an improved equation of state for the hard convex molecule system. *Chem. Phys. Lett.* 41:55.

NguyenHuynh, D., J.-P. Passarello, P. Tobaly, and J.-C. de Hemptinne. 2008. Application of GC-SAFT EOS to polar systems using a segment approach. *Fluid Phase Equilibr.* 264:62

Novenario, C. R., J. M. Caruthers, and K.-C. Chao. 1998. Chain-of-rotators equation of state for polar and non-polar substances and mixtures. *Fluid Phase Equilibr.* 142:83.

O'Lenick, R. and Y. C. Chiew. 1995. Variational theory for Lennard-Jones chains. *Mol. Phys.* 85:257.

Onsager, L. 1949. The effects of shape on the interaction of colloidal particles. *Ann. NY Acad. Sci.* 51:627.

Paras, E. P. A., C. Vega, and P. A. Monson. 1992. Application of cell theory to the thermodynamic properties of hard dumbbell solids. *Mol Phys.* 77:803.

Paredes, M. L. L., R. Nobrega, and F. W. Tavares. 2001. Square-well chain mixture: analytic equation of state and Monte Carlo simulation data. *Fluid Phase Equilibr.* 179:245.

Parsons, J. D. 1979. Nematic ordering in a system of rods. *Phys. Rev. A* 19:1225.

Peng, Y., K. D. Goff, M. C. dos Ramos, and C. McCabe. 2009. Developing a predictive group-contribution-based SAFT-VR equation of state. *Fluid Phase Equilibr.* 277:131.

Peng, Y., K. D. Goff, M. C. dos Ramos, and C. McCabe. 2010. Predicting the phase behavior of polymer systems with the GC-SAFT-VR approach. *Ind. Eng. Chem. Res.* 49:1378.

Perram, J. W. and E. R. Smith. 1975. A model for the examination of phase behaviour in multicomponent systems. *Chem. Phys. Lett.* 35:138.

Perram, J. W. and L. R. White. 1974. Perturbation theory for the angular correlation function. *Mol. Phys.* 28:527.

Phan, S., E. Kierlik, M. L. Rosinberg, H. Yu, and G. Stell. 1993. Equation of state for hard chain molecules *J. Chem. Phys.* 99:5326.

Phan, S., E. Kierlik, and M. L. Rosinberg. 1994. An equation of state for fused hard-sphere polyatomic molecules. *J. Chem. Phys.* 101:7997.

Pults, J. D., R. A. Greenkorn, and K.-C. Chao. 1989a. Chain-of-rotators group contribution equation of state. *Chem. Eng. Sci.* 44:2553.

Pults, J. D., R. A. Greenkorn, and K.-C. Chao. 1989b. Fluid phase equilibrium and volumetric properties from the chain-of-rotators group contribution equation of state. *Fluid Phase Equilibr.* 51:147.

Rigby, M. 1976. Scaled particle equation for hard non-spherical molecules. *Mol. Phys.* 32:575.

Rosenfeld, Y. 1988. Scaled particle theory of the structure and the thermodynamics of isotropic hard particle fluids. *J. Chem. Phys.* 89:4272.

Saager, B. and J. Fischer. 1992. Construction and application of physically based equations of state: Part II. The dipolar and quadrupolar contributions to the Helmholtz energy. *Fluid Phase Equilibr.* 72:67.

Saager, B., R. Hennenberg, and J. Fischer. 1992. Construction and application of physically based equations of state. Part I. Modification of the BACK equation *Fluid Phase Equilibr.* 72:41.

Scott, G. D. 1962. Radial distribution of the random close packing of equal spheres. *Nature* 194:956.

Sear, R. P. and G. Jackson. 1995. The gas, liquid, and solid phases of dimerizing hard spheres and hard-sphere dumbbells. *J. Chem. Phys.* 102:939.

Shukla, K. P. and W. G. Chapman. 2000. TPT2 and SAFTD equations of state for mixtures of hard chain copolymers. *Mol. Phys.* 98:2045.

Singer, S. J. and R. Mumaugh. 1990. Monte Carlo study of fluid-plastic crystal coexistence in hard dumbbells. *J. Chem. Phys.* 93:1278.

Song, Y., S. M. Lambert, and J. M. Prausnitz. 1994a. Equation of state for mixtures of hard-sphere chains including copolymers. *Macromolecules* 27:441.

Song, Y., S. M. Lambert, and J. M. Prausnitz. 1994b. A perturbed hard-sphere-chain equation of state for normal fluids and polymers. *Ind. Eng. Chem. Res.* 33:1047.

Strobrants, A., H. N. W. Lekkerkerker, and D. Frenkel. 1986. Evidence for smectic order in a fluid of hard parallel spherocylinders. *Phys. Rev. Lett.* 57:1452.

Strobrants, A., H. N. W. Lekkerkerker, and D. Frenkel. 1987. Evidence for one-, two-, and three-dimensional order in a system of hard parallel spherocylinders. *Phys. Rev. A* 36:2929.

Tamouza, S., J.-P. Passarello, P. Tobaly, and J.-C. de Hemptinne. 2004. Group contribution method with SAFT EOS applied to vapor liquid equilibria of various hydrocarbon series. *Fluid Phase Equilibr.* 222–223:67.

Tamouza, S., J.-P. Passarello, P. Tobaly, and J.-C. de Hemptinne. 2005. Application to binary mixtures of a group contribution SAFT EOS (GC-SAFT). *Fluid Phase Equilibr.* 228–229:409.

Tan, S. P., H. Adidharma, and M. Radosz. 2008. Recent advances and applications of statistical associating fluid theory. *Ind. Eng. Chem. Res.* 47:8063.

Tang, X. and J. Gross. 2010. Renormalization-group corrections to the perturbed-chain statistical associating fluid theory for binary mixtures. *Ind. Eng. Chem. Res.* 49:9436.

Tang, Y. and B. C.-Y. Lu. 1996. Direct calculation of radial distribution function for hard-sphere chains. *J. Chem. Phys.* 105:8262.

Tang, Y. and B. C.-Y. Lu. 1997a. Analytical representation of the radial distribution function for classical fluids. *Mol. Phys.* 90:215.

Tang, Y. and B. C.-Y. Lu. 1997b. Analytical description of the Lennard-Jones fluid and its application. *AIChE J.* 43:2215.

Tang, Y. and B. C.-Y. Lu. 2000. A study of associating Lennard-Jones chains by a new reference radial distribution function. *Fluid Phase Equilibr.* 171:27.

Tavares, F. W., J. Chang, and S. I. Sandler. 1995. Equation of state for the square-well chain fluid based on the dimer version of Wertheim's perturbation theory. *Mol. Phys.* 86:1451.

Tavares, F. W., J. Chang, and S. I. Sandler. 1997. A completely analytic equation of state for the square-well chain fluid of variable well width. *Fluid Phase Equilibr.* 140:129.

Tildesley, D. J. and W. B. Streett. 1980. An equation of state for hard dumbbell fluids. *Mol. Phys.* 41:85.

Tjipto-Margo, B. and G. T. Evans. 1990. The Onsager theory of the isotropic-nematic liquid-crystal transition: Incorporation of the higher virial coefficients. *J. Chem. Phys.* 93:4254.

Tumakaka, F. and G. Sadowski. 2004. Application of the perturbed-chain SAFT equation of state to polar systems. *Fluid Phase Equilibr.* 217:233.

Varga, S. and I. Szalai. 2000. Modified Parsons–Lee theory for fluids of linear fused hard sphere chains. *Mol. Phys.* 98:693.

Veerman, J. A. C. and D. Frenkel 1990. Phase diagram of a system of hard spherocylinders by computer simulation. *Phys. Rev. A* 41:3237.

Vega, C. and S. Lago. 1994. Isotropic-nematic transition of hard polar and nonpolar molecules. *J. Chem. Phys.* 100:6727.

Vega, C., S. Lago, and B. Garzón. 1994. Virial coefficients and equation of state of hard alkane models. *J. Chem. Phys.* 100:2182.

Vega, C. and L. G. MacDowell. 2001. Extending Wertheim's perturbation theory to the solid phase: The freezing of the pearl-necklace model. *J. Chem. Phys.* 114:10411.

Vega, C., L. G. MacDowell, and P. Padilla. 1996. Equation of state for hard n-alkane models: Long chains. *J. Chem. Phys.* 104:701.

Vega, C. and C. McBride. 2002. Scaling laws for the equation of state of flexible and linear tangent hard sphere chains. *Phys. Rev. E* 65:052501.

Vega, C., C. McBride, and L. G. MacDowell. 2001. Liquid crystal phase formation for the linear tangent hard sphere model from Monte Carlo simulations. *J. Chem. Phys.* 115:4203.

Vimalchand, P. and M. D. Donohue. 1985. Thermodynamics of quadrupolar molecules: The perturbed-anisotropic-chain theory. *Ind. Eng. Chem. Fundam.* 24:246.

von Solms, N., R. O'Lenick, and Y. C. Chiew. 1999. Lennard-Jones chain mixtures: Variational theory and Monte Carlo simulation results. *Mol. Phys.* 96:15.

Walsh, J. M. and K. E. Gubbins. 1990. A modified thermodynamic perturbation theory equation for molecules with fused hard sphere cores. *J. Phys. Chem.* 94:5115.

Walsh, J. M. and K. E. Gubbins. 1993. The liquid structure and thermodynamic properties of Lennard-Jones spheres with association sites. *Mol. Phys.* 80:65.

Wertheim, M. S. 1984a. Fluids with highly directional attractive forces. I. Statistical thermodynamics. *J. Stat. Phys.* 35:19.

Wertheim, M. S. 1984b. Fluids with highly directional attractive forces. II. Thermodynamic perturbation theory and integral equations. *J. Stat. Phys.* 35:35.

Wertheim, M. S. 1986a. Fluids with highly directional attractive forces. III. Multiple attraction sites. *J. Stat. Phys.* 42:459.

Wertheim, M. S. 1986b. Fluids of dimerizing hard spheres, and fluid mixtures of hard spheres and dispheres. *J. Chem. Phys.* 85:2929.

Wertheim, M. S. 1987. Thermodynamic perturbation theory of polymerization. *J. Chem. Phys.* 87:7323.

Wichert, J. M., H. S. Gulati, and C. K. Hall. 1996. Binary hard chain mixtures. I. Generalized Flory equations of state. *J. Chem. Phys.* 105:7669.

Wichert, J. M. and C. K. Hall. 1994. Generalized Flory equation of state for hard chain-hard monomer mixtures of unequal segment diameter. *Chem. Eng. Sci.* 49:2793.

Williamson, D. C. and F. del Río. 1998. The isotropic-nematic phase transition in a fluid of square well spherocylinders. *J. Chem. Phys.* 109:4675.

Williamson, D. C. and G. Jackson. 1995. Excluded volume for a pair of linear chains of tangent hard spheres with an arbitrary relative orientation. *Mol. Phys.* 86:819.

Williamson, D. C. and G. Jackson. 1998. Liquid crystalline phase behavior in systems of hard-sphere chains. *J. Chem. Phys.* 108:10294.

Yeom, M. S., J. Chang, and H. Kim. 2002. An equation of state for the hard-sphere chain fluid based on the thermodynamic perturbation theory of sequential polymerization. *Int. J. Thermophys.* 23:135.

Yethiraj, A. 1995. Monte Carlo simulations for the equation of state of athermal linear alkanes. *J. Chem. Phys.* 102:6874.

Yethiraj, A., J. G. Curro, K. S. Schweizer, and J. D. McCoy. 1993. Microscopic equations of state of polyethylene: Hard-chain contribution to the pressure. *J. Chem. Phys.* 98:1635.

Yethiraj, A. and C. K. Hall. 1990. Local structure of fluids containing chain-like molecules: Polymer reference interaction site model with Yukawa closure. *J. Chem. Phys.* 93:5315.

Yethiraj, A. and C. K. Hall. 1991a. Equations of state for star polymers. *J. Chem. Phys.* 94:3943.

Yethiraj, A. and C. K. Hall. 1991b. Square-well chains: Bulk equation of state using perturbation theory and Monte Carlo simulations of the bulk pressure and of the density profiles near walls. *J. Chem. Phys.* 95:1999.

Yethiraj, A. and C. K. Hall. 1991c. Generalized Flory equations of state for square-well chains. *J. Chem. Phys.* 95:8494.

Yethiraj, A. and C. K. Hall. 1992. Monte Carlo simulations and integral equation theory for microscopic correlations in polymeric fluids. *J. Chem. Phys.* 96:797.

Yethiraj, A., and C. K. Hall. 1993. On the equation of state for hard chain fluids. *Mol. Phys.* 80:469.

Zhou, Y., C. K. Hall, and G. Stell. 1995a. Thermodynamic perturbation theory for fused hard-sphere and hard-disk chain fluids. *J. Chem. Phys.* 103:2688.

Zhou, Y., S. W. Smith, and C. K. Hall. 1995b. Linear dependence on chain length for the thermodynamic properties of tangent hard-sphere chains. *Mol. Phys.* 86:1157.

9 Inhomogeneous Systems

This chapter summarizes the foundations of the density functional formalism, including a summary of the main perturbative and nonperturbative approximations in density functional theory. The results of some of these theories are discussed for some simple model systems with both isotropic and anisotropic interactions.

9.1 FUNDAMENTALS OF THE DENSITY FUNCTIONAL FORMALISM

Let us consider a system with volume V in contact with a heath reservoir at temperature T and with a molecule reservoir with chemical potential μ, and suppose that each particle is subjected to an external potential $\mathscr{V}(\mathbf{r})$, which depends on the position $\mathbf{r}$ of the particle and is the responsible for the inhomogeneity of the system. It can be proved (see Evans 1979) that, for a given intermolecular potential, there is a unique external potential $\mathscr{V}(\mathbf{r})$ compatible with a given grand canonical equilibrium single-particle density $\rho^{(1)}(\mathbf{r})$, Equation 2.50 for $n = 1$, which we will denote by $\rho(\mathbf{r})$, for simplicity, from now on. Therefore, all the thermodynamic functions may be expressed as functionals of the single-particle density, whence the name of *density functional theory* (DFT) given to this formalism. In particular, defining a local potential $\psi(\mathbf{r}) = \mu - \mathscr{V}(\mathbf{r})$, the grand potential $\Omega = F - \mu N = -k_B T \ln \Xi$ will take the form

$$\Omega\left[\rho\left(\mathbf{r}\right)\right] = \mathscr{F}\left[\rho\left(\mathbf{r}\right)\right] - \int \rho\left(\mathbf{r}\right)\psi\left(\mathbf{r}\right)d\mathbf{r}, \tag{9.1}$$

where $\mathscr{F}\left[\rho\left(\mathbf{r}\right)\right]$ is a free energy functional, which is related to the Helmholtz free energy of the system by means of the relationship

$$F\left[\rho\left(\mathbf{r}\right)\right] = \mathscr{F}\left[\rho\left(\mathbf{r}\right)\right] + \int \mathscr{V}\left(\mathbf{r}\right)\rho\left(\mathbf{r}\right)d\mathbf{r}, \tag{9.2}$$

where $\rho(\mathbf{r})$ is the equilibrium nonuniform density, which will be the one minimizing the grand potential (Equation 9.1). This yields (Lebowitz and Percus 1963a)

$$\frac{\delta\mathscr{F}\left[\rho\left(\mathbf{r}\right)\right]}{\delta\rho\left(\mathbf{r}\right)} = \psi\left(\mathbf{r}\right), \tag{9.3}$$

where δ means a functional derivative.

On the other hand, Lebowitz and Percus (1963a) also showed that

$$\frac{\delta\Omega}{\delta\psi\left(\mathbf{r}\right)} \equiv -\frac{k_B T\delta\ln\Xi}{\delta\psi\left(\mathbf{r}\right)} = -\rho\left(\mathbf{r}\right), \tag{9.4}$$

as can be easily seen from the definition of $\rho(\mathbf{r})$ in the grand canonical ensemble, Equation 2.50 for $n = 1$, and the properties of the functional derivative (see Evans 1979 for a detailed derivation). For an ideal gas, the intermolecular potential is $u(r) = 0$ and, from Equation 2.49, one easily obtains

$$\Omega_{id}\left[\rho\left(\mathbf{r}\right)\right] = -k_B T \Lambda^{-3} \int e^{\beta\psi(\mathbf{r})} d\mathbf{r}, \tag{9.5}$$

where Λ is the thermal wavelength. Then, Equation 9.4 yields for the equilibrium single-particle density of the ideal gas

$$\rho\left(\mathbf{r}\right) = \Lambda^{-3} e^{\beta\psi(\mathbf{r})}. \tag{9.6}$$

Taking into account that $\Omega = -PV$ and, therefore, for an ideal gas with average number of particles $\langle N \rangle$ we have $\Omega = -\langle N \rangle k_B T$, and that $\int \rho(\mathbf{r}) d\mathbf{r} = \langle N \rangle$, from Equation 9.1 the ideal gas free energy functional is

$$\mathscr{F}_{id}\left[\rho\left(\mathbf{r}\right)\right] = \int \rho\left(\mathbf{r}\right) f_{id}\left(\rho\left(\mathbf{r}\right)\right) d\mathbf{r} = k_B T \int \rho\left(\mathbf{r}\right)\left\{\ln\left(\Lambda^3 \rho\left(\mathbf{r}\right)\right) - 1\right\} d\mathbf{r}, \tag{9.7}$$

which is a quite obvious extension to an inhomogeneous ideal gas of expression (6.42) for the free energy of the homogeneous ideal gas, where $f_{id}(\rho)$ is the free energy per particle of an ideal gas with density ρ.

For an inhomogeneous system of interacting particles, the effect of the interactions may be accounted for by introducing an effective one-body dimensionless potential $\mathscr{C}(\mathbf{r})$ (Yang et al. 1976, Haymet and Oxtoby 1981), so that

$$\rho\left(\mathbf{r}\right) = \Lambda^{-3} e^{\beta\psi(\mathbf{r}) + \mathscr{C}(\mathbf{r})}, \tag{9.8}$$

from which

$$\psi\left(\mathbf{r}\right) = k_B T \left\{\ln\left(\rho\left(\mathbf{r}\right)\Lambda^3\right) - \mathscr{C}\left(\mathbf{r}\right)\right\}. \tag{9.9}$$

Introducing this result into Equation 9.1, and taking into account Equation 9.7, one obtains

$$\Omega\left[\rho\left(\mathbf{r}\right)\right] = \mathscr{F}^E\left[\rho\left(\mathbf{r}\right)\right] + k_B T \int \rho\left(\mathbf{r}\right)\left\{\mathscr{C}\left(r\right) - 1\right\} d\mathbf{r}, \tag{9.10}$$

where we have introduced the excess free energy functional $\mathscr{F}^E = \mathscr{F} - \mathscr{F}_{id}$. Then, minimization of the grand potential yields

$$\beta\frac{\delta\mathscr{F}^E}{\delta\rho\left(\mathbf{r}\right)} = -\mathscr{C}\left(\mathbf{r}\right). \tag{9.11}$$

The second functional derivative is

$$\frac{\beta\delta^2\mathscr{F}^E}{\delta\rho\left(\mathbf{r}\right)\delta\rho\left(\mathbf{r}'\right)} = -\frac{\delta\mathscr{C}\left(\mathbf{r}\right)}{\delta\rho\left(\mathbf{r}'\right)} = -\mathscr{C}\left(\mathbf{r}, \mathbf{r}'\right), \tag{9.12}$$

and, in general,

$$\frac{\beta \delta^n \mathscr{F}^E}{\delta \rho\left(\mathbf{r}_1\right) \delta \rho\left(\mathbf{r}_2\right) \cdots \delta \rho\left(\mathbf{r}_n\right)} = -\frac{\delta \mathscr{C}\left(\mathbf{r}_1, \mathbf{r}_2, \ldots, \mathbf{r}_{n-1}\right)}{\delta \rho\left(\mathbf{r}_n\right)} = -\mathscr{C}\left(\mathbf{r}_1, \mathbf{r}_2, \ldots, \mathbf{r}_n\right). \quad (9.13)$$

For a homogeneous fluid, $\mathscr{C}\left(\mathbf{r}_1, \mathbf{r}_2, \ldots, \mathbf{r}_n\right)$ is equal to the n-particle DCF, that is

$$\mathscr{C}_{homo}\left(\mathbf{r}_1, \mathbf{r}_2, \ldots, \mathbf{r}_n\right) = c^{(n)}\left(\mathbf{r}_1, \mathbf{r}_2, \ldots, \mathbf{r}_n\right). \quad (9.14)$$

In particular, for $n = 2$ and a spherically symmetric potential, $c^{(2)}\left(\mathbf{r}_1, \mathbf{r}_2\right)$ depends only on the relative distance r between the particles and is the (two-particle) DCF $c(r)$ as defined by the OZ equation (4.1).

Another useful relationship involving the excess free energy functional for a system with pair potential $u(\mathbf{r}, \mathbf{r}')$ is

$$\frac{\delta \mathscr{F}^E}{\delta u\left(\mathbf{r}, \mathbf{r}'\right)} = \frac{1}{2}\rho^{(2)}\left(\mathbf{r}, \mathbf{r}'\right), \quad (9.15)$$

(see, e.g., Evans 1979 for a detailed derivation). In this equation, which for a uniform fluid with spherically symmetric pair potential leads to Equation 6.104, $\rho^{(2)}(\mathbf{r}, \mathbf{r}')$ is the pair density distribution of a system with density $\rho(\mathbf{r})$.

9.2 SOME DENSITY FUNCTIONAL APPROXIMATIONS

9.2.1 PERTURBATION EXPANSIONS

The excess free energy functional can be obtained from integration of Equation 9.15 as

$$\mathscr{F}^E\left[\rho\left(\mathbf{r}\right)\right] = \frac{1}{2}\int\int \rho^{(2)}\left(\mathbf{r}, \mathbf{r}'\right) u\left(\mathbf{r}, \mathbf{r}'\right) d\mathbf{r} d\mathbf{r}'. \quad (9.16)$$

In particular, if the positions of the particles are uncorrelated, then $\rho^{(2)}(\mathbf{r}, \mathbf{r}') = \rho(\mathbf{r})\rho(\mathbf{r}')$, and Equation 9.16 reduces to the MFA

$$\mathscr{F}^E\left[\rho\left(\mathbf{r}\right)\right] = \frac{1}{2}\int\int \rho\left(\mathbf{r}\right) \rho\left(\mathbf{r}'\right) u\left(\mathbf{r}, \mathbf{r}'\right) d\mathbf{r} d\mathbf{r}', \quad (9.17)$$

which, according to Equation 9.12, is equivalent to take $\mathscr{C}(\mathbf{r}, \mathbf{r}') = -\beta u(\mathbf{r}, \mathbf{r}')$.

Many density functional approximations use a suitable reference system with known thermodynamic and structural properties, and the departure of the properties of the inhomogeneous system with respect to the reference one are treated within a perturbation scheme. Let us consider reference $u_0(\mathbf{r}, \mathbf{r}')$ and perturbation $u_1(\mathbf{r}, \mathbf{r}')$ potentials coupled by means of a parameter α to give rise to an α-dependent intermolecular pair potential of the form

$$u\left(\mathbf{r}, \mathbf{r}'; \alpha\right) = u_0\left(\mathbf{r}, \mathbf{r}'\right) + \alpha u_1\left(\mathbf{r}, \mathbf{r}'\right), \quad (9.18)$$

with $0 \leq \alpha \leq 1$ so that for $\alpha = 0$ reduces to the reference potential and for $\alpha = 1$ gives rise to the full potential $u(\mathbf{r}, \mathbf{r}')$. Then, the free energy functional for the actual system can be obtained from integration as

$$\mathscr{F}[\rho(\mathbf{r})] = \mathscr{F}_0[\rho(\mathbf{r})] + \frac{1}{2} \int_0^1 d\alpha \int \int \rho^{(2)}(\mathbf{r}, \mathbf{r}'; \alpha) u_1(\mathbf{r}, \mathbf{r}') d\mathbf{r} d\mathbf{r}', \qquad (9.19)$$

where $\rho^{(2)}(\mathbf{r}, \mathbf{r}'; \alpha)$ is the pair density distribution for a system with pair potential $u_1(\mathbf{r}, \mathbf{r}'; \alpha)$ and density $\rho(\mathbf{r})$. Equation 9.19, which may be considered as the extension of Equation 6.21 with $\alpha = 1$ to inhomogeneous systems, provides a route to obtain the free energy functional. From Equation 9.19 for an homogeneous fluid with density ρ, using Equation 9.12 with the derivatives evaluated at $\rho(\mathbf{r}) = \rho$, one can obtain the DCF $c(r)$ of the actual system in terms of that of the reference system $c_0(r)$. If, in addition, the correlation between the particles is negligible, so that $\rho^{(2)}(\mathbf{r}, \mathbf{r}'; \alpha) = \rho(\mathbf{r})\rho(\mathbf{r}')$, the RPA approximation, Equation 4.44, results.

In a similar way, given reference and actual inhomogeneous systems with densities $\rho_0(\mathbf{r})$ and $\rho(\mathbf{r})$, respectively, and a coupling parameter α with $0 \leq \alpha \leq 1$, and defining a family of α-dependent densities $\rho(\mathbf{r}; \alpha)$ such that $\rho(\mathbf{r}; 0) = \rho_0(\mathbf{r})$ and $\rho(\mathbf{r}; 1) = \rho(\mathbf{r})$, integration of Equations 9.11 and 9.12 yields

$$\beta \mathscr{F}^E[\rho(\mathbf{r})] = \beta \mathscr{F}^E[\rho_0(\mathbf{r})] - \int_0^1 d\alpha \int \rho'(\mathbf{r}; \alpha) \, \mathscr{C}(\mathbf{r}; [\rho(\mathbf{r}; \alpha)]) \, d\mathbf{r}, \qquad (9.20)$$

$$\mathscr{C}(\mathbf{r}; [\rho(\mathbf{r})]) = \mathscr{C}_0(\mathbf{r}; [\rho(\mathbf{r})]) + \int_0^1 d\alpha' \int \rho'(\mathbf{r}'; \alpha') \, \mathscr{C}(\mathbf{r}, \mathbf{r}'; [\rho(\mathbf{r}'; \alpha')]) \, d\mathbf{r}', \qquad (9.21)$$

where $\rho'(\mathbf{r}; \alpha) = d\rho(\mathbf{r}; \alpha)/d\alpha$. Replacing in the latter expression the upper limit of integration with α, we can obtain $\mathscr{C}(\mathbf{r}; [\rho(\mathbf{r}; \alpha)])$, which enters in Equation 9.20. In particular, if

$$\rho(\mathbf{r}, \alpha) = \rho_0(\mathbf{r}) + \alpha \Delta \rho(\mathbf{r}), \qquad (9.22)$$

with $\Delta \rho(\mathbf{r}) = \rho(\mathbf{r}) - \rho_0(\mathbf{r})$, then $\rho'(\mathbf{r}; \alpha) = \Delta \rho(\mathbf{r})$ is independent of α.

On the other hand, $\mathscr{C}(\mathbf{r})$ can be expanded in terms of the single-particle density $\rho(\mathbf{r})$ around the value ρ_0 corresponding to the uniform fluid, with the result

$$\mathscr{C}(\mathbf{r}_1) = c_0^{(1)}(\mathbf{r}_1, \rho_0) + \int c_0^{(2)}(\mathbf{r}_1, \mathbf{r}_2; \rho_0) \, \Delta \rho(\mathbf{r}_2) \, d\mathbf{r}_2$$

$$+ \frac{1}{2} \int \int c_0^{(3)}(\mathbf{r}_1, \mathbf{r}_2, \mathbf{r}_3; \rho_0) \, \Delta \rho(\mathbf{r}_2) \, \Delta \rho(\mathbf{r}_3) \, d\mathbf{r}_2 d\mathbf{r}_3 + \cdots, \qquad (9.23)$$

where $c_0^{(2)}(\mathbf{r}_1, \mathbf{r}_2; \rho_0) = c_0(|\mathbf{r}_1 - \mathbf{r}_2|; \rho_0)$ and $c_0^{(3)}(\mathbf{r}_1, \mathbf{r}_2, \mathbf{r}_3; \rho_0) = c_0^{(3)}(|\mathbf{r}_1 - \mathbf{r}_2|, |\mathbf{r}_1 - \mathbf{r}_3|; \rho_0)$ are the two- and three-particle DCFs of the uniform reference fluid, respectively, $\Delta \rho(\mathbf{r}) = \rho(\mathbf{r}) - \rho_0$, and for the reference fluid $\mathscr{V}(\mathbf{r}) = 0$ and so, from Equation 9.9 and the definition of $\psi(\mathbf{r})$, $\mathscr{C}(\mathbf{r}) = c_0^{(1)}(\mathbf{r}; \rho_0) = -\beta \mu^E(\rho_0)$.

A similar expansion can be done for the excess free energy functional, namely

$$\beta \mathscr{F}^E [\rho(\mathbf{r})]$$

$$= \beta F^E (\rho_0) - \int c_0^{(1)} (\mathbf{r}_1; \rho_0) \, \Delta\rho(\mathbf{r}_1) \, d\mathbf{r}_1$$

$$- \frac{1}{2} \int \int c_0^{(2)} (\mathbf{r}_1, \mathbf{r}_2; \rho_0) \, \Delta\rho(\mathbf{r}_1) \, \Delta\rho(\mathbf{r}_2) \, d\mathbf{r}_1 d\mathbf{r}_2$$

$$- \frac{1}{6} \int \int \int c_0^{(3)} (\mathbf{r}_1, \mathbf{r}_2, \mathbf{r}_3; \rho_0) \, \Delta\rho(\mathbf{r}_1) \times \Delta\rho(\mathbf{r}_2) \, \Delta\rho(\mathbf{r}_3) \, d\mathbf{r}_1 d\mathbf{r}_2 d\mathbf{r}_3 + \cdots ,$$

$$(9.24)$$

Alternatively, this equation can be rewritten as

$$\beta \Delta\mathscr{F} [\rho(\mathbf{r})]$$

$$= \int \rho(\mathbf{r}_1) \ln \left[\frac{\rho(\mathbf{r}_1)}{\rho_0} \right] d\mathbf{r}_1 - \int c_0^{(1)} (\mathbf{r}_1; \rho_0) \, \Delta\rho(\mathbf{r}_1) \, d\mathbf{r}_1$$

$$- \frac{1}{2} \int \int c_0^{(2)} (\mathbf{r}_1, \mathbf{r}_2; \rho_0) \, \Delta\rho(\mathbf{r}_1) \, \Delta\rho(\mathbf{r}_2) \, d\mathbf{r}_1 d\mathbf{r}_2$$

$$- \frac{1}{6} \int \int \int c_0^{(3)} (\mathbf{r}_1, \mathbf{r}_2, \mathbf{r}_3; \rho_0) \, \Delta\rho(\mathbf{r}_1) \, \Delta\rho(\mathbf{r}_2) \, \Delta\rho(\mathbf{r}_3) \, d\mathbf{r}_1 d\mathbf{r}_2 d\mathbf{r}_3 + \cdots ,$$

$$(9.25)$$

where $\Delta\mathscr{F} [\rho(\mathbf{r})] = \mathscr{F} [\rho(\mathbf{r})] - F(\rho_0)$ and we have made use of Equation 9.7. Equation 9.24, truncated at the second-order term, allows us to obtain the free energy functional provided that we know the DCF and the Helmholtz free energy of the uniform reference system and that $\Delta\rho(\mathbf{r})$ is small. In terms of the grand potential, and taking into account Equation 9.10, Equation 9.25 leads to

$$\beta \Delta\Omega [\rho(\mathbf{r})]$$

$$= \int \rho(\mathbf{r}_1) \ln \left[\frac{\rho(\mathbf{r}_1)}{\rho_0} \right] d\mathbf{r}_1 - \int \Delta\rho(\mathbf{r}_1) \, d\mathbf{r}_1$$

$$- \frac{1}{2} \int \int c_0^{(2)} (\mathbf{r}_1, \mathbf{r}_2; \rho_0) \, \Delta\rho(\mathbf{r}_1) \, \Delta\rho(\mathbf{r}_2) \, d\mathbf{r}_1 d\mathbf{r}_2$$

$$- \frac{1}{6} \int \int \int c_0^{(3)} (\mathbf{r}_1, \mathbf{r}_2, \mathbf{r}_3; \rho_0) \, \Delta\rho(\mathbf{r}_1) \, \Delta\rho(\mathbf{r}_2) \, \Delta\rho(\mathbf{r}_3) \, d\mathbf{r}_1 d\mathbf{r}_2 d\mathbf{r}_3 + \cdots \quad (9.26)$$

The excess free energy functional can be expressed as

$$\beta \mathscr{F}^E [\rho(\mathbf{r})] = \int \Phi [\rho(\mathbf{r})] \, d\mathbf{r}, \qquad (9.27)$$

where $\beta^{-1}\Phi\left[\rho(\mathbf{r})\right]$ is the excess free energy density of the nonuniform system. Among the different approximations proposed to obtain the excess free energy functional, the simplest one is known as the *local density approximation* (LDA), based on the assumption that the excess free energy density of the nonuniform system equals that of a uniform system with the density ρ replaced with the local density $\rho(\mathbf{r})$, that is $\Phi\left[\rho(\mathbf{r})\right] = \Phi_0\left(\rho(\mathbf{r})\right)$, a reasonable approximation only when the latter is slowly varying enough. For more quickly varying densities, the free energy density may be expressed as an expansion in series of the density gradients $\nabla\rho(\mathbf{r})$, the *gradient expansion*. Truncating the series after the first term yields the LDA. The next term involves the square gradient $(\nabla\rho(\mathbf{r}))^2$ and the corresponding coefficient is related to the DCF of the homogeneous system (see Evans 1979 for further details). Truncation of the series at this order results in the *square-gradient approximation* (SGA)

$$\Phi\left[\rho\left(\mathbf{r}\right)\right] = \Phi_0\left(\rho\left(\mathbf{r}\right)\right) + \frac{1}{12}\int c_0^{(2)}\left(\mathbf{r};\rho\left(\mathbf{r}\right)\right)\left(\nabla\rho\left(\mathbf{r}\right)\right)^2 r^2 d\mathbf{r}. \qquad (9.28)$$

Higher-order terms involve higher-order correlation functions, which makes difficult the calculations. Moreover, these higher-order terms do not exist for potentials varying as r^{-n} for large distances (see Evans 2009). Neither is this approach suitable to deal with systems with hard-core interactions. The LDA can also be improved by adding a mean field term of the form of Equation 9.17. The resulting approximation, called by Löwen (1994) *LDA plus MFA*, is applicable to moderately inhomogeneous fluids.

On the other hand, when the density is slowly varying it can be shown that

$$\beta\mathscr{F}^E\left[\rho\left(\mathbf{r}\right)\right] = \int \Phi_0\left(\rho\left(\mathbf{r}\right)\right)d\mathbf{r} + \frac{1}{4}\int\int c^{(2)}\left(\left|\mathbf{r}-\mathbf{r}'\right|;\bar{\rho}\right)\left(\rho\left(\mathbf{r}\right)-\rho\left(\mathbf{r}'\right)\right)^2 d\mathbf{r}d\mathbf{r}', \qquad (9.29)$$

where $\bar{\rho}$ is some average density, as for example $\bar{\rho} = (\rho(\mathbf{r})+\rho(\mathbf{r}'))/2$. The preceding result was derived by Ebner et al. (1976) from expanding the free energy functional in powers of the difference $\rho(\mathbf{r}) - \rho(\mathbf{r}')$ and by Evans (1979) from a partial summation of the gradient expansion.

9.2.2 WEIGHTED DENSITY APPROXIMATION AND RELATED THEORIES

Different theories have been designed with the aim of overcoming the limitations of perturbation approaches arising from the unavoidable truncation of the perturbation series, usually at second order, and are considered nonperturbative approaches, although they are closely related to the perturbation expansions. A number of them are discussed next.

Starting from Equation 9.27, one can approximate $\Phi\left[\rho(\mathbf{r})\right]$ with $\beta\rho(\mathbf{r})f^E(\bar{\rho}(\mathbf{r}))$ or, equivalently, the exact *local* excess free energy per particle $f^E\left[\rho(\mathbf{r})\right]$ with the approximate one $f^E(\bar{\rho}(\mathbf{r}))$, where $f^E(\rho)$ is the excess free energy per particle of a uniform fluid with density ρ and $\bar{\rho}(\mathbf{r})$ is a *weighted density*, that is

$$\mathscr{F}^E\left[\rho\left(\mathbf{r}\right)\right] = \int \rho\left(\mathbf{r}\right)f^E\left(\bar{\rho}\left(\mathbf{r}\right)\right)d\mathbf{r}. \qquad (9.30)$$

This is the basis of the *weighted density approximation* (WDA) (Tarazona 1984, Curtin and Ashcroft 1985), with a number of versions differing with each other in the way the weighted density $\bar{\rho}(\mathbf{r})$ is determined. Following Tarazona (1984), the weighted and unweighted densities are related by

$$\bar{\rho}(\mathbf{r}) = \int \rho(\mathbf{r}') w\left(|\mathbf{r} - \mathbf{r}'|; \bar{\rho}(\mathbf{r})\right) d\mathbf{r}', \qquad (9.31)$$

where $w(r)$ is the weighting function, which satisfies the normalizing condition

$$\int w(r)\, d\mathbf{r} = 1. \qquad (9.32)$$

The weighting function may be determined in such a way to yield, through Equations 9.12 and 9.14, a prescribed DCF for the homogeneous fluid, but this procedure is quite involved (see Baus 1990 for details).

Using the procedure just outlined, and starting from Equation 9.19 with the reference system replaced with the ideal gas, Curtin and Ashcroft (1985) derived a general relationship between the weighting function and the DCF of the homogeneous fluid. They also noted that, with w determined in this way, the expansion of the density functional in terms of the single-particle density $\rho(\mathbf{r})$ around the uniform fluid density ρ_0 is identical to the expansion (9.24) to second order in $\Delta\rho(\mathbf{r})$. A simplified version of the WDA, the *simple weighted density approximation* (SWDA), consisting in replacing the weighted density $\bar{\rho}(\mathbf{r})$ in the weighting function with the uniform fluid density ρ_0, was proposed by Kim and Suh (1996), with the weighting function determined from the exact low-density expansion of the two-particle DCF. The same approach was used by Zhou (1999), with $w(|\mathbf{r} - \mathbf{r}'|; \rho_0)$ determined as in Equation 9.78.

Denton and Ashcroft (1989a) introduced a *modified weighted density approximation* (MWDA) as

$$\mathscr{F}^E[\rho(\mathbf{r})] = Nf^E(\bar{\rho}), \qquad (9.33)$$

where $\bar{\rho}$ is a weighted density defined by

$$\bar{\rho} = \frac{1}{N} \int \rho(\mathbf{r})\, d\mathbf{r} \int \rho(\mathbf{r}') w\left(|\mathbf{r} - \mathbf{r}'|; \bar{\rho}\right) d\mathbf{r}', \qquad (9.34)$$

with a normalization condition for the weighting function w similar to Equation 9.32. Therefore, in this approximation, the *global* excess free energy per particle $\mathscr{F}^E[\rho(\mathbf{r})]/N$ is replaced with the approximate one $f^E(\bar{\rho})$, which is independent of the position. Using Equations 9.12 and 9.14 yields for the weighting function w the simple expression*

$$w\left(|r - r'|; \rho\right) = -\frac{1}{2f'^E(\rho)} \left\{ k_B Tc\left(|r - r'|; \rho\right) + \frac{1}{V}\rho f''^E(\rho) \right\}, \qquad (9.35)$$

---

* Note that we are using the same notation for different weighting functions and different weighted densities.

where $f^E(\rho)$ and $c(|\mathbf{r} - \mathbf{r}'|; \rho)$ are the excess free energy per particle and the DCF, respectively, of a uniform fluid with density ρ and volume V, and primed quantities mean derivatives with respect to the density ρ. Although simpler than the WDA, the MWDA has the drawback that it is not suitable for application to interfacial properties due to its nonlocal character, as it is based on the *global* excess free energy, instead of the *local* excess free energy per particle as in the WDA. As noted by Denton and Ashcroft (1989a), the self-consistent way in which is determined the weighting function in the MWDA guaranties that it matches the exact expansion (9.24) to second order, in addition to include the contribution of the higher-order terms in an approximate way. The MWDA was extended by Likos and Ashcroft (1992, 1993) by including the contribution of the third-order DCF $c_0^{(3)}(r)$ of the reference fluid, thus giving rise to the *extended modified weighted density approximation* (EMA).

The combination of the WDA expressions (Equations 9.30 and 9.31), replacing in the weighting function the effective density $\bar{\rho}(\mathbf{r})$ with $\bar{\rho}$ as given by the MWDA expression (9.34), gives rise to the *hybrid weighted-density approximation* (HWDA) devised by Leidl and Wagner (1993).

Khein and Ashcroft (1997) introduced a *generalized density functional theory* in which

$$\mathscr{F}^E[\rho(\mathbf{r})] = Nf^E(\rho) a(x, y), \tag{9.36}$$

where $x = \bar{\rho}/\rho$, $y = f^E(\bar{\rho})/f^E(\rho)$, $a(x, y)$ is a scaling function, and the effective density is introduced by means of another scaling function $b(x, y)$ in the form

$$\rho b(x, y) = \frac{1}{N} \int \rho(\mathbf{r}) \, d\mathbf{r} \int \rho(\mathbf{r}') \, w\left(|\mathbf{r} - \mathbf{r}'|; \bar{\rho}\right) d\mathbf{r}'. \tag{9.37}$$

The weighting function w is determined from the condition that $\mathscr{F}^E[\rho(\mathbf{r})]$ must give the right second-order DCF in the homogeneous limit, corresponding to $x = y = 1$. Different choices for the weighting functions lead to different density functional approximations, including the MWDA (for $a = y$), and allow for new approximations to be introduced. Khein and Ashcroft assumed

$$b(x) = b_0 x + (1 - b_0)x^2, \tag{9.38}$$

where b_0 is a parameter that must be determined from some condition and, in particular, for $b_0 = 1$ leads to the MWDA definition (9.34) of $\bar{\rho}$.

Baus and Colot (1985a) (see also Baus 1990) developed the *effective liquid approximation* (ELA), in which the nonuniform system is approximated by a uniform fluid with density ρ_0 with the two-particle DCF calculated at an effective density $\bar{\rho}$. Combining Equations 9.20 through 9.22, we have

$$\beta \mathscr{F}^E[\rho(\mathbf{r})] = \beta \mathscr{F}^E[\rho_0(\mathbf{r})] - \int \mathscr{C}(\mathbf{r}; [\rho_0(\mathbf{r})]) \, \Delta\rho(\mathbf{r}) \, d\mathbf{r}$$

$$- \int_0^1 (1 - \alpha) \, d\alpha \int\int \mathscr{C}(\mathbf{r}, \mathbf{r}'; [\rho(\mathbf{r}'; \alpha)]) \, \Delta\rho(\mathbf{r}) \, \Delta\rho(\mathbf{r}') \, d\mathbf{r} d\mathbf{r}', \tag{9.39}$$

where the double integration with respect to the coupling parameter has been simplified by means of an integration by parts. Now, taking a uniform fluid with density ρ_0 as the reference system, the first term of the right-hand side of Equation 9.39 reduces to $\beta F^E(\rho_0)$. On the other hand, as explained in Section 9.2, for a uniform fluid $\mathscr{C}(\mathbf{r}) = c^{(1)}(r) = -\beta\mu^E$, and so the second term of the right-hand side of Equation 9.39 vanishes if one takes $\rho_0 = (1/V)\int\rho(\mathbf{r})$, the average density of the nonuniform system, because in this case $\int\Delta\rho(\mathbf{r})d\mathbf{r} = 0$. Therefore, provided that the excess free energy $F^E(\rho_0)$ of the reference fluid is known, we only need to determine $\mathscr{C}(\mathbf{r},\mathbf{r}';[\rho(\mathbf{r},\alpha)])$, where $\rho(\mathbf{r},\alpha)$ is given by Equation 9.22. To this end, Baus and Colot (1985a) introduced an effective density $\bar{\rho}(\rho_0)$ and approximated $\mathscr{C}(\mathbf{r},\mathbf{r}';[\rho(\mathbf{r},\alpha)])$ with $c_0^{(2)}(|\mathbf{r}-\mathbf{r}'|;\bar{\rho}(\rho_0))$, where $c_0^{(2)}(r;\bar{\rho})$ is the DCF of the reference fluid at the effective density $\bar{\rho}$. Then, Equation 9.39 transforms into

$$\beta\mathscr{F}^E[\rho(\mathbf{r})] = \beta F^E(\rho_0) - \frac{1}{2}\int\int c_0^{(2)}(|\mathbf{r}-\mathbf{r}'|;\bar{\rho})\,\Delta\rho(\mathbf{r})\,\Delta\rho(\mathbf{r}')\,d\mathbf{r}d\mathbf{r}'. \quad (9.40)$$

The effective density $\bar{\rho}$ must be determined from some condition. In particular, using Equation 9.7, we can rewrite Equation 9.40 in the form

$$\beta\Delta\mathscr{F} = \int\rho(\mathbf{r})\ln\left(\frac{\rho(\mathbf{r})}{\rho_0}\right)d\mathbf{r} - \frac{1}{2}\int\int c_0^{(2)}(|\mathbf{r}-\mathbf{r}'|;\bar{\rho})\,\Delta\rho(\mathbf{r})\,\Delta\rho(\mathbf{r}')\,d\mathbf{r}d\mathbf{r}',$$
$$(9.41)$$

where $\Delta\mathscr{F} = \mathscr{F}[\rho(\mathbf{r})] - F(\rho_0)$, and ρ_0 is again the average density of the nonuniform system. Using this expression, Colot et al. (1986) determined the effective density $\bar{\rho}$ by minimizing $\mathscr{F}[\rho(\mathbf{r})]$.

If $\rho_0 = 0$ is chosen as the reference state in Equation 9.39, one has

$$\beta\mathscr{F}^E[\rho(\mathbf{r})] = -\int_0^1(1-\alpha)\,d\alpha\int\int\mathscr{C}(\mathbf{r},\mathbf{r}';[\alpha\rho(\mathbf{r}')])\,\rho(\mathbf{r})\,\rho(\mathbf{r}')\,d\mathbf{r}d\mathbf{r}'. \quad (9.42)$$

Introducing in this expression the ELA approximation leads to the *modified effective liquid approximation* (MELA)

$$\beta\mathscr{F}^E[\rho(\mathbf{r})] = -\int_0^1(1-\alpha)\,d\alpha\int\int c_0^{(2)}(|\mathbf{r}-\mathbf{r}'|;\alpha\bar{\rho})\,\rho(\mathbf{r})\,\rho(\mathbf{r}')\,d\mathbf{r}d\mathbf{r}', \quad (9.43)$$

proposed by Baus (1989). The effective density $\bar{\rho}$ is obtained from the self-consistency condition that the *local* excess free energies per particle f^E of the inhomogeneous system and the effective fluid must be equal. This yields $\bar{\rho} = \bar{\rho}(\mathbf{r})$, with

$$\bar{\rho}(\mathbf{r}) = \frac{\int_0^1(1-\alpha)\,d\alpha\int\int c_0^{(2)}(|\mathbf{r}-\mathbf{r}'|;\alpha\bar{\rho}(\mathbf{r}))\,\rho(\mathbf{r}')\,d\mathbf{r}'}{\int_0^1(1-\alpha)\,d\alpha\int\int c_0^{(2)}(|\mathbf{r}''|;\alpha\bar{\rho}(\mathbf{r}))\,d\mathbf{r}''}. \quad (9.44)$$

It is easy to see that the effective density $\bar{\rho}(\mathbf{r})$ obtained in this way can be put in the form of the weighted density (Equation 9.31), and so the MELA expression for $\mathscr{F}^E$ is of the form of the WDA Equation (9.30). If the same self-consistency condition is applied to the *global* excess free energy per particle $\mathscr{F}^E/N$, a uniform effective density results, with the expression

$$\bar{\rho} = \frac{\int_0^1 (1-\alpha)\, d\alpha \int \int c_0^{(2)} (|\mathbf{r} - \mathbf{r}'|\,;\alpha\bar{\rho})\, \rho(\mathbf{r})\, \rho(\mathbf{r}')\, d\mathbf{r} d\mathbf{r}'}{N \int_0^1 (1-\alpha)\, d\alpha \int \int c_0^{(2)} (|\mathbf{r}''|\,;\alpha\bar{\rho})\, d\mathbf{r}''}, \qquad (9.45)$$

then the MELA $\mathscr{F}^E$ is of the form of the MWDA equation (9.33). In both cases, for a uniform fluid with density ρ one obtains $\bar{\rho} = \rho$ (Baus 1989, 1990). Replacing $\alpha\bar{\rho}$ in Equation 9.43 with $\bar{\rho}\,[\alpha\rho(\mathbf{r})]$ and imposing the same self-consistency condition leads to the *generalized effective liquid approximation* (GELA), proposed by Lutsko and Baus (1990a,b), in which the weighted density

$$\bar{\rho}\,[\rho(\mathbf{r})] = \frac{\int_0^1 (1-\alpha)\, d\alpha \int \int c_0^{(2)} (|\mathbf{r} - \mathbf{r}'|\,;\bar{\rho}\,[\alpha\rho(\mathbf{r})])\, \rho(\mathbf{r})\, \rho(\mathbf{r}')\, d\mathbf{r} d\mathbf{r}'}{N \int_0^1 (1-\alpha)\, d\alpha \int \int c_0^{(2)} (|\mathbf{r}''|\,;\alpha\bar{\rho}\,[\rho(r)])\, d\mathbf{r}''}, \qquad (9.46)$$

is now a functional of $\rho(\mathbf{r})$.

9.2.3 REFERENCE FUNCTIONAL APPROACH

Oettel (2005) denoted in this way a generalization of an approach developed by Rosenfeld (1993), which connects the DFT with the RHNC theory. Although Rosenfeld derived the theory directly for mixtures, we will give an outline of the theory for pure fluids, as the extension to mixtures is straightforward. Let us consider again reference and perturbed systems, but now both of them being inhomogeneous. In this case, we can rewrite the expansion (9.24) in the form

$$\beta\mathscr{F}^E\,[\rho(\mathbf{r})] = \beta\mathscr{F}_0^E\,[\rho_0(\mathbf{r})] - \int \mathscr{C}_0\,(\mathbf{r};[\rho_0])\,\Delta\rho(\mathbf{r})\,d\mathbf{r}$$

$$- \frac{1}{2} \int \int \mathscr{C}_0\,(\mathbf{r},\mathbf{r}';[\rho_0])\,\Delta\rho(\mathbf{r})\,\Delta\rho(\mathbf{r}')\,d\mathbf{r} d\mathbf{r}' + \beta\mathscr{F}^B\,[\rho(\mathbf{r})], \quad (9.47)$$

where $\Delta\rho(\mathbf{r}) = \rho(\mathbf{r}) - \rho_0(\mathbf{r})$ and $\mathscr{F}^B\,[\rho(\mathbf{r})]$ is a functional that accounts for all the terms in the expansion beyond the second-order term. Applying Equation 9.8 to both fluids, in combination with Equations 9.11 and 9.47 one obtains

$$\ln\left[\frac{\rho(\mathbf{r})}{\rho_0(\mathbf{r})}\right] + \beta\Delta\mathscr{V}(\mathbf{r}) = \mathscr{C}(\mathbf{r};[\rho(\mathbf{r})]) - \mathscr{C}_0(\mathbf{r};[\rho_0(\mathbf{r})])$$

$$= \int \mathscr{C}_0\,(\mathbf{r},\mathbf{r}';[\rho_0(\mathbf{r})])\,\Delta\rho(\mathbf{r}')\,d\mathbf{r}' - \beta\frac{\delta\mathscr{F}^B\,[\rho(\mathbf{r})]}{\delta\rho(\mathbf{r})}. \qquad (9.48)$$

Let us now consider an homogeneous fluid and fix a particle, *the test particle*, at position $\mathbf{r} = 0$. The interaction potential between this particle and the rest of the fluid may be considered as an external potential that renders the fluid inhomogeneous. In this case, $\Delta \mathscr{V}(\mathbf{r}) = u(\mathbf{r})$, the intermolecular potential of the homogeneous fluid, $\rho_0 = \rho$, the bulk density of the fluid, $\rho(\mathbf{r}) = \rho g(\mathbf{r})$, the RDF of the homogeneous fluid (Lebowitz and Percus 1963b), $\Delta \rho(\mathbf{r}) = \rho(\mathbf{r}) h(\mathbf{r})$, because far from the influence of the test particle ($r \to \infty$) the fluid is homogeneous with density ρ and $\lim_{r \to \infty} g(r) = 1$, and $\mathscr{C}(\mathbf{r}, \mathbf{r}'; [\rho]) = c^{(2)}(\mathbf{r})$, the (two-particle) DCF of the homogeneous fluid. Then, it is easy to see that Equation 9.48 is equivalent to the exact closure (4.10) if we identify $B(r) = -\lim_{\rho(r) \to \rho} \beta \delta \mathscr{F}^B([\rho(\mathbf{r})])/\delta \rho(\mathbf{r})$. Therefore, Equation 9.48 may be viewed as the extension to inhomogeneous systems of the exact closure (4.10), with the last term defining the *bridge functional*

$$B[\rho(\mathbf{r})] = -\beta \frac{\delta \mathscr{F}^B[\rho(\mathbf{r})]}{\delta \rho(\mathbf{r})}. \tag{9.49}$$

Rosenfeld (1993) invoked the ansatz of the *universality of the bridge functional* to approximate the bridge functional of the actual inhomogeneous system with that of a reference system, with known free energy functional, as determined from (9.49). Self-consistency was introduced through the test particle limit. This means that with $c^{(2)}(r)$ obtained from Equation 9.12 in the homogeneous limit, with the bridge function obtained from the bridge functional (9.49) in the test particle limit, both Equation 9.48 and the OZ equation must give the same result for the RDF. Therefore, in this limit, the reference functional approach is equivalent to the RHNC theory, but for the fact that the bridge function in the reference functional approach is obtained from a bridge functional derived from an appropriate DFT.

Usually the reference system is the HS system and, therefore, some prescription must be established to determine the effective HS diameter. To this end, one may use the optimizing condition (4.55). A different optimizing condition was proposed by Oettel (2005), based on minimizing the difference $\Delta \mathscr{F}_0^B = \mathscr{F}_0^B[\rho_0 g(\mathbf{r}; d)] - \mathscr{F}_0^B[\rho_0 g_0(\mathbf{r}; d)]$ with respect to d.

9.3 FUNDAMENTAL MEASURE THEORY

A different kind of DFT was devised by Rosenfeld (1989) for HS mixtures by extending to inhomogeneous systems the fundamental measure concept (see Section 8.1), thus giving rise to the *fundamental measure theory* (FMT). We will not give here an exhaustive account of this and other related theories but only a brief summary of the most important results. The interested reader is addressed to the reviews by Tarazona et al. (2008), Evans (2009), and Roth (2010), apart from the literature that will be cited later, for further details.

For a homogeneous fluid, the density expansion of the excess free energy can be obtained from integration of the virial expansion of the EOS, like in Equation 8.156. In a similar way, taking into account the expressions (2.39) and (2.40), the low-density expansion of the excess free energy functional of an inhomogeneous m-component mixture of additive HS may be written as

$$\beta \mathscr{F}^E \left[\{ \rho_i \left(\mathbf{r} \right) \} \right]$$

$$= -\frac{1}{2} \sum_{i,j=1}^{m} \int \int \rho_i \left(\mathbf{r}_1 \right) \rho_j \left(\mathbf{r}_2 \right) f_{ij} \left(r_{12} \right) d\mathbf{r}_1 d\mathbf{r}_2$$

$$- \frac{1}{6} \sum_{i,j,k=1}^{m} \int \int \int \rho_i \left(\mathbf{r}_1 \right) \rho_j \left(\mathbf{r}_2 \right) \rho_k \left(\mathbf{r}_3 \right) f_{ij} \left(r_{12} \right) f_{ik} \left(r_{13} \right) f_{jk} \left(r_{23} \right) d\mathbf{r}_1 d\mathbf{r}_2 d\mathbf{r}_3 + \cdots,$$

$$(9.50)$$

where
  $\rho_i(\mathbf{r})$ is the density profile of component i
  $f_{ij}(r)$ is the Mayer function for the pair ij

The Mayer function for a pair of spheres is a step function that determines the pair excluded volume, namely, $f_{ij}(r_{ij}) = -\Theta[(R_i + R_j) - r_{ij}]$, where R_i and R_j are the radius of the spheres. For a pair of HCBs, the excluded volume can be expressed in terms of the individual measures of each body, as indicates Equation (8.8). Rosenfeld (1989) used the identity

$$\Theta \left[\left(R_i + R_j \right) - r_{ij} \right] = w_i^{(3)} * w_j^{(0)} + w_j^{(3)} * w_i^{(0)} + w_i^{(1)} * w_j^{(2)} + w_j^{(1)} * w_i^{(2)}$$

$$+ w_i^{(1)} * w_j^{(2)} + w_j^{(1)} * w_i^{(2)}, \qquad (9.51)$$

where the symbol "$*$" means a convolution, so that

$$w_i^{(a)} * w_j^{(b)} = \int w_i^{(a)} \left(\mathbf{r} - \mathbf{r}_i \right) w_j^{(b)} \left(\mathbf{r} - \mathbf{r}_j \right) d\mathbf{r}, \qquad (9.52)$$

with $a, b = 0, 1, 2, 3$, the boldface indicate vector quantities with implicit dot products, and

$$w_i^{(3)} \left(\mathbf{r} \right) = \Theta \left(R_i - r \right), \quad w_i^{(2)} \left(\mathbf{r} \right) = \frac{\mathbf{r}}{r} \delta \left(R_i - r \right), \quad w_i^{(2)} \left(\mathbf{r} \right) = \delta \left(R_i - r \right),$$

$$w_i^{(1)} \left(\mathbf{r} \right) = \frac{w_i^{(2)} \left(\mathbf{r} \right)}{4 \pi R_i}, \quad w_i^{(0)} \left(\mathbf{r} \right) = \frac{w_i^{(2)} \left(\mathbf{r} \right)}{4 \pi R_i^2}, \quad w_i^{(1)} \left(\mathbf{r} \right) = \frac{w_i^{(2)} \left(\mathbf{r} \right)}{4 \pi R_i}, \qquad (9.53)$$

where $\delta(x)$ is the Dirac delta function. An alternative to Equation 9.51 that avoids the use of the vector weights $w_i^{(l)}(\mathbf{r})$ was derived by Kierlik and Rosinberg (1990) and Phan et al. (1993) showed the equivalence of the two proposals. The weights of Equation 9.53 were used by Rosenfeld to define the set of weighted densities

$$n_l \left(\mathbf{r} \right) = \sum_{i=1}^{m} \int \rho_i \left(\mathbf{r}' \right) w_i^{(l)} \left(\mathbf{r} - \mathbf{r}' \right) d\mathbf{r}', \qquad (9.54)$$

either scalar or vectorial. In the uniform limit, the scalar densities $n_l(\mathbf{r})$ reduce to the composition-averaged measures $\xi^{(l)}$ of Equation 8.26, whereas the vector densities $\mathbf{n}_l(\mathbf{r})$ vanish. The excess free energy functional was expressed in a similar way as in Equation 9.27, namely,

$$\beta \mathscr{F}^E [\{\rho_i (\mathbf{r})\}] = \int \Phi [\{n_l (\mathbf{r})\}] \, d\mathbf{r}. \tag{9.55}$$

On the basis of dimensional analysis, Rosenfeld assumed for Φ the form

$$\Phi [\{n_l (\mathbf{r})\}] = \varphi_1 (n_3) \, n_0 + \varphi_2 (n_3) \, n_1 n_2 + \varphi_3 (n_3) \, n_2^3$$
$$+ \varphi_4 (n_3) \, \mathbf{n}_1 \cdot \mathbf{n}_2 + \varphi_5 (n_3) \, n_2 (\mathbf{n}_2 \cdot \mathbf{n}_2) . \tag{9.56}$$

The functions φ_i were determined from the requirement that the exact low-density expansion (9.50) must be fulfilled up to third order, that is to say, up to the contribution from the third virial coefficient, together with the exact condition (Reiss et al. 1959, Lebowitz and Rowlinson 1964)

$$\lim_{R_i \to \infty} \mu_i^E = P v_i, \tag{9.57}$$

for the uniform fluid, where $v_i = (4/3)\pi R_i^3$ is the volume of a sphere of radius R_i and μ_i^E the corresponding excess chemical potential. The latter is given by the relationship

$$\mu_i^E = k_B T \frac{\partial \Phi}{\partial \rho_i} = k_B T \sum_{a=0}^{3} \frac{\partial \Phi}{\partial n_l} \frac{\partial n_l}{\partial \rho_i}. \tag{9.58}$$

On the other hand, from Equations 9.53 and 9.54, it easy to see that $\partial n_3/\partial \rho_i = (4/3)\pi R_i^3 \equiv v_i$ and that $\partial n_l/\partial \rho_i \propto R_i^l$ for $l < 3$ and so, from Equation 9.58, one has

$$\lim_{R_i \to \infty} \left(\frac{\mu_i^E}{v_i} \right) = P = k_B T \frac{\partial \Phi}{\partial n_3}. \tag{9.59}$$

The pressure of the homogeneous fluid can be obtained from $\Omega^E/V = -P^E = k_B T(\Phi - \sum_i \rho_i \partial \Phi/\partial \rho_i)$. This expression remains valid for the inhomogeneous fluid, provided that we replace ρ_i with n_l and the homogeneous pressure with the inhomogeneous local pressure, that is

$$P^E = k_B T \left(-\Phi + \sum_{l=0}^{3} n_l \frac{\partial \Phi}{\partial n_l} \right). \tag{9.60}$$

Combining Equations 9.59 and 9.60, one finally obtains

$$\frac{\partial \Phi}{\partial n_3} = -\Phi + \sum_{l=0}^{3} n_l \frac{\partial \Phi}{\partial n_l} + n_0, \tag{9.61}$$

where n_0 accounts for the ideal gas contribution to the pressure.

Proceeding in this way Rosenfeld obtained

$$\varphi_1 = -\ln(1 - n_3), \quad \varphi_2 = -\varphi_4 = \frac{1}{1 - n_3}, \quad \varphi_3 = -\frac{\varphi_5}{3} = \frac{1}{24\pi} \frac{1}{(1 - n_3)^2},$$
$$(9.62)$$

and consequently

$$\Phi(\{n_l(\mathbf{r})\}) = -n_0 \ln(1 - n_3) + \frac{n_1 n_2 - \mathbf{n}_1 \mathbf{n}_2}{1 - n_3} + \frac{1}{8\pi} \frac{n_2^3/3 - n_2(\mathbf{n}_2 \cdot \mathbf{n}_2)}{(1 - n_3)^2}, \quad (9.63)$$

which for the homogeneous fluid reduces to the SPT result. Using Equation 9.13, Rosenfeld also derived a general expression for the DCFs of the homogeneous fluid in terms of the weight functions $w_i^{(l)}(\mathbf{r})$ and showed that the second-order DCF is identical to the PY result.

The Rosenfeld FMT represents an improvement over previous DFTs for some properties, but not for others (see Tarazona et al. 2008 for a more detailed discussion). In particular, it fails to predict the freezing transition of a pure HS fluid. The situation much improves when *dimensional crossover* constraints are incorporated into the FMT. This is based on the fact that a DFT for a three-dimensional $(D = 3)$ HS fluid should give account of the density distributions of lower dimensional systems: a two-dimensional $(D = 2)$ hard-disk fluid and a one-dimensional $(D = 1)$ hard-rod system. This idea was first used by Tarazona et al. (1987) within the context of the WDA. The dimensional crossover concept was extended by Rosenfeld et al. (1996, 1997) to a zero-dimensional $(D = 0)$ cavity, a cavity that can contain at most a single sphere, for which the excess free energy can be determined exactly. Consequently, these authors proposed an empirical modification, which will be denoted here E-FMT, of the Rosenfeld expression (9.63) to give the correct $D = 0$ result. Tarazona and Rosenfeld (1997) further developed the cavity theory by introducing a procedure to derive free-energy functionals within the FMT on the basis of the exact free energy for the zero-dimensional cavity.

However, evidence was found by Rosenfeld et al. (1996, 1997) that the last term in Equation 9.63 should include tensor weighted densities, instead of only the scalar and vector weighted densities (9.54). To account for this fact, Tarazona (2000, 2002) proposed a *dimensional interpolation fundamental measure theory* (DI-FMT), which reproduces the exact density functional for hard rods derived by Percus (1976), is approximately correct for a zero-dimensional cavity, and reproduces the PY DCF and compressibility EOS for the homogeneous monocomponent HS fluid $(D = 3)$.

A procedure to derive a free energy functional that gives rise to an improved DCF of the homogeneous HS fluid was devised by González and White (2001) on the basis of the *generating function approach* (GFA) proposed by González et al. (1997). This is based, within the spirit of the SPT, on the ansatz

$$\Phi(\rho(\mathbf{r})) = \sum_{i=1}^{3} \frac{a_i}{R^{3-i}} \frac{\partial^i G(R; \rho(\mathbf{r}))}{\partial R^i}, \quad (9.64)$$

where

$$G(R; \rho(\mathbf{r})) = (1 - n_3)(\ln(1 - n_3) - 1) \qquad (9.65)$$

is a generating function inspired in the form of Φ for an ideal gas, as results from Equations 9.7 and 9.27. The coefficients in Equation 9.64 were determined by González et al. (1997) from the condition that in the uniform limit it must give rise to a prescribed EOS, such as the PY-c or the CS equations, and by González and White (2001) by imposing the right low-density behavior of the EOS and the DCF of the homogeneous fluid. In the first case, the $D = 3$ functional provides fairly satisfactory dimensional crossover behavior, in that it yields quite accurate results for the EOS of the $D = 2$ and $D = 1$ homogeneous systems and, when using the PY EOS, the exact $D = 0$ limit, although in some other aspects the theory provides worse results than the Rosenfeld FMT. In the second case, an improved DCF for the homogeneous HS fluid, compatible with the CS EOS through the compressibility route, is obtained.

The *White Bear (WB)* version of the FMT for HS mixtures, proposed by Roth et al. (2002),[*] is to some extent related to the DI-FMT and the GFA approaches. In the WB-FMT, the function Φ is considered of the form of Equation 9.56 with $\varphi_2 = -\varphi_4$ and $\varphi_3 = -\varphi_5/3$, like in the Rosenfeld FMT. Equation 9.60 is used to determine the functions φ_i with the pressure given by the BMCSL equation (7.91), whereas condition (9.61) is removed. Proceeding in this way the same expressions as in the Rosenfeld theory are obtained for φ_1 and φ_2, whereas φ_3 changes to

$$\varphi_3 = \frac{1}{36\pi} \frac{n_3 + (1 - n_3)^2 \ln(1 - n_3)}{n_3^2 (1 - n_3)^2}. \qquad (9.66)$$

Removing condition (9.61) gives rise to inconsistency in that the pressure obtained from $\partial\Phi/\partial n_3$ differs from that of the prescribed BMCSL equation. In order to improve consistency in this respect, Hansen-Goos and Roth (2006) combined the Rosenfeld procedure with a generalization of the PY EOS depending on a number of parameters that were determined from Equation 9.61 and the condition of recovering the CS equation for the pure HS fluid. The resulting EOS for HS mixtures was claimed to be more accurate than the BMCSL equation (7.91).

When applied to the monocomponent HS system, the WB-FMT suffers from similar problems as the Rosenfeld FMT. To partially overcome them, Roth et al. (2002) proposed either to use the empirical modification introduced by Rosenfeld et al. (1996, 1997) mentioned earlier in this section or to follow the same procedure as in the DI-FMT derived by Tarazona (2000, 2002). In the latter case, the WB-FMT for the pure HS system is equivalent to the CS version of the DI-FMT.

Thus far we have considered additive HS mixtures. A few comments on the extension of the FMT to NAHS mixtures, which are much more challenging than their additive counterpart, are relevant. Several extensions of the FMT have been

---

[*] The same approach, which is also known as *modified FMT* or *MFMT*, was independently derived by Yu and Wu (2002).

developed for particular nonadditive hard sphere (NAHS) mixtures, some of which will be discussed in Section 10.3. A generalization of the Rosenfeld FMT for the general case of NAHS mixtures with moderate nonadditivity has been developed by Schmidt (2004). The excess free energy functional is written as

$$\beta \mathscr{F}^E [\rho_1 (\mathbf{r}), \rho_2 (\mathbf{r})] = \int \sum_{i,j=0}^{3} K_{ij}^{(12)} (|\mathbf{r} - \mathbf{r}'|) \, \Phi_{ij} (\{n_l^{(1)} (\mathbf{r})\}, \{n_l^{(2)} (\mathbf{r}')\}) \, d\mathbf{r} d\mathbf{r}',$$

$$(9.67)$$

where $l = 0, 1, 2, 3$, $K_{ij}^{(12)}(r)$ is a function controlling the range on nonlocality between unlike species, and

$$n_l^{(k)} (\mathbf{r}) = \int \rho_k (\mathbf{r}) \, w_l^{(k)} (|\mathbf{r} - \mathbf{r}''|) \, d\mathbf{r}'', \quad k = 1, 2, \qquad (9.68)$$

with the Kierlik and Rosinberg (1990) scalar form adopted for the weights $w_l^{(l)}(r)$. Both sets of functions $w_l^{(k)}(r)$ and $K_{ij}^{(12)}(r)$ are determined in such a way that Equation 9.50 is reproduced exactly up to the second-order term, and the theory reduces to the Rosenfeld FMT for nonadditivity parameter $\Delta = 0$.

To end this section, we will mention the *soft fundamental measure theory* (SFMT) developed by Schmidt (1999, 2000a,b) as an extension of the FMT to soft interactions, by introducing temperature-dependent weighted densities. An improved version of the SFMT as well as a unified description for both hard and soft interactions was devised by Sweatman (2002).

9.4 SIMPLE FLUIDS AND SOLIDS

One of the earlier successful applications of the DFT was the prediction of the fluid–solid transition of the HS and other simple fluids. To this end, one has to first determine the equilibrium density of the solid. This can be achieved by minimizing the difference between the grand potentials in both phases, that is

$$\frac{\delta \Delta \Omega [\rho (\mathbf{r})]}{\delta \rho (\mathbf{r})} = 0. \qquad (9.69)$$

Then, the coexistence densities can be determined from the condition of equal chemical potentials and pressures in both phases at a given temperature or, equivalently, $\Delta \Omega = 0$ for fixed T and μ. This requires an expression for the density $\rho(\mathbf{r})$ of the solid as well as, at least, the second-order DCF $c_0^{(2)}(r)$ of the fluid.

9.4.1 LOCAL DENSITY OF THE SOLID

The local density of the solid is often approximated by a sum of Gaussians like in Equation 5.96, which we will rewrite here as

$$\rho (\mathbf{r}) = \sum_i \tilde{\rho} (\mathbf{r} - \mathbf{r}_i), \quad \tilde{\rho} (\mathbf{r} - \mathbf{r}_i) = \left(\frac{\gamma}{\pi}\right)^{3/2} e^{-\gamma (\mathbf{r} - \mathbf{r}_i)^2}, \qquad (9.70)$$

with the parameter γ determined from minimizing the free energy of the solid for a given density. At high densities, the Gaussians become narrow (γ large) and so neighboring Gaussians do not overlap. In this case, the first term in the right-hand side of Equations 9.25 and 9.26 takes the particularly simple form

$$\int \rho\left(\mathbf{r}\right) \ln\left[\frac{\rho\left(\mathbf{r}\right)}{\rho_0}\right] d\mathbf{r} = N\left[\frac{3}{2}\ln\left(\frac{\gamma}{\pi}\right) - \ln\rho_0 - \frac{3}{2}\right],$$ (9.71)

where ρ_0 is the density of the homogeneous fluid.

A more general method is based on the Fourier expansion of the local density of the solid

$$\rho\left(\mathbf{r}\right) = \rho_S + \sum_{\mathbf{k}\neq 0}\rho_{\mathbf{k}}e^{i\mathbf{k}\cdot\mathbf{r}},$$ (9.72)

where
  $\{\mathbf{k}\}$ is the set of reciprocal lattice vectors for the particular lattice structure considered
  $\rho_{\mathbf{k}}$ are the Fourier components of the density
  ρ_S is the average density of the solid

Some care must be taken in performing the sum (see Harrowell et al. 1985). Expressions (9.70) and (9.72) not only can deal with different lattice structures by introducing the appropriate lattice vectors $\mathbf{r}_i$ in the former and the reciprocal lattice vectors $\{\mathbf{k}\}$ in the latter, but also can be used for the local density of amorphous (glassy) solids provided that we introduce the vectors corresponding to the random structure appropriate for the amorphous solid of interest. However, it is to be noted that, whereas for any substance there is a limited number of crystalline structures, there is an unlimited variety of amorphous structures.

9.4.2 *n*-PARTICLE DIRECT CORRELATION FUNCTIONS OF THE HOMOGENEOUS FLUID

Concerning the (two-particle) DCF of the homogeneous fluid, for the HS fluid apart from the PY expression (4.15), there are available a number of analytical expressions, such as the accurate semi-empirical expression reported by Henderson and Grundke (1975) and the Tang (2003) expression derived from the FMSA. In some situations, higher-order DCFs may be needed, as we have seen in Section 9.2. Barrat et al. (1987a) approximate the third-order DCF by

$$c_0^{(3)}\left(\mathbf{r}_1, \mathbf{r}_2, \mathbf{r}_3\right) \equiv c_0^{(3)}\left(\mathbf{r}, \mathbf{r}'\right) = l\left(r\right)l\left(r'\right)l\left(\left|\mathbf{r} - \mathbf{r}'\right|\right),$$ (9.73)

where
  $\mathbf{r} = \mathbf{r}_1 - \mathbf{r}_2$
  $\mathbf{r}' = \mathbf{r}_1 - \mathbf{r}_3$

$l(r)$ is an arbitrary function to be determined from the exact condition

$$\frac{\partial c_0^{(n-1)}\left(\mathbf{r}_1,\ldots,\mathbf{r}_{n-1}\right)}{\partial \rho} = \int c_0^{(n)}\left(\mathbf{r}_1,\ldots,\mathbf{r}_n\right)d\mathbf{r}_n, \qquad (9.74)$$

which is easily derived from Equations 9.13 and 9.14 and, for $n=3$ with the approximation (9.73) gives

$$\frac{\partial c_0^{(2)}\left(\mathbf{r}\right)}{\partial \rho} = \int c_0^{(3)}\left(\mathbf{r},\mathbf{r}'\right)d\mathbf{r}' = l\left(r\right)\int l\left(r'\right)l\left(\left|\mathbf{r}-\mathbf{r}'\right|\right)d\mathbf{r}', \qquad (9.75)$$

whose solution allows us to obtain the three-particle DCF of the reference fluid if we know the second-order one.

Iyetomi and Ichimaru (1988) used the approximation

$$c_0^{(3)}\left(r_1,r_2,r_3\right) = h\left(r\right)h\left(r'\right)h\left(\left|r-r'\right|\right). \qquad (9.76)$$

Curtin and Ashcroft (1987) proposed to use Equations 9.13 and 9.14, with the WDA approximation for $\mathscr{F}^E\left[\rho(\mathbf{r})\right]$, to obtain $c_0^{(n)}(\mathbf{r}_1,\ldots,\mathbf{r}_n)$, with $n>2$, from any known $c_0^{(2)}(\left|\mathbf{r}_1-\mathbf{r}_2\right|)$. A related procedure was derived by Denton and Ashcroft (1989b), based on the approximation $\mathscr{C}(\mathbf{r}) \approx c_0^{(1)}(\mathbf{r};\bar{\rho}(\mathbf{r}))$, with $\bar{\rho}(\mathbf{r})$ given by the WDA Equation 9.31 with the weighting function determined from the condition that Equation 9.12 must be satisfied in the uniform limit, that is

$$\lim_{\rho(\mathbf{r})\to\rho_0}\left[\frac{\delta\mathscr{C}\left(\mathbf{r}\right)}{\delta\rho\left(\mathbf{r}'\right)}\right] = c_0^{(2)}\left(\left|\mathbf{r}-\mathbf{r}'\right|;\rho_0\right). \qquad (9.77)$$

This yields the simple result

$$w\left(r;\rho_0\right) = \frac{c_0^{(2)}\left(r,\rho_0\right)}{\partial c_0^{(1)}\left(\rho_0\right)/\partial\rho_0}. \qquad (9.78)$$

The same result was derived by Zhou and Ruckenstein (2000a) with $\bar{\rho}(\mathbf{r})$ given by the SWDA instead of the WDA. Once $\mathscr{C}(\mathbf{r})$ is known, higher-order DCFs can be obtained from successive functional derivatives with respect to $\rho(\mathbf{r})$ using Equations 9.13 and 9.14. With this procedure Zhou and Ruckenstein derived the general expression

$$c_0^{(n)}\left(\mathbf{r},\mathbf{r}_1,\ldots,\mathbf{r}_{n-1};\rho_0\right) = \frac{\partial^{n-1}c_0^{(1)}\left(\rho_0\right)/\partial\rho_0^{n-1}}{\left[\partial c_0^{(1)}\left(\rho_0\right)/\partial\rho_0\right]^n}$$

$$\times\int c_0^{(2)}\left(\mathbf{r}',\mathbf{r};\rho_0\right)c_0^{(2)}\left(\mathbf{r}',\mathbf{r}_1;\rho_0\right)\cdots c_0^{(2)}\left(\mathbf{r}',\mathbf{r}_{n-2};\rho_0\right)d\mathbf{r}'$$
$$(9.79)$$

for the n-particle DCFs of the uniform fluid with $n \geq 3$.

In Fourier space, Equation 9.74 takes the simple form

$$\frac{\partial \hat{c}_0^{(n-1)}(\mathbf{k}_1, \dots, \mathbf{k}_{n-1})}{\partial \rho} = \hat{c}_0^{(n)}(\mathbf{k}_1, \dots, \mathbf{k}_{n-1}, 0). \qquad (9.80)$$

Khein and Ashcroft (1999a) proposed to use this condition, instead of Equation 9.74, to obtain the higher-order DCFs, thus avoiding the need of solving an integral equation like Equation 9.75. Using suitable ansatzs for the mathematical form of $\hat{c}_0^{(3)}(\mathbf{k}, \mathbf{k}')$, they derived two accurate analytical expressions for the third-order DCF in terms of the second-order DCF.

Next we will give a brief survey of the results provided by some of the density functional approximations described in the previous sections for the solid–fluid coexistence densities and the thermodynamic properties of the solid phase for some simple systems.

9.4.3 FREEZING OF HARD SPHERES

Haymet and Oxtoby (1981) derived a DFT, which in particular reproduces the results of a non-DFT approach previously developed by Ramakrishnan and Yussouff (1979), for the interfacial properties at the crystal–liquid interface. Based on similar grounds, Haymet (1983, 1985) developed a theory for the freezing of HS. The starting point in the latter is the ratio of the density of the solid to that of the uniform fluid at coexistence in the absence of external potential which, from Equation 9.8, can be expressed as

$$\ln\left[\frac{\rho(\mathbf{r})}{\rho_0}\right] = \mathscr{C}(\mathbf{r}) - c_0^{(1)}(r). \qquad (9.81)$$

Combining this equation with Equation 9.23 and the expansion (9.72) for the density of the solid, one obtains a set of coupled implicit equations for ρ_S and the parameters $\rho_\mathbf{k}$. Solutions with $\rho_S = \rho_0$ and $\rho_\mathbf{k} = 0$ for $\mathbf{k} \neq 0$ will correspond to the fluid phase. For some temperature and density ranges, there may be solutions with $\rho_S \neq \rho_0$ and $\rho_\mathbf{k} \neq 0$ for $\mathbf{k} \neq 0$, which will correspond to crystal phases. Inserting Equation 9.23 into Equation 9.26 allows us to obtain the coexistence densities from the condition of equal pressures in both phases, as obtained from Equation 9.26, provided that we know at least the two-particle DCF of the homogeneous fluid.

Haymet and Oxtoby (1986) used this theory, truncated at first order in expansion (9.23), to analyze the freezing of HS. They considered both the PY DCF with the PY compressibility EOS and the Henderson and Grundke (1975) semi-empirical DCF with the CS EOS (5.21). Very good agreement with simulation data was obtained for the coexistence densities, especially with the second choice, as shown in Table 9.1.

Tarazona (1984) applied a WDA to the freezing of HS. To this end, the excess free energy per particle was obtained from the CS equation (5.21) and a

TABLE 9.1

Coexistence Properties for the Hard-Sphere Fluid-to-fcc-Solid Transition

	η_F	η_S	$\Delta\eta/\eta_F$	P^*	
Sim[a]	0.494	0.545	0.103	11.70	
HO[b]	0.506	0.541	0.069		(PY)
HO[b]	0.495	0.543	0.091		(CS)
WDA[c]	0.494	0.555	0.125		(CS)
WDA[d]	0.480	0.547	0.141		(PY)
MWDA[e]	0.476	0.542	0.138		(PY)
MWDA[f]	0.461	0.538	0.168		(CS)
OMWDA3[g]	0.500	0.555	0.111		(CS)
EMA[f]	0.479	0.531	0.109		(PY)
EMA[f]	0.517	0.519	0.126		(CS)
ELA[h]	0.520	0.567	0.090	16.09	(k,PY)
ELA[i]	0.518	0.564	0.089		(m,PY)
ELA[i]	0.513	0.561	0.094		(k,CS)
ELA[i]	0.511	0.556	0.088		(m,CS)
MELA[j]	0.484	0.538	0.112	11.2	(PY)
MELA[j]	0.508	0.560	0.102	13.3	(CS)
GELA[j]	0.472	0.522	0.106	10.3	(PY)
GELA[j]	0.495	0.545	0.101	11.9	(CS)
E-FMT[k]	0.491	0.540	0.100	11.9	
DI-FMT[l]	0.467	0.516	0.104		(PY)
DI-FMT[l]	0.489	0.536	0.0953		(CS)

$P^* = P\sigma^3/k_B T$ is the reduced pressure at coexistence. PY refers to the use of the Percus–Yevick DCF and EOS. k indicates that the effective density $\bar{\rho}$ was determined from the condition $k_{min}(\rho) = k(\bar{\rho})$ and m that was determined by minimizing $\Delta\mathscr{F}$. CS indicates the use of an accurate semi-empirical DCF and the CS equation (5.21). The effective density (9.45) was used in the MELA.

[a] Hoover and Ree (1968).
[b] Haymet and Oxtoby (1986).
[c] Tarazona (1985).
[d] Curtin and Ashcroft (1985) (see also ref. e).
[e] Denton and Ashcroft (1989a).
[f] Likos and Ashcroft (1993).
[g] Khein and Ashcroft (1999b).
[h] Baus and Colot (1985a).
[i] Colot et al. (1986).
[j] Lutsko and Baus (1990b).
[k] Rosenfeld et al. (1996).
[l] Tarazona (2002).

density-independent step function, previously introduced by Nordholm et al. (1980), was used for $w(r)$, namely,

$$w(r) = \frac{3}{4\pi\sigma^3}\Theta(\sigma - r).\qquad(9.82)$$

The local density of the solid was assumed to be of the form of Equation 9.70 with the parameter γ determined by minimizing the free energy per particle at a given density, with densities for which $\gamma = 0$ corresponding to the fluid phase and those with $\gamma \neq 0$ to the solid phase. For certain density range, two solutions appeared: one with $\gamma = 0$ and another with $\gamma \neq 0$, and the coexistence properties were determined from the usual conditions of equal pressures and chemical potentials in both phases. The coexistence densities obtained in this way were only in modest agreement with the simulation data because of the simple ansatz adopted for $w(r)$.

A much better performance was achieved by Tarazona (1985) by using a density-dependent weighting function of the form

$$w(r) = w_0(r) + w_1(r)\rho + w_2(r)\rho^2.\qquad(9.83)$$

$w_0(r)$ and $w_1(r)$ were determined from the condition that Equation 9.13 for $n = 2$, in combination with Equations 9.14, 9.30, and 9.31, must give the exact zero- and first-order terms in the density expansion of the DCF of the HS fluid. The expression for $w_0(r)$ thus obtained is the same as in Equation 9.82, whereas $w_1(r)$ was calculated numerically and fitted to a simple expression. $w_2(r)$ was empirically adjusted to give a good description of $c(r)$ for the whole density range of the HS fluid. As shown in Table 9.1, the predicted coexistence densities with this version of the WDA are in close agreement with the simulations. Comparably accurate results, also included in the table, were obtained by Curtin and Ashcroft (1985) (see also Denton and Ashcroft 1989a) from their general relationship between $w(r)$ and $c^{(2)}(r)$, and by Denton and Ashcroft (1989a) from the MWDA; in both cases, the PY expression (5.15) was used for the DCF and the corresponding EOS (5.17) for the fluid. In contrast, the calculations of Denton and Ashcroft (1989a) reveal that both the WDA and MWDA theories, and especially the latter, considerably underestimate the compressibility factor along the solid branch near melting, as illustrated in Figure 9.1.

The extended MWDA with the third-order DCF $c_0^{(3)}(r)$ of the reference fluid as calculated by the Denton and Ashcroft (1989b) procedure (see earlier in this section), was applied by Likos and Ashcroft (1993) to the study of the freezing of HS. They used two approximate expressions for $c_0^{(2)}$: the PY solution, and the corresponding PY-c EOS, and the accurate semi-empirical expression, derived by Henderson and Grundke (1975), and the CS EOS (5.21). The results of this EMA are in poorer agreement with the simulations for the coexistence densities than those of the MWDA, as we can see in Table 9.1, but in contrast it considerably improves the predicted values of the EOS of the solid at low densities, as illustrated in Figure 9.1, especially when using the PY DCF. The introduction of the Barrat et al. (1987a) approximation, Equation 9.73, for the third-order DCF within the EMA resulted in the failure of predicting a fluid to fcc-solid transition.

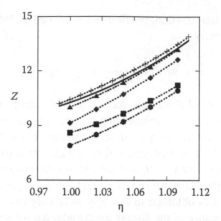

FIGURE 9.1 Compressibility factor Z for the fcc HS solid near melting from different density functional theories as compared with the Hall (1972) parametrization of the simulation data (continuous curve). Circles: MWDA with the CS equation, from Likos and Ashcroft (1993). Squares: *Optimized modified weighted density approximation* (OMWDA), from Khein and Ashcroft (1997). Triangles: OMWDA3, from Khein and Ashcroft (1999b). Crosses: DI-FMT, from Tarazona (2000; personal communication, 2012). The dotted curves are guides for the eye.

By determining b_0 in Equation 9.38 from the exact condition $\beta F^E/N = 1$ in the close packing limit $\rho^* = \sqrt{2}$, some improvement was achieved by Khein and Ashcroft (1997) with the generalized density functional theory particularized for the MWDA and optimized in this way (OMWDA), as compared with the predictions of the ordinary MWDA. The generalized density functional theory was carried out to third order in the DCF by Khein and Ashcroft (1999b), using an accurate $c_0^{(3)}(r)$ derived by them (Khein and Ashcroft 1999a; see earlier in this section), giving rise to a remarkable improvement in the predicted coexistence densities (see Table 9.1) and the pressure near melting (see Figure 9.1).

Baus and Colot (1985a) applied the ELA to the freezing of HS with the effective density $\bar{\rho}$ determined from equating the lowest reciprocal lattice vector $k_{min}(\rho)$ of the solid with density ρ to the position $k(\bar{\rho})$ of the main peak of the static structure factor of the reference fluid with density $\bar{\rho}$. For the DCF of the HS fluid, they used the PY solution (see Section 5.3) and for the local density of the solid they assumed a form like in Equation 9.70 with the parameter γ determined by minimizing the grand potential. The results obtained in this way for the coexistence properties are in fairly good agreement with the simulation data, as shown in Table 9.1, with more accurate prediction for the coexistence pressure using the PY virial EOS than using the PY compressibility one. As a matter of fact, the latter choice gives rise to too high values for the pressure along the whole solid branch (Colot and Baus 1985). Instead, Colot et al. (1986) applied Equation 9.41 with the effective density $\bar{\rho}$ determined by minimizing $\mathscr{F}[\rho(\mathbf{r})]$. The results obtained in this way for the coexistence densities with the PY solution for the DCF were nearly the same as those obtained from the condition $k_{min}(\rho) = k(\bar{\rho})$. Slightly better agreement with simulation data was achieved by using for the DCF a semi-empirical expression, compatible with the CS EOS (5.21), instead of the PY one (see Table 9.1).

The performance of the MELA for the coexistence properties is better than that of the ELA, but still better is the accuracy of the GELA; as a matter of fact, the results provided with the latter are nearly in perfect agreement with the simulations, as seen in Table 9.1, when the Carnahan–Starling equation (5.21) is used for the fluid. MELA and GELA also provide excellent agreement with the simulation data for the pressure along the solid branch near melting.

On the other hand, as mentioned previously, the Rosenfeld FMT fails to predict the freezing transition of a pure HS fluid. Instead, Rosenfeld et al. (1996, 1997) obtained with the E-FMT very good predictions for the solid–fluid coexistence densities and pressure of the monocomponent HS system, as shown in Table 9.1. Moreover, although the calculated pressures for the fcc solid near melting were somewhat high, the theory proved to be able to predict the divergence of the pressure at close packing, in contrast with previous WDAs.

The DI-FMT provides excellent accuracy for the EOS of the fcc HS solid, as depicted in Figure 9.1, and also yields the correct behavior near close packing (the first two terms of the right-hand side of Equation 5.95), but fails to give accurate estimates of the coexistence densities (see Table 9.1). The reason for this shortcoming is precisely the very accurate description of the solid combined with a less accurate description of the fluid as given by the PY compressibility EOS (Tarazona 2002). Forcing the DI-FMT to reproduce the CS EOS instead of the PY-c one results in a considerable improvement in the predicted coexistence densities (see Table 9.1), but at the expense of a considerable worsening in the quality of the crystal description, as found by Tarazona (2002). Neither is the predicted DCF improved in this way.

9.4.4 RELATIVE STABILITY OF THE SC, BCC, FCC, AND HCP PHASES OF HARD SPHERES

In the preceding discussion, we had focused on the fluid-to-fcc-solid transition and the thermodynamic properties of the fcc solid. However, the DFTs considered can be applied to the study of other crystalline phases provided that we introduce the appropriate structure through the Gaussian approximation (Equation 9.70) or the expansion (Equation 9.72) for the solid density. Thus, for example, the ELA was used by Baus and Colot (1985b) and Colot and Baus (1985) to analyze the stability of sc, bcc, fcc, and hcp phases of HS. They concluded that the sc phase is always mechanically unstable; the bcc phase is mechanically stable for $0.62 \leq \eta \leq 0.68$, the latter being the close packing fraction for the bcc lattice, but is always thermodynamically metastable with respect to the fluid; the fcc phase is mechanically stable for $\eta \geq 0.50$, thermodynamically metastable for $0.504 < \eta < 0.548$, and thermodynamically stable for $\eta > 0.548$ up to the fcc close packing fraction $\eta = 0.74048$. The theory was not able to clearly determine which of the two compact lattices, fcc or hcp, is the most stable one, partially due to the use of the PY DCF in addition to the fact that the simulation data reveal that the difference in the free energies of the two lattices is very small.

In contrast, using the WDA, Curtin and Runge (1987) found the bcc phase at high densities to be stable with respect to the fluid but metastable with respect to the fcc phase, whereas at lower densities the fluid is more stable than the bcc solid, which in turn is more stable than the fcc solid. Similar conclusion was reached by Lutsko and Baus (1990b) from the GELA. The latter authors also analyzed the relative stability of the sc lattice, with the result that at high densities the sc phase is metastable with respect to both the bcc and fcc phases but stable with respect to the fluid phase, whereas at low densities the sc phase is the most stable of the solid phases but metastable relative to the fluid. However, these results must be considered with caution because of the scarcity of the simulation data available for the bcc HS solid, although both theories are in close agreement with some simulation data for the free energy of the bcc solid reported by Curtin and Runge (1987). Moreover, as noted by Lutsko and Baus, the theoretical results for the bcc solid are more sensitive to the accuracy of the EOS used for the fluid than those for the fcc phase, and it is to be expected that the sensitivity of the thermodynamic properties of the sc solid will be even greater. In any case, it may be of interest to know the availability of DFTs capable of describing solid phases other than the fcc one for the HS system, as these phases may be useful as reference systems in perturbation theory for the thermodynamic properties of more realistic solids with stable sc or bcc structures.

On the other hand, Lutsko (2006) analyzed the performance of the E-FMT, DI-FMT, and WB-FMT versions for the structure and thermodynamic properties of the fcc, hcp, bcc, and sc HS crystals. For the fcc phase, it was found that the three theories predict a minimum in the free energy for $\gamma = 0$, which corresponds to the fluid phase, even for $\eta > 0.6366$, corresponding to the RCP density (see Section 5.1). This unphysical behavior was attributed to the use of the PY and CS equations of state that diverge at $\eta = 1$, a packing fraction unreachable to the HS system. Another unphysical behavior was found in the E-FMT, for nonzero values of γ, consisting in the presence of two minima in the free energy within certain range of densities near the random close packing density. In any case, the predicted Lindemann parameter, which is the ratio of the root-mean-square displacement of the particles to the nearest neighbor distance, was seen to decrease toward zero for all the three theories as the density approached the fcc close-packing density, corresponding to $\eta = 0.74048$, as it should be. None of these theories was able to clearly distinguish between the free energies of the fcc and hcp phases. For the bcc phase, all the three theories predicted more than one minimum for $\gamma \neq 0$ within some high-density ranges, with the Lindemann parameter for the low-γ solution first decreasing with increasing density and then increasing again in approaching the bcc close packing, corresponding to $\eta = 0.6802$, and the high-γ solution displaying the right behavior by vanishing at close packing. However, all of these theories again showed some unphysical behavior, as in both the E-FMT and the WB-FMT the low-γ solution resulted to be the most stable one and in the DI-FMT the Lindemann parameter was found to increase with density at intermediate densities and to present a discontinuous jump at $\eta \sim 0.60$. Finally, for the sc phase, the calculations revealed that none of these theories predicts multiple minima, the E-FMT gives stable solutions for $\eta \gtrsim 0.45$ up to the neighborhood of the sc close packing density, corresponding to $\eta = 0.5236$, the DI-FMT only provides stable solutions with $\gamma \neq 0$ for densities very near close packing, and no stable solutions are found with the WB-FMT.

The picture that emerges from the preceding discussion is that, although the different versions of the FMT for the HS fluid represent a considerable improvement over other DFTs, they are not completely satisfactory in some aspects. The shortcomings and limitations of the different FMT versions have been discussed by Cuesta et al. (2002) and Tarazona et al. (2008).

9.4.5 HARD-SPHERE GLASS

The HS glass and the glass transition have been studied by means of DFT by several authors. Thus, Singh et al. (1985) used expansion (9.24) truncated at second order to describe this transition. For the DCF of the HS fluid, they considered the PY solution, the Henderson and Grundke (1975) semi-empirical approximation, and a modification of the latter to improve the tail of the DCF. For the density of the glass, they assumed the Gaussian form (Equation 9.70) with the lattice sites corresponding to a random close packing of HS obtained by Bennett (1972) from computer simulation. With the PY DCF, the theory failed to give a minimum in the free energy for any reasonable value of γ in Equation 9.70, in contrast with the situation when using a more accurate approximation for the DCF. In the latter case, minima appeared for $\eta \gtrsim 0.539$. The free energy per particle at the minima was determined as a function of the density and the curve was found to cross that for the uniform fluid at $\eta = 0.597$, reasonably close to the value $\eta \approx 0.571$ at which is located the glass transition in some simulations, as mentioned in Section 5.1. Using this same approach, but with the Henderson and Grundke (1975) DCF and the CS EOS for the homogeneous fluid, Kaur and Das (2001) noticed the appearance of two different inhomogeneous noncrystalline structures: one, the ordinary glass, with high values of γ (highly localized particles), and the other with lower values of γ (weakly localized particles), the latter being more stable than the homogeneous fluid for $\eta > 0.576$ and the former for $\eta > 0.597$.

Another approach was derived by Baus and Colot (1986) within the framework of the ELA with the Gaussian density (Equation 9.70) and the high-density approximation (9.71). Their expression for the free energy difference between the glass and the uniform fluid is

$$\frac{\beta \Delta \mathscr{F} [\rho(\mathbf{r})]}{N} = \left[\frac{3}{2} \ln \left(\frac{\gamma}{\pi} \right) - \ln \rho_0 - \frac{3}{2} \right] + \frac{\rho_0}{2} \int c_0^{(1)} (|\mathbf{r}| ; \bar{\rho}) \, d\mathbf{r}$$

$$- \frac{1}{2} \int \int c_0^{(2)} (|\mathbf{r} - \mathbf{r}'| ; \bar{\rho}) \, \tilde{\rho}(\mathbf{r}) \, \tilde{\rho}(\mathbf{r}') \, d\mathbf{r} d\mathbf{r}'$$

$$- \frac{\rho_0}{2} \int g(|\mathbf{R}|) \, d\mathbf{R} \int \int c_0^{(2)} (|\mathbf{r} - \mathbf{r}'| ; \bar{\rho}) \, \tilde{\rho}(\mathbf{r}') \, \tilde{\rho}(\mathbf{r} - \mathbf{R}) \, d\mathbf{r} d\mathbf{r}'. \tag{9.84}$$

Here, ρ_0 is the density of the homogeneous fluid, equal to the average density of the glass, and $g(|\mathbf{R}|)$ is the site–site pair correlation function of the glass; the term between brackets in the right-hand side is the high-density approximation (9.71), the next term arises from the difference in the free energy of the fluid of density ρ_0 and the effective fluid of density $\bar{\rho}$, the third term is the contribution of any lattice site chosen as the central site, and the last term is the contribution from two different sites. Baus and Colot (1986) used the PY approximation for the DCF of the reference fluid and the Bennett (1972) simulation data for the site–site correlation function $g_B(|\mathbf{R}|)$ of the glass at random close packing. The site–site pair correlation function of the glass was obtained by scaling as $g(|\mathbf{R}|) = g_B[(\eta/\eta_{RCP})^{1/3}R]$, with the random close packing density considered as an adjustable parameter. They found that the glass is mechanically stable for $\eta \gtrsim 0.56$ and is thermodynamically stable with respect to the fluid for $\eta > 0.62$, provided that the packing fraction corresponding to the random close packing is $\eta_{RCP} > 0.69$ (see Section 5.1 for more details about random packings of HS), but in any case the glass is metastable with respect to the fcc solid, which seems to be in agreement with the simulations.

Löwen (1990) applied a similar treatment to the glass within the MWDA theory. The weighting function was determined from the condition of reproducing exactly the DCF of the homogeneous system as given by the PY approximation. As before, it was found that the fcc solid is always more stable than the glass, but the latter becomes stable at high densities for high values of η_{RCP}. The fluid–glass transition was found to be first order and to take place for $\eta_{RCP} > 0.672$. Using this same DFT, but with the Henderson and Grundke (1975) DCF and the CS EOS for the homogeneous fluid, Kaur and Das (2002) found that for $\eta_{RCP} \sim 0.70$ two different minima appeared in the free energy per particle: one for high values of γ, corresponding to a system with highly localized particles, the ordinary glass, and the other for low values of γ, corresponding to a much more weakly localized structure, thus confirming the previous findings of the same authors on the basis of expansion (9.24), as mentioned earlier in this section. The weakly localized structure only appeared in the disordered phase, but not in the crystal phase, and showed to be more stable than the ordinary glass up to a packing fraction $\eta \approx 0.539$. Another remarkable finding was the existence of a crossover value of the average density of the inhomogeneous system below which only the minimum corresponding to low γ was observed and this crossover density increased with the value of η_{RCP} considered.

9.4.6 HARD-SPHERE MIXTURES

Equations 9.24 through 9.26 are readily extended to m-component mixtures. Thus, for example, the latter equation takes the form

$$\beta \Delta \Omega [\{\rho_i(\mathbf{r})\}] = \sum_{i=1}^{m} \int \rho_i(\mathbf{r}_1) \ln \left[\frac{\rho_i(\mathbf{r}_1)}{\rho_{0i}} \right] d\mathbf{r}_1 - \sum_{i=1}^{m} \int \Delta \rho_i(\mathbf{r}_1) d\mathbf{r}_1$$

$$- \frac{1}{2} \sum_{i,j=1}^{m} \int \int c_{0ij}^{(2)}(\mathbf{r}_1, \mathbf{r}_2; \rho_0) \Delta \rho_i(\mathbf{r}_2) \Delta \rho_j(\mathbf{r}_1) d\mathbf{r}_1 d\mathbf{r}_2 + \cdots$$

$$- \frac{1}{6} \sum_{i,j,k=1}^{m} \int \int \int c_{0ijk}^{(3)}(\mathbf{r}_1, \mathbf{r}_2, \mathbf{r}_3; \rho_0) \Delta \rho_i(\mathbf{r}_1) \Delta \rho_j(\mathbf{r}_2)$$

$$\times \Delta \rho_k(\mathbf{r}_3) d\mathbf{r}_1 d\mathbf{r}_2 d\mathbf{r}_3 + \cdots, \qquad (9.85)$$

where

$\rho_i(\mathbf{r})$ is the local density of component i in the inhomogeneous mixture

ρ_{0i} that in the homogeneous reference mixture, $\Delta\rho_i(\mathbf{r}) = \rho_i(\mathbf{r}) - \rho_{0i}$

$c_{0ij}^{(2)}(\mathbf{r}_1, \mathbf{r}_2; \rho_0)$ and $c_{0ijk}^{(3)}(\mathbf{r}_1, \mathbf{r}_2, \mathbf{r}_3; \rho_0)$ are the second- and third-order DCF of the reference mixture, respectively

When the inhomogeneous mixture is a crystalline solid, the local density of component i may be approximated by a sum of Gaussians in a similar way as for a monocomponent solid, with the parameters γ_i again determined from minimizing the free energy. The equilibrium densities $\rho_i(\mathbf{r})$ of the inhomogeneous phase and the coexistence densities are determined in a similar way as for the monocomponent systems.

Using Equation 9.85 truncated at third order, Smithline and Haymet (1987) studied the freezing of equimolar binary mixtures of HS, with the PY expression for the second-order DCFs $c_{0ij}^{(2)}(\mathbf{r}, \mathbf{r}'; \rho_0)$ and an approximation for the third-order DCFs. Different choices for the crystalline structures, required as an input for the theory, were considered, and the coexistence densities as well as the crystal phases were determined as a function of the relative diameters. Denton and Ashcroft (1990) extended to binary mixtures of HS the WDA and MWDA approximations and applied the latter theory to predict the fluid–solid phase transition as a function of the concentration for several diameter ratios and crystalline structures in very good agreement with the simulation data.

On the other hand, the performance of the Schmidt (2004) extension of the FMT to NAHS mixtures was analyzed by the same author for several mixtures of this kind. Schmidt obtained the partial DCFs $c_{ij}^{(2)}(r)$ from the equivalent of Equation 9.12 for mixtures in the homogeneous limit, and then the partial RDFs $g_{ij}(r)$ by means of the OZ equation (7.47) for mixtures. A remarkable agreement with the simulations for the $g_{ij}(r)$ obtained in this way was reported for the diameter ratio $\sigma_1/\sigma_2 = 2$ and nonadditivity parameter $\Delta = 0.2$ at supercritical conditions. The theory was also found to predict the critical density of the demixing transition in reasonable agreement with simulations for symmetric mixtures with $\Delta > 0$, and even for quite extreme diameter ratios the results were qualitatively correct. The predictions of this theory for the fluid–fluid coexistence for NAHS mixtures with size ratio $\sigma_1/\sigma_2 = 10$ were analyzed by Hopkins and Schmidt (2010) for $\Delta = 0.2, 0.3, 0.4$, and 0.5, with the result that, although qualitatively correct, the theory considerably underestimates the coexistence pressure and the deviation from simulations increases with Δ.

Rosenfeld (1994) tested the capability of the self-consistent reference functional approach to predict the phase separation of AHS mixtures with the reference functional given by the FMT. Kahl et al. (1996) applied the same theory to obtain the partial RDFs $g_{ij}(r)$ of equal-sized binary NAHS mixtures with different compositions and nonadditivities. They used an iterative procedure starting from the HNC solution for the reference HS fluid, imposing test particle self-consistency, and using the optimizing condition (7.117) for the reference system parameters, obtaining in this way good agreement with simulations.

9.4.7 Systems with Soft and Attractive Potentials

In a similar way as in the free energy perturbation theory for homogeneous fluids, the simplest approach to deal with inhomogeneous systems with soft and attractive potentials is to split the potential into reference $u_0(r)$ and perturbation $u_1(r)$ contributions, approximate the free energy functional for the reference system with that of a system of HS with suitable diameters, and treat the perturbation in the MFA, as given by Equation 9.17 with $u(r)$ replaced with $u_1(r)$. Introducing the resulting functional into Equation 9.12, in the homogeneous limit one obtains the RPA approximation (Equation 4.44). When more accuracy is needed, one may resort to expanding Equation 9.19 in power series of the coupling parameter α, in a similar way as for homogeneous fluids (see Section 6.1) but for the fact that now inhomogeneous correlation functions are involved. The latter may be approximated with those for the homogeneous reference fluid, eventually calculated at an effective density. It is worth to comment some of the results achieved with different extensions and applications of the DFT to the phase equilibria and bulk properties of systems with potentials other than the HS potential. In the analysis next, the cases for the SW, HCY, SS, and LJ potentials are considered.

9.4.7.1 Square-Well Potential

A *perturbation weighted density approximation* (PWDA) was proposed by Mederos et al. (1993) in the form

$$\mathscr{F}[\rho(\mathbf{r})] = \mathscr{F}_0[\rho(\mathbf{r})] + \frac{1}{2} \int \int \rho_0^{(2)}(\mathbf{r}, \mathbf{r}'; [\rho(\mathbf{r})]) u_1(|\mathbf{r} - \mathbf{r}'|) d\mathbf{r} d\mathbf{r}', \qquad (9.86)$$

that results from expanding the last term of Equation 9.19 in power series of the coupling parameter α and truncating at first order. The HS fluid was chosen as the reference one with free energy functional as given by the WDA, and the pair density distribution was approximated with that of the reference fluid with effective density $\hat{\rho}$, which in general will depend on the position, that is

$$\rho_0^{(2)}(\mathbf{r}, \mathbf{r}'; [\rho(\mathbf{r})]) \approx \rho(\mathbf{r}) \rho(\mathbf{r}') g_0(|\mathbf{r} - \mathbf{r}'|; \hat{\rho}). \qquad (9.87)$$

The effective density $\hat{\rho}$ is obtained from the exact *lc equation* (Mederos et al. 1993)

$$1 + \int \rho(\mathbf{r}') [g(\mathbf{r}, \mathbf{r}') - 1] d\mathbf{r}' = \frac{k_B T}{\rho(\mathbf{r})} \frac{d\rho(\mathbf{r})}{d\mu}, \qquad (9.88)$$

which is the equivalent for inhomogeneous systems of the compressibility equation (2.54), because $\rho \kappa_T = (1/\rho)(\partial \rho / \partial \mu)_T$ for an homogeneous fluid with density ρ. Replacing in Equation 9.88 $g(\mathbf{r}, \mathbf{r}')$ with $g_0(|\mathbf{r} - \mathbf{r}'|; \hat{\rho}(\mathbf{r}))$ and, in the right-hand side, $\rho(\mathbf{r})$ with $\bar{\rho}(\mathbf{r})$, the WDA density, one can solve the resulting equation to obtain $\hat{\rho}(\mathbf{r})$, which completes the PWDA. A simplified version of the PWDA, denoted SPWDA, was proposed by Mederos et al. (1994) for inhomogeneous solids by using instead of the lc equation (Equation 9.88) the *global compressibility equation*

$$1 + \frac{1}{N} \int \rho\left(\mathbf{r}\right) d\mathbf{r} \int \rho\left(\mathbf{r}'\right) \left[g\left(\mathbf{r}, \mathbf{r}'\right) - 1\right] d\mathbf{r}' = \rho k_B T \kappa_T, \tag{9.89}$$

where ρ is the average density, which leads to a global uniform density $\hat{\rho}$ instead of a position-dependent local density $\hat{\rho}(\mathbf{r})$. With respect to the PWDA, the SPWDA has the disadvantage that only can be applied to macroscopically homogeneous systems, like solids, but the advantage that it is much simpler to calculate while providing comparable accuracy.

The SPWDA was applied by Rascón et al. (1995a) to short-ranged SW systems. They found that for reduced well ranges $\lambda \lesssim 1.06$ the phase diagram exhibits a fcc–fcc isostructural transition, which ends at a critical point, and a triple point where the two fcc phases coexist with the fluid, whereas the liquid phase is metastable, as the gas–liquid coexistence curve is below the fluid–solid one in the $T - \rho$ plane. As the well width increases, the solid–solid transition evolves toward lower densities while the critical temperature remains nearly constant. For $1.06 < \lambda < 1.25$ both the solid–solid and the gas–liquid transitions are metastable and for $\lambda \gtrsim 1.25$ the gas–liquid transition becomes stable. This behavior is in qualitative agreement with the simulations. The calculated critical temperatures for the solid–solid transition were considerably higher than the simulation ones, whereas the critical densities were accurately predicted.

The fcc–fcc isostructural transition in the SW solid with $1.04 \leq \lambda \leq 1.09$ was also studied by Likos et al. (1994) using the MWDA for the repulsive part of the potential and the MFA for the attractive one. For $\lambda \leq 1.08$, the theory was found to predict that the transition is stable, and metastable for larger ranges. On the other hand, the reported results show that the theory gives values considerably higher than the simulation data for the critical temperature, which was attributed to the MFA, whereas the critical densities and the triple-point temperatures are very accurately predicted.

Fu and Wu (2004) combined the MFMT for the repulsive contribution with the quadratic approximation (9.29) for the dispersive one. The latter may be written as

$$\beta \mathscr{F}_{dis}\left[\rho\left(\mathbf{r}\right)\right] = \beta \int \varphi_{dis}\left[\rho\left(\mathbf{r}\right)\right] d\mathbf{r} + \frac{1}{4} \int \int c_{dis}^{(2)}\left(\left|\mathbf{r} - \mathbf{r}'\right| ; \rho_m\right) \left[\rho\left(\mathbf{r}\right) - \rho\left(\mathbf{r}'\right)\right]^2 d\mathbf{r} d\mathbf{r}', \tag{9.90}$$

where
  $\varphi_{dis}(\rho)$ is the dispersive contribution to the free energy per unit volume of a
    uniform fluid with density ρ
  $c_{dis}^{(2)}(\left|\mathbf{r} - \mathbf{r}'\right|)$ is the corresponding contribution to the DCF
  $\rho_m = [\rho(\mathbf{r}) + \rho(\mathbf{r}')]/2$

Good agreement with simulations was achieved in this way for the liquid–vapor coexistence densities and pressures of SW fluids with variable range, except near the

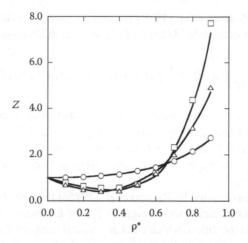

FIGURE 9.2 Compressibility factor Z for SW fluids as a function of the reduced density ρ^*. Points: simulation data from Largo et al. (2005) for $\lambda = 1.05$ and $T^* = 0.5$ (circles), for $\lambda = 1.5$ and $T^* = 1.5$ (squares), and for $\lambda = 2.0$ and $T^* = 3.0$ (triangles). Curves: MFMT-FMT+FMSA (see explanation in the text) from Tang (2008; personal communication, 2012).

critical point. Introducing corrections to account for the long-range fluctuations in the critical region, accurate results were obtained also in this region. The MFMT was also used by Tang (2008) for the repulsive contribution, whereas the attractive one was approximated with a perturbation expansion similar to Equation 9.24 truncated at second order, with the attractive contributions to the bulk fluid free energy and DCF as given by the analytical expressions derived by Tang and Lu (1994) and by Tang (2007), respectively, from the FMSA. This MFMT+FMSA-DCF combination resulted in a considerable improvement over the FMSA in the prediction of the RDF for SW fluids, especially for very short-ranged potentials, including the values at discontinuities σ^+, $\lambda\sigma^-$, and $\lambda\sigma^+$ which, according to Equation 2.44, determine the EOS. As a consequence, the latter property is accurately predicted even for very narrow potential wells except near the critical point, as illustrated in Figure 9.2 for several examples.

9.4.7.2 Hard-Core Yukawa Potential

The SPWDA was used by Rascón et al. (1995b) to study the phase diagram of short-ranged HCY systems, with the solid phase restricted to the fcc structures. The theory predicted a solid–solid isostructural transition, with the critical temperature varying approximately linearly with the parameter κ of the potential and with a slope in agreement with the simulations. Also, the critical densities obtained from the theory were in close agreement with the simulations. However, the calculated values for the critical temperature were some 13.5% higher than those found in the simulations. In contrast with other authors, they argued that most of the solid correlation structure is accounted for by the MFA, contrarily to the case of fluids, and so this cannot be the reason for the discrepancy. Instead, they considered that the solid–solid transition involves a delicate balance between the different contributions to the solid phase free energy, and so it is very sensitive to the accuracy of the theory.

The fcc–fcc isostructural transition in HCY solids was studied also by Német and Likos (1995) by means of MWDA+MFA. The found that with this approximation the transition takes place for $\kappa \gtrsim 25$, with critical temperatures in close agreement with the simulation data, whereas for lower values it is preempted by melting. However, the predicted variation of the critical temperature with κ is the opposite of that observed in simulations.

The MFMT plus quadratic approximation mentioned earlier in this section was applied by Fu and Wu (2004) also to HCY systems with similar accuracy as for the SW. Satisfactory results were also obtained by Tang (2004) from the combination MFMT+FMSA-DCF, with the analytical expressions for the attractive contributions to the free energy and the DCF derived by Tang (2003) from the FMSA. Ayadim et al. (2009) used the self-consistent reference functional approach with the Rosenfeld FMT for the reference fluid and the Oettel (2005) optimizing condition (see Section 9.2.3) to predict the liquid–vapor coexistence and the pressure for a short-ranged HCY fluid, with very good accuracy.

9.4.7.3 Soft-Sphere Potential

Most of the DFTs considered in this chapter give poor results for the freezing transition of SS systems. This is the case, among others, for the WDA, ELA, and related approximations (see e.g., Barrat et al. 1987b, Laird and Kroll 1990, de Kuijper et al. 1990). At the best case, these theories predict values for the coupling parameter Γ for the fluid-to-fcc-solid transition considerably higher than the simulation data, and the situation worsens with increasing the softness of the potential. In particular, Laird and Kroll (1990) found that the MELA and GELA theories are unable to predict the freezing into either fcc or bcc solid phases.

The one-component plasma, an extreme case of the SS potential, has been studied by different DFT approximations. Some of the earlier analyses of the freezing of the OCP were based on the expansion (9.24) truncated at second or third order. To this respect, Iyetomi and Ichimaru (1988) noted that higher-order terms involving n-particle DCFs with $n \geq 3$ are of crucial importance. Moreover, even using a three-particle DCF, the predicted value of the coupling parameter Γ at freezing is very sensitive to the accuracy of the theory used to determine the structure of the homogeneous fluid. Thus, Bagchi et al. (1984) considered the expansion (9.24) up to third order together with the expansion (9.72) for the density of the OCP solid, an accurate analytical expression for the structure factor of the homogeneous fluid derived by Chaturvedi et al. (1981) on the basis of the self-consistent MSA, and a three-particle DCF obtained from the two-particle DCF. The theory predicted a fluid-to-bcc phase transition without volume change at a value of Γ in very good agreement with simulation. Also, the EMA was successfully applied by Likos and Ashcroft (1992) to this system, with either approximations (9.73) or (9.76) for the three-particle DCF. With both choices very good agreement with simulations was achieved for the coupling parameter Γ and the Lindemann ratio at melting, as well as for the change in entropy on melting.

9.4.7.4 Lennard–Jones Potential

The freezing of an LJ system has been studied by Marshall et al. (1985) from the third-order expansion (9.26), with the Fourier expansion (9.72) for the solid density and the two-particle DCF determined from a fitting of the simulation data for $g(r)$. The predicted coexistence densities were in quite good agreement with the simulation data at low temperatures, although at higher temperatures the theoretical results increasingly deviated from the simulations with rising temperature. Using a similar approach, Laird et al. (1987) compared the results obtained from the Fourier expansion (9.72) and the Gaussian approximation (9.70) for the solid density obtaining nearly the same results from the two choices.

Mederos et al. (1993) obtained the phase diagram of the LJ system from the PWDA, with the WCA splitting of the potential, in close agreement with the simulations for $T^* \lesssim 2.75$. Similar accuracy for the solid–fluid coexistence densities was achieved by Mederos et al. (1994) with the SPWDA and even with a further simplified mean field version. Based on a different perturbed WDA is the approach devised by Curtin and Ashcroft (1986), who write

$$\beta \mathscr{F} \left[\rho \left(\mathbf{r} \right) \right] = \beta F_{HS}^{WDA} \left[\rho \left(\mathbf{r} \right) ; d \right] + \frac{N\beta}{2} \int \int \rho_{HS}^{(2)} \left(r; d \right) u \left(r \right) d\mathbf{r}$$

$$- \frac{1}{2} \int \int \Delta c^{(2)} \left(\left| \mathbf{r} - \mathbf{r}' \right| \right) \Delta \rho \left(\mathbf{r} \right) \Delta \rho \left(\mathbf{r}' \right) d\mathbf{r} d\mathbf{r}'. \tag{9.91}$$

$\Delta c^{(2)} (|\mathbf{r} - \mathbf{r}'|)$, the difference between the two-particle DCFs of the actual and reference fluids, is approximated by

$$\Delta c^{(2)} \left(r \right) = \begin{cases} 0, & 0 < r < r_s/2 \\ -\beta \varepsilon, & r_s/2 < r < r_m , \\ -\beta u \left(r \right), & r > r_m \end{cases} \tag{9.92}$$

where
   r_s is the interparticle spacing
   $-\varepsilon$ is the maximum potential depth
   r_m is the position of the minimum of the potential

Using the Gaussian approximation for the density of the solid and replacing the LJ potential by a sum of two Yukawa potentials closely mimicking the LJ potential, very good agreement with simulation data was found for the phase diagram of the LJ system for $T^* \lesssim 1.35$.

de Kuijper et al. (1990) analyzed the performance of the expansion (9.26) truncated at second order, the MWDA, and the MWDA+MF approximations, the latter consisting in a combination of the MWDA for the repulsive contribution plus a MFA for the attractive one, for the solid–fluid coexistence of the LJ system. The needed input for the homogeneous fluid was obtained from the HMSA closure (see Section 4.2). It was found that the second-order expansion (9.26) provides too high values of the coexistence densities and pressures and the MWDA fails to predict the phase transition for $T^* < 5.0$, whereas the MWDA+MF approximation yields close agreement with the simulations. Ohnesorge et al. (1991) used the MWDA for the reference HS fluid, with the DCF and the free energy as given by the PY theory, and a perturbation term like the one in Equation 9.86, with the Verlet and Weis (1972) expression for the HS RDF. With this approach, the predicted liquid–vapor coexistence densities are in good agreement with the simulations; the same applies to the solid densities in the solid–fluid coexistence, but not for the fluid densities that are underestimated.

Rosenfeld (1998) applied the self-consistent reference functional approach, with the FMT free energy functional for the reference inhomogeneous HS fluid, to calculate the thermodynamic and structural properties of the LJ fluid, and obtained very good agreement with the simulations for the RDF, the DCF, and the bridge function near the triple point, as well as for the pressure, the energy, and the isothermal compressibility as functions of density and temperature. Similar accuracy was also achieved by Kahl et al. (1996) for LJ mixtures and binary alloys using the self-consistent reference functional approach with the HNC approximation for the reference fluid and the optimizing condition (7.117). Fu and Wu (2004) combined the MFMT with the quadratic approximation (9.90) to determine the liquid–vapor coexistence of the LJ fluid, obtaining very good results for the coexistence densities and pressures when properly accounting for the contribution due to long-scale fluctuations near the critical point. Tang (2005) used the FMSA theory instead of Equation 9.90 for the perturbation contribution. Excellent results for the EOS, the energy, and the liquid–vapor coexistence were also obtained by Oettel (2005) for a cut-off and shifted LJ fluid, using the Rosenfeld FMT for the reference HS system and the optimizing condition $\partial \Delta \mathscr{F}_0^B / \partial d = 0$ (see Section 9.2.3).

An ansatz for the DCF of simple fluids has been proposed by Lutsko (2007) in the form

$$c^{(2)}(r; \rho) = c_{HS}^{(2)}(r; d, \rho) + \left(a_0 + a_1 \frac{r}{d}\right) \Theta(d - r) + c_{tail}(r; d, \rho). \qquad (9.93)$$

The first term of the right-hand side is the DCF of a fluid of HS with diameter d, the effective diameter of the molecules of the actual fluid, the second term is a core correction depending on two parameters a_0 and a_1, which are determined from the condition that the DCF must be compatible with a prescribed EOS for the uniform fluid through the free energy functional in the uniform limit, and the last term is a tail correction. The only input required by the theory is the free energy, or equivalently the EOS, of the homogeneous fluid, which may be obtained from any suitable theory, together with a simple ansatz for the tail contribution.

Expression (9.93), with a density-independent tail correction $c_{tail}(r; d, \rho) = \exp(-\beta u(r)) - 1 + \Theta(d - r)$, was used by Lutsko (2007) with a density functional which may be derived from Equation 9.39 applied to the inhomogeneous LJ and HS fluids, both with density $\rho(\mathbf{r})$, considering the two reference inhomogeneous fluids with the same density $\rho_0(\mathbf{r})$, and subtracting, with the result

$$\beta \mathscr{F}^E [\rho(\mathbf{r})] = \beta \mathscr{F}^E_{HS} [\rho(\mathbf{r};d)] + \beta \Delta \mathscr{F}^E [\rho_0(\mathbf{r};d)] - \int \Delta \mathscr{C}(\mathbf{r};[\rho_0(\mathbf{r})]) \Delta \rho(\mathbf{r}) \, d\mathbf{r}$$

$$- \int_0^1 (1-\alpha) \, d\alpha \int \int \Delta C(\mathbf{r},\mathbf{r}';[\rho(\mathbf{r}';\alpha)]) \Delta \rho(\mathbf{r}) \Delta \rho(\mathbf{r}') \, d\mathbf{r} d\mathbf{r}', \tag{9.94}$$

where d is the effective HS diameter. Then, the reference system for both the LJ and HS inhomogeneous fluids were taken to be the corresponding homogeneous fluids, both with density ρ_0, and the difference $\Delta \mathscr{C}(\mathbf{r},\mathbf{r}',[\rho(\mathbf{r})])$ was approximated by $\Delta \bar{c}^{(2)}(|\mathbf{r}-\mathbf{r}'|)$, where $\bar{c}^{(2)}(|\mathbf{r}-\mathbf{r}'|)$ is some intermediate value between $c^{(2)}(|\mathbf{r}-\mathbf{r}'|;\rho(\mathbf{r}))$ and $c^{(2)}(|\mathbf{r}-\mathbf{r}'|;\rho(\mathbf{r}'))$.

Instead, Lutsko (2008) assumes $c_{tail}(r;d) = -\beta\Theta(r-d)u(r)$ and a free energy functional consisting of a term of the form of Equation 9.27 for the core contribution, due to the hard spherical core plus the core correction, and a mean field contribution of the form of Equation 9.17, that is

$$\beta \mathscr{F}^E [\rho(\mathbf{r})] = \int (\Phi_{HS}[\rho(\mathbf{r})] + \Phi_{CC}[\rho(\mathbf{r})]) \, d\mathbf{r}$$

$$+ \frac{1}{2} \int \int \rho(\mathbf{r}) \rho(\mathbf{r}') \Theta(|\mathbf{r}-\mathbf{r}'|-d) u(|\mathbf{r}-\mathbf{r}'|) d\mathbf{r} d\mathbf{r}', \tag{9.95}$$

where $\Phi_{HS}[\rho(\mathbf{r})]$ is given by the FMT expression and for the contribution $\Phi_{CC}[\rho(\mathbf{r})]$ of the core correction it is assumed the FMT form with coefficients φ_i determined from the condition that the DCF (9.93) must be recovered in the homogeneous limit.

9.5 SURFACES AND INTERFACES

Other practical applications of the density functional formalism refer to the structure and thermodynamic properties of fluids near surfaces and interfaces. Some of this kind of phenomena are briefly considered next.

9.5.1 DENSITY PROFILES NEAR A HARD WALL

The density profile $\rho(\mathbf{r})$ of a nonuniform fluid near the surface can be related to the bulk density ρ_0 of the fluid applying Equation 9.8 to both densities, which gives

$$\rho(\mathbf{r}) = \rho_0 \exp\left\{-\beta \mathscr{V}(\mathbf{r}) + \mathscr{C}(\mathbf{r};[\rho(\mathbf{r})]) - c_0^{(1)}(r;\rho_0)\right\}. \tag{9.96}$$

In the absence of any external potential, the inhomogeneity is caused by the presence of the surface. If the surface is planar and parallel to the xy plane, the density profile will vary rapidly along the z direction, and we can obtain the corresponding density profile $\rho(z)$ by averaging the density profile $\rho(\mathbf{r})$ on the xy plane, that is $\rho(z) = (1/A) \int_A \rho(\mathbf{r}) dx dy$, where A is the surface area. In the case of a planar hard wall, the interaction with the wall can be treated as an external potential of the form

$$\mathscr{V}(z) = \begin{cases} \infty, & z < d/2 \\ 0, & z > d/2 \end{cases}, \tag{9.97}$$

where d is the effective diameter. For a fluid in a spherical cavity with hard wall

$$\mathcal{V}(\mathbf{r}) = \begin{cases} \infty, & |\mathbf{r}| < d/2 \\ 0, & |\mathbf{r}| > d/2 \end{cases}. \tag{9.98}$$

Expressions for external potentials corresponding to other situations, like fluids confined between two hard walls, or in contact with a big HS, or in contact with soft or attractive walls, are straightforward to set up.

Computer simulations and experiment reveal an oscillatory character of the density profiles of fluids in contact with hard walls. One of the earliest DFT approximations succeeding in reproducing, at least qualitatively, this behavior was developed by Tarazona and Evans (1984) for HS and LJ fluids, on the basis of the WDA for the repulsive contribution plus a mean field term to account for the attractive one, using the WCA splitting of the LJ potential.

More accurate results for HS were obtained by Denton and Ashcroft (1991) with a variant of the WDA based on the approximation

$$\mathcal{C}(\mathbf{r}; [\rho(\mathbf{r})]) \approx c_0^{(1)}(\bar{\rho}(\mathbf{r})), \tag{9.99}$$

where the weighted density $\bar{\rho}(\mathbf{r})$ is given by Equation 9.31 and $\mathcal{C}(\mathbf{r}; [\rho(\mathbf{r})])$ is determined from the condition that through Equations 9.12 and 9.14 must give a prescribed two-particle DCF for the uniform fluid with density ρ_0, which yields for the weighting function the result (9.78). For this purpose, the PY solution (see Section 5.3) was used for the uniform HS fluid. To determine the density profiles of the HS fluid near a hard wall, the approximation (9.99) is introduced into Equation 9.96, with the external potential (9.97) with $d = \sigma$, with the result

$$\rho(z) = \rho_0 \exp\left\{c_0^{(1)}(\bar{\rho}(z)) - c_0^{(1)}(\rho_0)\right\}, \tag{9.100}$$

with

$$\bar{\rho}(z) = \int \rho(z') w\left(|z - z'|; \bar{\rho}(z)\right) dz', \tag{9.101}$$

in which $w(z)$ is the average of $w(r)$ over the plane xy. The theory was also extended by the same authors to binary HS mixtures, with the Lebowitz (1964) solution of the PY theory (see Section 7.5.2) for one- and two-particle DCFs of the uniform fluid, also with quite satisfactory results. Similar accuracy for this kind of mixtures was achieved by Patra (1999) with the Zhou (1999) version of the SWDA.

Choudhury and Ghosh (1999) applied the expansion (9.23) up to second order to obtain the density profiles of HS mixtures near a hard wall. They used the PY theory for the two-particle DCFs of HS mixtures and an approximate ansatz for the three-particle DCFs. The theory was found to compare favorably with the WDA of Denton and Ashcroft (1991). Zhou (2001) also used the expansion (9.23) truncated at second order in $\Delta\rho(\mathbf{r})$, with the n-particle DCFs given by Equation 9.79 and the coefficient $\left[\partial^2 c_0^{(1)}(\rho_0)/\partial\rho_0^2\right] / \left[\partial c_0^{(1)}(\rho_0)/\partial\rho_0\right]^3$ in $c_0^{(3)}(\mathbf{r}, \mathbf{r}_1, \mathbf{r}_2; \rho_0)$ replaced with a parameter determined from the exact condition that βP must be equal to the average density $\rho(\sigma/2)$ of HS in contact with the wall, as shown by Reiss et al. (1959), that is

$$\beta P = \rho \left(\frac{\sigma}{2} \right). \tag{9.102}$$

The theory was shown to provide good agreement with simulations for the density profiles of LJ and SHS fluids confined between two hard walls.

Several approximations, based on truncating either the perturbation series (9.23) or the (9.24) one and introducing in the last considered term a density, different from the bulk density ρ_0, to correct in some way for the neglected terms, have been devised by Zhou (2002, 2003a,b, 2004a,b). Thus, Zhou (2002) starts from the expansion (9.23) truncated at first order and replaces the bulk density ρ_0 with a weighted density $\bar{\rho}(\bar{\mathbf{r}})$. This yields

$$\mathscr{C}(\mathbf{r};[\rho(\mathbf{r})]) \approx c_0^{(1)}(\rho_0) + \int c_0^{(2)} \left(|\mathbf{r} - \mathbf{r}'| ; \bar{\rho}(\bar{\mathbf{r}}) \right) \Delta\rho\left(\mathbf{r}'\right) d\mathbf{r}', \tag{9.103}$$

where $\Delta\rho(\mathbf{r}) = \rho(\mathbf{r}) - \rho_0$, $\bar{\mathbf{r}} = (\mathbf{r} + \mathbf{r}')/2$, and $\bar{\rho}(\mathbf{r})$ is defined in the form

$$\bar{\rho}(\mathbf{r}) = \int \widetilde{\rho}\left(\mathbf{r}'\right) w\left(|\mathbf{r} - \mathbf{r}'| ; \bar{\rho}(\mathbf{r}) \right) d\mathbf{r}'. \tag{9.104}$$

Here, $\widetilde{\rho}(\mathbf{r}) = \rho_0 + \chi\left(\rho(\mathbf{r}) - \rho_0\right)$, in which χ is some quantity that is determined from the exact condition (9.102), is an intermediate density introduced to account in an approximate way for the effect of the higher-order terms in the expansion (9.23), and $w(|\mathbf{r} - \mathbf{r}'|; \bar{\rho}(\mathbf{r}))$ is a weighting function. To define the latter, the bulk fluid two-particle DCF $c_0^{(2)}(r; \rho_0)$ is split into short-range $c_{0S}^{(2)}(r; \rho_0)$ and long-range $c_{0L}^{(2)}(r; \rho_0)$ contributions at some distance d, and then the weighting function is taken to be

$$w\left(|\mathbf{r} - \mathbf{r}'| ; \bar{\rho}(\mathbf{r}) \right) = \frac{c_{0S}^{(2)}\left(|\mathbf{r} - \mathbf{r}'| ; \rho_0 \right)}{\int c_{0S}^{(2)}\left(|\mathbf{r} - \mathbf{r}'| ; \rho_0 \right) d\mathbf{r}}. \tag{9.105}$$

Zhou (2004a) also starts from Equation 9.23, truncated at first order, and rewrites it in the form

$$\mathscr{C}(\mathbf{r};[\rho(\mathbf{r})]) \approx c_0^{(1)}(\rho_0) + \frac{\partial c_0^{(1)}(\rho_0)}{\partial \rho_0} \left[\frac{\int c_0^{(2)}\left(|\mathbf{r} - \mathbf{r}'| ; \rho_0 \right) \rho\left(\mathbf{r}'\right) d\mathbf{r}'}{\partial c_0^{(1)}(\rho_0)/\partial \rho_0} - \rho_0 \right]$$

$$= c_0^{(1)}(\rho_0) + \frac{\partial c_0^{(1)}(\rho_0)}{\partial \rho_0} \left(\bar{\rho}(\mathbf{r}) - \rho_0 \right), \tag{9.106}$$

where $\bar{\rho}(\mathbf{r})$ is the weighted density as defined in the SWDA, Equation 9.31 with $\bar{\rho}(\mathbf{r})$ in the weighting function replaced with ρ_0, and we have taken into account the expression (9.78) of the weighted density in the SWDA approximation. Based on the similarity of Equation 9.106 with a truncated Taylor series expansion, Zhou (2004a) writes

$$\mathscr{C}(\mathbf{r};[\rho(\mathbf{r})]) = c_0^{(1)}(\rho_0) + \left.\frac{\partial c_0^{(1)}(\rho_0)}{\partial \rho_0}\right|_{\rho_0 \to \rho_0 + \chi(\bar{\rho}(\mathbf{r}) - \rho_0)} (\bar{\rho}(\mathbf{r}) - \rho_0), \qquad (9.107)$$

where the derivative is calculated at the density $\rho_0 + \chi(\bar{\rho}(\mathbf{r}) - \rho_0)$ to account for the higher-order terms neglected in Equation 9.106, and parameter χ is determined from condition (9.102).

Instead, the approximation proposed by Zhou (2004b) is based on the expansion (9.24) truncated at first order with the bulk fluid one-particle DCF replaced with the nonuniform $\mathscr{C}(\mathbf{r};[\tilde{\rho}(\mathbf{r})])$, to account for the contribution of the suppressed higher-order terms. This yields

$$\beta \mathscr{F}^E[\rho(\mathbf{r})] = \beta F^E(\rho_0) - \int \mathscr{C}(\mathbf{r};[\tilde{\rho}(\mathbf{r})]) \Delta\rho(\mathbf{r}) d\mathbf{r}, \qquad (9.108)$$

where $\tilde{\rho}(\mathbf{r})$ has the same meaning as before with $\chi = 1/2$. Introducing the approximation $\mathscr{C}(\mathbf{r};[\tilde{\rho}(\mathbf{r})]) \approx c_0^{(1)}(\bar{\rho}(\mathbf{r}))$, where $\bar{\rho}(\mathbf{r})$ is a weighted density defined in a similar way as in the SWDA, namely,

$$\bar{\rho}(\mathbf{r}) = \int \tilde{\rho}(\mathbf{r}') w(|\mathbf{r} - \mathbf{r}'|; \rho_0) d\mathbf{r}', \qquad (9.109)$$

with the weighting function given by Equation 9.78, and performing the functional derivative of Equation 9.108 with respect to $\rho(\mathbf{r})$, one obtains

$$\mathscr{C}(r;[\rho(\mathbf{r})]) \approx c_0^{(1)}(\bar{\rho}(\mathbf{r})) + \int \frac{\partial c_0^{(1)}(\bar{\rho}(\mathbf{r}))}{\partial \rho(\mathbf{r})} w(|\mathbf{r} - \mathbf{r}'|; \rho_0) \Delta\rho(\mathbf{r}') d\mathbf{r}'. \quad (9.110)$$

These three and other related approximations have shown to perform satisfactorily for different potential models. As an example, Figure 9.3 illustrates the performance of the Zhou (2004a) theory for the density profiles of the HS fluid near a hard wall.

A novel strategy for fluids other than the HS fluid has been introduced by Zhou (2003a,b), based on dividing the two-particle DCF of the bulk fluid into an HS-like part, strongly varying with density, and a tail, only weakly dependent on density. Then, expansion (9.23) truncated at first order will be enough accurate for the contribution of the tail to $\mathscr{C}(\mathbf{r})$, whereas a higher-order truncation will be needed for the HS-like contribution. The reason for doing so is that there are available accurate DFTs for the inhomogeneous HS fluid, whereas for the tail contribution we usually can obtain accurate results only for the two-particle DCF of the homogeneous fluid. Treating the HS-like part with the approach synthesized in Equations 9.106 and 9.107, this *partitioned* DFT provides excellent agreement with simulations for the density profiles of the LJ fluid near a hard wall, as shown in Figure 9.4. This partitioning of the two-particle DCF has been further exploited by Zhou and Jamnik (2005a,b) by considering third- and second-order perturbation approximations for the HS-like and tail contributions, respectively.

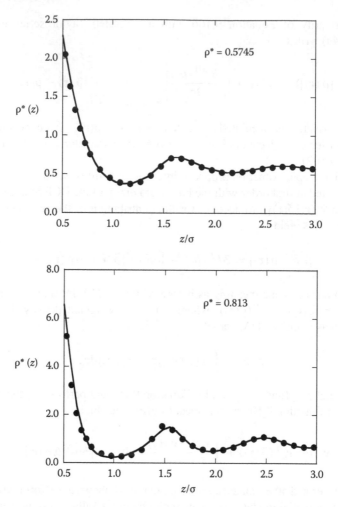

FIGURE 9.3 Density profiles $\rho^*(z)$ of the HS fluid near a hard wall for two different bulk densities ρ^*. Points: simulation data from Groot et al. (1987). Curves: results from Equation 9.23 with approximation (9.107), from Zhou (2004a; personal communication, 2012).

Closely related to the WDA of Denton and Ashcroft (1991) is the approach devised by Choudhury et al. (2002). It is based on the series expansion of the exponent of Equation 9.100 in terms of the difference $\bar{\rho}(z) - \rho_0$ truncated at third order, with the third-order term scaled by a parameter to account for the contribution of the higher-order terms. The scaling parameter is again determined from the exact condition (9.102). With the bulk fluid pressure and n-particle DCFs, with $n = 1, 2, 3, 4$, derived from the PY theory with the help of Equations 9.13 and 9.14, the results obtained for the density profiles near a hard wall were found to be in close agreement with the simulations for the pure HS fluid as well as for several HS mixtures.

Based on similar grounds is another approach proposed by Choudhury and Ghosh (2003), consisting in introducing the exact expansion (9.23) in the exponent of

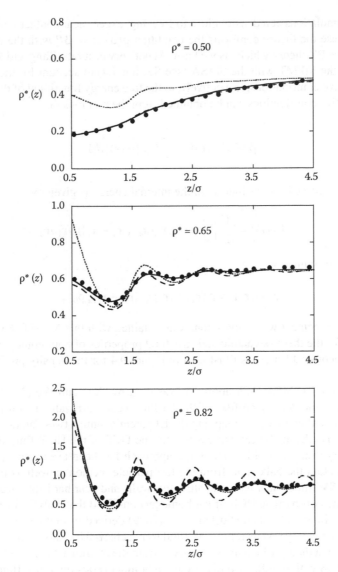

FIGURE 9.4 Density profiles $\rho^*(z)$ of the LJ fluid near a hard wall at $T^* = 1.35$ for three different bulk densities ρ^*. Points: simulation data from Balabanic et al. (1989). Curves: partitioned DFT (continuous), from Zhou (2003, private communication), MFMT+FMSA-RDF (dashed) and MFMT+MF (dotted), from Tang and Wu (2003; personal communication, 2012).

Equation 9.96, truncating the series at second order, with the involved three-particle DCF calculated at the weighted density (9.101), and using for the three-particle DCF an approximate ansatz depending on a parameter determined from the same condition as before.

Kierlik and Rosinberg (1990) applied their version of the FMT to obtain the density profiles of the HS fluid near a hard wall, obtaining nearly perfect agreement with the simulations at a density close to freezing, except near contact with the hard wall, because the theory conforms the condition $\rho(\sigma/2) = \beta P$ with the pressure P given by the PY theory which, as we know, is not very accurate. Tang and Wu (2003) combined the MFMT with the FMSA (see Section 4.4) to account for the repulsive and attractive contributions, respectively, to the free energy functional of the LJ fluid. The attractive contributions can be obtained from the relationship

$$\beta \mathscr{F}_{att}\left[\rho\left(\mathbf{r}\right)\right] = \int_{0}^{\beta} U_{att}\left[\rho\left(\mathbf{r}\right)\right] d\beta, \tag{9.111}$$

in which the attractive contribution to the internal energy is given by

$$U_{att}\left[\rho\left(\mathbf{r}\right)\right] = \frac{1}{2} \int\int \rho\left(\mathbf{r}_1, \mathbf{r}_2\right) u_{att}\left(|\mathbf{r}_1 - \mathbf{r}_2|\right) d\mathbf{r}_1 d\mathbf{r}_2. \tag{9.112}$$

With the approximation

$$\rho\left(\mathbf{r}_1, \mathbf{r}_2\right) \approx \rho\left(\mathbf{r}_1\right)\rho\left(\mathbf{r}_2\right) g\left(|\mathbf{r}_1 - \mathbf{r}_2|; \rho_0\right), \tag{9.113}$$

very good agreement with simulations was obtained with this MFMT+FMSA-RDF approach for the thermodynamic and structural properties of both homogeneous and inhomogeneous LJ fluids, as illustrated in Figure 9.4 for the density profiles near a hard wall.

The MFMT+FMSA-DCF approach was used by Tang and Wu (2004) to accurately predict the density profiles of the LJ fluid near hard walls, in slit pores, and around colloidal particles, by mapping the LJ potential onto a two-Yukawa potential, in order to use the analytical expressions for the DCF of the HCY fluid derived by Tang (2003) from the FMSA. The same approach has been used also by Jin et al. (2011) to obtain the bulk RDF from the test particle method as well as the density profiles of SW fluids near hard and attractive walls and near hard spherical cavities, using for the two-particle DCF of the homogeneous fluid the analytical expressions reported by Tang (2007) and Hlushak et al. (2009) derived from the FMSA.

Very satisfactory results for the HS fluid were achieved by Zhou and Ruckenstein (2000b) from a theory which may be derived from the reference functional approach, Equation 9.48 with (9.49), by replacing the inhomogeneous reference fluid with the corresponding homogeneous one and the bridge functional with the bridge function of the reference fluid, that is

$$\ln\left[\frac{\rho\left(\mathbf{r}\right)}{\rho_0}\right] + \beta \Delta \mathscr{V}\left(\mathbf{r}\right) = \mathscr{C}\left(\mathbf{r}; [\rho\left(\mathbf{r}\right)]\right) - c_0^{(1)}\left(r; \rho_0\right)$$

$$= \int c_0^{(2)}\left(|\mathbf{r} - \mathbf{r}'|; \rho_0\right) \Delta\rho\left(\mathbf{r}'\right) d\mathbf{r}' + B\left[\gamma\left(\mathbf{r}\right)\right]. \tag{9.114}$$

The required DCF $c_0^{(2)}(r; \rho_0)$ and bridge functional $B[\gamma(\mathbf{r})]$ of the homogeneous HS fluid was obtained from IET with the VM closure (Equation 4.23). The procedure was successfully applied by Zhou and Ruckenstein (2000c) to LJ fluids, using IET with a closure proposed by Duh and Henderson (1996) for the homogeneous fluid and a modified VM closure for the bridge functional, and by Zhou (2000) to binary HS mixtures with the version for mixtures of the VM closure. Choudhury and Ghosh (2001) also analyzed the performance of this approach for HS pure fluid and binary mixtures as well as for binary LJ mixtures. Again the VM closure was used for both the pure HS fluid and HS mixtures, but with parameter b in Equation 4.23 determined by forcing the contact values of the RDF of the pure HS fluids to agree with the CS expression (8.41), and those for HS mixtures to agree with the BMCSL expression (7.90). The theory not only showed very good agreement with simulations for all these fluids but also for the two former kinds of fluids was found to compare favorably with the Denton and Ashcroft version of the WDA near contact with the hard wall. Still better agreement with simulations near contact was achieved by Patra and Ghosh (2002) for HS mixtures by using the mixture version of the RY closure (4.30) instead of the VM one.

A general approach for inhomogeneous fluids has been proposed by Yu (2009) from combining the MFMT for the repulsive contribution with a mean field term of the form of Equation 9.17, with $u(\mathbf{r}, \mathbf{r}')$ replaced by $u_1(|\mathbf{r}-\mathbf{r}'|)$, plus an additional term to account for the effect of correlations. The latter is treated within the framework of the WDA, namely,

$$\mathscr{F}_{COR}[\rho(\mathbf{r})] = \int \bar{\rho}(\mathbf{r}) f_{COR}(\bar{\rho}(\mathbf{r})) d\mathbf{r}, \qquad (9.115)$$

where f_{COR} is the contribution to the free energy per particle due to correlations, which can be obtained from any suitable EOS, and $\bar{\rho}(\mathbf{r})$ is a weighted density defined as

$$\bar{\rho}(\mathbf{r}) = \int \rho(\mathbf{r}') w_{COR}(|\mathbf{r} - \mathbf{r}'|) d\mathbf{r}', \qquad (9.116)$$

with

$$w_{COR}(r) = \Theta(d - r)\frac{4}{3}\pi d^3, \qquad (9.117)$$

where d is the effective diameter. With f_{COR} obtained from the FMSA and d determined from the BH prescription (6.68), Yu (2009) obtained very accurate results for the density profiles of LJ fluids near hard walls and in slitlike pores. A similar approach for the perturbation contribution, namely, the combination of mean field and correlation terms, was independently devised by Zhou (2009).

Another approach developed with the aim of improving the MFA is the one proposed by Zhou (2010a,b). For inhomogeneous fluids with potentials consisting of a hard spherical core plus a tail $u_1(r)$, within the MFA the excess free energy functional can be expressed in the form

$$\mathscr{F}^E[\rho(\mathbf{r})] = \mathscr{F}^E_{HS}[\rho(\mathbf{r})] + \frac{1}{2}\int\int \rho(\mathbf{r})\rho(\mathbf{r}') u_1(|\mathbf{r} - \mathbf{r}'|) d\mathbf{r} d\mathbf{r}'. \qquad (9.118)$$

By applying Equations 9.13 and 9.14 for $n = 2$, the corresponding bulk fluid two-particle DCF is

$$c_{MFA}^{(2)}\left(\left|\mathbf{r} - \mathbf{r}'\right|; \rho_0\right) = c_{HS}^{(2)}\left(\left|\mathbf{r} - \mathbf{r}'\right|; \rho_0\right) - \beta u_1\left(\left|\mathbf{r} - \mathbf{r}'\right|\right), \qquad (9.119)$$

which allows us to rewrite Equation 9.118 in the form

$$\beta\mathscr{F}^E\left[\rho\left(\mathbf{r}\right)\right] = \beta\mathscr{F}_{HS}^E\left[\rho\left(\mathbf{r}\right)\right]$$

$$+ \frac{1}{2}\int\int \rho\left(\mathbf{r}\right)\rho\left(\mathbf{r}'\right)\left(c_{HS}^{(2)}\left(\left|\mathbf{r} - \mathbf{r}'\right|; \rho_0\right) - c_{MFA}^{(2)}\left(\left|\mathbf{r} - \mathbf{r}'\right|; \rho_0\right)\right)d\mathbf{r}d\mathbf{r}'.$$
$$(9.120)$$

This suggests that a way to improve the MFA results is to replace in the latter equation $c_{MFA}^{(2)}(\left|\mathbf{r} - \mathbf{r}'\right|)$ with a more accurate one as obtained from a suitable IET. The effective HS diameter is obtained from the self-consistent condition that the bulk fluid compressibility factor obtained from this DFT must be equal to that obtained from the IET. Note that with this choice for the effective HS diameter the theory can be applied to soft-core potentials and the potential is not split in advance into reference and perturbation parts.

9.5.2 SURFACE MELTING

Melting usually does not take place in the bulk crystalline solid, but at the surfaces. Between the bulk crystalline solid and liquid, a slab, several molecular diameters thick, appears when the crystal starts melting. The density profile through the slab is determined by minimizing the difference in the grand potential between the interface and the bulk phase, which is given by (Oxtoby and Haymet 1982)

$$\Delta\Omega\left[\rho\left(\mathbf{r}\right)\right] = \mathscr{F}\left[\rho\left(\mathbf{r}\right)\right] - \mu\int \rho\left(\mathbf{r}\right)d\mathbf{r} + PV, \qquad (9.121)$$

as is easily derived from the thermodynamic relationship $F = \mu N + \Omega = \mu N + \Omega_0 + \Delta\Omega$, where $\Omega_0 = -PV$ is the grand potential of the homogeneous system and $\Delta\Omega = \Omega - \Omega_0$ is the difference between the interfacial and bulk, either solid or fluid, grand potentials. Once the density profile is determined, the interfacial tension is $\gamma = \Delta\Omega/A$ (Yang et al. 1976), where A is the area of the interface.

The average density in the slab will vary slowly between those of the solid and liquid, as the difference between them is relatively small. Therefore, the square-gradient approximation is expected to be applicable to this situation with the density $\rho(\mathbf{r})$ as given by Equation 9.72 with parameters that now will vary along the z direction, supposed to be perpendicular to the surface. This kind of approach was worked out by Haymet and Oxtoby (1981), who reported the expressions of the interfacial density profile $\rho(z)$ and the grand potential and noted that the theory predicts different density profiles for different crystal faces, as for example the [100], [110], and [111] faces of the fcc lattice, because of the change in the projections of the lattice vectors {$\mathbf{k}$} perpendicular to the face. The same approach was used by Oxtoby and Haymet (1982) to determine the density profiles and surface free energies of bcc lattices. Löwen et al. (1989) and Ohnesorge et al. (1991) split the potential into short-range and long-range parts and used a square-gradient approximation for the contribution of the former to the free energy functional, whereas the contribution of the tail was treated in the MFA.

The WDA was used by Curtin (1987, 1989) for the HS free energy functional, together with a two-parameter simple ansatz for the density profile in the interfacial slab, to analyze the HS [100] and [111] fcc–fluid interface of the HS system, as well as the [111] fcc–liquid interface of the LJ system, with the effect of the attractive tail in the latter introduced through a first-order perturbation theory like in Equation 9.86. The width Δz of the interfacial slab is determined from minimization of $\Delta \Omega$ with respect to Δz and the interfacial tension is then $\gamma = \Delta \Omega / A$ at the minimum. In order to reduce the computational effort required by the WDA, Marr and Gast (1993) used the planar-averaged density $\rho(z)$ to define a planar weighted density $\bar{\rho}(z)$, which was introduced into the WDA expression (9.30) of the excess free energy functional. Combined with the Curtin (1987) ansatz for the density profile in the interfacial slab, the interfacial tension obtained for the [111] fcc–fluid interface of the HS system was in close agreement with the result obtained using the three-dimensional WDA theory. The procedure was also applied to the LJ system by determining an effective diameter from the BH prescription (6.68) with the attractive contribution determined from first-order perturbation theory. With the aim of further simplifying the calculations, Choudhury and Ghosh (1998) used a procedure similar to that of Marr and Gast (1993), but based on the MWDA instead of the WDA. To this end, the interfacial slab was divided into a number of atomic layers and a weighted density within the MWDA was determined for each of the layers. Again the same ansatz as in the Curtin treatment was adopted for the density profile in the interfacial slab.

A different way of reducing the computational effort involved in the WDA was adopted by Ohnesorge et al. (1994). Based on the fact that it depends smoothly on the density, they expanded the weighting function for HS in Taylor series around $\rho^* = 0.5$ truncated at the second-order term, which allows to analytically solve Equation 9.31 to obtain the weighted density $\bar{\rho}(\mathbf{r})$. It was found that the difference in the weighting function obtained in this way with the exact one is negligible for $0.15 \lesssim \rho^* \lesssim 0.9$. The expansion (9.72) was adopted for the density, with z-dependent order parameters determined from minimization. For the LJ system again the potential was split into repulsive and attractive parts, with the free energy functional for the former contribution determined from the WDA for HS with an effective BH diameter and the tail contribution treated in the MFA.

Also, the FMT has been applied by Warshavsky and Song (2006) to study the solid–fluid interfacial properties of the HS system. They used both the E-FMT and the WB-FMT versions (see Section 9.3 for details) with the Curtin (1987) parametrization of the density profile in the interfacial slab.

The results for the interfacial tension obtained from several of the theories cited earlier in this section for the fcc solid–fluid interfaces are compared with the simulation data in Table 9.2 for the HS system and in Table 9.3 for the LJ system. It is to be noted the remarkable differences between the predictions from different, but closely related, approaches.

TABLE 9.2

Interfacial Tension, in Units of $k_B T / \sigma^2$, for the HS fcc Solid–Fluid Interface from Several DFTs and Molecular Dynamics (MD) and Monte Carlo (MC) Simulations

	γ_{100}	γ_{110}	γ_{111}	γ	n
MD[a]	0.62 ± 0.01	0.64 ± 0.01	0.58 ± 0.01	0.61 ± 0.01	
MC[b]	0.64 ± 0.02	0.62 ± 0.02	0.61 ± 0.02	0.62 ± 0.02	
WDA[c]	0.66		0.63		~4
WDA[d]			0.60		~4
WDA[e]	0.35	0.30	0.26	0.30	~7
MWDA[f]			0.33		~8
E-FMT[g]	0.79	0.89	0.87	0.85	
WB-FMT[g]	0.68	0.84	0.82	0.78	

γ is the average interfacial tension and n is the number of atomic layers in the interfacial slab.

[a] Davidchack and Laird (2000).

[b] Mu et al. (2005).

[c] Curtin (1987).

[d] Marr and Gast (1993).

[e] Ohnesorge et al. (1994).

[f] Choudhury and Ghosh (1998).

[g] Warshavsky and Song (2006).

TABLE 9.3

Interfacial Tension, in Units of ε / σ^2, for the LJ fcc Solid–Fluid Interface at the Triple Point from Several DFTs and Molecular Dynamics (MD) Simulation

	γ_{100}	γ_{110}	γ_{111}	γ	n
MD[a]	0.36 ± 0.02	0.34 ± 0.02	0.35 ± 0.02	0.35 ± 0.02	~5
WDA[b]			0.43		~4
WDA[c]			0.51		~4
WDA[d]	0.29	0.27	0.23	0.26	~7

γ is the average interfacial tension and n is the number of atomic layers in the interfacial slab.

[a] Broughton and Gilmer (1986).

[b] Curtin (1989).

[c] Marr and Gast (1993).

[d] Ohnesorge et al. (1994).

9.5.3 Liquid–Vapor Interface

Several perturbation theories for the equilibrium properties of homogeneous fluids have been extended to inhomogeneous fluids and used to determine the liquid–vapor surface tension. Thus, Toxvaerd (1971) applied the BH theory with the mc approximation and the PY compressibility solution for the reference HS fluid, together with a simple ansatz for the density profile in the interface, to determine the surface tension of the LJ liquid with fairly satisfactory results, as shown in Figure 9.5. Also, the WCA was extended to this kind of application by Upstill and Evans (1977) and Singh and Abraham (1977).

Concerning DFTs, contrarily to the case of the crystal–melt interface, the SGA is not reliable for the liquid–vapor interface because the density varies rapidly through the interface, except near the critical point, and so truncating the gradient expansion at second order is not justified. In fact, numerical calculations by Telo da Gama and Evans (1979) for the LJ fluid reveal that the SGA considerably overestimates the surface tension at low temperatures whereas approaches the right behavior near the critical point. Instead, using the approximation (9.29), which as said before can be obtained from a partial summation of the gradient expansion, with the properties of the homogeneous fluid determined from the PY theory and a three-parameter ansatz for the density profile at the interface, quite good results, also shown in Figure 9.5, were achieved by Ebner et al. (1976) for the surface tension of the LJ system.

Wadewitz and Winkelmann (2000) determined the surface tension of the LJ fluid from a perturbation theory with the WCA splitting of the potential. The reference

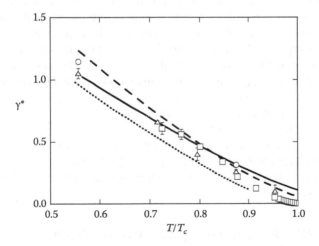

FIGURE 9.5 Reduced surface tension $\gamma^* = \gamma \sigma^2 / \varepsilon$ for the LJ liquid. Points: simulation data from Mecke et al. (1997) (circles), Potoff and Panagiotopoulos (2000) (squares), and Duque et al. (2004) (triangles). Continuous curve: Equation 9.94, from Lutsko (2007). Dashed curve: quadratic approximation (9.29), from Ebner et al. (1976). Dotted curve: BH theory, from Toxvaerd (1971).

free energy functional was treated in the LDA and the contribution of the attractive forces was considered as a first-order perturbation. The free energy functional is then

$$\mathscr{F}^E \left[\rho \left(\mathbf{r} \right) \right] = \mathscr{F}_0^E \left[\rho \left(\mathbf{r} \right) \right] + \frac{1}{2} \int \int \rho \left(\mathbf{r}_1 \right) \rho \left(\mathbf{r}_2 \right) g_0 \left(\mathbf{r}_1, \mathbf{r}_2; \rho \left(\mathbf{r} \right) \right) u_1 \left(\mathbf{r}_1, \mathbf{r}_2 \right) d\mathbf{r}_1 d\mathbf{r}_2$$

$$\approx \mathscr{F}_{HS}^E \left[\rho \left(\mathbf{r} \right) \right] + \frac{1}{2} \int \int \rho \left(\mathbf{r}_1 \right) \rho \left(\mathbf{r}_2 \right) g_{HS} \left(r_{12}; \bar{\rho}_m \left(\mathbf{r} \right) \right) u_1 \left(r_{12} \right) d\mathbf{r}_1 d\mathbf{r}_2.$$

$$(9.122)$$

In this expression, $g_0(\mathbf{r}_1, \mathbf{r}_2; \rho(\mathbf{r}))$ is the pair correlation function of an inhomogeneous reference system with density $\rho(\mathbf{r})$ and, following the approach devised by Sokolowski and Fischer (1992), has been replaced with the RDF of a homogeneous HS fluid with mean weighted density $\bar{\rho}_m(\mathbf{r}) = [\bar{\rho}(\mathbf{r}_1) + \bar{\rho}(\mathbf{r}_2)]/2$, where

$$\bar{\rho} \left(\mathbf{r} \right) = \int \rho \left(\mathbf{r}' \right) w \left(\left| \mathbf{r} - \mathbf{r}' \right| \right) d\mathbf{r}',$$

$$(9.123)$$

with

$$w \left(\left| \mathbf{r} - \mathbf{r}' \right| \right) = \begin{cases} 3/ \left(4\pi \left| \mathbf{r} - \mathbf{r}' \right|^3 \right), & \left| \mathbf{r} - \mathbf{r}' \right| \leq r_0 \\ 0, & \left| \mathbf{r} - \mathbf{r}' \right| > r_0 \end{cases}.$$

$$(9.124)$$

The cut-off distance for the weighting function was settled to $r_0 = 0.8\sigma$ and the properties of the reference system were obtained from the PY theory. The density profile at the interface was determined by minimizing the grand potential and the resulting equation was solved iteratively starting from a step function profile until achieving convergence. A careful analysis led to the conclusion that the cut-off distance for the potential needs to be settled to $r_c = 5.5\sigma$, because for lower values both the theoretical calculations and the simulations show an increase of surface tension with r_c. The results obtained in this way for the surface tension of the LJ fluid were in excellent agreement with the simulations.

Several other approximations based on the expansion (9.24) have been developed by Tang (2005). To this end, the n-particle DCFs of the bulk fluid, with $n \leq 2$, were determined from the FMSA and the higher-order DCFs were approximated by those given by the LDA. The resulting approximation, which was termed LDA+DCF, may be expressed as

$$\beta \mathscr{F}^E \left[\rho \left(\mathbf{r} \right) \right] = \beta \mathscr{F}_{LDA}^E \left(\rho \left(\mathbf{r} \right) \right) - \int \Delta c_0^{(1)} \left(\mathbf{r}_1; \rho_0 \right) \Delta \rho \left(\mathbf{r}_1 \right) d\mathbf{r}_1$$

$$- \frac{1}{2} \int \int \Delta c_0^{(2)} \left(\left| \mathbf{r}_1 - \mathbf{r}_2 \right|; \rho_0 \right) \Delta \rho \left(\mathbf{r}_1 \right) \Delta \rho \left(\mathbf{r}_2 \right) d\mathbf{r}_1 d\mathbf{r}_2, \quad (9.125)$$

where $\Delta c_0^{(n)}$, with $n = 1, 2$, denotes the difference between the n-particle DCFs of the bulk fluid calculated from the FMSA and from the LDA. If $\Phi(\rho(\mathbf{r}))$ in the LDA is determined from the FMSA, then $\Delta c_0^{(1)} = 0$ and, after some algebra, the preceding equation takes the simple form

$$\beta \mathscr{F}^E \left[\rho\left(\mathbf{r}\right) \right] = \beta \mathscr{F}_{LDA}^E \left(\rho\left(\mathbf{r}\right)\right) + \left[2\pi \int_0^{\infty} r^2 c_0^{(2)}\left(r\right) dr \right] \int \left(\rho\left(\mathbf{r}_1\right)\right)^2 d\mathbf{r}_1$$

$$- \frac{1}{2} \int c_0^{(2)}\left(\left|\mathbf{r}_1 - \mathbf{r}_2\right|\right) \rho\left(\mathbf{r}_1\right) \rho\left(\mathbf{r}_2\right) d\mathbf{r}_1 d\mathbf{r}_2, \qquad (9.126)$$

where $c_0^{(2)}(r)$ is the FMSA solution for the two-particle DCF of the bulk fluid.

Alternatively, the free energy functional can be split into repulsive and attractive contributions, with the former given by the MFMT and the latter by the earlier-mentioned in this section expansion applied only to the attractive part of the potential. This approximation was denoted by Tang as MFMT+DCF. Both the LDA+DCF and the MFMT+DCF were found to provide nearly identical results, and in excellent agreement with the simulations, for the surface tension of the LJ liquid. A third approximation based on replacing the last two terms in Equation 9.126 with those provided by the SGA gave poorer results.

Lutsko (2007) compared the results obtained from Equation 9.94, with the DCF (9.93) and two simple approximations for $\Delta \bar{c}^{(2)}(r_{12})$, with those obtained using Equation 9.90, and from the SGA, for the surface tension of the LJ liquid. It was found that the SGA considerably overestimates the simulation data at low temperatures, whereas the other approximations provide nearly the same results, all of them in close agreement with the simulation data, except perhaps near the critical point (see Figure 9.5).

9.6 INHOMOGENEOUS SYSTEMS WITH ANISOTROPIC INTERACTIONS

Density functional theories can be extended to inhomogeneous systems with anisotropic interactions, for which one must take into account that the density profiles and correlation functions will depend on the orientation of the molecules in addition to their position. To end this chapter, we will give a short account of some of these extensions.

Considering first hard-body molecules, one may use perturbation expansions around a reference homogeneous system, like those in Equations 9.23 through 9.26. This kind of approach has been frequently used to analyze the phase transitions in HB systems, with the expansion truncated at second order and the two-particle DCF related in one way or another to that of the HS fluid. This may be achieved by expanding the two-particle DCF in spherical harmonics, but most frequently it is approximated by that of an HS fluid with an orientation-dependent scaled distance and eventually using an effective averaged density. Several approximations to this end have been proposed in the literature (see, e.g., Pynn 1974a,b, Wulf 1977, Singh and Singh 1986, Marko 1988, 1989). Concerning the single-particle density of the (partially) ordered phase, one may resort to Fourier expansions, like in Equation 9.72, or to Gaussian approximations, like in Equation 9.70, generalized to include the angular dependence, or to other kind of parametrizations. Thus, Lipkin and Oxtoby (1983) developed a density functional approach for smectic phases with the single-particle density expressed as an expansion in both Fourier series and spherical harmonics and the two-particle DCF also expanded in spherical harmonics. The isotropic-plastic and isotropic-nematic transitions in a system of hard ellipsoids were studied by Singh and Singh (1986) with generalized Fourier expansions for both the nematic and plastic phases and the Pynn (1974a,b) approximation for the two-particle DCF of the isotropic phase, which decouples the orientational and positional degrees of freedom. To describe the same transitions, Marko (1988, 1989) used a Gaussian approximation for the density of the plastic phase, a three-parameter ansatz for the density of the nematic phase, and an improved DCF for the isotropic phase obtained from the DCF of the HS fluid with angle-dependent scaled distance multiplied by a factor depending on a parameter variationally optimized. The freezing of hard ellipsoids of revolution and hard diatomics was analyzed by Smithline et al. (1988), considering a single-particle density with independent translational and oriental contributions, using the Gaussian approximation for the former contribution and a simple ansatz for the orientational one, and a spherical harmonic expansion of the two-particle DCF of the isotropic fluid.

Alternatively, one can start from a generalization of Equation 9.39, to account for the dependence on orientation of the involved quantities, as done by Baus et al. (1987) and Colot et al. (1988) to study the isotropic-nematic transition of hard ellipsoids. The local density of the nematic phase was expressed as the product of the averaged density ρ by the probability density of orientations $f(\Omega)$, that is $\rho(\mathbf{r}) = \rho f(\Omega)$, using a parametric ansatz for $f(\Omega)$. The two-particle DCF of both the nematic and isotropic phases was decoupled into the product of angular and translational contributions, with the latter approximated by the PY solution for the HS fluid with a scaled distance and an effective density. This approximation and the Parsons–Lee one (see Section 8.8) are essentially equivalent, as shown by Vroege and Lekkerkerker (1992).

The WDA was extended to anisotropic hard-body molecular systems independently by Somoza and Tarazona (1988, 1989a,b) and by Poniewierski and Hołist (1988) and Hołist and Poniewierski (1989), by introducing orientation-dependent weighting functions. Somoza and Tarazona chose a system of parallel ellipsoids of revolution (PHE) as the reference system and the Mayer function as the weighting function. Then, the excess free energy functional was constructed in the form

$$\mathscr{F}_{HB}^{E}\left[\rho\left(\mathbf{r}, \Omega\right)\right] = \int \int \rho\left(\mathbf{r}, \Omega\right) f_{PHE}^{E}\left(\bar{\rho}\left(\mathbf{r}\right)\right) d\mathbf{r} d\Omega$$

$$\times \frac{\int \int \rho\left(\mathbf{r}', \Omega'\right) f_{HB}\left(\mathbf{r} - \mathbf{r}', \Omega, \Omega'\right) d\mathbf{r}' d\Omega'}{\int \int \rho\left(\mathbf{r}'\right) f_{PHE}\left(\mathbf{r} - \mathbf{r}'\right) d\mathbf{r}'}, \qquad (9.127)$$

where
$f_{PHE}^{E}(\bar{\rho}(\mathbf{r}))$ is the excess free energy per particle of the reference PHE system
$f_{HB}(\mathbf{r} - \mathbf{r}', \Omega, \Omega')$ and $f_{PHE}(\mathbf{r} - \mathbf{r}')$ are the Mayer functions of the HB and PHE systems, respectively

As all the molecules in the PHE system are aligned along the same direction, the corresponding Mayer function is independent of the molecular orientations, but depends on the direction of $\mathbf{r} - \mathbf{r}'$. By construction the theory satisfies the Onsager limit at low densities and the Parsons–Lee approximation for uniform fluids. With $f_{PHE}^E(\bar{\rho}(\mathbf{r}))$ determined from the WDA, the theory was successfully applied to determine the critical density for the nematic-smectic-A transition and the EOS of both phases in a system of parallel hard spherocylinders with variable length, as well as the coexistence densities for the smectic-A-columnar phase transition.

Instead, Poniewierski and Hołyst considered as the reference system the isotropic HB fluid, and then the excess free energy functional $\mathscr{F}_{HB}^E[\rho(\mathbf{r}, \Omega)]$ was expressed like in Equation 9.30. For the weighting function, they chose

$$w\left(\mathbf{r} - \mathbf{r}', \Omega, \Omega'\right) = -\frac{f_{HB}\left(\mathbf{r} - \mathbf{r}', \Omega, \Omega'\right)}{2B_2^{HB}(\text{iso})}, \qquad (9.128)$$

where $B_2^{HB}(\text{iso})$ is the second virial coefficient for the isotropic fluid. The weighted density was assumed to be

$$\bar{\rho}(\mathbf{r}) = \int\int \rho\left(\mathbf{r}', \Omega'\right)\frac{\rho(\mathbf{r}, \Omega)}{\rho(\mathbf{r})}w\left(\mathbf{r} - \mathbf{r}', \Omega, \Omega'\right)d\mathbf{r}'d\Omega', \qquad (9.129)$$

and excess free energy per particle of the reference fluid

$$\beta f_{HB}^E(\rho) = \rho B_2^{HB} + \left[\beta f_{CS}^E - 4\eta\right], \qquad (9.130)$$

where f_{CS}^E is the excess free energy per particle of the HS fluid as given by the CS equation (5.21). Defined in this way, f_{HB}^E satisfies the Onsager limit for infinitely long molecules and reduces to the CS result for HS.

A considerable effort has been conducted to extend the FMT to fluids with hard anisotropic particles (see Tarazona et al. 2008 for a review), but with limited success up to now, and so we will not extend on this point. Much better achievements have been reached with the extensions to inhomogeneous systems of the Wertheim and related perturbation theories for hard polyatomic molecules analyzed in Chapter 8. Thus, a DFT for nonuniform fluids was derived by Kierlik and Rosinberg (1992, 1993) from the Wertheim first-order perturbation theory for rigid and flexible molecules, respectively, but the calculations are computationally demanding. Segura et al. (1997) combined the WDA with the Wertheim TPT1 to obtain the density profiles of a fluid of associating HS near a hard wall at different temperatures and densities.

Concerning fluids with both repulsive and attractive anisotropic interactions, Blas et al. (2001) developed a simple SAFT-MF-DFT in which the repulsive contribution to the free energy functional is treated in the LDA and a MF approximation is adopted for the attractive contribution, so that for $\mathscr{V}(\mathbf{r}) = 0$

$$F[\rho(\mathbf{r})] = \int \rho(\mathbf{r})f_{rep}(\rho(\mathbf{r}))d\mathbf{r} + F_{MF}[\rho(\mathbf{r})], \qquad (9.131)$$

where $\rho(\mathbf{r})$ is the segment density profile. The repulsive contribution to the free energy per particle is given by the sum

$$f_{rep}\left(\rho\left(\mathbf{r}\right)\right) = f_I\left(\rho\left(\mathbf{r}\right)\right) + f_{HS}\left(\rho\left(\mathbf{r}\right)\right) + f_{chain}\left(\rho\left(\mathbf{r}\right)\right) + f_{assoc}\left(\rho\left(\mathbf{r}\right)\right). \tag{9.132}$$

The ideal gas contribution is given by Equation 9.7; the HS contribution is readily obtained from integration of the CS EOS (5.21) as

$$f_{HS}\left(\rho\left(\mathbf{r}\right)\right) = k_B T \frac{4\eta\left(\mathbf{r}\right) - 3\eta^2\left(\mathbf{r}\right)}{\left(1 - \eta\left(\mathbf{r}\right)\right)^2}, \tag{9.133}$$

where $\eta(\mathbf{r}) = v\rho(\mathbf{r})$ is the packing fraction for segments of volume v; the chain term is obtained from the Wertheim TPT1 expression (see Equation 8.111) in the form

$$f_{chain}\left(\rho\left(\mathbf{r}\right)\right) = k_B T\left(1 - n\right)\ln y_{HS}\left(\sigma; \rho\left(\mathbf{r}\right)\right), \tag{9.134}$$

where

$$y_{HS}\left(\sigma; \rho\left(\mathbf{r}\right)\right) \equiv g_{HS}\left(\sigma; \rho\left(\mathbf{r}\right)\right) = \frac{1 - \eta\left(\mathbf{r}\right)/2}{\left(1 - \eta\left(\mathbf{r}\right)\right)^3}, \tag{9.135}$$

in agreement with the CS equation; and the association contribution (see Equation 8.86) is

$$f_{assoc}\left(\rho\left(\mathbf{r}\right)\right) = k_B T \sum_A \left(\ln X_A\left(\mathbf{r}\right) - \frac{X_A\left(\mathbf{r}\right)}{2}\right) + \frac{1}{2}M, \tag{9.136}$$

and, finally, the mean field term is given by Equation 9.17 with $u(\mathbf{r})$ replaced with $u_{att}(\mathbf{r})$.

The theory was able to correlate the surface tension as a function of the temperature for water and two refrigerants. The SW version of the SAFT-VR approximation (see Section 8.7) was used by Gloor et al. (2004) instead of the ordinary SAFT, the SAFT-HS, to incorporate the effects of the attractions beyond the MFA. This SAFT-VR-DFT showed a strong improvement in the prediction of the interfacial properties at the liquid–vapor coexistence of the SW fluid with respect to the SAFT-MF-DFT version. Fairly good agreement with simulations was also found for the surface tension of the dimerizing SW fluid as a function of the temperature. The theory also proved to be useful for correlating the surface tension of several organic compounds either linear or branched, associating and nonassociating. A closely related theory was applied by Kahl and Winkelmann (2008) to inhomogeneous nonassociating LJ chains. In this theory, the LJ potential is split according to the WCA prescription and the repulsive interaction is approximated by that of HS with an effective diameter determined from the BH equation (6.68) with the upper limit of integration replaced with r_m. The HS contribution to the SAFT free energy functional is treated in the LDA, the chain contribution is given by the TPT1 expression, like in Equation 9.134 with y_{HS} replaced with y_{LJ}, and the attractive contribution is incorporated as a first-order perturbation simplified as in the SAFT-VR,

Equation 8.151. The theory was used to predict the surface tension of the LJ fluid, in excellent agreement with the simulations, and to correlate the experimental data on the liquid–vapor equilibrium and the surface tension of several alkane and aromatic compounds.

Another approximation was derived by Tripathi and Chapman (2005a,b) for homonuclear chains within the framework of the SAFT and was denoted *inhomogeneous/interfacial SAFT (i-SAFT)*. The theory, in which the HS contribution is approximated by the Rosenfeld FMT and the attractive contribution is again treated in the MF approximation, reduces to the SAFT for the bulk fluid. It was used by Tripathi and Chapman (2005a,b) to predict the segment density profiles and surface properties of athermal flexible n-mer fluids, either linear or branched, and blends in overall excellent agreement with the simulations. Dominik et al. (2006) showed that this i-SAFT provides satisfactory agreement with the experimental data for the surface tension of the n-alkanes when using the chain length n and segment diameter σ parameters regressed by Gross and Sadowski (2001) with the PC-SAFT (see Section 8.7), and the segment dispersion energy parameter ε is regressed from experimental bulk properties. The theory has been extended to heteronuclear chains by Jain et al. (2007), under the name of *modified i-SAFT*, and to associating mixtures of polyatomic molecules by Bymaster and Chapman (2010).

REFERENCES

Ayadim, A., M. Oettel, and S. Amokrane. 2009. Optimum free energy in the reference functional approach for the integral equations theory. *J. Phys.: Condens. Matter* 21:115103.

Bagchi, B., C. Cerjan, U. Mohanty, and S. A. Rice. 1984. Crystallization of the classical one-component plasma. *Phys. Rev. B* 29:2857.

Balabanic, C., B. Borštnik, R. Milčič, A. Rubčič, and F. Sokolic. 1989. In *Static and Dynamic Properties of Liquids*, ed. M. Davidovic and A. K. Soper, *Springer Proceedings in Physics* 40:70. Berlin, Germany: Springer.

Barrat, J.-L., J.-P. Hansen, and G. Pastore. 1987a. Factorization of the triplet direct correlation function in dense fluids. *Phys. Rev. Lett.* 58:2075.

Barrat, J.-L., J.-P. Hansen, G. Pastore, and E. M. Waisman. 1987b. Density functional theory of soft sphere freezing. *J. Chem. Phys.* 86:6360.

Baus, M. 1989. Density functional theory of freezing with a self-consistent effective liquid. *J. Phys.: Condens. Matter* 1:3131.

Baus, M. 1990. The present status of the density-functional theory of the liquid-solid transition. *J. Phys.: Condens. Matter* 2:2111.

Baus, M. and J. L. Colot. 1985a. The freezing of hard spheres. The density functional theory revisited. *Mol. Phys.* 55:653.

Baus, M. and J. L. Colot. 1985b. Stability of the high-density hard-sphere solid in the density functional theories of freezing. *J. Phys. C: Solid State Phys.* 18:L365.

Baus, M. and J.-L. Colot. 1986. The hard-sphere glass: Metastability versus density of random close packing. *J. Phys. C* 19:L135.

Baus, M., J.-L. Colot, X.-G. Wu, and H. Xu. 1987. Finite-density Onsager-type theory for the isotropic-nematic transition of hard ellipsoids. *Phys. Rev. Lett.* 59:2184.

Bennett, C. H. 1972. Serially deposited amorphous aggregates of hard spheres. *J. Appl. Phys.* 43:2727.

Blas, F. J., E. Martín del Río, E. de Miguel, and G. Jackson. 2001. An examination of the vapour-liquid interface of associating fluids using a SAFT-DFT approach. *Mol. Phys.* 99:1851.

Broughton, J. Q. and G. H. Gilmer. 1986. Molecular dynamics investigation of the crystal-fluid interface. VI. Excess surface energies of crystal-liquid systems. *J. Chem. Phys.* 84:5759.

Bymaster, A. and W. G. Chapman. 2010. An *i*SAFT density functional theory for associating polyatomic molecules. *J. Phys. Chem. B* 114:12298.

Chaturvedi, D. K., G. Senatore, and M. P. Tosi. 1981. Structure of the strongly coupled classical plasma in the self-consistent mean spherical approximation. *Nuovo Cimento B* Ê62:375.

Choudhury, N. and S. K. Ghosh. 1998. Modified weighted density-functional approach to the crystal-melt interface. *Phys. Rev. E* 57:1939.

Choudhury, N. and S. K. Ghosh. 1999. A perturbative density functional theory of inhomogeneous fluid mixture. *J. Chem. Phys.* 110:8628.

Choudhury, N. and S. K. Ghosh. 2001. Density functional theory of inhomogeneous fluid mixture based on bridge function. *J. Chem. Phys.* 114:8530.

Choudhury, N. and S. K. Ghosh. 2003. Structure of an inhomogeneous fluid mixture: A new weighted density-functional theory within a perturbative approach. *J. Chem. Phys.* 118:1237.

Choudhury, N., C. N. Patra, and S. K. Ghosh. 2002. A new perturbative weighted density functional theory for an inhomogeneous hard-sphere fluid mixture. *J. Phys.: Condens. Matter* 14:11955.

Colot, J. L. and M. Baus. 1985. The freezing of hard spheres. II. A search for structural (f.c.c.-h.c.p.) phase transition. *Mol. Phys.* 56:807.

Colot, J. L., M. Baus, and X. Xu. 1986. The freezing of hard spheres. III. Testing the approximations. *Mol. Phys.* 57:809.

Colot, J.-L., X.-G. Wu, H. Xu, and M. Baus. 1988. Density-functional, Landau, and Onsager theories of the isotropic-nematic transition of hard ellipsoids. *Phys. Rev. A* 38:2022.

Cuesta, J. A., Y. Martínez-Ratón, and P. Tarazona. 2002. Close to the edge of fundamental measure theory: A density functional for hard-sphere mixtures. *J. Phys.: Condens. Matter* 14:11965.

Curtin, W. A. 1987. Density-functional theory of the solid-liquid interface. *Phys. Rev. Lett.* 59:1228.

Curtin, W. A. 1989. Density-functional theory of the crystal-melt interface. *Phys. Rev. B* 39:6775.

Curtin, W. A. and N. W. Ashcroft. 1985. Weighted-density-functional theory of inhomogeneous liquids and the freezing transition. *Phys. Rev. A* 32:2909.

Curtin, W. A. and N. W. Ashcroft. 1986. Density-functional theory and freezing of simple liquids. *Phys. Rev. Lett.* 56:2775.

Curtin, W. A. and N. W. Ashcroft. 1987. Triplet and higher-order direct correlation functions in dense fluids. *Phys. Rev. Lett.* 59:2385.

Curtin, W. A. and K. Runge. 1987. Weighted-density-functional and simulation studies of the bcc hard-sphere solid. *Phys. Rev. A* 35:4755.

Davidchack, R. L. and B. B. Laird. 2000. Direct calculation of the hard-sphere crystal/melt interfacial free energy. *Phys. Rev. Lett.* 85:4751.

de Kuijper, A., W. L. Vos, J.-L. Barrat, J.-P. Hansen, and J. A. Schouten. 1990. Freezing of simple systems using density functional theory. *J. Chem. Phys.* 93:5187.

Denton, A. R. and N. W. Ashcroft. 1989a. Modified weighted-density-functional theory of nonuniform classical liquids. *Phys. Rev. A* 39:4701.

Denton, A. R. and N. W. Ashcroft. 1989b. High-order direct correlation functions of uniform classical liquids. *Phys. Rev. A* 39:426.

Denton, A. R. and N. W. Ashcroft. 1990. Weighted-density-functional theory of nonuniform fluid mixtures: Application to freezing of binary hard-sphere mixtures. *Phys. Rev. A* 42:7312.

Denton, A. R. and N. W. Ashcroft. 1991. Weighted-density-functional theory of nonuniform fluid mixtures: Application to the structure of binary hard-sphere mixtures near a hard wall. *Phys. Rev. A* 44:8242.

Dominik, A., S. Tripathi, and W. G. Chapman. 2006. Bulk and interfacial properties of polymers from interfacial SAFT density functional theory. *Ind. Eng. Chem. Res.* 45:6785.

Duh, D.-M. and D. Henderson. 1996. Integral equation theory for Lennard-Jones fluids: The bridge function and applications to pure fluids and mixtures. *J. Chem. Phys.* 104:6742.

Duque, D., J. C. Pàmies, and L. F. Vega. 2004. Interfacial properties of Lennard-Jones chains by direct simulation and density gradient theory. *J. Chem. Phys.* 121:11395.

Ebner, C., W. F. Saam, and D. Stroud. 1976. Density-functional theory of simple classical fluids. I. Surfaces. *Phys. Rev. A* 14:2264.

Evans, R. 1979. The nature of the liquid-vapour interface and other topics in the statistical mechanics of non-uniform, classical fluids. *Adv. Phys.* 28:143.

Evans, R. 2009. Density functional theory for inhomogeneous fluids I: Simple fluids in equilibrium. *In Lectures at 3rd Warsaw School of Statistical Physics*. Kazimierz Dolny, 27 June–3 July 2009. http://www.phy.bris.ac.uk/people/evans_r/papers/Evans194.pdf (accessed November 18, 2011).

Fu, D. and J. Wu. 2004. A self-consistent approach for modelling the interfacial properties and phase diagrams of Yukawa, Lennard-Jones and square-well fluids. *Mol. Phys.* 102:1479.

Gloor, G. J., G. Jackson, F. J. Blas, E. Martín del Río, and E. de Miguel. 2004. An accurate density functional theory for the vapor-liquid interface of associating chain molecules based on the statistical associating fluid theory for potentials of variable range. *J. Chem. Phys.* 121:12740.

González, A. and J. A. White. 2001. Generating function density functional theory: Free-energy functionals and direct correlation functions for hard-spheres. *Physica A* 296:347.

González, A., J. A. White, and R. Evans. 1997. Density functional theory for hard-sphere fluids: A generating function approach. *J. Phys.: Condens. Matter* 9:2375.

Groot, R. D., N. M. Faber, and J. P. van der Eerden. 1987. Hard sphere fluids near a hard wall and a hard cylinder. *Mol. Phys.* 62:861.

Gross, J. and G. Sadowski. 2001. Perturbed-chain SAFT: An equation of state based on a perturbation theory for chain molecules. *Ind. Eng. Chem. Res.* 40:1244.

Hall, K. H. 1972. Another hard sphere equation of state. *J. Chem. Phys.* 57:2252.

Hansen-Goos, H. and R. Roth. 2006. A new generalization of the Carnahan-Starling equation of state to additive mixtures of hard spheres. *J. Chem. Phys.* 124:154506.

Harrowell, P. R., D. W. Oxtoby, and A. D. J. Haymet. 1985. On the positivity of the density in molecular theories of freezing. *J. Chem. Phys.* 83:6058.

Haymet, A. D. J. 1983. A molecular theory for the freezing of hard spheres. *J. Chem. Phys.* 78:4641.

Haymet, A. D. J. 1985. The density functional theory of freezing: Results and high-density artifacts. *J. Phys. Chem.* 89:887.

Haymet, A. D. J. and D. W. Oxtoby. 1981. A molecular theory for the solid-liquid interface. *J. Chem. Phys.* 74:2559.

Haymet, A. D. J. and D. W. Oxtoby. 1986. A molecular theory for freezing: Comparison of theories, and results for hard spheres. *J. Chem. Phys.* 84:1769.

Henderson, D. and E. W. Grundke. 1975. Direct correlation function: Hard sphere fluid. *J. Chem. Phys.* 63:601.

Hlushak, S., A. Trokhymchuk, and S. Sokolowsky. 2009. Direct correlation function of the square-well fluid with attractive well width up to two particle diameters. *J. Chem. Phys.* 130:234511.

Hołist, R. and A. Poniewierski. 1989. Nematic-smectic-*A* transition for perfectly aligned spherocylinders: Application of the smoothed-density approximation. *Phys. Rev. A* 39:2742.

Hoover, W. G. and F. H. Ree. 1968. Melting transition and communal entropy for hard spheres. *J. Chem. Phys.* 49:3609.

Hopkins, P. and M. Schmidt. 2010. Binary non-additive hard sphere mixtures: Fluid demixing, asymptotic decay of correlations and free fluid interfaces. *J. Phys.: Condens. Matter* 22:325108.

Iyetomi, H. and S. Ichimaru. 1988. Nonlinear density-functional approach to the crystallization of the classical one-component plasma. *Phys. Rev. B* 38:6761.

Jain, S., A. Dominik, and W. G. Chapman. 2007. Modified interfacial statistical associating fluid theory: A perturbation density functional theory for inhomogeneous complex fluids. *J. Chem. Phys.* 127:244904.

Jin, Z., Y. Tang, and J. Wu. 2011. A perturbative density functional theory for square-well fluids. *J. Chem. Phys.* 134:174702.

Kahl, G., B. Bildstein, and Y. Rosenfeld. 1996. Structure and thermodynamics of binary liquid mixtures: Universality of the bridge *functional*. *Phys. Rev. E* 54:5391.

Kahl, H. and J. Winkelmann. 2008. Modified PT-LJ-SAFT density functional theory I. Prediction of surface properties and phase equilibria of non-associating fluids. *Fluid Phase Equilibr.* 270:50.

Kaur, C. and S. P. Das. 2001. Heterogeneities in supercooled liquids: A density-functional study. *Phys. Rev. Lett.* 86:2062.

Kaur, C. and S. P. Das. 2002. Metastable structures with modified weighted density-functional theory. *Phys. Rev. E* 65:026123.

Khein, A. and N. W. Ashcroft. 1997. Generalized density functional theory. *Phys. Rev. Lett.* 78:3346.

Khein, A. and N. W. Ashcroft. 1999a. Symmetry based approach to triplet correlation function. *Phys. Rev. E* 59:1803.

Khein, A. and N. W. Ashcroft. 1999b. Generalized density-functional theory: Extended weighted density approaches. *Phys. Rev. E* 60:2875.

Kierlik, E. and M. L. Rosinberg. 1990. Free-energy density functional for the inhomogeneous hard-sphere fluid: Application to interfacial adsorption. *Phys. Rev. A* 42:3382.

Kierlik, E. and M. L. Rosinberg. 1992. A perturbation density-functional theory for polyatomic fluids. I. Rigid molecules. *J. Chem. Phys.* 97:9222.

Kierlik, E. and M. L. Rosinberg. 1993. A perturbation density functional theory for polyatomic fluids. II. Flexible molecules. *J. Chem. Phys.* 99:3950.

Kim, S.-C. and S.-H. Suh. 1996. Weighted-density approximation and its application to classical fluids. *J. Chem. Phys.* 104:7233.

Laird, B. B. and D. M. Kroll. 1990. Freezing of soft spheres: A critical test for weighted-density-functional theories. *Phys. Rev. A* 42:4810.

Laird, B. B., J. D. McCoy, and D. J. Haymet. 1987. Density functional theory of freezing: Analysis of crystal density. *J. Chem. Phys.* 87:5449.

Largo, J., J. R. Solana, S. B. Yuste, and A. Santos. 2005. Pair correlation function of short-ranged square-well fluids. *J. Chem. Phys.* 122:084510.

Lebowitz, J. L. 1964. Exact solution of generalized Percus-Yevick equation for a mixture of hard spheres. *Phys. Rev.* 133:A895.

Lebowitz, J. L. and J. K. Percus. 1963a. Statistical thermodynamics of nonuniform fluids. *J. Math. Phys.* 4:116.

Lebowitz, J. L. and J. K. Percus. 1963b. Asymptotic behavior of the radial distribution function. *J. Math. Phys.* 4:248.

Lebowitz, J. L. and J. S. Rowlinson. 1964. Thermodynamic properties of mixtures of hard spheres. *J. Chem. Phys.* 41:133.

Leidl, R. and H. Wagner. 1993. Hybrid WDA: A weighted-density approximation for inhomogeneous fluids. *J. Chem. Phys.* 98:4142.

Likos, C. N. and N. W. Ashcroft. 1992. Self-consistent theory of freezing of the classical one-component plasma. *Phys. Rev. Lett.* 69:316.

Likos, C. N. and N. W. Ashcroft. 1993. Density-functional theory of nonuniform classical liquids: An extended modified weighted-density approximation. *J. Chem. Phys.* 99:9090.

Likos, C. N., Zs. T. Német, and H. Löwen. 1994. Density-functional theory of solid-to-solid isostructural transitions. *J. Phys.: Condens. Matter* 6:10965.

Lipkin, M. D. and D. W. Oxtoby. 1983. A systematic density functional approach to the mean field theory of smectics. *J. Chem. Phys.* 79:1939.

Löwen, H. 1990. Elastic constants of the hard-sphere glass: A density functional approach. *J. Phys.: Condens. Matter* 2:8477.

Löwen, H. 1994. Melting, freezing and colloidal suspensions. *Phys. Rep.* 237:249.

Löwen, H., T. Beier, and H. Wagner. 1989. Van der Waals theory of surface melting. *Europhys. Lett.* 9:791.

Lutsko, J. F. 2006. Properties of non-fcc hard-sphere solids predicted by density functional theory. *Phys. Rev. E* 74:021121.

Lutsko, J. F. 2007. Density functional theory of inhomogeneous liquids. I. The liquid-vapor interface in Lennard-Jones fluids. *J. Chem. Phys.* 127:054701.

Lutsko, J. F. 2008. Density functional theory of inhomogeneous liquids. II. A fundamental measure approach. *J. Chem. Phys.* 128:184711.

Lutsko, J. F. and M. Baus. 1990a. Can the thermodynamic properties of a solid be mapped onto those of a liquid? *Phys. Rev. Lett.* 64:761.

Lutsko, J. F. and M. Baus. 1990b. Nonperturbative density-functional theories of classical nonuniform systems. *Phys. Rev. A* 41:6647.

Marko, J. F. 1988. Accurate calculation of isotropic-plastic and isotropic-nematic transitions in the hard-ellipsoid fluid. *Phys. Rev. Lett.* 60:325.

Marko, J. F. 1989. First-order phase transitions in the hard-ellipsoid fluid from variationally optimized direct pair correlation. *Phys. Rev. A* 39:2050.

Marr, D. W. and A. P. Gast. 1993. Planar density-functional approach to the solid-fluid interface of simple liquids. *Phys. Rev. E* 47:1212.

Marshall, C., B. B. Laird, and A. D. J. Haymet. 1985. Freezing of the Lennard-Jones liquid. *Chem. Phys. Lett.* 122:320.

Mecke, M., J. Winkelmann, and J. Fischer. 1997. Molecular dynamics simulation of the liquid-vapor interface: The Lennard-Jones fluid. *J. Chem. Phys.* 107:9264.

Mederos, L., G. Navascués, P. Tarazona, and E. Chacón. 1993. Perturbation weighted-density approximation: The phase diagram of a Lennard-Jones system. *Phys. Rev. E* 47:4284.

Mederos, L., G. Navascués, and P. Tarazona. 1994. Perturbation theory applied to the freezing of classical systems. *Phys. Rev. E* 49:2161.

Mu, Y., A. Houk, and X. Song. 2005. Anisotropic interfacial free energies of the hard-sphere crystal-melt interfaces. *J. Phys. Chem. B* 109:6500.

Német, Zs. T. and C. N. Likos. 1995. Solid to solid isostructural transitions: The case of attractive Yukawa potentials. *J. Phys.: Condens. Matter* 7:L537.

Nordholm, S., M. Johnson, and B. C. Freasier. 1980. Generalized van der Waals theory. III. The prediction of hard sphere structure. *Aust. J. Chem.* 33:2139.

Oettel, M. 2005. Integral equations for simple fluids in a general reference functional approach. *J. Phys.: Condens. Matter* 17:429.

Ohnesorge, R., H. Löwen, and H. Wagner. 1991. Density-functional theory of surface melting. *Phys. Rev. A* 43:2870.

Ohnesorge, R., H. Löwen, and H. Wagner. 1994. Density functional theory of crystal-fluid interfaces and surface melting. *Phys. Rev. E* 50:4801.

Oxtoby, D. W. and A. D. J. Haymet. 1982. A molecular theory of the solid-liquid interface. II. Study of bcc crystal-melt interfaces. *J. Chem. Phys.* 76:6262.

Patra, C. N. 1999. Structure of binary hard-sphere mixtures near a hard wall: A simple weighted-density-functional approach. *J. Chem. Phys.* 111:6573.

Patra, C. N. and S. K. Ghosh. 2002. Structure of nonuniform fluid mixtures: A self-consistent density-functional approach. *J. Chem. Phys.* 117:8933.

Percus, J. K. 1976. Equilibrium state of a classical fluid of hard rods in an external field. *J. Stat. Phys.* 15:505.

Phan, S., E. Kierlik, M. L. Rosinberg, B. Bildstein, and G. Kahl. 1993. Equivalence of two free-energy models for the inhomogeneous hard-sphere fluid. *Phys. Rev. E* 48:618.

Poniewierski, A. and R. Hołyst. 1988. Density-functional theory for nematic and smectic-A ordering of hard spherocylinders. *Phys. Rev. Lett.* 61:2461.

Potoff, J. J. and A. Z. Panagiotopoulos. 2000. Surface tension of the three-dimensional Lennard-Jones fluid from histogram-reweighting Monte Carlo simulations. *J. Chem. Phys.* 112:6411.

Pynn, R. 1974a. Theory of static correlations in a fluid of linear molecules. *Solid State Commun.* 14:29.

Pynn, R. 1974b. Density and temperature dependence of the isotropic-nematic transition. *J. Chem. Phys.* 60:4579.

Ramakrishnan, T. V. and M. Yussouff. 1979. First-principles order-parameter theory of freezing. *Phys. Rev. B* 19:2775.

Rascón C., L. Mederos, and G. Navascués. 1995b. Solid-to-solid isostructural transition in the hard sphere/attractive Yukawa system. *J. Chem. Phys.* 103:9795.

Rascón C., G. Navascués, and L. Mederos. 1995a. Phase transitions in systems with extremely short-ranged attractions: A density-functional theory. *Phys. Rev. B* 51:14899.

Reiss, H., H. L. Frisch, and J. L. Lebowitz. 1959. Statistical mechanics of rigid spheres. *J. Chem. Phys.* 31:369.

Rosenfeld, Y. 1989. Free-energy model for the inhomogeneous hard-sphere fluid mixture and density-functional theory of freezing. *Phys. Rev. Lett.* 63:980.

Rosenfeld, Y. 1993. Free energy model for inhomogeneous fluid mixtures: Yukawa-charged hard spheres, general interactions, and plasmas. *J. Chem. Phys.* 98:8126.

Rosenfeld, Y. 1994. Phase separation of asymmetric binary hard-sphere fluids: Self-consistent density functional theory. *Phys. Rev. Lett.* 72:3831.

Rosenfeld, Y. 1998. Self-consistent density functional theory and the equation of state for simple fluids. *Mol. Phys.* 94:929.

Rosenfeld, Y., M. Schmidt, H. Löwen, and P. Tarazona. 1996. Dimensional crossover and the freezing transition in density functional theory. *J. Phys.: Condens. Matter* 8:L577.

Rosenfeld, Y., M. Schmidt, H. Löwen, and P. Tarazona. 1997. Fundamental-measure free-energy density functional for hard spheres: Dimensional crossover and freezing. *Phys. Rev. E* 55:4245.

Roth, R. 2010. Fundamental measure theory for hard-sphere mixtures: A review. *J. Phys.: Condens. Matter* 22:063102.

Roth, R., R. Evans, A. Lang, and G. Khal. 2002. Fundamental measure theory for hard-sphere mixtures revisited: The White Bear version. *J. Phys.: Condens. Matter* 14:12063.

Schmidt, M. 1999. An *ab initio* density functional for penetrable spheres. *J. Phys.: Condens. Matter* 11:10163.

Schmidt, M. 2000a. Density functional for additive mixtures. *Phys. Rev. E* 62:3799.

Schmidt, M. 2000b. Fluid structure from density-functional theory. *Phys. Rev. E* 62:4976.

Schmidt, M. 2004. Rosenfeld functional for non-additive hard spheres. *J. Phys.: Condens. Matter* 16:L351.

Segura, C. J., W. G. Chapman, and K. P. Shukla. 1997. Associating fluids with four bonding sites against a hard wall: Density functional theory. *Mol. Phys.* 90:759.

Singh, Y. and F. F. Abraham. 1977. Statistical mechanical theory for nonuniform fluids: Properties of the hard-sphere system and a perturbation theory for nonuniform simple fluids. *J. Chem. Phys.* 67:537.

Singh, U. P. and Y. Singh. 1986. Molecular theory for freezing of a system of hard ellipsoids: Properties of isotropic-plastic and isotropic-nematic transition. *Phys. Rev. A* 33:2725.

Singh, Y., J. P. Stoessel, and P. G. Wolynes. 1985. Hard-sphere glass and density-functional theory of aperiodic crystals. *Phys. Rev. Lett.* 54:1059.

Smithline, S. J. and A. D. J. Haymet. 1987. Density functional theory for the freezing of 1:1 hard sphere mixtures. *J. Chem. Phys.* 86:6486.

Smithline, S. J., S. W. Rick, and A. D. J. Haymet. 1988. Density functional theory of freezing for molecular liquids. *J. Chem. Phys.* 88:2004.

Sokolowski, S. and J. Fischer. 1992. The role of attractive intermolecular forces in the density functional theory of inhomogeneous fluids. *J. Chem. Phys.* 96:5441.

Somoza, A. M. and P. Tarazona. 1988. Nematic-smectic-*A*-smectic-*C* transitions in systems of parallel hard molecules. *Phys. Rev. Lett.* 61:2566.

Somoza, A. M. and P. Tarazona. 1989a. Density functional approximation for hard-body liquid crystals. *J. Chem. Phys.* 91:517.

Somoza, A. M. and P. Tarazona. 1989b. Columnar liquid crystal of parallel hard spherocylinders. *Phys. Rev. A* 40:4161.

Sweatman, M. B. 2002. Fundamental measure theory for pure systems with soft, spherically repulsive interactions. *J. Phys.: Condens. Matter* 14:11921.

Tang, Y. 2003. On the first-order mean spherical approximation. *J. Chem. Phys.* 118:4140.

Tang, Y. 2004. First-order mean spherical approximation for inhomogeneous fluids. *J. Chem. Phys.* 121:10605.

Tang, Y. 2005. First-order mean-spherical approximation for interfacial phenomena: A unified method from bulk-phase equilibria study. *J. Chem. Phys.* 123:204704.

Tang, Y. 2007. Direct correlation function for the square-well potential. *J. Chem. Phys.* 127:164504.

Tang, Y. 2008. An accurate theory for the square-well potential: From long to short well width. *Mol. Phys.* 106:2431.

Tang, Y. and B. C.-Y. Lu. 1994. An analytical analysis of the square-well fluid behaviors. *J. Chem. Phys.* 100:6665.

Tang, Y. and J. Wu. 2003. A density functional theory for bulk and inhomogeneous Lennard-Jones fluids from the energy route. *J. Chem. Phys.* 119:7388.

Tang, Y. and J. Wu. 2004. Modeling inhomogeneous van der Waals fluids using an analytical direct correlation function. *Phys. Rev. E* 70:011201.

Tarazona, P. 1984. A density functional theory of melting. *Mol. Phys.* 52:81.

Tarazona, P. 1985. Free-energy density functional for hard spheres. *Phys. Rev. A* 31:2672. See also Tarazona, P. 1985. Erratum: Free-energy density functional for hard spheres. *Phys. Rev. A* 32:3148.

Tarazona, P. 2000. Density functional for hard sphere crystals: A fundamental measure approach. *Phys. Rev. Lett.* 84:694.

Tarazona, P. 2002. Fundamental measure theory and dimensional interpolation for the hard spheres fluid. *Physica A* 306:243.

Tarazona, P., J. A. Cuesta, and Y. Martínez Ratón. 2008. Density functional theories of hard particle systems. In *Theory and Simulation of Hard-Sphere Fluids and Related Systems*, ed. A. Mulero, *Lect. Notes Phys.* 753:247–341. Berlin, Germany: Springer-Verlag.

Tarazona, P. and R. Evans. 1984. A simple density functional theory for inhomogeneous liquids. Wetting by gas at a solid-liquid interface. *Mol. Phys.* 52:847.

Tarazona, P. and Y. Rosenfeld. 1997. From zero-dimension cavities to the free-energy functionals for hard disks and hard spheres. *Phys. Rev. E* 55:R4873.

Tarazona, P., U. Marini Bettolo Marconi, and R. Evans. 1987. Phase equilibria of fluid interfaces and confined fluids. Non-local versus local density functionals. *Mol. Phys.* 60:573.

Telo da Gama, M. M. and R. Evans. 1979. The density profile and surface tension of a Lennard-Jones fluid from a generalized van der Waals theory. *Mol. Phys.* 38:367.

Toxvaerd, S. 1971. Perturbation theory for nonuniform fluids: Surface tension. *J. Chem. Phys.* 55:3116.

Tripathi, S. and W. G. Chapman. 2005a. Microstructure and thermodynamics of inhomogeneous polymer blends and solutions. *Phys. Rev. Lett.* 94:087801.

Tripathi, S. and W. G. Chapman. 2005b. Microstructure of inhomogeneous polyatomic mixtures from a density functional formalism for atomic mixtures. *J. Chem. Phys.* 122:094506.

Upstill, C. E. and R. Evans. 1977. The surface tension and density profile of simple liquids. *J. Phys. C: Solid State Phys.* 10:2791.

Verlet, L. and J.-J. Weis. 1972. Equilibrium theory of simple liquids. *Phys. Rev. A* 5:939.

Vroege, G. J. and H. N. W. Lekkerkerker. 1992. Phase transitions in lyotropic colloidal and polymer liquid crystals. *Rep. Prog. Phys.* 55:1241.

Wadewitz, T. and J. Winkelmann. 2000. Application of density functional perturbation theory to pure liquid-vapor interfaces. *J. Chem. Phys.* 113:2447.

Warshavsky, V. B. and X. Song. 2006. Fundamental-measure density functional theory study of the crystal-melt interface of the hard sphere system. *Phys. Rev. E* 73:031110.

Wulf, A. 1977. Short-range correlations and the effective orientational energy in liquid crystals. *J. Chem. Phys.* 67:2254.

Yang, A. J. M., P. D. Fleming, and J. H. Gibbs. 1976. Molecular theory of surface tension. *J. Chem. Phys.* 64:3732.

Yu, Y.-X. 2009. A novel weighted density functional theory for adsorption, fluid-solid interfacial tension, and disjoining properties of simple liquid films on planar solid surfaces. *J. Chem. Phys.* 131:024704.

Yu, Y.-X. and J. Wu. 2002. Structures of hard-sphere fluids from a modified fundamental-measure theory. *J. Chem. Phys.* 117:10156.

Zhou, S. 1999. A simple weighted-density-functional method: Test and its application to hard sphere fluid in spherical cavity. *J. Chem. Phys.* 110:2140.

Zhou, S. 2000. Inhomogeneous mixture system: A density functional formalism based on the universality of the free energy density functional. *J. Chem. Phys.* 113:8719.

Zhou, S. 2001. Density functional theory based on the universality principle and third-order expansion approximation for adhesive hard-sphere fluid near surfaces. *J. Phys. Chem. B* 105:10360.

Zhou, S. 2002. Perturbation density functional theory for density profile of a nonuniform and uniform hard core attractive Yukawa model fluid. *J. Phys. Chem. B* 106:7674.

Zhou, S. 2003a. Mean spherical approximation-based partitioned density functional theory. *Commun. Theor. Phys. (Beijing, China).* 40:721.

Zhou, S. 2003b. Partitioned density functional approach for a Lennard-Jones fluid. *Phys. Rev. E* 68:061201.

Zhou, S. 2004a. Perturbative density functional approximation in the view of weighted density concept and beyond. *Chem. Phys. Lett.* 385:208.

Zhou, S. 2004b. Formally exact truncated nonuniform excess Helmholtz free energy density functional: Test and application. *J. Phys. Chem.* 108:3017.

Zhou, S. 2009. A new scheme for perturbation contribution in density functional theory and application to solvation force and critical fluctuations. *J. Chem. Phys.* 131:134702.

Zhou, S. 2010a. New free energy density functional and application to core-softened fluid. *J. Chem. Phys.* 132:194112.

Zhou, S. 2010b. Going beyond the mean field approximation in classical density functional theory and application to one attractive core-softened model fluid. *J. Stat. Mech.* 210:11039.

Zhou, S. and A. Jamnik. 2005a. Analysis of the validity of perturbation density functional theory: Based on extensive simulation for simple fluid at supercritical and subcritical temperature under various external potentials. *J. Chem. Phys.* 122:064503.

Zhou, S. and A. Jamnik. 2005b. Global and critical test of the perturbation density-functional theory based on extensive simulation of Lennard-Jones fluid near an interface and in confined systems. *J. Chem. Phys.* 123:124708.

Zhou, S. and E. Ruckenstein. 2000a. High-order direct correlation functions of uniform fluids and their application to the high-order perturbative density-functional theory. *Phys. Rev. E* 61:2704.

Zhou, S. and E. Ruckenstein. 2000b. A density functional theory based on the universality of the free energy density functional. *J. Chem. Phys.* 112:8079.

Zhou, S. and E. Ruckenstein. 2000c. A new density functional approach to nonuniform Lennard-Jones fluids. *J. Chem. Phys.* 112:5242.

10 Overview to Perturbation Theories for More Complex Systems

In this chapter, we give a quick look at the theoretical description of more complex systems, such as fluids near the critical point, electrolyte solutions, liquid metals, polymers, colloids, and protein solutions. The complexity of the fluids near the critical point arises from the large-scale fluctuations involved. Most of the theories analyzed in Chapters 4, 6 through 9 fail to provide accurate results near the critical point unless appropriate corrections are introduced, as will be explained in Section 10.1. In the remaining systems considered in this chapter, the complexity arises from the interactions involved. In particular, in the mixtures discussed in Sections 10.3 and 10.4 of this chapter, the complexity is partially due to the large asymmetric sizes of the components of the mixtures. In these complex mixtures, the problem is largely simplified by introducing an effective potential allowing to map the mixture onto an effective one-component fluid. Therefore, a considerable part of Section 10.3 is devoted to the introduction of several effective potential models. The application of the effective one-component approach to a class of particularly challenging systems, namely, globular proteins in solution, is illustrated in Section 10.4.

10.1 FLUIDS NEAR THE CRITICAL POINT

As mentioned in Section 3.6, near the critical point, several thermodynamic properties either diverge or tend to zero according to power laws. Thus, for a fluid near the critical point

$$C_V \sim |T - T_c|^{-\alpha}, \quad \rho_L - \rho_G \sim |T - T_c|^{\beta}, \quad \kappa_T \sim |T - T_c|^{-\gamma},$$

$$|P - P_c| \sim |\rho - \rho_c|^{\delta}, \quad \xi \sim |T - T_c|^{-\nu}, \tag{10.1}$$

where ξ is the *correlation length*, which is a measure of the size of the density fluctuations and is related to the asymptotic (large r) behavior of the total correlation function

$$h(r) \sim \frac{1}{r^{1+\eta}} e^{-r/\xi}, \tag{10.2}$$

and α, β, γ, δ, ν, and η are *critical exponents*. Classical (mean field) calculations yield $\alpha = 0$, $\beta = 1/2$, $\gamma = 1$, $\delta = 3$, $\nu = 1/2$, and $\eta = 0$ (see, e.g., Wilson and

Kogut 1974), in disagreement with the experimentally measured values $\alpha \sim 0.11$, $\beta \sim 0.33$, $\gamma \sim 1.2$, $\delta \sim 4.3$, $\nu \sim 0.62$, and $\eta \sim 0.042$ (Pelissetto and Vicari 2002).

The reason for this breakdown of the classical behavior is the strong increase in the correlation length near the critical point. Far from the critical point ξ is of the order of several molecular diameters and the thermodynamic properties of the system are determined by the interaction potential. Near the critical point ξ strongly increases and the thermodynamic properties are mainly determined by cooperative effects arising from a huge number of particles within the correlation length, whereas the system-dependent interaction potential plays a minor role. As a consequence, systems with very different interactions behave in a similar way near the critical point, which leads to the concept of *universality*, as mentioned in Section 3.6. This means that systems within the same *universality class* have the same scaling functions and critical exponents, and only the amplitudes, that is the proportionality coefficients in the scaling laws (10.1), are system-dependent.

The *renormalization group* (RG) approach (Wilson 1971a,b, Wilson and Kogut 1974) provides a way to deal with systems near the critical point. First formulated for the Ising model, the results can be extended to ordinary fluids as they belong to the same universality class. Let us consider for simplicity a square lattice like the one in Figure 10.1a, with lattice parameter a. Following a previous idea of Kadanoff (1966), Wilson (1971a,b) groups the lattice sites into blocks, like in Figure 10.1b, and then replaces the particles in each block with an effective particle, thus giving rise to the lattice in Figure 10.1c with lattice spacing $2a$. If the interactions in the initial lattice are local, that is to say is limited to the neighboring particles, with Hamiltonian $\mathcal{H}_0$, the effective interactions in the resulting lattice will be also local with Hamiltonian $\mathcal{H}_1$. The process can be repeated as many times as needed until the lattice space of the effective lattice reaches the value ξ, so that, if this occurs after n transformations, then $2^n a \sim \xi$. If we denote by T the transformation from an effective Hamiltonian to the next one, we will have $T(\mathcal{H}_0) = \mathcal{H}_1$, $T(\mathcal{H}_1) = \mathcal{H}_2,..., T(\mathcal{H}_{n-1}) = \mathcal{H}_n$. At each step we only need to consider a few number of particles and after the n steps we are considering a subsystem with size $\sim \xi$, but the effective Hamiltonian $\mathcal{H}_n$ is obtained from the effective Hamiltonian $\mathcal{H}_{n-1}$ and only a small number of effective particles

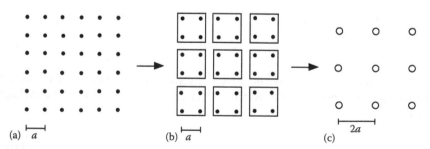

(a) a (b) a (c) $2a$

FIGURE 10.1 Illustration of the blocking procedure involved in the RG approach. The sites in the original lattice (a) are grouped into blocks (b) and the particles in each block are replaced by an effective particle (c). The procedure is repeated as many times as needed until the lattice spacing of the effective lattice reaches a value of the order of the correlation length.

are involved. Eventually, at the end of the process the transformation will reach a fixed point such that $T(\mathcal{H}_n) = \mathcal{H}_n$. A system may have more than one fixed point, and the fixed point attained at the end of the iteration will depend on the initial $\mathcal{H}_0$.

The key question in the iteration is to determine the equations governing the change in the parameters that define the interaction, and thus the Hamiltonian, at each renormalization step. Kadanoff applied the procedure to the Ising model assuming a discrete scaling parameter a and introducing simple ansatzs for the scaling of the interaction parameters with a. Wilson generalized the procedure by allowing a to vary continuously and deriving the differential equations governing the dependence of the interaction parameters with a.

The RG theory predicts that the critical exponents α, γ, and ν are the same; either the critical point is approached from above or from below the critical temperature. The theory also predicts several relationships between the critical exponents for the Ising universality class. For a three-dimensional system, they are

$$\alpha + 2\beta + \gamma = 2, \quad \beta(\delta + 1) + \alpha = 2, \quad \gamma = \nu(2 - \eta), \quad 3\nu + \alpha = 2. \quad (10.3)$$

For a three-dimensional Ising model, the critical exponents cannot be determined exactly, but very accurate estimations have been obtained from a number of procedures (see, e.g., Pelissetto and Vicari 2002 for a review). They give $\alpha \simeq 0.110$, $\beta \simeq 0.3265$, $\gamma \simeq 1.237$, $\delta \simeq 4.79$, $\nu \simeq 0.630$, and $\eta \simeq 0.036$, in very good agreement with the earlier-mentioned experimental values.

The RG theory was extended to fluids by White and Zhang (1993, 1995) by applying the procedure to the free energy rather than to the Hamiltonian. To this end, the interaction potential is separated into repulsive (short-ranged) and attractive (long-ranged) parts. The short-ranged interactions do not contribute significantly to the long-range correlations, so that their contribution to the free energy is obtained from integration of any suitable EOS. The renormalization procedure is then applied only to the attractive interactions. The effect of the density fluctuations on the attractive interactions is considered recursively for successively larger wavelengths, covering all the wavelengths greater than the range of the attractive interactions. The attractive contribution to the free energy density $f^*_{att}(T, \rho)$ is written as the sum of fluctuation and mean field contributions

$$f^*_{att}(T, \rho) = f^*_F(T, \rho) + f^*_{MF}(T, \rho). \quad (10.4)$$

The fluctuation contribution is obtained from the RG procedure as

$$f^*_F(T, \rho) = \lim_{n \to \infty} f^*_n(T, \rho), \quad (10.5)$$

with

$$f^*_n(T, \rho) = f^*_{n-1}(T, \rho) + \delta f^*_n(T, \rho), \quad (10.6)$$

and the MF contribution is of the vdW form

$$f^*_{MF}(T, \rho) = -a\rho^2, \quad (10.7)$$

with a, the vdW constant, given by Equation 6.45.

The iterative process starts at $n = 0$ with the repulsive free energy density $f_0^*(T, \rho)$, as determined from an appropriate EOS, and the effect of density fluctuations on $f_F^*(T, \rho)$ is incorporated step by step. At each step n, the effects of density fluctuations with wavelengths shorter than λ_n are incorporated into $f_n^*(T, \rho)$ through Equation 10.6, and $\delta f_n^*(T, \rho)$ is determined from the Wilson (1971b) phase-space cell method, with the result

$$\delta f_n^*(T, \rho) = -\frac{1}{\beta V_n} \ln \left\{ \frac{\int_0^\rho \exp\left[-\beta V_n \mathscr{Q}(x, D_n)\right] dx}{\int_0^\rho \exp\left[-\beta V_n \mathscr{Q}(x, D_0)\right] dx} \right\}, \qquad (10.8)$$

in which $\beta = 1/k_B T$, and

$$\mathscr{Q}(x, D_n) = \frac{f_{n-1}^*(T, \rho + x) + f_{n-1}^*(T, \rho - x)}{2} - f_{n-1}^*(T, \rho) + u_n(x, D_n). \qquad (10.9)$$

White and Zhang assumed a sinusoidal form for the density fluctuations and obtained

$$V_n = \left(\frac{z\lambda_n}{2}\right)^3, \qquad (10.10)$$

where $z \simeq 1$ is a parameter, $\lambda_n = t^n \lambda_0$, t is a factor,

$$\lambda_0 = \frac{2\pi}{\sqrt{2}} R, \qquad (10.11)$$

R is the effective range of the attractive part of the potential defined from

$$R^2 = \frac{\int (\mathbf{n} \cdot \mathbf{r})^2 u_{att}(r) \, d\mathbf{r}}{\int u_{att}(r) \, d\mathbf{r}}, \qquad (10.12)$$

where $\mathbf{n}$ is a unit vector pointing in an arbitrary direction, and

$$u_n(x, D_n) = -a(1 - D_n)x^2, \qquad (10.13)$$

with

$$D_n = \frac{1}{t^{2n}}. \qquad (10.14)$$

Alternatively, Equation 10.10 can be rewritten as

$$V_n = ct^{3n}. \qquad (10.15)$$

Using for $f_0^*(T, \rho)$ the free energy density derived from the CS EOS for HS, taking $t = 2$, and considering c as an adjustable parameter determined from the experimental value of the compressibility factor at the critical point, White and Zhang (1993) obtained excellent agreement with the experimental data for the pressure isotherms of pentane above and below the critical point up to fairly high pressures. Similar accuracy was achieved by White and Zhang (1995) for hydrogen with the SW potential using Equation 10.10 with fixed $z = 1.06$. An improved version of the theory was developed by White (2000) and applied to the LJ fluid with excellent results for the coexistence densities, vapor pressures, critical constants, and pressure along isotherms. The theory has been extended by Mi et al. (2004a, 2005a) to molecular fluids and mixtures, respectively, in combination with the SAFT.

The White and Zhang theory was reformulated by Tang (1998) in order to render it applicable with any theory for the fluid, instead of the vdW mean field theory, thus allowing to obtain more accurate results outside of the critical region without using adjustable parameters. In this case, the effective range of the potential is defined in the form

$$R^2 (T, \rho) = \frac{\int_0^\beta d\beta' \int \left[g(r) + \rho \frac{\partial g(r)}{\partial \rho} + \frac{\rho^2}{4} \frac{\partial g^2(r)}{\partial \rho^2} \right] (\mathbf{n} \cdot \mathbf{r})^2 u_{att}(r) \, d\mathbf{r}}{\int_0^\beta d\beta' \int g(r) u_{att}(r) \, d\mathbf{r}}, \tag{10.16}$$

where $\beta = 1/k_B T$ and $g(r)$ is the (temperature- and density-dependent) RDF of the fluid, and the function $u_n(x, D_n)$ in Equation 10.9 is replaced with

$$u'_n (x, D_n) = -\frac{x^2 D_n}{\rho^2} f^*_{att} (T, \rho). \tag{10.17}$$

The theory that reduces to the White and Zhang theory when combined with the MF approximation $g(r) = 1$ was applied to the LJ fluid, using $z = 1$ in Equation 10.10 and the FMSA for the RDF and the free energy density. This FMSA+RG theory resulted in a much better prediction of the liquid–vapor coexistence densities near the critical point as compared with the FMSA. The calculated critical temperature and density were found to be in close agreement with the simulation results, and the value 0.336 obtained for the critical exponent β also closely agrees with the earlier-quoted values found from RG theory and from experiment.

Based on similar grounds is the approach devised by Lue and Prausnitz (1998a), in which the homogeneous fluid is treated in the MSA and the short-wavelength attractive contribution is incorporated into the reference free energy density after subtracting the long-wavelength contribution treated in the MF approximation, that is,

$$f^*_0 (T, \rho) = f^*_{MSA} (T, \rho) + a\rho^2. \tag{10.18}$$

Now, the recursive relations become

$$\delta f^*_n(T, \rho) = -K_n \ln \frac{I_{n,s}(T, \rho)}{I_{n,l}(T, \rho)}, \quad 0 \le \rho < \frac{\rho_{max}}{2},$$

$$= 0, \quad \frac{\rho_{max}}{2} \le \rho < \rho_{max}, \tag{10.19}$$

with

$$K_n = \frac{k_B T}{(2^n \lambda_0)^3},$$ (10.20)

$$I_{n,i} = \int_0^\rho \exp\left[\frac{-G_{n,i}(x,\rho)}{K_n}\right] dx, \quad i = s, l,$$ (10.21)

$$G_{n,i}(x,\rho) = \frac{\bar{f}_{n,i}^*(T,\rho+x) + \bar{f}_{n,i}^*(T,\rho-x)}{2} - \bar{f}_{n,i}^*(T,\rho), \quad i = s, l,$$ (10.22)

$$\bar{f}_{n,l}^*(T,\rho) = f_{n-1}^*(T,\rho) + a\rho^2,$$ (10.23)

$$\bar{f}_{n,s}^*(T,\rho) = f_{n-1}^*(T,\rho) + b\frac{a}{2}\left(\frac{R}{2^n \lambda_0}\right)^2 \rho^2,$$ (10.24)

where

ρ_{max} is the highest density allowed to the fluid

$f_n^*(T,\rho)$ is given by Equation 10.6

R is the effective range of the attractive interaction obtained from

$$R^2 = \frac{1}{3}\frac{\int u_{att}(r) r^2 d\mathbf{r}}{\int u_{att}(r) d\mathbf{r}},$$ (10.25)

and b is a parameter.

The theory was applied by Lue and Prausnitz (1998a,b) to pure fluids and mixtures, respectively, with λ_0 and b treated as adjustable parameters, and extended by Jiang and Prausnitz (1999, 2000) to chain molecular fluids and mixtures, as mentioned in Section 8.7. The same RG approach was used by Fu and Wu (2004) in combination with the MFMT (see Section 9.3) plus the quadratic approximation (9.90) to obtain the coexistence densities and pressures of SW, HCY, and LJ fluids, and by Llovell et al. (2004) in combination with the soft-SAFT (see Section 8.7) to fit the liquid–vapor coexistence densities of LJ chains and n-alkanes. Based on the Lue and Prausnitz RG approach is also the PC-SAFT+RG approximation, with three RG adjustable parameters, used by Bymaster et al. (2008) for n-alkanes and extended by Tang and Gross (2010) to binary mixtures, as also cited in Section 8.7.

Other improved approximations, free from adjustable parameters, based on the use of the White and Zhang RG approximation within the context of the SAFT in combination with the FMSA, have been reported by Mi et al. (2004b, 2005b, 2006). The FMSA-SAFT+RG approximation was successfully applied to predict the liquid–vapor coexistence densities, pressure–density isotherms, and critical parameters of LJ chains, n-alkanes, and mixtures.

10.2   LIQUID METALS, MOLTEN SALTS, AND ELECTROLYTE SOLUTIONS

10.2.1   LIQUID METALS

In a liquid metal, the conduction electrons move more or less freely through the volume of the liquid, so that the system can be considered as a fluid mixture of ions and electrons. However, in contrast with ordinary mixtures, whereas the ion fluid behaves classically, the dense electron gas is a strongly degenerate Fermi gas and must be treated quantically. On the other hand, whereas the interactions in simple fluids are short-ranged, long-range Coulomb interactions are present in liquid metals. The situation becomes more complex because of the simultaneous ion–ion, ion–electron, and electron–electron interactions. The core electrons of each ion prevent the conduction electrons to become closer to the ion center than some distance r_c, the radius of the ion core, because of the Pauli exclusion principle. On the other hand, the conduction electrons produce a screening effect that largely compensates the ion–electron attractions, thus weakening the interaction and giving rise to the so-called *pseudopotential*, which may be considered as a perturbation of the electron gas. Each ion together with the associated screening charge may be regarded as a *pseudoatom* that interacts with each other by means of an effective potential and moves in a background potential due to the presence of the electron gas. The total potential energy of the system is then (Hasegawa and Watabe 1972)

$$\mathcal{U}_N = Nu_0(n) + \sum_{i=1}^{N-1} \sum_{j=i+1}^{N} u\left(r_{ij}; n\right), \tag{10.26}$$

where $n = z\rho$ is the number density of electrons, in which z is the number of conduction electrons per ion and $\rho = N/V$ is the number density of ions, and $u(r_{ij}; n)$ is the effective interaction between two pseudoatoms. As a consequence, the energy (2.46) and pressure (2.36) equations must be modified in the form

$$U^E = Nu_0(n) + 2\pi N\rho \int_0^\infty g(r)\, u(r; n)\, r^2 dr, \tag{10.27}$$

and

$$Z \equiv \frac{P}{\rho k_B T} = 1 + \frac{n}{k_B T} \frac{du_0(n)}{dn} - \frac{2}{3}\pi \frac{\rho}{k_B T} \int_0^\infty \left[r \frac{\partial u(r; n)}{\partial r} - 3n \frac{\partial u(r; n)}{\partial n} \right] g(r) r^2 dr, \tag{10.28}$$

respectively.

According to the pseudopotential theory (see, e.g., Young 1987, 1992), the Fourier transform $\hat{u}(k; n)$ of the effective potential $u(r; n)$ can be expressed as

$$\hat{u}(k; n) = \hat{u}_d(k) + \hat{u}_i(k; n), \tag{10.29}$$

where

$$\hat{u}_d (k) = \frac{4\pi z^2 e^2}{k^2} \qquad (10.30)$$

is the direct Coulomb ion–ion interaction, and

$$\hat{u}_i (k; n) = \frac{k^2}{4\pi e^2} \left[\frac{1}{\epsilon (k; n)} - 1 \right] \hat{u}_p^2 (k) \qquad (10.31)$$

is the indirect interaction due to the presence of the conduction electrons. In the latter expression, $\epsilon(k; n)$ is the *dielectric screening function* for which Ichimaru and Utsumi (1981) reported a frequently used expression, and $\hat{u}_p (k)$ is the Fourier transform of the pseudopotential

$$u_p (r) = u_c (r) - \frac{ze^2}{r}, \qquad (10.32)$$

in which $u_c(r)$ represents the interaction with the ion core and the second term of the right-hand side is the electrostatic attraction outside of the core. A simple approximation is the Aschroft (1966) empty-core model

$$u_p (r) = \begin{cases} 0 & r \leq r_c \\ -ze^2/r & r > r_c \end{cases}, \qquad (10.33)$$

from which

$$\hat{u}_p (k) = -\frac{4\pi z^2 e^2}{k^2} \cos (kr_c), \quad k \neq 0. \qquad (10.34)$$

Finally, from Hasegawa and Watabe (1972)

$$u_0 (n) = z \left(1 - \frac{1}{2} V^2 \frac{d^2}{dV^2} \right) u_{el} + \frac{1}{2} \int \hat{u}_i (k; n) \, d\mathbf{k}, \qquad (10.35)$$

where u_{el} is the energy per electron in the uniform electron gas in a neutralizing background.

The general shape of the potential $u(r; n)$ presents a sharp decay with distance at short distances, due to the repulsive core, and damped oscillations (the Friedel oscillations) for larger distances. The specific shape depends on three parameters: the radius r_c of the ion core, the screening length λ_{sc}, and the Friedel wavelength λ_F (Hafner and Heine 1983). Figure 10.2 shows two examples of the effective potential of metals as calculated from the pseudopotential theory by Hafner and Heine (1983). These authors used the theory to explain the trends in the crystal structures of the elements along rows of the periodic table.

Once the effective potential is determined, any theory suitable for simple fluids may, in principle, be used to determine the structure and thermodynamic properties

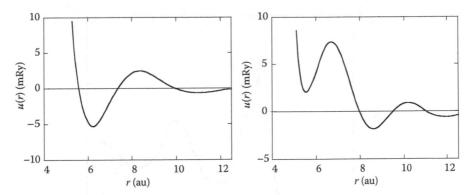

FIGURE 10.2 Two examples of the effective potential energy of metals, in mRy (1 Ry $=$ 2.18 $\times$ 10^{-18} J), as a function of the distance in au (1 au $=$ 5.29 $\times$ 10^{-11} m), after Hafner and Heine (1983).

of liquid metals. Two of the preferred approaches are the RPA and VPT. Thus, for example, Evans and Sluckin (1981) determine the structure factor of liquid alkali metals from the RPA expression (4.45). Two approximations were adopted by these authors for the reference structure factor. In the first of them, the softness of the repulsive reference potential was accounted for by means of a blip function expansion about the HS fluid, yielding (Jacobs and Andersen 1975, Telo da Gama and Evans 1980)

$$S_0(k) = \frac{S_{HS}(k)}{1 - \rho S_{HS}(k)\, B(k)}, \tag{10.36}$$

where $B(k)$ is the Fourier transform of the blip function defined by the integrand of Equation 6.73. Using the latter equation to determine the effective diameter d, they showed that for low k $B(k)$ is negligible and the expression of the structure factor reduces to Equation 4.45 with $S_0(k)$ replaced by $S_{HS}(k)$.

In the second approximation, instead of splitting the effective potential into short-range (HS-like) and long-range parts, it was separated into ion–ion Coulomb repulsion and screening contributions, and the structure factor of the OCP was used for the reference system. With this choice, very good agreement with experiment was achieved for the $S(0)$ of the liquid alkali metals near melting. For lower densities, the importance of the screening effects increases and the RPA becomes inaccurate.

Hafner and Kahl (1984) used the ORPA with the WCA reference system to obtain the structure factor $S(k)$ of liquids Rb and Ge near melting in very good agreement with the experimental data. The theory was shown by Kahl and Hafner (1984) to be accurate for the structure factor of expanded liquid Rb up to $T \approx 1400$ K (melting temperature $T_m = 312.65$ K). The structure factor of Ga at the melting point was calculated by Bretonnet and Regnaut (1985) by means of the ORPA with a modified blip function. They obtained excellent agreement with the simulation data (see Figure 10.3) performed with an empirical potential, with little difference

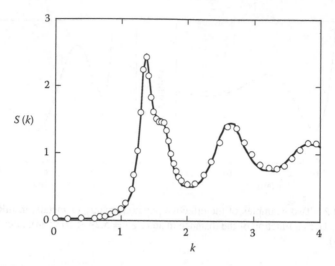

FIGURE 10.3 Structure factor of liquid Gallium at the melting point. Points: simulation data; curve: ORPA, from Bretonnet and Regnaut (1985).

between using the WCA or the SS as the reference systems, and showed that this theory provides much better performance than the RPA and SS approximations.

On the other hand, the VPT, with the HS structure factor as given by the PY theory, was applied by Umar and Young (1974) to obtain the structure factor of a number of metals near melting, with partial success. Mon et al. (1981) analyzed the performance of the HS and OCP as the reference systems within the VPT for the structure factor, and concluded that the second choice is better suited for alkali metals whereas the HS reference is more appropriate for polyvalent metals. The same reference systems, as well as the SS one, were considered by Ross et al. (1981) in their VPT calculations of the Coulomb contribution to the compressibility factor and the excess energy of a model of Li, with the better agreement achieved with the OCP reference, especially for the first of these thermodynamic quantities.

A problem that arises in using the OCP as the reference system in VPT for liquid metals is that the value $\Gamma \sim 130$ obtained for the OCP coupling parameter for liquid alkali metals is much lower than the value $\Gamma \sim 170$ expected from the structural data. To attempt to overcome this drawback, other reference systems, such as the one consisting of CHS in a uniform neutralizing background, have been considered. This reference system was adopted by Lai et al. (1990), together with a pseudopotential model developed by Li et al. (1986), with either two variational parameters, the packing fraction η and Γ, or the variational parameter η, with Γ determined from the structure, to obtain the structure and thermodynamic properties of liquid alkali metals. The analytical solution derived by Palmer and Weeks (1973) within the MSA was used for the reference system. It was found that the one-parameter VPT provides nearly perfect agreement with the experimental data for the liquid structure factor near melting, as illustrated in Figure 10.4 for cesium, whereas the accuracy of the two-parameter VPT is somewhat lower.

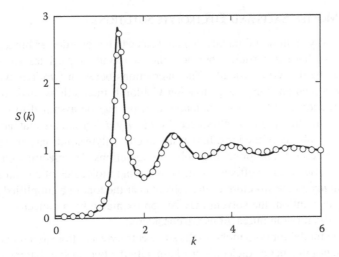

FIGURE 10.4 Structure factor of liquid cesium near the melting point. Points: experimental data from Waseda (1980); curve: VPT from Lai et al. (1990).

Both approaches yield nearly equal, and very accurate, results for the excess energy, but the predicted excess entropy deviates more appreciably from the experiment, with the worse results achieved with the one-parameter approximation. On the other hand, the values $\Gamma \sim 50$ obtained from the two-parameter VPT are much lower than the earlier-quoted expected ones. A similar approach was applied by Bretonnet et al. (1992) to obtain the structure factor of 3d liquid transition metals, with an effective potential derived by Bretonnet and Silbert (1992), in close agreement with the experiment.

The VPT-CHS approach was further worked out by Badirkhan et al. (1992), who attributed the failure of the Lai et al. approximation in obtaining the right value of Γ to the lack of thermodynamic consistency of the MSA solution for the CHS fluid. In addition, Badirkhan et al. noted that the theory fails completely in the region of the CHS fluid close to the OCP limit (low η and high Γ), as it predicts negative values of the RDF near contact. Therefore, Badirkhan et al. (1992) replaced the MSA solution for the CHS fluid by that obtained from RHNC, which in the limits $\Gamma = 0$ and $\eta = 0$ accurately predicts the thermodynamic and structural properties of the HS and OCP fluids, respectively. The resulting VPT predicts two minima in the free energy near melting. For the liquid alkali metals, one of them corresponds to $\eta \approx 0.42$ and $\Gamma \approx 120$ and the other to $\eta \approx 0.05$ and $\Gamma \approx 150$, with the latter providing the lowest free energy. For polyvalent metals, the two minima appear at $\eta \approx 0.42$ and $\Gamma \approx 30$ and at $\eta \approx 0$ and $\Gamma \approx 160$, respectively, with the latter again having slightly lower free energy than the former. In all cases, the differences in the free energies of the two minima are very small and the two solutions provide comparable accuracy in the calculated energy, entropy, and structure factor. Therefore, near melting both the HS and OCP reference systems are nearly equally suitable. At higher temperatures, the results indicate that OCP provides better results.

10.2.2 MOLTEN SALTS AND ELECTROLYTE SOLUTIONS

In a similar way as in liquid metals, molten salts can be regarded as binary mixtures of oppositely charged particles, but now the two kind of particles are of similar size and can be treated classically. The interactions between two ions will include short-range repulsion, long-range ion–ion Coulomb interactions, and ion-induced dipole interactions, and the total potential energy may be assumed to be pairwise additive. A simplification is introduced by the *rigid ion model* that neglects the induced polarizability of the ions. In the case of electrolyte solutions, the presence of the solvent introduces additional ion–solvent interactions. The electric field of an ion will produce orientational effects on the surrounding molecules of a polar solvent as well as changes on the structure of the solvent near the ion. In a simplified picture of an electrolyte solution, the solvent may be approximated by a dielectric continuum and so only the ion–ion interactions are considered.

A simple model for both molten salts and electrolyte solutions is a binary mixture of CHS in a homogeneous dielectric medium with the two species having charges of equal magnitude ze but opposite sign, for which the interaction potential is of the form of the RPM or PM potentials, Equations 1.16 and 1.17, respectively. The reduced variables characterizing the state of this system are the reduced density $\rho^* = \rho\sigma^3$, where $\sigma = (\sigma_1 + \sigma_2)/2$ is the average diameter of the spheres, and $\beta^* = \beta(ze)^2/\epsilon\sigma$, where $\beta = 1/k_bT$ and ϵ is the dielectric constant of the medium. Low values of these parameters correspond to the electrolyte-solution regime and high values to the molten-salt regime.

The thermodynamic and structural properties of this kind of systems are usually obtained from integral equation theory with different closures. Thus, for example, Larsen (1978) solved the HNC and RHNC theories for the RPM in concentrated electrolyte-solution and molten-salt regimes with quite satisfactory results, as compared with the simulation data, for the RDF $g_{++}(r)$ and $g_{+-}(r)$, as well as for the structure factor, but not for the thermodynamic properties in the molten-salt regime. Stell and Hafskjold (1981) also obtained satisfactory results for the RDFs in the molten-salt regime from a self-consistent GMSA (a version of the SCOZA) and the HNC approximations. The predicted structure for the RPM from the HNC and RHNC theories may be improved by introducing suitably chosen effective HS diameters, instead of diameters equal to those of the CHS, as shown by Caccamo et al. (1984). The performance of the MSA, a partially self-consistent GMSA, and the RHNC for the structure and thermodynamic properties of the PM with several diameter ratios was analyzed by Abramo et al. (1983) in the molten-salt regime. The best overall agreement with simulations was provided by the GMSA and comparable accuracy was achieved also with the RHNC, whereas the results supplied by the MSA were considerably worse. The structure and thermodynamic properties of molten alkali chlorides were studied by Ballone et al. (1984). They concluded that to obtain satisfactory agreement with experimental data from a modified HNC, the appropriate reference system for the bridge function should be a mixture of strongly NAHS. Close agreement with experimental data was reported by these authors for the structure of molten NaCl, using an empirical bridge function, with the ratios η_{++}/η_{+-} and η_{--}/η_{+-} of the effective HS packing fractions determined from the peak positions of the corresponding RDFs solved in the HNC approximation, and η_{+-} determined by imposing self-consistency between the virial and compressibility routes.

With regard to the electrolyte-solution regime, Abramo et al. (1984) carried out extensive simulations for the thermodynamic and structural properties of the PM of aqueous 1:1 electrolytes with concentrations in the range 0.1–0.8 M and diameter ratios in the range 0.1–0.8, and compared the simulation data with the results from the HNC, MSA, and EXP approximations. Close agreement with the simulations was found for the excess internal energy calculated from both the HNC and the MSA, except in the latter for the diameter ratio 0.1 at the highest concentration, as well as for the osmotic pressure from the HNC and also from the energy route of the MSA. Concerning the structure, very satisfactory results for the RDFs were obtained from the EXP approximation, whereas the MSA provided too low values, especially near contact. Weis and Levesque (2001) also performed computer simulations for the thermodynamic and structural properties of a PM electrolyte with diameter ratio 0.4 covering a wide range of temperatures and densities, and confirmed the accuracy of the HNC theory in the density range 0.4–0.8. For low temperatures and densities, the theory fails to yield accurate results, or even to yield a solution, as a consequence of its inability of accounting for the clustering of the ions. The importance of an accurate description of the bridge function was tested by Duh and Haymet (1992) by comparing the predicted structure obtained from IET with different closures for a RPM of aqueous 2:2 electrolyte. The best overall results were obtained with a modified HNC incorporating a bridge function derived by these authors by inverting the simulation data with an optimized splitting of the potential. The clustering in the RPM near the critical region was analyzed by MC simulation by Bresme et al. (1995), who found that in approaching the critical temperature an increasing fraction of the ions group into neutral clusters formed by 4–6 ions. They also used an inversion procedure to extract the "exact" bridge function from the simulation data and compared the results with those of the Duh and Haymet (1992) closure. It was found that the closure is accurate for $\beta^* < 10$, but is inaccurate for higher values (lower temperatures) or even fails to converge, and it was concluded that a different splitting of the potential and a reparametrization of the bridge function might remedy these deficiencies.

A different way to improve upon the HNC results is based on the inhomogeneous Ornstein–Zernike approach, devised by Attard (1989a) and extended to mixtures by Jorge et al. (2001), which includes the three-particle correlations. This kind of approach was applied by Jorge et al. (2002) to models of 3:3 and 1:3 electrolyte solutions with different concentrations and size ratios. The HNC3 approximation was shown to appreciably improve the predicted structure as compared with the HNC.

10.3 COLLOIDS AND COLLOID-POLYMER MIXTURES

10.3.1 DEPLETION POTENTIALS

In colloids and colloid polymer mixtures are involved species with very disparate sizes. In a solution consisting in large solute particles and smaller solvent particles, a purely entropic interaction, the *depletion interaction*, appears. This interaction

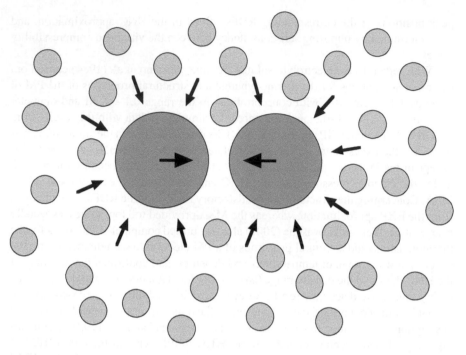

FIGURE 10.5 Illustration of the depletion forces present in a solution in which the diameter of the solute particles is much larger than that of the solvent particles. When two solute particles become close enough to each other, a region between them is free of solvent particles. This causes an imbalance in the pressure forces, indicated by the thin arrows, around the larger particles that gives rise to an effective attraction, indicated by the thick arrows.

arises from excluded volume effects when two solute particles become so close to each other that the separation between their surfaces is smaller than the effective diameter of the solvent particles. In this situation, the imbalance in the pressure around the solute particles gives rise to an effective attractive interaction, as illustrated in Figure 10.5.

A simple model of a colloidal suspension is a binary HS mixture with large values of the diameter ratio $R = \sigma_1/\sigma_2$. In this situation, the usual theories for HS mixtures (see Section 7.5) in general fail to provide accurate results. Computer simulation of very asymmetric HS mixtures is also challenging and requires special techniques (Lue and Woodcock 1999). For this model, the depletion interaction is attractive when the separation between the surfaces of the larger particles is $\delta < \sigma_2$, increases monotonically with δ, is zero for $\delta \geq \sigma_2$, and is proportional to the packing fraction of the solvent. From simple geometrical arguments, Asakura and Oosawa (1954, 1958) and Vrij (1976) determined the depletion force $F_d(\delta)$ acting between two large HS at infinite dilution in the mixture, $\rho_1 \to 0$ and $\rho_2 \equiv \rho$, with their centers separated a distance $r = \sigma_1 + \delta$, assuming that the solvent fluid behaves as an ideal gas, with the result

$$F_d(\delta) = -\frac{1}{4}\pi\rho k_B T \left[(\sigma_1 + \sigma_2)^2 - (\sigma_1 + \delta)^2\right], \qquad \delta \leq \sigma_2,$$

$$0, \qquad\qquad\qquad\qquad\qquad\qquad\qquad\qquad \delta > \sigma_2.$$

(10.37)

The corresponding *Asakura–Ossawa* (AO) potential is

$$u_d(\delta) = -\int_{\sigma_2}^{\delta} F_d(\delta')\, d\delta'$$

$$= -\frac{1}{4}\pi\rho k_B T (\sigma_2 - \delta)\left[\sigma_2\left(\sigma_1 + \frac{2\sigma_2}{3}\right) - \left(\sigma_1 + \frac{\sigma_2}{3}\right)\delta - \frac{\delta^2}{3}\right], \qquad \delta \leq \sigma_2,$$

$$= 0, \qquad\qquad\qquad\qquad\qquad\qquad\qquad\qquad\qquad\qquad\qquad\qquad \delta > \sigma_2.$$

(10.38)

For $\sigma_1 \gg \sigma_2$, the preceding expression reduces to

$$u_d(\delta) = -\frac{1}{4}\pi\rho k_B T \sigma_1 (\sigma_2 - \delta)^2, \qquad \delta \leq \sigma_2,$$

$$0, \qquad\qquad\qquad\qquad\qquad\qquad\qquad \delta > \sigma_2.$$

(10.39)

This approach was carried to third order in the solvent density by Mao et al. (1995).

The ideal gas assumption is valid only at low solvent densities. At higher densities, the force acting on a sphere immersed in a fluid of smaller spheres depends on the density profile $\rho(\mathbf{r})$ of solvent particles around the solute particle in the form (Attard 1989b, Dickman et al. 1997).

$$\mathbf{F} = -k_B T \int_S \rho(\mathbf{r})\, \mathbf{n}\, dA,$$

(10.40)

where

S is the surface of the exclusion sphere for the center of a solvent particle around the solute particle

$\mathbf{n}$ is the unit vector normal to the surface outward

The density profile of the solvent particles is influenced by the presence of a second large sphere close to the first one. If $\mathbf{n}$ is the unit vector along the line joining the centers of a large and a small sphere, and θ is the angle between $\mathbf{n}$ and the line joining the centers of the two large spheres (see Figure 10.6), from Equation 10.40 the depletion force is readily obtained to be

$$F_d(\delta) = -\frac{1}{2}\pi k_B T (\sigma_1 + \sigma_2)^2 \int_0^{\pi} \rho(\sigma_{12}; \theta)\sin\theta \cos\theta\, d\theta,$$

(10.41)

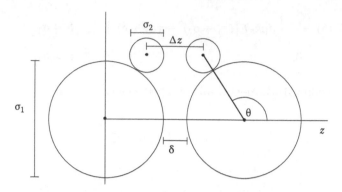

FIGURE 10.6   Geometry involved in the derivation of Equation 10.41.

where $\sigma_{12} = \frac{1}{2}(\sigma_1 + \sigma_2)$ and $\rho(\sigma_{12}; \theta)$ is the density of the solvent fluid at contact distance between the large and small spheres. Equation 10.41 reduces to the Asakura–Oosawa expression (10.37) if the solvent fluid is considered an ideal gas as shown by Attard (1989b).

On the other hand, Equation 10.41 may be viewed as the local pressure $\rho(\mathbf{r})k_B T$ integrated over the area of the exclusion sphere for a sphere of species 2 around a sphere of species 1. Götzelmann et al. (1998) rewrite Equation 10.41 as

$$F_d(\delta) = \frac{1}{2}\pi k_B T (\sigma_1 + \sigma_2)^2 \int\limits_{\pi/2}^{\pi} \Delta\rho\,(\sigma_{12}; \theta)\sin\theta\,(-\cos\theta)\,d\theta, \qquad (10.42)$$

where $\Delta\rho(\sigma_{12}; \theta) = \rho(\sigma_{12}; \theta) - \rho(\sigma_{12}; \pi - \theta)$, with $\pi/2 < \theta < \pi$, is the difference between the contact densities in the left and right hemispheres of the sphere in the right of Figure 10.6, which results in a difference in the corresponding pressures. In the limit $R \to \infty$, the larger spheres can be considered as planar hard walls and the solvent fluid between them behaves like a HS fluid confined in a slit with planar hard walls, separated a distance Δz that depends on the position of the contact point between the large and small spheres (see Figure 10.6). In this case, the pressure difference can be replaced by the solvation force per unit area $f_s(\Delta z)$, that is, the excess pressure of the solvent due to its confinement between two planar hard walls separated a distance x. Then, after some algebra, from Equation 10.42 Götzelmann et al. (1998) obtain

$$F_d(\delta) = \pi\frac{\sigma_1 + \sigma_2}{2} \int\limits_{\delta}^{\infty} f_s(\Delta z)\,d\,(\Delta z), \qquad (10.43)$$

which is the Derjaguin (1934) approximation.

By definition, the solvation force per unit area is

$$f_s(\Delta z) = -\frac{1}{A}\left(\frac{\partial\Omega}{\partial\Delta z}\right)_{\mu,T} - P = -\left(\frac{\partial\gamma\,(\Delta z)}{\partial\Delta z}\right)_{\mu,T}, \qquad (10.44)$$

where
  Ω is the grand potential
  P is the bulk solvent pressure
  $\gamma(\Delta z)$ is the contribution to the grand potential per unit area due to the slit hard walls

In particular, when $0 < \Delta z < \sigma_2$ the solvent particles cannot pass through the slit, so that the contact density is zero within the slit and $f_s(\Delta z) = -P$. Introducing these results into Equation 10.43, and taking into account that $\gamma(\sigma_2) = 0$ (Götzelmann and Dietrich 1997), Götzelmann et al. (1998) obtained the expression

$$F_d = \pi \frac{\sigma_1 + \sigma_2}{2} [P(\delta - \sigma_2) - 2\gamma], \quad \delta \leq \sigma_2, \tag{10.45}$$

which coincides with the result previously derived by Attard et al. (1991), where $\gamma = \gamma(\infty)/2$ is the surface tension of the solvent at a planar wall in a semi-infinite system. Therefore, the depletion force depends on the pressure and surface tension of a monodisperse fluid of HS with diameter σ_2. The pressure can be obtained from the CS equation (5.21) whereas for the surface tension one can use the Henderson and Plischke (1987) expression

$$\gamma = -\frac{9}{2\pi} \frac{k_B T}{\sigma_2^2} \frac{\eta^2 \left(1 + \frac{44}{35}\eta - \frac{4}{5}\eta^2\right)}{(1 - \eta)^3}. \tag{10.46}$$

From Equation 10.45, the depletion potential in the Derjaguin approximation is

$$u_d(\delta) = u_d(\sigma_2) - \pi \frac{\sigma_1 + \sigma_2}{2}(\delta - \sigma_2)\left[\frac{1}{2}P(\delta - \sigma_2) - 2\gamma\right], \quad \delta \leq \sigma_2, \tag{10.47}$$

with $u_d(\sigma_2) \simeq 0$.

The Derjaguin approximation is exact in the limit $R \to \infty$, but for finite values of R the large particles cannot be considered as planar walls. More appropriate for this case is the so-called *wedge approximation* proposed by Götzelmann et al. (1998)

$$F_d = \pi \frac{\sigma_1 + \sigma_2}{2} [P(\delta - \sigma_2) - 2\gamma \cos \alpha], \quad \delta \leq \sigma_2, \tag{10.48}$$

where $\cos \alpha = (\sigma_1 + \delta)/(\sigma_1 + \sigma_2)$. The depletion potential in this case is

$$u_d(\delta) = u_d(\sigma_2) - \pi \frac{\sigma_1 + \sigma_2}{4}(\delta - \sigma_2)\left[P(\delta - \sigma_2) - 2\gamma \frac{2\sigma_1 + \sigma_2 + \delta}{\sigma_1 + \sigma_2}\right], \quad \delta \leq \sigma_2. \tag{10.49}$$

Another expression for the depletion potential was also derived by Götzelmann et al. (1998) based on its exact expression in the limit $R = 1$. In this case, within the context of the test particle method (see Section 9.2), the potential of mean force $\Psi(r)$ in Equation 2.31 can be written as the sum of the pair potential $u(r) \equiv u(\sigma + \delta)$, considered as an external potential due to the particle at the origin, and the depletion potential $u_d(\delta)$. As for HS $u(\sigma + \delta) = 0$ for $\delta > 0$, from Equation 2.31 the depletion potential is given by

$$u_d(\delta) = -k_B T \ln g_{HS}(\sigma + \delta). \tag{10.50}$$

In particular, for $\delta = \sigma$ the preceding expression yields $u_d(\sigma) = -k_B T \ln g_{HS}(2\sigma)$, and we can write

$$u_d(\delta) = u_d(\sigma) - k_B T \ln \left[\frac{g_{HS}(\sigma + \delta)}{g_{HS}(2\sigma)} \right]. \tag{10.51}$$

For $R > 1$ this result is no longer valid and Götzelmann et al. (1998) proposed the ansatz

$$u_d(\delta) = u_d(\sigma_2) - k_B T \frac{\sigma_1 + \sigma_2}{2\sigma_2} \ln \left[\frac{g_{HS}(\sigma_2 + \delta)}{g_{HS}(2\sigma_2)} \right], \tag{10.52}$$

which reduces to the exact result (10.51) for $R = 1$ and is expected to be accurate for R close to 1.

To end this short account on the depletion potential for additive HS mixtures, we will mention the approach devised by Almarza and Enciso (1999) based on a density expansion of the grand potential for the solvent. To second order in the solvent packing fraction η_2 they obtain

$$\beta u_d(\delta) = a_1(\delta)\eta_2 + a_2(\delta)\eta_2^2, \tag{10.53}$$

in which

$$a_1(\delta) = -\frac{1}{2}\left(1 - \frac{\delta}{\sigma_2}\right)^2\left[3R + 2 + \frac{\delta}{\sigma_2}\right], \qquad \delta \le \sigma_2, \tag{10.54}$$
$$0, \qquad\qquad\qquad\qquad\qquad\qquad\qquad\quad \delta > \sigma_2,$$

$$a_2(\delta) = 8a_1(\delta) + 2R^3\left[\left(1 + \frac{1}{R}\right)a_0 - \frac{\delta}{\sigma_1} - 1\right]^2$$
$$\left[2\left(1 + \frac{1}{R}\right)a_0 + \frac{\delta}{\sigma_1} + 1\right], \quad \delta \le \sigma_1\left[\left(1 + \frac{1}{R}\right)a_0 - 1\right], \tag{10.55}$$

and $a_2(\delta) = 0$ for larger δ, with

$$a_0 = \left[1 + \frac{9}{8}\frac{1}{1+R} - \frac{1}{4}\frac{1}{(1+R)^3}\right]^{1/3}. \tag{10.56}$$

The effective depletion potential between two large spheres immersed in a fluid of smaller spheres can be determined from computer simulations, as done by Dickman

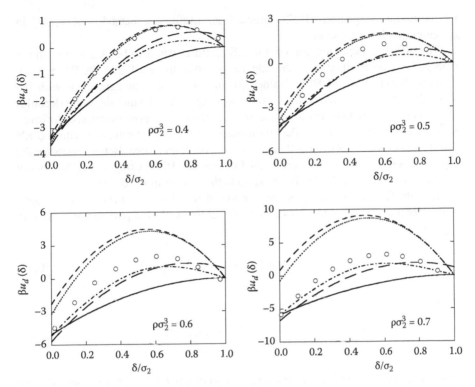

FIGURE 10.7 Depletion potential for the diameter ratio $R = 10$ and several densities of the smaller spheres. Points: simulation data from Biben et al. (1996). Continuous curve: Asakura–Oosawa approximation, Equation 10.38. Short dashed curve: Derjaguin approximation, Equation 10.47. Dotted curve: wedge approximation, Equation 10.49. Dash-dotted curve: Götzelmann et al. approximation, Equation 10.52. Long-dashed curve: Almarza–Enciso approximation, Equation 10.53.

et al. (1997) for diameter ratios $R = 5$ and 10 and by Ashton et al. (2011) for $R = 10$, 20, 50, and 100. These simulations reveal that the depletion potential is nearly negligible for $\delta > \sigma_2$, in agreement with most of the theoretical predictions. For shorter distances, the analytical theories described earlier are only partially successful, as illustrated in Figure 10.7 for $R = 10$ and several reduced densities $\rho\sigma_2^3$. At low densities, the Derjaguin and wedge approximations yield fairly accurate results, but as density increases both tend to increasingly overestimate the depletion potential, except for $\delta = \sigma_2$. All the other analytical approximations considered predict too low values of the depletion potential, except in the limits $\delta = 0$ and $\delta = \sigma_2$, with the better performance at moderate densities provided by the Götzelmann et al., Equation 10.52, and Almarza and Enciso, Equation 10.53, approximations. In contrast with these results, the Mao et al. (1995) third-order density expansion was found by Biben et al. (1996) to be in close agreement with the simulations for $\delta \leq \sigma_2$ and all the densities considered in Figure 10.7. This theory, as well as the Götzelmann et al. (1998)

approximation (Equation 10.52), predicts an oscillatory behavior of $u_d(\delta)$, in agreement with the simulations.

The effective depletion potential in HS mixtures can also be obtained from the RDF $g_{12}(r_{12})$ determined from the integral equation theory. Using the PY and RY closures, Biben et al. (1996) obtained in this way excellent agreement with the simulations for the depletion force corresponding to $R = 10$ and the same densities as in Figure 10.7, with little difference between these two approximations. However, the accuracy in the predicted depletion potential for $\delta \leq \sigma_2$ achieved with the RY closure was considerably lower than that for the depletion force. Better results for the depletion potential were reported by Dickman et al. (1997) using an MHNC theory incorporating the exact bridge function up to third order in the density.

Density functional theory is another useful tool for this purpose. Götzelmann et al. (1999) and Roth et al. (2000) showed that the depletion potential in a mixture is given by

$$\beta u_d(\mathbf{r}) = \lim_{\rho_1 \to 0} \left(\mathscr{C}_1(\mathbf{r} \to \infty) - \mathscr{C}_1(\mathbf{r}) \right), \tag{10.57}$$

where

$$\mathscr{C}(\mathbf{r}) = -\beta \frac{\delta \mathscr{F}^E \left[\rho_1(\mathbf{r}), \rho_2(\mathbf{r}) \right]}{\delta \rho_1(\mathbf{r})}. \tag{10.58}$$

These authors used the FMT to obtain the depletion potential for $R = 5$ and 10 with excellent results, but perhaps near contact. The performance of the FMT and two versions of the VB-FMT has been ascertained by Ashton et al. (2011) by comparing with the simulation data for $R = 10, 20, 50$, and 100, although some deviations were observed for large solvent packing fractions. Highly successful in the context of the DFT is also the reference functional approach, introduced at the end of Section 9.2, as shown by Amokrane et al. (2005).

10.3.2 Effective One-Component Approach for a Hard-Sphere Colloidal Model

The knowledge of the depletion potential allows us to treat the larger particles in a binary mixture of AHS with large diameter ratio R as an effective fluid of particles with an effective hard-core pair potential

$$u(r) = u_0(r) + u_d(r), \tag{10.59}$$

where

$u_0(r)$ is the HS potential for spheres with diameter σ_1

$u_d(r)$ is the depletion potential acting on them due to the presence of the small spheres

One may wonder whether the depletion potential, determined in the low-density limit, is appropriate for higher densities and to what extent the properties of the actual mixture can be obtained from those of the effective fluid. These questions were analyzed by Malherbe and Amokrane (2001) for mixtures with $R = 5, 10$, and 20 from simulations performed for the true mixture and for the effective fluid, with a potential of the form (10.59) determined from computer simulation and assuming pairwise additivity. The results showed that the effective one-component approximation overestimates the contact values $g_{11}(\sigma_{11})$ of the RDF, with a relative deviation nearly independent of the partial packing fraction $\eta_1 = (\pi/6)\rho_1\sigma_1^3$ of the solute but increasing with the partial packing fraction η_2 of the solvent and with the diameter ratio R. To further test the reliability of the pairwise additivity, the mean force acting on two, in a set of four, large spheres in a bath of smaller ones was determined directly from simulation and compared with that obtained from the superposition approximation on the basis of simulations for two large spheres in the solvent bath. This confirmed that the superposition approximation overestimates the actual force, as a consequence of changes in the solvent density, and the deviation increases with the solvent packing fraction.

In spite of the earlier-mentioned limitations, the effective one-component approach may be useful to gain insight into the behavior of strongly asymmetric binary mixtures of AHS, for which theoretical approaches for the true mixtures usually fail to provide reliable results and simulations are difficult to perform accurately. In particular, this approach was used to analyze the phase coexistence of this kind of mixtures from computer simulation by Dijkstra et al. (1999), using for the depletion potential an expression derived by Götzelmann et al. (1998) on the basis of an expansion of Equation 10.47 to third order in the packing fraction. They reported evidence of fluid–fluid and isostructural solid–solid transitions for $R \geq 10$, with the fluid–fluid coexistence being metastable for all values of R and the solid–solid coexistence being stable for $R \geq 20$. Similar conclusions were reached by Almarza and Enciso (1999) using the depletion potential (10.53) in their simulations for mixtures with $R = 10$ and $R = 20$, but for the fact that they found the solid–solid isostructural transition to be stable also for $R = 10$.

With the same end, Velasco et al. (1999) used perturbation theory to first order in the HTE with the same two depletion potentials used in these simulations for $R \geq 5$. The theory was found to predict with both depletion potentials a metastable fluid–fluid demixing transition, although shifted to higher densities as compared with the simulations, as well as a stable solid–solid transition for $R \geq 10$ in fairly good agreement with the simulations, and a fluid–solid transition with an accuracy increasing with R. Similar results were obtained with the AO potential, except for the fact that for $R = 10$ the solid–solid transition in this case resulted to be still metastable with respect to the fluid–solid transition. However, Germain and Amokrane (2002) questioned the reliability of the first-order perturbation theory to study the phase diagram of systems with very-short-ranged potentials in general and, in particular, of the one-component approximation with a depletion potential for HS mixtures. Comparing the results of the first-order perturbation theory with those of the RHNC, using in both theories the third-order density expansion of Götzelmann et al. (1998) for the depletion potential, they concluded that the former of these theories is appropriate for the solid–fluid coexistence, but inappropriate for the fluid–fluid transition, which is predicted to take place at unphysically high densities. For the solid, the first-order perturbation theory was shown to be accurate near the close packing limit, but not for the low-density solid because the short range of the attractive depletion potential gives rise to a considerable increase in the contact value of the RDF, as compared with the HS solid, and so the latter is not a suitable reference system. As a consequence, only partial success can be expected for the prediction of the isostructural solid–solid transition. These conclusions are in agreement with the results obtained by Velasco et al. (1999). The partial failure of the first-order perturbation theory to describe the phase diagram of AHS mixtures in the one-component approach can be explained from the short-range nature of the depletion potential, which differs markedly from the shape of the potentials of ordinary atomic fluids. The resulting effects on the phase diagram and structure of these effective one-component fluids have been discussed by Louis (2001). Interestingly, Zhou (2005) showed that the AO potential can be satisfactorily mapped onto a HCY potential. This allows us to use higher-order perturbation theories for the latter, such as those cited at the end of Section 4.4.

On the other hand, the phase diagram of AHS mixtures, within the effective one-component approach, was analyzed by Germain and Amokrane (2007) for $R = 4$ and 12.5 by means of the RHNC theory with the AO depletion potential as well as a depletion potential obtained from the RHNC. They showed that, whereas the two choices yield the same general features of the phase diagram, the detailed characteristics of the latter depend on the depletion potential used. In particular, the isostructural solid–solid transition for $R = 12.5$ was found to be stable only for the second of them, and for $R = 4$ only with the Asakura–Oosawa approximation was obtained a fluid–fluid metastable transition. Therefore, the conclusions obtained about the behavior of AHS mixtures using simple models of the depletion potential need to be considered with caution.

10.3.3 Effective One-Component Approach in Simple Colloid-Polymer Mixtures

In mixtures of colloids and nonadsorbing polymers, the colloidal and polymeric particles may be considered as spheres with effective diameters σ_1 and $\sigma_2 = 2R_G$, respectively, where R_G is the radius of gyration of the polymer molecules. However, whereas the closest distance of approach between two colloidal particles is $\sigma_{11} = \sigma_1$, for the polymer molecules $\sigma_{22} < \sigma_2$, because the latter can interpenetrate to each other. In this situation, a simple model of colloid–polymer mixtures will be an asymmetric nonadditive binary mixture of HS, with the nonadditivity giving rise to an enhancement in the depletion interaction with respect to that arising in additive HS mixtures (Dijkstra et al. 1999). For NAHS mixtures, $\sigma_{11} = \sigma_1$, $\sigma_{22} = \sigma_2$, and $\sigma_{12} = \frac{1}{2}(\sigma_1 + \sigma_2)(1 + \Delta)$, as we know, and, as mentioned in Section 7.6, values $\Delta > 0$ favor the existence of stable fluid–fluid demixing transitions, in contrast with the additive case corresponding to $\Delta = 0$. In the simplest case, the polymer molecules are considered as an ideal gas, so that $\sigma_{22} = 0$, and the AO depletion potential applies. This corresponds to a NAHS mixture with $\sigma_{22} = 0$ and $\Delta = 1/R > 0$. On the contrary, some colloidal mixtures exhibit negative nonadditivity. Moreover, as shown by Louis and Roth (2001), a wide variety of depletion potentials arising in more realistic mixtures, and the corresponding phase diagrams, can be mapped onto those for NAHS mixtures by properly adjusting the nonadditivity parameter. Therefore, we will discuss briefly the case of very asymmetric NAHS mixtures from the viewpoint of the effective one-component approximation.

The problem of determining the depletion potential for asymmetric NAHS was addressed by Roth and Evans (2001) by means of the DFT procedure previously developed by Roth et al. (2000) for AHS, based on Equation 10.57. Roth and Evans showed that in the limit $\rho_1 \to 0$ the binary NAHS mixture can be mapped exactly onto an AHS mixture, and so the FMT for the latter was used to determine the depletion potential. The resulting depletion potential displays an enhanced oscillatory behavior for $\Delta < 0$, and damped for $0 < \Delta < 1/R$, with respect to that corresponding to the AHS mixtures ($\Delta = 0$), and becomes indistinguishable of the AO potential for $\Delta = 1/R$. In the range $\delta \leq \sigma_2$ the depletion potential for $0 < \Delta < 1/R$ is lower (more negative) than the AO potential. An analytical expression for the depletion potential was also derived by Louis et al. (2000) on the basis of a generalization of the earlier-mentioned density expansions of the depletion potential for AHS mixtures derived by Mao et al. (1995) and Götzelmann et al. (1998).

From the theoretical viewpoint, first-order perturbation theory in the HTE is not expected to be accurate for the phase diagram of very asymmetric NAHS mixtures, in a similar way as for the additive case. Instead, Louis et al. (2000), using a first-order perturbation theory of the form of Equation 7.102 with their depletion potential, showed that the theory predicts a fluid–fluid transition for $R = 5$ at low-packing fractions η_1 of the bigger spheres in good agreement with the results of IET with the BPGG closure. Moreover, the predicted fluid–solid transition from the perturbation theory was found in close agreement with the simulation data. The stability of the fluid–fluid demixing transition in NAHS mixtures was studied by Pellicane et al. (2006) from the RY and RHNC integral equation theories for the actual mixtures, and from the same first-order perturbation theory and depletion potential as before, for $R = 4/3, 5/3, 2$, and $10/3$, and nonadditivities $\Delta = 0.05$ and 0.1 in all cases. It was found that all these theories yield similar results for the fluid–fluid separation curves and for the predicted critical points, and suggest that the fluid–fluid coexistence is stable for $\Delta = 0.1$ and metastable for $\Delta = 0.05$. The Louis et al. and AO potentials, in combination with the HRT and RHNC theories for the fluid and the second-order BH perturbation theory for the solid, was applied by Lo Verso et al. (2005) to the same problem. With the former of these potentials, for $R = 5$ the HRT was found to predict a stable fluid–fluid transition for $\Delta = 0.25$ and metastable for $\Delta = 0.1$. The results of the RHNC were in qualitative agreement with those of the HRT, but providing lower values for the critical packing fraction of the bigger spheres. For colloid–polymer mixtures with the AO potential, the HRT was found to predict metastable fluid–fluid coexistence for $R = 4$ and 5, and stable for $R = 5/3$. In the two latter cases, the theory also agreed closely with simulation data for both the fluid–fluid and fluid–solid coexistence. In contrast, the earlier-mentioned first-order perturbation theory predicted too low values for the fluid–fluid coexistence, although quite satisfactory for the fluid–solid coexistence.

10.3.4 CHARGE-STABILIZED COLLOIDAL SUSPENSIONS

Real colloidal mixtures are more complex than considered thus far. On the one hand, the big particles behave as soft, rather hard, spheres. On the other hand, other kinds of interactions, such as vdW attractions and electrostatic repulsions, may be present. Moreover, the big particles in real colloids are not equal to each other as they exhibit some degree of polydispersity in size, shape, or charge. These characteristics may have a strong influence on the phase equilibria of the mixture. Thus, for example, from computer simulation of a HS mixture in the effective one-component approach, Largo and Wilding (2006) showed that the presence of polydispersity gives rise to a displacement of the fluid–fluid critical point deeper into the percolation region, thus favoring dynamical arrest against fluid–fluid phase separation. Concerning the effect of the softness of the colloidal particles, Cinacchi et al. (2007) studied a binary mixture with soft repulsive interactions and found that the depletion potential, obtained from Equation 10.57 as an expansion to second order in the density as well as from computer simulation, was increasingly more attractive with increasing softness. Moreover, the PWDA (see Equation 9.86) solved within the effective one-component approach suggested the possible existence of a stable fluid–fluid transition for enough soft potentials. As for the electrostatic repulsions, they contribute to stabilize the suspension at conditions for which otherwise a phase transition would take place.

Charge-stabilized colloidal suspensions are complex mixtures usually including colloidal macroions, microions (salt), and a polar solvent. This gives rise to complex interactions between the constituent species (see, e.g., Dijkstra 2001 for a discussion on this subject). The direct computation, either from theory or from computer simulations, of the thermodynamic properties and phase behavior of these mixtures is extremely challenging, and so usually the problem is faced from the effective one-component approach. The most frequently used approximation to map the charge stabilized colloids into an effective one-component fluid is the *Derjaguin–Landau–Verwey–Overbeek* (DLVO) *theory* (Derjaguin and Landau 1941, Verwey and Overbeek 1948), which leads to a depletion potential consisting in a hard-core vdW attraction plus a screened electrostatic repulsion of the HCY form, that is,

$$u_d(r) = u_{rep}(r) + u_{att}(r). \tag{10.60}$$

For equal-sized colloidal particles with effective diameter σ and number density ρ, the vdW contribution can be expressed in the Hamaker (1937) theory as

$$u_{att}(r) = -\frac{C_H}{12}\left[\frac{\sigma^2}{r^2 - \sigma^2} + \frac{\sigma^2}{r^2} + 2\ln\left(1 - \frac{\sigma^2}{r^2}\right)\right], \tag{10.61}$$

where $C_H = \pi^2 C \rho^2$ is the Hamaker constant in which C is a constant that determines the strength of the interaction. Assuming London dispersive forces, the constant C is equal to the constant c_6 defined in Equation 1.2.

The screened electrostatic contribution is of the form of Equation 1.17, with plus sign. The range parameter of the potential is $\kappa = \kappa_D$, the Debye screening parameter

$$\kappa_D \equiv \lambda_D^{-1} = \left(4\pi\lambda_B \sum_i \rho_i z_i^2 \right)^{1/2}, \tag{10.62}$$

in which $\lambda_B = e^2/4\pi\epsilon k_B T$ is the Bjerrum length for a medium with dielectric constant ϵ, and ρ_i and z_i are the average number density and charge number, respectively, of the particles of species i. The potential energy at contact for colloidal particles with charge number z is given by

$$\varepsilon = \frac{z^2 e^2}{\sigma\epsilon \left(1 + \kappa_D\sigma/2 \right)^2}. \tag{10.63}$$

The competition between repulsive and attractive contributions results in a shape like the one in Figure 10.8. In this example, the DLVO potential is strongly attractive at very short distances, exhibits a repulsive barrier at intermediate distances, followed by a secondary minimum, and decays to zero at large distances. The existence of a sharp repulsive barrier prevents the aggregation (coagulation) of the colloidal particles, thus stabilizing the colloidal suspension. However, the existence of a secondary minimum may give rise to reversible coagulation. For a given colloidal mixture, the height of the barrier strongly depends on the concentration of salt. As the concentration increases the height decreases, eventually allowing aggregation. For enough high concentrations, the barrier vanishes and the potential becomes strongly attractive, giving rise to the formation of an amorphous aggregate.

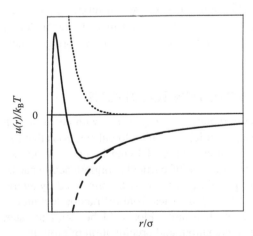

FIGURE 10.8 Illustration of the DLVO potential (continuous curve) showing the repulsive (dotted curve) and attractive (dashed curve) contributions.

A drawback of the DLVO potential is that it diverges to minus infinity at contact distance, so that near contact the attraction always prevails over repulsion at short distances and the thermodynamically stable state is always the aggregate state, although with high repulsive barriers the relaxation of the system to the equilibrium state may be slow enough as to remain fluid for long time. The problem may be avoided by defining an effective HS diameter $\sigma \geq r_{max}$. This was done by Victor and Hansen (1985) by splitting the DLVO potential in the form of the WCA and applying condition (6.73). Then, first-order perturbation theory predicted reversible coagulation to take place under conditions in qualitative agreement with experiment. A similar strategy was adopted by Morales et al. (2003), who used a modified DLVO potential, with the potential for $r < r_{max}$ replaced by a HS core, to analyze the phase equilibria from simulation and theory. Simulation showed phase separation for Debye screening parameters κ_D above a certain critical value. The RHNC theory was found to provide accurate results for the RDF and fairly good predictions for the phase equilibria, although underestimating the critical value of κ_D, but in any case the RHNC proved to perform better than perturbation theory. The comparison with experiment revealed the considerable importance of interactions, like hydrophobic attractions, not accounted for in the DLVO theory.

To overcome the limitations of the DLVO theory, the effective potential can be determined from IET. This is the approach adopted by Anta and Lago (2002), who determined the effective potential from the HNC approximation and then the effective one-component fluid was solved in the RHNC approximation for size and charge asymmetric hard-core colloidal mixtures. The colloid–colloid total correlation functions obtained in this way were in excellent agreement with simulations. A remarkable fact of the derived effective potential is that, in contrast with the DLVO potential, it accounts for the attraction between the counterions surrounding colloidal particles and other colloidal particles, which eventually compensates the repulsive Coulombic repulsions between the colloids. Subsequently, Anta et al. (2003) showed that this *coarse grained HNC* (CGHNC) is much more efficient than the ordinary HNC applied to the actual mixture, and that the former theory provides a solution for some regions for which the latter does not. These authors also reported evidence of the inadequacy of the HS bridge function for this kind of mixtures.

10.4 AQUEOUS PROTEIN SOLUTIONS

To end this look on complex systems, let us examine in brief the case of protein solutions. They are a particular type of colloidal solutions that form an essential part of the constitution and functioning of living beings. The knowledge of the phase properties of these solutions is of particular importance because, on the one hand, crystallization of the proteins, which may be prevented by aggregation, is essential to determine some of their properties from diffraction techniques and, on the other hand, certain diseases, such as amyloidosis, Alzheimer, and cataracts, are related to the self-assembling, aggregation, and precipitation of proteins.

Proteins in nature are frequently present in solutions that contain water and salty substances. Globular proteins have polar groups placed at the surface and the net charge depends on the pH of the solution. This gives rise to complex inter molecular protein–protein, in addition to protein–salt and protein–water, interactions. These interactions are different in nature: electrostatic, hydrophobic, etc. The interactions and, therefore, the protein behavior, depends on temperature, protein concentration, salt type and concentration, etc. By adjusting the pH of the solution, the surface net charge eventually will vanish, which corresponds to the *isoelectric point* (pI) of the solution. As the interactions between the surface charges are repulsive, they help to stabilize the solution, so that at the pI the solubility abruptly drops and the effect of the relatively strong protein–protein attractions favors aggregation to form an amorphous precipitate, rather than a crystalline solid. Therefore, to achieve crystallization it is necessary that the solution is away its pI and to partially compensate the attractions by means of the electrostatic repulsions. From the experiment it is known that when crystallization takes place the average center-to-center distance between the protein molecules in the solution is comparable to that in the crystal, which requires that the range of the repulsive forces be much smaller than the diameter of the protein. The strength of the repulsive forces can be "tuned" by screening the surface charges with the counterions of an appropriate added salt. The type and concentration of the latter must be determined empirically (see Rosenberger 1996 for a short review on these subjects).

Experimental evidence found by Guo et al. (1999) shows that the solubility of globular proteins is correlated with the osmotic second virial coefficient B_{11}, independently of the specific substances added to the solution, and this correlation has been confirmed from theory by Haas et al. (1999) and Ruppert et al. (2001). Moreover, conditions favorable to crystallization correspond to a narrow range -1×10^{-4} mol mL $g^{-2} \lesssim B_{11} \lesssim -8 \times 10^{-4}$ mol mL g^{-2}, called the *crystallization slot*, as observed by George and Wilson (1994) and confirmed by George et al. (1997) from experiment. For higher values the solubility is too high and the solution remains stable, whereas for lower values the attractive interactions are too strong and an amorphous aggregate forms. On the other hand, Lekkerkerker (1997) showed that, for systems with hard-core potentials, the liquid–vapor coexistence is metastable with respect to the solid–fluid coexistence when the ratio of the range of the attractive interaction to that of the repulsive one is 1.25 or, equivalently, the effective width of the attractive well is 0.25 in units of the effective diameter. This is the situation in some globular protein solutions, for which there is a liquid–liquid phase coexistence between dilute and concentrated protein solutions, which is metastable with respect to the solid–liquid coexistence. From the analysis of experimental data, Lekkerkerker (1997) and Haas and Drenth (1998) concluded that the crystallization slot in some protein solutions lies near the liquid–liquid critical point, which suggests that the strong concentration fluctuations near the transition favor the crystal nucleation. In any case, these findings state which conditions are suitable for nucleation, but tell us nothing about the time needed for the growth of the crystal to a usable size, which eventually might require several months. Pullara et al. (2005, 2007) measured by light scattering the size and lifetime of the large fluctuations occurring near the liquid–liquid transition and determined the spinodal temperatures T_s for some protein solutions. They found that the induction time, the time needed for the appearance of sizeable crystals, depends only on the ratio

$|T - T_s|/T_s$ over many orders of magnitude, and not on the specific values of the parameters, like temperature, pH, protein concentration, salt type and concentration, etc., that determine the ratio. On this basis, they determined a window of this ratio for which the conditions for crystallization are optimal. This seems to provide a more restrictive criterion for crystallization than the one based on the second virial coefficient.

The preceding discussion highlights the complexity of the interactions involved in globular protein solutions and evidence the difficulty of a reliable theoretical prediction of the phase diagram and of the conditions for fast crystal growth, in spite of which some useful results have been derived from theory. This will be illustrated for some protein solutions, especially lysozyme solutions that are particularly suitable for this purpose because for them there is available a considerable amount of experimental information. Lysozyme is a protein present in hen egg white and in different human secretions. It is formed by 129 amino acids, its mass is $\sim 14,300$ Daltons, and its shape is approximately ellipsoidal with dimensions $\sim 40 \times 30 \times 30$ Å. The pI of lysozyme is at pH $= 11.4$ and its net charge is ~ 15 at pH $= 3$, ~ 9 at pH $= 6$, and ~ 7 at pH $= 9$, as determined by Tanford and Roxby (1972).

As the interactions between globular proteins in solution include a hard core, screened electrostatic repulsions, and van der Waals attractions, a good candidate for the interaction potential seems to be the DLVO potential discussed in Section 10.3. However, the situation here is more complex because, on the one hand, there is shape and charge anisotropy and, on the other hand, hydrophobic (attractive) effects are important and strongly temperature-dependent. The patchiness in the distribution of charges and hydrophobic sites is an added complication. Notwithstanding, simple potential models have been used, in combination with theory, to correlate experimental results or in simulation of simple models of protein solutions. We will give next a short review of some of the advances in this field from the simplest potential models to the most complex ones.

Moon et al. (2000) obtained good agreement between experimental data for the osmotic pressure of aqueous solutions of lysozyme and those calculated from the compressibility equation (8.36) for SHS, with parameter τ determined by equating the experimental second virial coefficient of the solution to the theoretical one. The cloud-point temperatures, the temperatures for which no more protein can be solved and start to precipitate, were measured by Grigsby et al. (2001) for lysozyme solutions under different conditions of salt type, salt concentration, and pH. To correlate the experimental data they applied the mean field approximation, Equations 6.45 and 6.46, with the CS equation for the HS reference fluid and the potential of the mean force approximated by the SW potential with $\lambda = 1.2$. To account for the specific (salt-dependent) effects, the potential depth ε_{NaCl} of the solution with NaCl salt was taken as the reference and, for each ionic strength, the deviations of the cloud point temperature observed for any other added salt, with respect to that with NaCl, were attributed to specific effects and included by adding a contribution ε_{sp} to the total potential depth ε, so that $\varepsilon = \varepsilon_{NaCl} + \varepsilon_{sp}$. The dependence of ε_{sp} with the ionic strength was found to be linear for monovalent salts and quadratic for divalent cationic salts.

The SW potential was used also by Lomakin et al. (1996) to analyze the liquid–liquid phase separation of aqueous solutions of γ-crystallin. They found that the coexistence curve for $\lambda = 1.25$ obtained from computer simulation can be mapped onto the experimental one using a temperature-dependent potential depth ε. Following the ideas of Sanchez and Balazs (1989), the temperature dependence of the potential depth has been attributed to the entropic contribution arising from the specific effect of a particular salt on the structure of water molecules around the protein surface. As a consequence, the nonspecific energetic potential depth ε must be replaced by a free energy parameter $\varepsilon - Ts$ incorporating also the entropic contribution. Going ahead along these lines, Chang and Bae (2003) incorporated a temperature-dependent potential depth into the mean field approximation plus a correction to account for the unequal distribution of the small ions between the two chambers of the osmometer, and obtained in this way better accuracy with respect to the mean field approximation of Grigsby et al. (2001) for the predicted osmotic pressure of lysozyme solutions.

The potential of the mean force between two protein molecules in protein solutions was approximated by Jin et al. (2006) by a hard core plus two Yukawa tails, one attractive to account for the dispersive contribution and the other repulsive to account for the electrostatic repulsion. Combined with the analytical theory derived by Tang et al. (2005) for multi-Yukawa fluids within the framework of the FMSA, with the energy parameter ε_1 of the dispersive Yukawa term adjusted to fit the experimental data, very good overall results were achieved for the osmotic pressure of a variety of protein solutions.

As said before, a more appropriate potential model for the protein–protein interactions in dilute aqueous electrolyte solutions seems to be, at first sight, the DLVO potential. However, Moon et al. (2000) showed that the DLVO potential is unable to reproduce the observed dependence of the osmotic second virial coefficient with pH. Neither is applicable to concentrated salt solutions, in which the solvation forces, not accounted for in the DLVO theory, constitute the main contribution to the potential of the mean force. In spite of this, Pellicane et al. (2004a,b) used the DLVO potential, with the potential parameters fixed to the values known from experiment, except for the Hamaker constant that was fitted to the second virial coefficient, to determine the phase diagram of lysozyme and γ-crystallin solutions by means of the HMSA IET, obtaining in this way good agreement with the experimental data for both the fluid–fluid demixing and crystallization curves.

To attempt to overcome some of the limitations of the DLVO potential, several improvements, as well as alternative approaches, have been developed. Thus, Ho and Middelberg (2003) added to the DLVO potential osmotic and solvation contributions. The osmotic contribution is a depletion attraction of the Asakura–Oosawa type, arising from the exclusion of the ions from a volume between two colloidal particles when they are close enough, and is negligible for low salt concentrations, but becomes important for concentrated salt solutions. The solvation term was included to account for other kinds of interactions, such as hydration and hydrophobic forces and hydrogen bonds, as obtained from the experimental osmotic second virial coefficient by means of the procedure devised by Blanch et al. (2002), namely

$$B_{11}^{exp} = B_{11}^{DLVO+O} - \frac{2\pi N_A}{M^2} \int_d^{d_h} \left[e^{-\beta u_s(r)} - 1 \right] r^2 dr, \tag{10.64}$$

where
  B_{11}^{exp} is the experimental osmotic second virial coefficient
  B_{11}^{DLVO+O} is the one obtained in the usual way, Equation 2.38, from the DLVO
    potential plus the osmotic contribution
  M is the molecular weight
  $u_s(r)$ is the solvation potential sought for
  d and d_h are, respectively, the effective diameter and the hydrated effective
    diameter of the protein

By comparing the experimental data for the osmotic second virial coefficient of lysozyme with the results obtained from the DLVO + osmotic potential, Ho and Middelberg showed that the solvation contribution has a considerable importance even for moderate salt concentrations.

Thus far, we have not considered explicitly the nonuniform distribution of the interactions onto the protein surface, although its effect is accounted for in an averaged way when the effective potential is determined from the experimental osmotic second virial coefficient. However, an accurate description of the phase diagram as well as of certain properties, such as protein aggregation and self-assembling, requires to take into account the anisotropy in the shape and interactions of the protein molecules. To study the conditions under which the potential is averageable, Lomakin et al. (1999) considered a simple model of a globular protein consisting in a HS with a number of nonoverlapping attractive spots distributed at fixed positions onto the surface. A SW potential with temperature-dependent depth $\varepsilon_{eff}(T) = 2k_B T \ln[a \exp(\beta \varepsilon/2) + (1-a)]$, where a is the fractional area covered by the spots, was assumed for them. The interactions between the particles were defined in such a way that the energy of each particle was dependent of its orientation an relative position with respect to the other particles, but not on the orientations of the latter. To this end, a particle i was considered in contact with particle j when the vector $\mathbf{r}_{ij}$ was passing through an attractive spot of particle i. The pair potential for particles i and j was then $u(\Omega_i, \Omega_j, \mathbf{r}_{ij}) = u(\Omega_i, \mathbf{r}_{ij}) + u(\Omega_j, \mathbf{r}_{ij})$, where Ω_i defines the orientation of molecule i, and $u(\Omega_i, \mathbf{r}_{ij}) = -c\varepsilon/2$, where c is the number of contacts of particle i. Considering 25 randomly distributed spots with $a = 0.01$, it was found that the interactions were averageable when the number of neighbors was $n = 3$, but not for $n = 5$, because under these conditions a particle can never make five contacts, that is, $c < n = 5$, and so on averaging we are overestimating the absolute value of the energy. These authors concluded that the interactions are averageable near the critical point, for which $n \approx 3$, but not in the crystal phase, for which $n > 5$, so that different potentials need to be used for the fluid and solid phases. This simple model was able to satisfactorily reproduce the experimental solid–fluid and fluid–fluid coexistence curves of a γ-crystallin solution.

Curtis et al. (2001) analyzed the phase diagram of aqueous lysozyme solutions approximating the protein molecules by HS with sticky SW sites distributed on the surface, and using the Wertheim TPT1 for the fluid and a free volume model for the crystal. The effective potential depth was obtained by equating the experimental osmotic second virial coefficient with the one calculated with the angle-averaged Mayer function. For the systems studied, double bonding between two molecules was not allowed. The theory was used to correlate the experimental data of the lysozyme solubility with the osmotic second virial coefficient. The best correlations were those corresponding to six protein–protein contacts in the crystal, in agreement with the experimental results. The theory also predicted the dependence of the location of the liquid–liquid critical point with the number of attractive sites per protein molecule, in contrast with the case of spherically symmetric potentials for which the location of the critical point depends on the range of the attractive potential.

Liu et al. (2007) combined a SW isotropic interaction with a SW-patchy model, introduced by Kern and Frenkel (2003), as a model for lysozyme and γ-crystallin solutions. By means of computer simulation, and considering 4–7 patches per protein molecule, these authors showed that the experimental curves for the liquid–liquid transitions in these solutions can be reproduced by the model with SW ranges $\lambda_i = 1.15$ and $\lambda_p = 1.05$ for the isotropic and patchy interactions, respectively, with little influence of the number of patches considered. In contrast, the purely isotropic SW interaction provided a too narrow coexistence curve.

The effect of the location of the charges in spots (patchiness) and of the charge fluctuations was considered by Grant (2001). They showed that both mechanisms contribute to reduce the electrostatic repulsions, the former being significant at distances smaller than the Debye length and the latter acting at all distances, and together give rise to appreciable deviations of the effective potential from that corresponding to uniformly charged protein molecules.

The phase equilibria of aqueous lysozyme solutions were also determined by Gögelein et al. (2008) by means of the BH second-order perturbation theory, for both the fluid and solid phases, within the mc approximation. The protein molecule was considered to be spherical and the interaction potential was assumed to consists of a HS core, a repulsive screened Coulomb interaction, and an attractive patchy contribution, closely related to the earlier-mentioned patchy model of Kern and Frenkel (2003). The results obtained in this way were found to closely agree with experiment for the liquid–liquid and solid–liquid equilibrium curves.

Lund et al. (2008) devised a realistic structure of the lysozyme molecule with each amino acid approximated by a neutral sphere placed at positions according to the real structure, as determined by x-ray, and explicit inclusion of the polar and hydrophobic residues and cationic and anionic sites. The solvent was considered as a dielectric continuum. The interaction potential included explicitly the interactions between ions of the solution and the protein surfaces, and the dispersive interaction forces between two protein molecules approximated by the LJ potential. The potential of the mean force was calculated by simulation with the described potential model for several salt types and concentrations, and used to determine the second virial coefficient in very good agreement with experiment. Subsequently, the Lund et al. potential of mean force was fitted by Lettieri et al. (2009) to obtain from simulation the phase diagrams of aqueous solutions of lysozyme and analyze the effects of the hydrophobic forces.

From this discussion, it emerges that the theoretical prediction of the thermodynamic properties and the phase diagram of globular proteins in aqueous solutions is an extremely challenging task. This is perhaps not so much because of the lack of suitable theoretical tools, but rather due to the insufficient description of the interactions available at present. Many effects, still incompletely understood, are involved and, whereas simple potential models can give account, at least qualitatively, of some of the observed properties, a more complete and quantitative approach requires a detailed description of the interactions.

REFERENCES

Abramo, M. C., C. Caccamo, G. Malescio, G. Pizzimenti, and S. A. Rodge. 1984. Equilibrium properties of charged hard spheres of different diameters in the electrolyte solution regime: Monte Carlo and integral equation results. *J. Chem. Phys.* 80:4396.

Abramo, M. C., C. Caccamo, and G. Pizzimenti. 1983. MSA and GMSA for charged hard spheres of different diameters: Comparison with RHNC and MC calculations. *J. Chem. Phys.* 78:357.

Almarza, N. G. and E. Enciso. 1999. Phase equilibria of asymmetric hard sphere mixtures. *Phys. Rev. E* 59:4426.

Amokrane, S., A. Ayadim, and J. G. Malherbe. 2005. Structure of highly asymmetric hard-sphere mixtures: An efficient closure of the Ornstein–Zernike equations. *J. Chem. Phys.* 123:174508.

Anta, J. A., F. Bresme, and S. Lago. 2003. Integral equation studies of charged colloids: Non-solution boundaries and bridge functions. *J. Phys.: Condens. Matter* 15:S3491.

Anta, J. A. and S. Lago. 2002. Self-consistent effective interactions in charged colloidal suspensions. *J. Chem. Phys.* 116:10514.

Asakura, S. and F. Oosawa. 1954. On the interaction between two bodies immersed in a solution of macromolecules. *J. Chem. Phys.* 22:1255.

Asakura, S. and F. Oosawa. 1958. Interaction between particles suspended in solutions of macromolecules. *J. Polym. Sci.* 33:183.

Ashcroft, N. W. 1966. Electron-ion pseudopotentials in metals. *Phys. Lett.* 23:48.

Ashton, D. J., N. B. Wilding, R. Roth, and R. Evans. 2011. Depletion potentials in highly size-asymmetric binary hard-sphere mixtures: Comparison of simulation results with theory. *Phys. Rev. E* 84:061136.

Attard, P. 1989a. Spherically inhomogeneous fluids. I. Percus-Yevick hard spheres: Osmotic coefficients and triplet correlations. *J. Chem. Phys.* 91:3072.

Attard, P. 1989b. Spherically inhomogeneous fluids. II. Hard-sphere solute in a hard-sphere solvent. *J. Chem. Phys.* 91:3083.

Attard, P., D. R. Bérard, C. P. Ursenbach, and G. N. Patey. 1991. Interaction free energy between planar walls in dense fluids: An Ornstein–Zernike approach with results for hard-sphere, Lennard-Jones, and dipolar systems. *Phys. Rev. A* 44:8224.

Badirkhan, Z., O. Akinlade, G. Pastore, and M. P. Tosi. 1992. Thermodynamics and structure of liquid metals: A critical assessment of the charged-hard-sphere reference system. *J. Phys.: Condens. Matter* 4:6173.

Ballone, P., G. Pastore, and M. P. Tosi. 1984. Structure and thermodynamic properties of molten alkali chlorides. *J. Chem. Phys.* 81:3174.

Biben, T., P. Blandon, and D. Frenkel. 1996. Depletion effects in binary hard-sphere fluids. *J. Phys.: Condens. Matter* 8:10799.

Blanch, H. W., J. M. Prausnitz, R. A. Curtis, and D. Bratko. 2002. Molecular thermodynamics and bioprocessing: From intracellular events to bioseparations. *Fluid Phase Equilibr.* 194–197:31.

Bresme, F., E. Lomba, J. J. Weis, and J. L. F. Abascal. 1995. Monte Carlo simulation and integral-equation studies of a fluid of charged hard spheres near the critical region. *Phys. Rev. E* 51:289.

Bretonnet, J. L. and C. Regnaut. 1985. Determination of the structure factor of simple liquid metals from the pseudopotential theory and optimized random-phase approximation: Application to Al and Ga. *Phys. Rev. B* 31:5071.

Bretonnet, J. L. and M. Silbert. 1992. Interionic interactions in transition metals. Application to vanadium. *Phys. Chem. Liq.* 24:169.

Bretonnet, J. L., G. M. Bhuiyan, and M. Silbert. 1992. Gibbs-Bogoliubov variational scheme calculations for the liquid structure of 3d transition metals. *J. Phys.: Condens. Matter* 4:5359.

Bymaster, A., C. Emborsky, A. Dominik, and W. G. Chapman. 2008. Renormalization-group corrections to a perturbed-chain statistical associating fluid theory for pure fluids near to and far from the critical region. *Ind. Eng. Chem. Res.* 47:6264.

Caccamo, C., G. Malescio, and L. Reatto. 1984. Modified hypernetted chain approximation for a two component charged hard sphere system. *J. Chem. Phys.* 81:4093.

Chang, B. H. and Y. C. Bae. 2003. Lysozyme-lysozyme and lysozyme-salt interactions in the aqueous saline solution: A new square-well potential. *Biomacromolecules* 4:1713.

Cinacchi, G., Y. Martínez-Ratón, L. Mederos, G. Navascués, A. Tani, and E. Velasco. 2007. Large attractive depletion interactions in soft repulsive-sphere binary mixtures. *J. Chem. Phys.* 127:214501.

Curtis, R. A., H. W. Blanch, and J. M. Prausnitz. 2001. Calculation of phase diagrams for aqueous protein solutions. *J. Phys. Chem. B* 105:2445.

Derjaguin, B. 1934. Untersuchungen über die reibung und adhäsion, IV. Theorie des anhaftens kleiner teilchen. *Kolloid-Z* 69:155.

Derjaguin, B. and L. Landau. 1941. Theory of the stability of strongly charged lyophobic sols and of the adhesion of strongly charged particles in solutions of electrolytes. *Acta Physicochim. (URSS)* 14:633.

Dickman, R., P. Attard, and V. Simonian. 1997. Entropic forces in binary hard sphere mixtures: Theory and simulation. *J. Chem. Phys.* 107:205.

Dijkstra, M. 2001. Computer simulations of charge and steric stabilised colloidal suspensions. *Curr. Opin. Colloid Interface Sci.* 6:372.

Dijkstra, M., R. van Roij, and R. Evans. 1999. Phase diagram of highly asymmetric binary hard-sphere mixtures. *Phys. Rev. E* 59:5744.

Duh, D.-M. and A. D. J. Haymet. 1992. Integral equation theory for charged liquids: Model 2-2 electrolytes and the bridge function. *J. Chem. Phys.* 97:7716.

Evans, R. and T. J. Sluckin. 1981. The long-wavelength behaviour of the structure factor of liquid alkali metals. *J. Phys. C: Solid State Phys.* 14:3137.

Fu, D. and J. Wu. 2004. A self-consistent approach for modelling the interfacial properties and phase diagrams of Yukawa, Lennard-Jones and square-well fluids. *Mol. Phys.* 102:1479.

George, A., Y. Chiang, B. Guo, A. Arabshahi, Z. Cai, and W. W. Wilson. 1997. Second virial coefficient as predictor in protein crystal growth. *Methods Enzymol.* 276:100.

George, A. and W. W. Wilson. 1994. Predicting protein crystallization form a dilute solution property. *Acta Cryst.* D50:361.

Germain, Ph. and S. Amokrane. 2002. Validity of the perturbation theory for hard particle systems with very-short-range attraction. *Phys. Rev. E* 65:031109.

Germain, Ph. and S. Amokrane. 2007. Equilibrium and glassy states of the Asakura–Oosawa and binary hard sphere mixtures: Effective fluid approach. *Phys. Rev. E* 76:031401.

Gögelein, C., G. Nägele, R. Tuinier, T. Gibaud, A. Stradner, and P. Schurtenberger. 2008. A simple patchy colloid model for the phase behavior of lysozyme dispersions. *J. Chem. Phys.* 129:085102.

Götzelmann, B. and S. Dietrich. 1997. Density profiles and pair correlation functions of hard spheres in narrow slits. *Phys. Rev. E* 55:2993.

Götzelmann, B., R. Evans, and S. Dietrich. 1998. Depletion forces in fluids. *Phys. Rev. E* 57:6785.

Götzelmann, B., R. Roth, S. Dietrich, M. Dijkstra, and R. Evans. 1999. Depletion potential in hard-sphere fluids. *Europhys. Lett.* 47:398.

Grant, M. L. 2001. Nonuniform charge effects in protein-protein interactions. *J. Phys. Chem. B* 105:2858.

Grigsby, J. J., H. W. Blanch, and J. M. Prausnitz. 2001. Cloud-point temperatures for lysozyme in electrolyte solutions: Effect of salt type, salt concentration and pH. *Biophys. Chem.* 91:231.

Guo, B., S. Kao, H. McDonald, A. Asanov, L. L. Combs, and W. W. Wilson. 1999. Correlation of second virial coefficients and solubilities useful in protein crystal growth. *J. Cryst. Growth* 196:424.

Haas, C. and J. Drenth. 1998. The protein-water phase diagram and the growth of protein crystals from aqueous solution. *J. Phys. Chem. B* 102:4226.

Haas, C., J. Drenth, and W. W. Wilson. 1999. Relation between the solubility of proteins in aqueous solutions and the second virial coefficient of the solution. *J. Phys. Chem. B* 103:2808.

Hafner, J. and V. Heine. 1983. The crystal structures of the elements: Pseudopotential theory revisited. *J. Phys. F: Met. Phys.* 13:2479.

Hafner, J. and G. Kahl. 1984. The structure of the elements in the liquid state. *J. Phys. F: Met. Phys.* 14:2259.

Hamaker, H. C. 1937. The London-van der Waals attraction between spherical particles. *Physica* 4:1058.

Hasegawa, M. and M. Watabe. 1972. Theory of compressibility of simple liquid metals. *J. Phys. Soc. Jpn.* 32:14.

Henderson, D. and M. Plischke. 1987. Sum rules for the pair-correlation functions of inhomogeneous fluids: Results for the hard-sphere-hard-wall system. *Proc. R. Soc. London A* 410:409.

Ho, J. G. S. and A. P. J. Middelberg. 2003. The influence of molecular variation on protein interactions. *Biotechnol. Bioeng.* 84:611.

Ichimaru, S. and K. Utsumi. 1981. Analytic expression for the dielectric screening function of strongly coupled electron liquids at metallic and lower densities. *Phys. Rev. B* 24:7385.

Jacobs, R. E., and H. C. Andersen. 1975. The repulsive part of the effective interatomic potential for liquid metals. *Chem. Phys.* 10:73.

Jiang, J. and J. M. Prausnitz. 1999. Equation of state for thermodynamic properties of chain fluids near-to and far-from the vapor-liquid critical region. *J. Chem. Phys.* 111: 5964.

Jiang, J. and J. M. Prausnitz. 2000. Phase equilibria for chain-fluid mixtures near to and far from the critical region. *AIChE J.* 46:2525.

Jin, L., Y.-X. Yu, and G.-H. Gao. 2006. A molecular-thermodynamic model for the interactions between globular proteins in aqueous solutions: Applications to bovine serum albumin

(BSA), lysozyme, α-chymotrypsin, and immuno-gamma-globulins (IgG) solutions. *J. Colloid Interface Sci.* 304:77.

Jorge, S., E. Lomba, and J. L. F. Abascal. 2001. An inhomogeneous integral equation for the triplet structure of binary liquids. *J. Chem. Phys.* 114:3562.

Jorge, S., E. Lomba, and J. L. F. Abascal. 2002. Study of the triplet and pair structure of strong electrolytes modeled via truncated Coulomb interactions. *J. Chem. Phys.* 117:3763.

Kadanoff, L. P. 1966. Scaling laws for Ising models near T_c. *Physics* 2:263.

Kahl, G. and J. Hafner. 1984. Optimized random-phase approximation for the structure of expanded fluid rubidium. *Phys. Rev. A* 29:3310.

Kern, N. and D. Frenkel. 2003. Fluid-fluid coexistence in colloidal systems with short-ranged strongly directional attraction. *J. Chem. Phys.* 118:9882.

Lai, S. K., O. Akinlade, and M. P. Tosi. 1990. Thermodynamic and structure of liquid alkali metals from the charged-hard-sphere reference fluid. *Phys. Rev. A* 41:5482.

Largo, J. and N. B. Wilding. 2006. Influence of polydispersity on the critical parameters of an effective potential model for asymmetric hard-sphere mixtures. *Phys. Rev. E* 73: 036115.

Larsen, B. 1978. Studies in statistical mechanics of Coulombic systems. III. Numerical solutions of the HNC and RHNC equations for the restricted primitive model. *J. Chem. Phys.* 68:4511.

Lekkerkerker, H. N. W. 1997. Strong, weak and metastable liquids. *Physica A* 244:227.

Lettieri, S., X. Li, and J. D. Gunton. 2009. Ion specific effects on phase transitions in protein solutions. *Phys. Rev. E* 79:031904.

Li, D. H., X. R. Li, and S. Wang. 1986. Variational calculation of Helmholtz free energies with applications to the sp-type liquid metals. *J. Phys. F: Met. Phys.* 16:309.

Liu, H., S. K. Kumar, and F. Sciortino. 2007. Vapor-liquid coexistence of patchy models: Relevance to protein phase behavior. *J. Chem. Phys.* 127:084902.

Llovell, F., J. C. Pàmies, and L. F. Vega. 2004. Thermodynamic properties of Lennard-Jones chain molecules: Renormalization-group corrections to a modified statistical associating fluid theory. *J. Chem. Phys.* 121:10715.

Lomakin, A., N. Asherie, and G. B. Benedek. 1996. Monte Carlo study of phase separation in aqueous protein solutions. *J. Chem. Phys.* 104:1646.

Lomakin, A., N. Asherie, and G. B. Benedek. 1999. Aeolotopic interactions of globular proteins. *Proc. Natl. Acad. Sci. USA* 96:9465.

Louis, A. A. 2001. Effective potentials for polymers and colloids: Beyond the van der Waals picture of fluids? *Phil. Trans. R. Soc. Lond. A* 359:939.

Louis, A. A., R. Finken, and J. P. Hansen. 2000. Crystallization and phase separation in nonadditive binary hard-sphere mixtures. *Phys. Rev. E* 61:R1028.

Louis, A. A. and R. Roth. 2001. Generalized depletion potentials. *J. Phys.: Condens. Matter* 13:L777.

Lo Verso, F., D. Pini, and L. Reatto. 2005. Fluid-fluid and fluid-solid phase separation in nonadditive asymmetric binary hard-sphere mixtures. *J. Phys.: Condens. Matter* 17:771.

Lue, L. and J. M. Prausnitz. 1998a. Renormalization-group corrections to an approximate free-energy model for simple fluids near to and far from the critical region. *J. Chem. Phys.* 108:5529.

Lue, L. and J. M. Prausnitz. 1998b. Thermodynamics of fluid mixtures near to and far from the critical region. *AIChE J.* 44:1455.

Lue, L. and L. V. Woodcock. 1999. Depletion effects and gelation in a binary hard-sphere fluid. *Mol. Phys.* 96:1435.

Lund, M., P. Jungwirth, and C. E. Woodward. 2008. Ion specific protein assembly and hydrophobic surface forces. *Phys. Rev. Lett.* 100:258105.

Malherbe, J. G. and S. Amokrane. 2001. True mixture versus effective one component fluid models of asymmetric binary hard sphere mixtures: A comparison by simulation. *Mol. Phys.* 99:355.

Mao, Y., M. E. Cates, and H. N. W. Lekkerkerker. 1995. Depletion force in colloidal systems. *Physica A* 222:10.

Mi, J., C. Zhong, and Y.-G. Li. 2005a. Renormalization group theory for fluids including critical region. II. Binary mixtures. *Chem. Phys.* 312:31.

Mi, J., Y. Tang, C. Zhong, and Y.-G. Li. 2005b. Prediction of global vapor-liquid equilibria for mixtures containing polar and associating components with improved renormalization group theory. *J. Phys. Chem. B* 109:20546.

Mi, J., C. Zhong, Y.-G. Li, and J. Chen. 2004a. Renormalization group theory for fluids including critical region. I. Pure fluids. *Chem. Phys.* 305:37.

Mi, J., C. Zhong, Y.-G. Li, and Y. Tang. 2004b. An improved renormalization group theory for real fluids. *J. Chem. Phys.* 121:5372.

Mi, J., C. Zhong, Y.-G. Li, and Y. Tang. 2006. Prediction of global VLE for mixtures with improved renormalization group theory. *AIChE J.* 52:342.

Mon, K. K., R. Gann, and D. Stroud. 1981. Thermodynamics of liquid metals: The hard-sphere versus one-component-plasma reference system. *Phys. Rev. A* 24:2145.

Moon, Y. U., R. A. Curtis, C. O. Anderson, H. W. Blanch, and J. M. Prausnitz. 2000. Protein-protein interactions in aqueous ammonium sulfate solutions. Lysozyme and bovine serum albumin (BSA). *J. Solution Chem.* 29:699.

Morales, V., J. A. Anta, and S. Lago. 2003. Integral equation prediction of reversible coagulation in charged colloidal suspensions. *Langmuir* 19:475.

Palmer, R. G. and J. D. Weeks. 1973. Exact solution of the mean spherical model for charged hard spheres in a uniform neutralizing background. *J. Chem. Phys.* 58:4171.

Pelissetto, A. and E. Vicari. 2002. Critical phenomena and renormalization-group theory. *Phys. Rep.* 368:549.

Pellicane, G., D. Costa, and C. Caccamo. 2004a. Theory and simulation of short-range models of globular protein solutions. *J. Phys.: Condens. Matter* 16:S4923.

Pellicane, G., D. Costa, and C. Caccamo. 2004b. Microscopic determination of the phase diagrams of lysozyme and γ-crystallin solutions. *J. Phys. Chem. B* 108:7538.

Pellicane, G., F. Saija, C. Caccamo, and P. V. Giaquinta. 2006. Thermodynamic stability of fluid–fluid phase separation in binary athermal mixtures: The role of nonadditivity. *J. Phys. Chem. B* 110:4359.

Pullara, F., A. Emanuele, M. B. Palma-Vittorelli, and M. U. Palma. 2005. Lysozyme crystallization rates controlled by anomalous fluctuations. *J. Cryst. Growth* 274:536.

Pullara, F., A. Emanuele, M. B. Palma-Vittorelli, and M. U. Palma. 2007. Protein aggregation/crystallization and minor structural changes: Universal versus specific aspects. *Biophys. J.* 93:3271.

Rosenberger, F. 1996. Protein crystallization. *J. Cryst. Growth* 166:40.

Ross, M., H. E. DeWitt, and W. B. Hubbard. 1981. Monte Carlo and perturbation-theory calculations for liquid metals. *Phys. Rev. A* 24:1016.

Roth, R. and R. Evans. 2001. The depletion potential in non-additive hard-sphere mixtures. *Europhys. Lett.* 53:271.

Roth, R., R. Evans, and S. Dietrich. 2000. Depletion potential in hard-sphere mixtures: Theory and applications. *Phys. Rev. E* 62:5360.

Ruppert, S., S. I. Sandler, and A. M. Lenhoff. 2001. Correlation between the osmotic second virial coefficient and the solubility of proteins. *Biotechnol. Prog.* 17:182.

Sanchez, I. C. and A. C. Balazs. 1989. Generalization of the lattice-fluid model for specific interactions. *Macromolecules* 22:2325.

Stell, G. and B. Hafskjold. 1981. Appropriate input for the charged-sphere GMSA. *J. Chem. Phys.* 74:5278.

Tanford, C. and R. Roxby. 1972. Interpretation of protein titration curves. Application to lysozyme. *Biochemistry* 11:2192.

Tang, Y. 1998. Outside and inside the critical region of the Lennard-Jones fluid. *J. Chem. Phys.* 109:5935.

Tang, X. and J. Gross. 2010. Renormalization-group corrections to the perturbed-chain statistical associating fluid theory for binary mixtures. *Ind. Eng. Chem. Res.* 49:9436.

Tang, Y., Y.-Z. Lin, and Y.-G. Li. 2005. First-order mean spherical approximation for attractive, repulsive, and multi-Yukawa potentials. *J. Chem. Phys.* 122:184505.

Telo da Gama, M. M. and R. Evans. 1980. Theory of the liquid-vapour interface of a binary mixture of Lennard-Jones fluids. *Mol. Phys.* 41:1091.

Umar, I. H. and W. H. Young. 1974. Hard sphere structure factors for liquid metals. *J. Phys. F: Metal Phys.* 4:525.

Velasco, E., G. Navascués, and L. Mederos. 1999. Phase behavior of binary hard-sphere mixtures from perturbation theory. *Phys. Rev. E* 60:3158.

Verwey, E. J. W. and J. Th. G. Overbeek. 1948. *Theory of Stability of Lyophobic Colloids*, Amsterdam, the Netherlands: Elsevier.

Victor, J.-M. and J.-P. Hansen. 1985. Spinodal decomposition and the liquid-vapour equilibrium in charged colloidal dispersions. *J. Chem. Soc. Faraday Trans. 2* 81:43.

Vrij, A. 1976. Polymers at interfaces and the interactions in colloidal dispersions. *Pure Appl. Chem.* 48:471.

Waseda, Y. 1980. *The Structure of Non-crystalline Materials*. New York: McGraw-Hill.

Weis, J.-J. and D. Levesque. 2001. Thermodynamic and structural properties of size-asymmetric charged hard spheres. *Chem. Phys. Lett.* 336:523.

White, J. A. 2000. Global renormalization calculations compared with simulations for Lennard-Jones fluid. *J. Chem. Phys.* 112:3236.

White, J. A. and S. Zhang. 1993. Renormalization group theory for fluids. *J. Chem. Phys.* 99:2012.

White, J. A. and S. Zhang. 1995. Renormalization theory of nonuniversal thermal properties of fluids. *J. Chem. Phys.* 103:1922.

Wilson, K. G. 1971a. Renormalization group and critical phenomena. I. Renormalization group and the Kadanoff scaling picture. *Phys. Rev. B* 4:3174.

Wilson, K. G. 1971b. Renormalization group and critical phenomena. II. Phase-space cell analysis of critical behavior. *Phys. Rev. B* 4:3184.

Wilson, K. G. and J. Kogut. 1974. The renormalization group and the ε expansion. *Phys. Rep.* 12:75.

Young, W. H. 1987. Thermodynamic theory of simple liquid metals and alloys. *Can. J. Phys.* 65:241.

Young, W. H. 1992. Structural and thermodynamic properties of NFE liquid metals and binary alloys. *Rep. Prog. Phys.* 55:1769.

Zhou, S. 2005. Investigation about suitability of hard core attractive Yukawa potential as a model potential for short-range attractive interactions in colloidal dispersions. *Colloids Surf. A* 262:187.

Index